Nuclear Shapes and Nuclear Structure at Low Excitation Energies

NATO ASI Series

Advanced Science Institutes Series

A series presenting the results of activities sponsored by the NATO Science Committee, which aims at the dissemination of advanced scientific and technological knowledge, with a view to strengthening links between scientific communities.

The series is published by an international board of publishers in conjunction with the NATO Scientific Affairs Division

A	**Life Sciences**	Plenum Publishing Corporation
B	**Physics**	New York and London
C	**Mathematical and Physical Sciences**	Kluwer Academic Publishers
D	**Behavioral and Social Sciences**	Dordrecht, Boston, and London
E	**Applied Sciences**	
F	**Computer and Systems Sciences**	Springer-Verlag
G	**Ecological Sciences**	Berlin, Heidelberg, New York, London,
H	**Cell Biology**	Paris, Tokyo, Hong Kong, and Barcelona
I	**Global Environmental Change**	

Recent Volumes in this Series

Volume 282—Quantum Measurements in Optics
edited by Paolo Tombesi and Daniel F. Walls

Volume 283—Cluster Models for Surface and Bulk Phenomena
edited by Gianfranco Pacchioni, Paul S. Bagus, and Fulvio Parmigiani

Volume 284—Asymptotics beyond All Orders
edited by Harvey Segur, Saleh Tanveer, and Herbert Levine

Volume 285—Highlights in Condensed Matter Physics and Future Prospects
edited by Leo Esaki

Volume 286—Frontiers of High-Pressure Research
edited by Hans D. Hochheimer and Richard D. Etters

Volume 287—Coherence Phenomena in Atoms and Molecules in Laser Fields
edited by André D. Bandrauk and Stephen C. Wallace

Volume 288—Intersubband Transitions in Quantum Wells
edited by Emmanuel Rosencher, Børge Vinter, and Barry Levine

Volume 289—Nuclear Shapes and Nuclear Structure at Low Excitation Energies
edited by Michel Vergnes, Jocelyne Sauvage, Paul-Henri Heenen, and Hong Tuan Duong

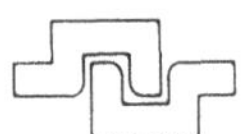

Series B: Physics

Nuclear Shapes and Nuclear Structure at Low Excitation Energies

Edited by

Michel Vergnes and Jocelyne Sauvage

Institut de Physique Nucléaire
Orsay, France

Paul-Henri Heenen

Université Libre de Bruxelles
Bruxelles, Belgium

and

Hong Tuan Duong

Laboratoire Aimé Cotton
Orsay, France

Springer Science+Business Media, LLC

Proceedings of a NATO Advanced Research Workshop on
Nuclear Shapes and Nuclear Structure at Low Excitation Energies,
held June 3-7, 1991,
in Cargèse, France

NATO-PCO-DATA BASE

The electronic index to the NATO ASI Series provides full bibliographical references (with key-words and/or abstracts) to more than 30,000 contributions from international scientists published in all sections of the NATO ASI Series. Access to the NATO-PCO-DATA BASE is possible in two ways:

—via online FILE 128 (NATO-PCO-DATA BASE) hosted by ESRIN, Via Galileo Galilei, I-00044 Frascati, Italy.

Additional material to this book can be downloaded from http://extra.springer.com.

Library of Congress Cataloging-in-Publication Data

Nuclear shapes and nuclear structure at low excitation energies / edited by Michel Vergnes ... [et al.].
p. cm. -- (NATO ASI series. Series B, Physics ; v. 289)
"Proceedings of a NATO Advanced Research Workshop on Nuclear Shapes and Nuclear Structure at Low Excitation Energies, held June 3-7, 1991, in Cargèse, France"--T.p. verso.
"Published in cooperation with NATO Scientific Affairs Division."
Includes bibliographical references and index.
ISBN 978-0-306-44195-0 ISBN 978-1-4615-3342-9 (eBook)
DOI 10.1007/978-1-4615-3342-9
1. Nuclear shapes--Congresses. 2. Nuclear structure--Congresses. 3. Nuclear spectroscopy--Congresses. I. Vergnes, Michel. II. North Atlantic Treaty Organization. Scientific Affairs Division. III. NATO Advanced Research Workshop on Nuclear Shapes and Nuclear Structure at Low Excitation Energies (1991 : Cargèse, France) IV. Series.
QC793.3.N83N83 1992
539.7'4--dc20 92-7776
CIP

ISBN 978-0-306-44195-0

Originally published by Plenum Press, New York in 1992

International Committee

H.G. BÖRNER *(Grenoble)*
C. BRIANCON *(Orsay)*
P.R. CHASMAN *(Argonne)*
A. FAESSLER *(Tübingen)*
A. GIZON *(Grenoble)*
W.D. HAMILTON *(Brighton)*
P.G. HANSEN *(Aarhus)*
R.W. HOFF *(Livermore)*
V.S. LETHOKOV *(Moscow)*
T. MARUMORI *(Tsukuba)*
A. RICHTER *(Darmstadt)*
H. SERGOLLE *(Orsay)*
R.K. SHELINE *(Tallahassee)*
V.G. SOLOVIEV *(Dubna)*
N.J. STONE *(Oxford)*
M. VENERONI *(Orsay)*
T. VON EGIDY *(München)*

Organizing Committee

R. BABINET *(Saclay)*
H.T. DUONG *(Orsay)*
D. GOUTTE *(Saclay)*
P.H. HEENEN *(Bruxelles)*
J. LIBERT *(Orsay)*
E.W. OTTEN *(Mainz)*
J. PINARD *(Orsay)*
P. QUENTIN *(Orsay)*
B. ROUSSIERE *(Orsay)*
J. SAUVAGE *(Orsay)*
M. VERGNES *(Orsay)*

Workshop Secretariat

F. DYKSTRA (Orsay)

PREFACE

The present NATO Advanced Research Workshop held in Cargèse (Corsica) from June 3rd to June 7th, 1991, was devoted to Nuclear Shapes and Nuclear Structure at Low Excitation Energies.

We tried to organize the Workshop to facilitate the exchange of information in a rapidly moving field, where theorists and experimentalists are continuously developing and implementing new and powerful techniques in order, both to improve our knowledge and understanding of already known areas and to open completely new and fascinating frontier domains, as for example in the case of the recent discovery of Superdeformations.

The informal atmosphere of Cargèse contributed to easy contacts and scientific exchanges and to relaxed - although fruitful and sometimes passionate - discussions.

We would like to express our gratitude to NATO for its financial support which made this Workshop possible.

We acknowledge the support of the Institut de Physique Nucléaire et de Physique des Particules (France), the Commissariat à l'Energie Atomique (France), and the Centre National de la Recherche Scientifique - Mathématiques et Physique de Base (France).

Our special appreciation is due to Frédérique Dykstra and Josepha Nsair for their outstanding organizational work throughout the preparation and duration of this conference. We want to acknowledge at this occasion the help of many people from the departments of the Institut de Physique Nucléaire of Orsay.

It is also a pleasure to thank the Université de Nice for making available the facilities of the Cargèse Scientific Institute.

Last but not least we would like to express our gratitude to Marie-France Hanseler, Director of the Cargèse Institute and to the staff of the Institute, in particular Chantal Ariano and Joseph Antoine Ariano for making our stay in Cargèse extremely enjoyable.

J. SAUVAGE
M. VERGNES

CONTENTS

I- INTRODUCTION

Introductory Remarks 3
E.W. Otten

Panoramic New Vistas of Nuclei in and near Their Ground States 17
J.H. Hamilton

Nuclear Shapes and Nuclear Structure at Low Excitation Energies: Theoretical Aspects 39
K. Heyde

Atomic Physics and Nuclear Properties 55
J. Bauche

II - ATOMIC METHODS IN NUCLEAR SPECTROSCOPY

Progress in Atomic Physics Experiments on Nuclear Properties 69
R. Neugart

Resonance Ionization Spectroscopy of Rare-Earth Elements at Iris Facility 81
G.D. Alkhazov, A.E. Barzakh, V.N. Panteleyev, E.P. Sudentas, V.N. Fedoseyev, V.S. Letokhov, V.I. Mishin and S.K. Sekatsky

Atomic Spectroscopy as a Probe of Parity Non-Conserving Effects in the Nucleus 87
A.M. Mårtensson-Pendrill

Isotope Shifts and Hyperfine Structure in Polonium Isotopes by Atomic Beam Laser Spectroscopy 93
D. Kowalewska, K. Bekk, S. Göring, A. Hanser, W. Kälber, G. Meisel and H. Rebel

III- SINGLE PARTICLES ASPECT

Single-Particle Response Function:Complementary Aspects of (e,e'p) and Transfer Reactions 101
S. Galès

Intruder States and Low Energy Nuclear Spectroscopy 123
R. Bengtsson

Rotational Structures and Residual Interactions in Doubly Odd Nuclei 143
A.J. Kreiner

Collective and Quasi-Particle Degrees of Freedom in ^{178}Hf 163
Ph. Quentin, S.J. Krieger, J. Libert and M.S. Weiss

Diabatic Effects and Shape Coexistence in Neutron Deficient Pb Isotopes 169
N. Tajima, P. Bonche, J. Dobaczewski, H. Flocard and P.-H. Heenen

IV - MULTIPHONON STATES

Low-Energy Multiphonon States in Deformed Nuclei 179
R. Piepenbring

Low-Lying Vibrational States in Deformed Nuclei 195
V.G. Soloviev, A.V. Sushkov and N.Yu. Shirikova

V - SHAPES AND COEXISTENCE

Algebraic Approaches to Nuclear Structure 203
R.F. Casten

Shape Coexistence in the Pb Region 219
M. Huyse, P. Decrock, P. Dendooven, G. Reusen, P. Van Duppen and J. Wauters

Population of the Oblate and Prolate Systems in ^{187}Au Produced at Various Beam Energies in Two Different Reactions 227
A. Korichi, N. Perrin, C. Bourgeois, F. Hannachi, D. Hojman, D.G. Popescu, M.G. Porquet and H. Sergolle

The Shell Model Distribution of Nuclear Shapes 233
J. Carvalho and D.J. Rowe

Magnetic Dipole Excitations of Heavy Nonspherical Nuclei 239
D. Zawischa and J. Speth

VI - OCTUPOLES

Reflection-Asymmetric Shapes in Atomic Nuclei 247
W. Nazarewicz

Electric Transition Strengths in Octupole Nuclei Measured by Coulomb Excitation 267
P.A. Butler; N. Clarkson, T.Czosnyka, R. Ibbotson, C.A. White, B. Kotlinski, D. Cline, R.A. Cunningham, M. Devlin, K.G. Helmer, T.H. Hoare, J.R. Hugues, G.D. Jones, A.E. Kavka, I-Y. Lee, R.J. Poynter, J. Srebrny, W. Urban, E.G. Vogt, R. Wadsworth,D.L. Watson and C.-Y. Wu

Open Questions on the Low-Energy Negative-Parity Excitations 271
I.N. Mikhailov and Ch. Briançon

VII - SUPERDEFORMATION

Microscopic Description of Superdeformation at Low Spin 281
P. Bonche, H. Flocard, P.-H. Heenen, J. Meyer, J. Dobaczewski, S.J. Krieger and M.S. Weiss

Superdeformation:An Experimental Review 299
R.V.F. Janssens

Thallium Isotopes:A Set of "Identical" Superdeformed Bands 315
F. Azaiez, C.W. Beausang, J.V. de Boer, M.A. Deleplanque, R.M. Diamond, J.E. Draper, W.H. Kelly, W. Korten, A.O. Macchiavelli, E. Rubel, M. Rohn, F.S. Stephens, J.A. Becker, M.J. Brinkman, J.A. Cizewski, E.A. Henry, A. Kuhnert, T.F. Wang and S.W. Yates

Microscopic Description of Quadrupole and Octupole Modes in the Superdeformation Region of ^{192}Hg 323
J.P. Delaroche, J.F. Berger, M. Girod, J. Libert, I. Deloncle and L.M. Robledo

VIII - EXOTIC NUCLEI

Isospin Forbidden β Transitions to Low-Lying 0^+ States and the Effect on Superallowed Decay Rates 331
E. Hagberg, V.T. Koslowsky, J.C. Hardy, and I.S. Towner

Nuclei Far from Stability and the R-Process Waiting-Point Concept 339
K.L. Kratz, H. Gabelmann, P. Möller, B. Pfeiffer, H.L. Ravn, A. Wöhr and F.-K. Thielemann

Emission of Neutrons and the Neutron Halo of ^{11}Li 347
M. Lewitowicz, R. Anne, S.E. Arnell, R. Bimbot , H. Emling, D. Guillemaud-Mueller, P.G. Hansen, L. Johannsen, B. Jonson, S. Mattsson, A.C. Mueller, R. Neugart, G. Nyman, F. Pougheon, A. Richter, K. Riisager, M.G. Saint-Laurent, O. Sorlin and K. Wilhelmsen

Search for New Radioactivities at the Proton Drip-Line 353
F. Pougheon, V. Borrel, D. Guillemaud-Mueller, A.C. Mueller, R. Anne, C. Détraz, M. Lewitowicz, D. Bazin, J.P. Dufour, A. Fleury, F. Hubert and M.S. Pravikoff

Internal Conversion Anomaly around A=187 359
B. Roussière, P. Kilcher, F. Le Blanc, J. Oms, J. Sauvage, H. Dautet and the ISOCELE COLLABORATION

Exploration of the Neutron-Rich Mass Surface from ^{11}Li to ^{66}Fe 365
D.J. Vieira, X.G. Zhou, X.L. Tu, J.M. Wouters, H.L. Seifert, K.E.G. Löbner, Z.Y. Zhou, V.G. Lind and G.W. Butler

IX - CHAOS

Statistical Properties of Complete Level Schemes 373
T. von Egidy

X - EXPERIMENTAL TECHNIQUES

Nuclear Moments by Orientation Methods 383
H. Postma

High-Accuracy Mass Determination of Unstable Rb, Sr, Cs, Ba, Fr and Ra Isotopes with a Penning Trap Mass Spectrometer 399
G. Bollen, H.-J. Kluge, H. Hartmann, T. Otto, G. Savard, L. Schweikhard, H. Stolzenberg, G. Audi, R.B. Moore, G. Rouleau and the ISOLDE COLLABORATION

γ-Ray Spectroscopy of Relativistic Fragments:A Novel Technique 407
R. Holzmann for the LAND COLLABORATION

Lifetime Measurements in Medium Heavy Nuclei Using the Gamma Ray Induced Doppler Broadening (GRID) Technique 415
J. Jolie and H.G. Börner

Post Acceleration of Radioactive Beams:A New Tool for Nuclear Physics Studies 423
P. Decrock, T. Delbar, P. Duhamel, W. Galster, M. Huyse, P. Leleux, I. Licot, E. Lienard, P. Lipnik, M. Loiselet, C. Michotte, G. Ryckewaert, P. Van Duppen, J. Vanhorenbeeck and J. Vervier

In Beam Nuclear Polarization of Radioactive Ion Beams 429
J. Wouters, W. Vanderpoorten, P. de Moor, P. Schuurmans, N. Severijns, R. Siebelink, J. Vanhaverbeke, L. Vanneste and J. Vervier

XI - CONCLUSION

Concluding Remarks 437
Ph. Quentin

Participants 447

Index 455

I - INTRODUCTION

INTRODUCTORY REMARKS

E.W. Otten

Institut für Physik der Universität Mainz
Postfach 3980 - D-6500 Mainz

Dear friends of nuclear structure and nuclear spectroscopy, dear friends of atomic structure and atomic spectroscopy!

This address sounds strange and incoherent and might raise the suspicion in some of your minds whether you have come to the wrong conference. It's my task to convince you that you did not, that atomic and nuclear physics are still on speaking terms as they were always, although Rutherfords "division" of the atom into a very small nucleus and a large electron cloud has separated their scales of dimension and energy by some 5 orders of magnitude.

Looking around the audience, I discover many faces which I saw here already 15 years ago on the occasion of the 3rd International Conference on Nuclei far off Stability, smiling and happy faces, in fact, because they are back to this beautiful place. So let me also address myself to the

Dear friends of Cargèse!

Rereading the proceedings of the 1976 Cargèse meeting [1] one finds them to be an outstanding milestone on our scientific route, from which we can measure the progress achieved, as well as identify the problems which remained open until today. I will return later to this theme, but first fulfill my task of tying together nuclear and atomic spectroscopy and spectroscopists by going through some simple, although a bit exotic examples of their unseparability in contradiction to the commom belief.

EXAMPLES OF "UNSEPARABLE" ATOMIC AND NUCLEAR SPECTROSCOPY

Dynamic Muonic Hyperfine Interaction

The first example is hyperfine interaction in muonic atoms. There the quadrupole interaction in the 2p-doublet, e.g., is of the same order as the finestructure splitting due to the high mass of this heavy electron and consequently small extension of its wavefunction (Fig.1). For a well deformed nucleus its of the order of some 10 keV and thus comparable to the rotational excitation energy of the nucleus itself. The usual hierarchy of interactions is lost and hence atomic and nuclear levels are wildly mixed. Not only that one has carefully measured this reshuffled level scheme, one even has observed γ-rays following the muonic cascade from nuclei left over in excited states. This was a big issue about 20 years ago, called dynamic hfi. [2]

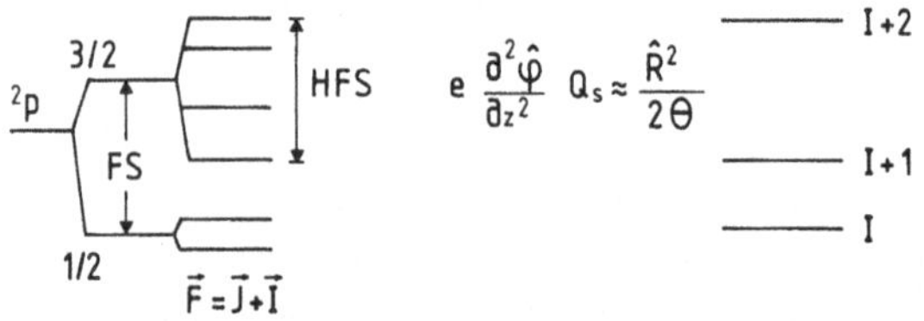

Fig.1. Left: μ-atomic level scheme for a p-doublet including fine- and hyperfinestructure;
right: nuclear rotational level scheme. Since the splittings are of same order, all the levels get mixed

Dynamic Electronic Hyperfine Interaction

My second example I owe to Jan Zylicz who presented it a few days ago at the ISOLDE workshop [3]. It is the same game as above but rescaled to the region of a few eV, and is a true pecularity (Fig. 2): In these days hydrogenlike ions of even the

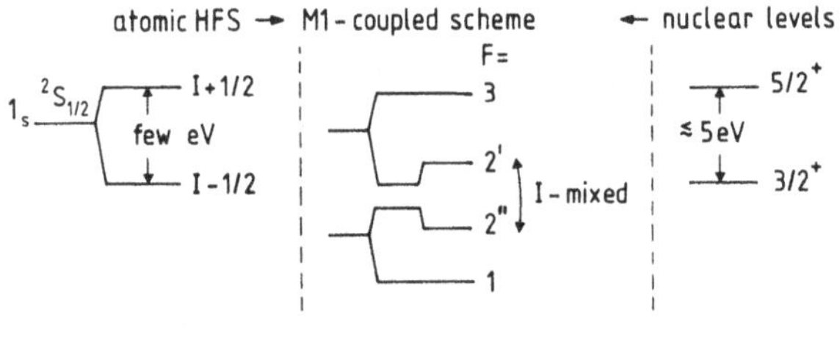

Fig.2. Nuclear spin mixing of ^{229}Ac in the magnetic hyperfine field of the hydrogenlike ion

heaviest elements become available at relativistic heavy ion accelerators and storage rings like at GSI, for example. In this case the hyperfine splitting (hfs) of the 1s electron, which is 1420 MHz for the case of atomic hydrogen, is scaled up by Z^3; thus it is pushed from the microwave up to the optical region of few eV. Still this energy doesn't impress an ordinary nucleus, unless it is extremely soft to magnetic forces due to closely spaced nuclear states. This seems to be the case for ^{229}Ac, where two levels $I^{P} = {}^{3}/_{2}{}^{+}$ and ${}^{5}/_{2}{}^{+}$ are claimed to be degenerated groundstates within 5eV. Again the usual order of matters is completely disturbed and the nucleus being excited to another spin state for a good fraction of time. The verification of this very exotic case by some laserspectroscopic method is certainly not easy. But it is out of question that the atomic spectroscopy of highly ionized atoms will be a central issue in future heavy ion research.

Nuclear Polarization Effect in the Isotope Shift

Next let me draw your attention to a less dramatic case, which plays a certain role in ordinary isotope shift (IS),however. The leading term in the contribution of nuclear volume to electronic binding is the diagonal matrix element of the mean squared nuclear charge radius times the electron density at the site of the nucleus (first term in eq. 1):

$$\Delta E_{vol} = -\frac{2\pi}{3} \cdot Ze^2 \cdot |\Psi^{(e)}(0)|^2 \cdot \delta < r^2 >^{(n)} + \quad \text{(Eq. 1)}$$

$$\sum_{i,j,k} \cdot \frac{|< \Psi_i^{(e)} \Psi_j^{(n)} | \hat{F}_k^{(e)} \otimes \hat{M}_k^{(n)} | \Psi_o^{(e)} \Psi_o^{(n)} >|^2}{E_i^{(e)} - E_o^{(e)} + E_j^{(n)} - E_o^{(n)}}$$

Reiner and Wilets have pointed out in 1962 already [4] that one should include non diagonal matrix elements in the electronic as well as in the nuclear wave funtion when one aims for an accuracy of better than a few per cent. The quantitative treatment of this problem was left aside and the whole subject was almost forgotten, until Hoffmann and Baur took it up again in 1984 and carried through the second order perturbation calculation, summing up over all multipole orders and states with realistic nuclear matrix elements [5]. Surprisingly enough, most of the electronic strength lies far out in the continuum at excitation energies up to 100MeV, as compared to which the nuclear excitation energy may be even neglected (see Fig. 3).

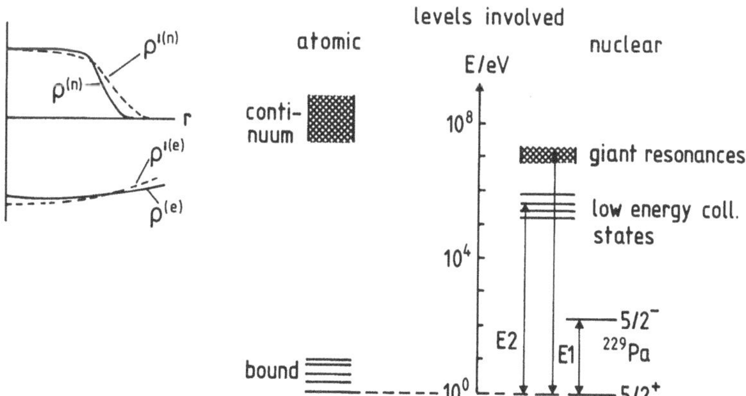

Fig. 3. Mutual polarization of nuclear $\rho^{(n)} \rightarrow \rho'^{(n)}$ and electronic $\rho^{(e)} \rightarrow \rho'^{(e)}$ charge distribution (left) and contributing levels (right).

Therefore the nuclear strength is exhausted in most cases by the giant resonances at excitation energies of 10 to 15MeV. (I admit that those high nuclear excitation energies are explicitely out of sight of our workshop, except for those, who still keep in mind, that the low energy collective modes are something like the Goldstone bosons of the giant resonances pushed down in energy by spontaneous symmetry breaking.) The second order nuclear and electronic polarization effect adds up to about 5 % of the ordinary IS in Pb. But since the giant resonances vary very smoothly with A, not much of it can be seen in the IS measurements.

Polarization Effect of Parity Doublets in Octupole Shaped Nuclei

The situation changes, however, if we conceive a very special case, and this leads to my last example and right into the centre of interest of this conference. Take the octupole deformed nucleus ^{229}Pa which exhibits a closely spaced ground state doublet $I^P = {}^5/_2{}^+, {}^5/_2{}^-$ separated by about 200eV only [6], (see Fig. 3). If this is the characteristic parity doublet of an inversion asymmetric nucleus, then it should be connected by a huge electric dipole transition matrix element of 0.1 - 1 Weißkopf units. In this case the matrix elements with low total excitation energy, which implies low electron excitation as well, explode and should cause a considerable contribution to the IS of this particular nucleus. One should carry through the calculation and try the experiment! It would be a textbook case.

I hope that by now I have convinced you of the intimate connection between atomic and nuclear spectroscopy, although the story was quite different from the higly developed canonical treatment of the problem which deals with the factorization of nucleus-electron interaction into diagonal matrix elements of nuclear moments and hyperfine fields and will be presented by J. Bauche at this meeting.

FROM CARGÈSE I (1976) TO CARGÈSE II (1991)

Shape Coexistence in the Hg-Region - or from RADOP to Laserspectroscopy

I turn now to the second part of my INTRODUCTORY REMARKS drawing some lines of important developments between the two Cargèse meetings in 1976 and 1991. First let us have a look to a figure which H.-J. Kluge has shown at this place 15 years ago (see Fig. 4) [7]. It displays the change of root mean squared charge radii in Hg-isotopes as known by that time from the so called RADOP-experiments (RADOP stands for RAdioactive Detection of Optical Pumping). This method represents again an interface between atomic and nuclear spectroscopy, but this time not on the theoretical, but on the experimental side of the problem. The remarkable discovery evidenced by this plot was the sudden increase of the nuclear charge radius for light odd Hg-isotopes interpreted as a shape transition from small oblate to large prolate deformation.

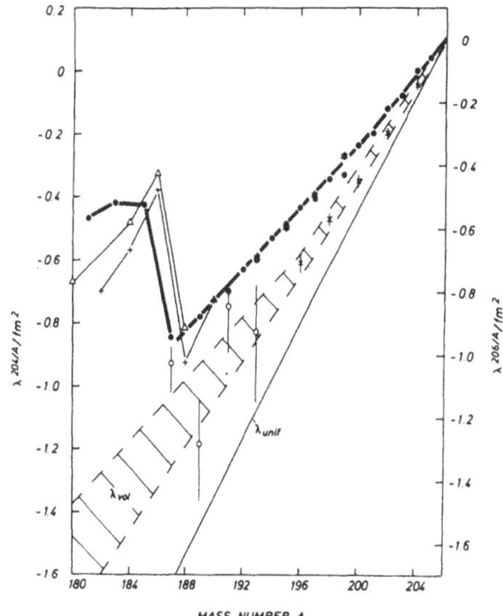

Fig.4. Change of mean squared charge radii in Hg-isotopes (full points and stars) including first on-line RADOP measurements on the neutron deficient side [7].

By that time one knew already that this picture showed only half of the truth, the other half lying with the even isotopes which don't undergo this shape transition, but maintain their slightly oblate shape. This second half of the truth was discovered by nuclear spectroscopy. It was deduced from the level scheme and BE2-quadrupole strengths observed in even Hg-nuclei. The situation as by 1976 is shown in Fig. 5 which the UNISOR group presented here at the Cargèse I meeting [8]. The spectra are characterized by two bands of quadrupole transitions. One of more vibrational character, typical for spherical nuclei; it starts from the 0^+ ground state. The other band is a deformed one, starting from a second, isomeric 0^+-state. These phenomena were then called shape coexistence and odd/even shape staggering or simply "flip flop nuclei".

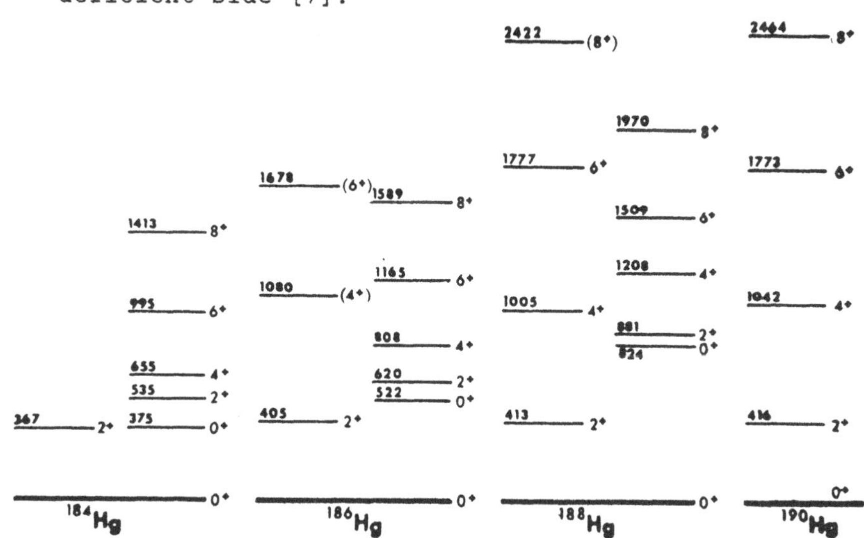

Fig.5. Coexistence of spherical and deformed bands in light, even Hg-isotopes as reported at the Cargèse conference in 1976 [8].

The RADOP method was certainly quite intricate but also testified that it was impossible at that time to achieve such measurements by atomic physics methods alone. This circumstance seriously limited their scope of application off stability. The future was clearly lying with laser spectroscopy which knocked already at the door of the Cargèse-meeting. The ORSAY collaboration between the institutes Aimé Cotton and René Bernas introduced the LIOP method (which stands for Laser Induced Optical Pumping) and presented the first series of results. It was obtained at the Orsay cyclotron and concerned Na-isotopes, whose hfs and IS are shown in Fig. 6 [9]. By now the method is well known to everybody in the audience and I need not explain it, nor will I discuss all the other experiments and methods in this field which followed, they have provided us with moments spins and charge radii of about 500 nuclides [10].

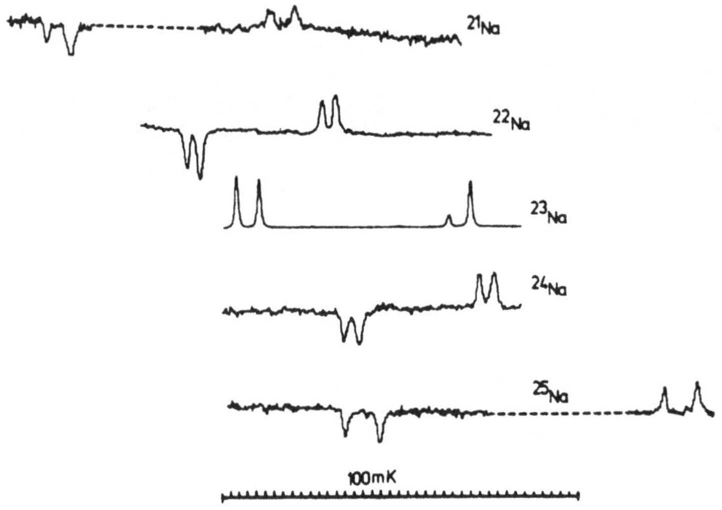

Fig.6. First on line laserspectroscopic results, obtained by the LIOP-method; The experiment concerns the hyperfine splitting of the resonance line in a series of Na-isotopes [9].

Comments on Recent Results from Laser Spectroscopy

Coming back to the problem of shape transitions in the Hg-region, Fig. 7 [14] shows how the chart of nuclear charge radii around Hg looks in our days, completed by the chains of Au, Pt, Tl and Pb isotopes. Its rich structure is a reflex of the complicated collective dynamics governing this region, whose microscopic basis we are going to understand better and better thanks to the numerous efforts which nuclear spectroscopists and theoreticians have invested in this problem. Quite a number of contributions to this meeting witness the progress of these activities.

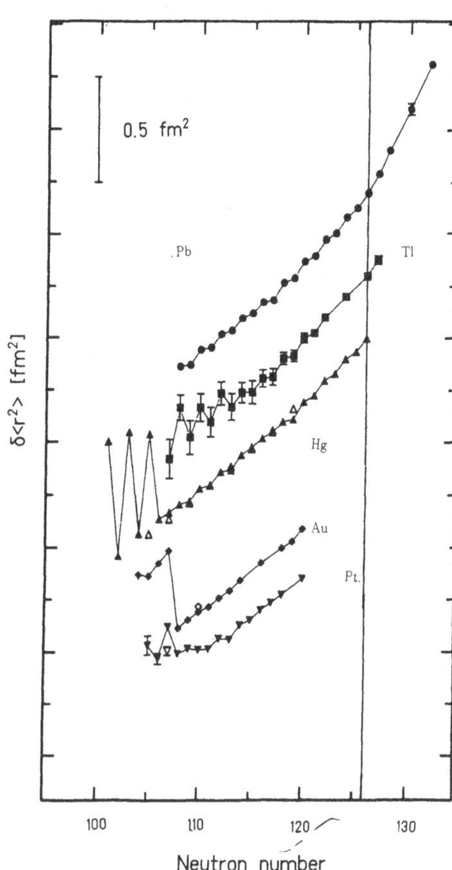

Fig.7. Present status of isotope shifts in the region of shape coexistence around Hg. (The different chains have been separated by more than the isotonic shift in order to avoid confusion.) [14]

A picture quite different from that in the Hg region is shown by the charge radii of In (Fig. 8) and its neighbourhood, although the proton number of these nuclei is also close to shell closure. The course of the IS is very smooth and regular and fits to a parabola over almost the entire range of the $50 \leq N \leq 82$ neutron shell [15]. Still this very regular shape is lacking a firm interpretation. The spherical droplet model would predict pretty much a straight line between the magic isotopes N = 50 and 82. The Zamick Talmi model, a schematic approach to core polarization based on shell model algebra, leads to such a parabolic shape, including odd/even staggering. Therefore, its parameters can be fitted to the experimental curve. However, compared to other experimental, collective observables, like spectroscopic quadrupole moments and BE2-values, only half of the parabolic bump can be accounted for. Since this phenomenon is so regular, it would be rewarding to search for the roots of these enhanced ground state correlations.

At that point I would like to leave the subject of spins, moments and radii; you will hear and see much more about it in the course of the workshop.

ISOTOPIC DEPENDENCE OF THE WEAK NEUTRAL NUCLEAR CHARGE

Can laserspectroscopy along isotopic chains serve purposes other than clearing up the electromagnetic structure of the nucleus? At the ISOLDE workshop at Leysin, last week, C. Wieman presented an exciting, though difficult and demanding proposal to measure

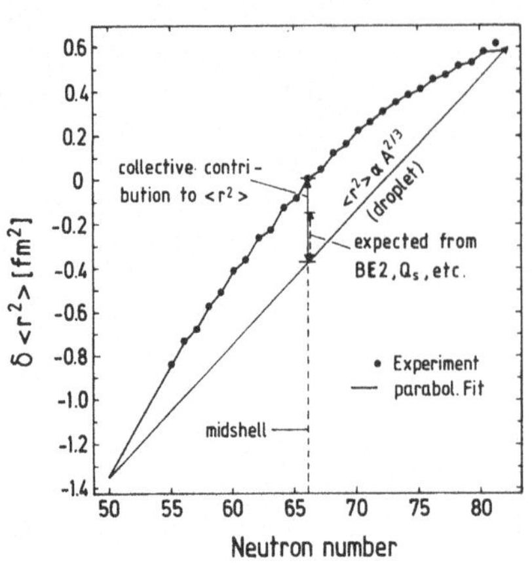

Fig.8. Isotope shift in the In-series covering most of the $50 \leq N \leq 82$ neutron shell. The parabolic bumps over the straight line is the collective contribution to $\langle r^2 \rangle$, only half of which can be explained by other collective observables like BE2-values or spectroscopic quadrupole moments [15].

the isotopic dependence of the weak neutral charge of the nucleus from the observed change of the parity violating asymmetry in the absorption of an atomic resonance line along an isotopic chain [16]. Favourable cases are Cs and Fr being produced at ISOLDE with high yields over a large span of isotopes. Still the experiment requires to make an economic use of these unstable atoms. For that purpose the atomic vapor in the resonance cell is cooled to a temperature of the order of μK by the action of 6 orthogonal resonant laser beams and is trapped to a volume of about 1 mm^3 (see Fig. 9 taken from ref.[17]). One hopes to achieve sufficient efficiency and density in trapping this vapor in order to observe the tiny parity violation effect of about 10^{-5} with an accuracy on the per mille level. The role which precision measurements of atomic parity violation plays in testing the limits of validity of the standard model will be discussed by A.M. Martensson-Pendrill at this workshop. Let me mention at this point only, that the PNC effect in atoms is proportional to the overlap of the electronic and nuclear wave function. Thus its analysis requires more than anything else a precise knowledge of atomic as well as nuclear structure.

NUCLEAR MASSES

Turning over the pages of the 76-Cargèse proceedings further, one meets the paper by Thibault [18], which reports on the first series of direct mass measurements with an on-line mass spectrometer. Table 1 and Fig.[10].reproduced from that paper show the mass excess of neutron rich Li and Na isotopes and 2-neutron-separation energies, respectively. I would like to draw your

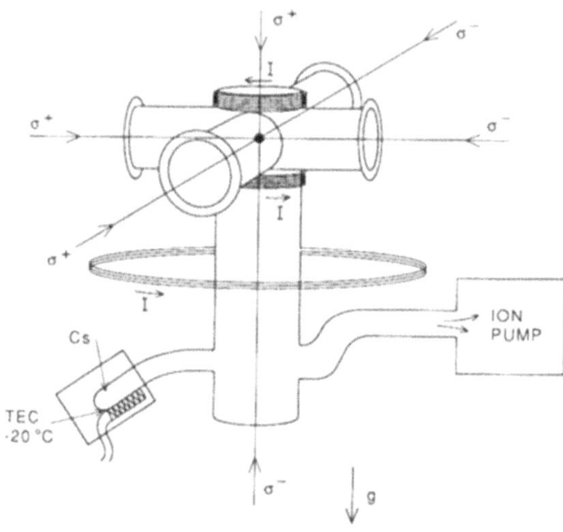

Fig.9. Setup for trapping Cs-atoms by 6. orthogonal, contrary polarized laser beams within their intersection [17].

attention in particular to the case of ^{11}Li. Its low S_{2n} value of about 200 keV only, is in fact the key number to the neutron halo [19] of this nucleus which was seen in nuclear reaction cross sections [20]. Precision mass measurements far off stability and the related question of nuclear structure at the drip lines form central issues of this workshop again.

RADIOACTIVE BEAMS AND TARGETS

The case of ^{11}Li also demonstrates that spectroscopic experiments far off stability have to be supplemented by nuclear reaction studies requiring beams and targets of radioactive

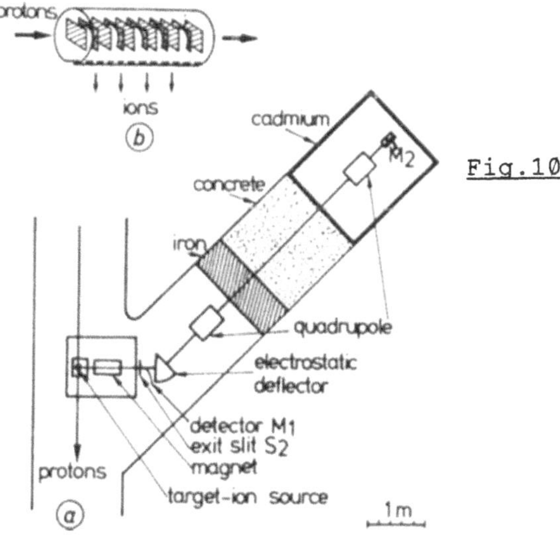

Fig.10. Two neutron separation energies of Na - isotopes (and their heavier isotones), obtained by on line mass spectrometry [18).

Isotope	Mass Excess (keV)	Cross section(μ barn)
		Ir + p
^{11}Li	40940 ± 80	∿ 50
		U + p
^{26}Na	- 6901 ± 25	∿ 5400
^{27}Na	- 5620 ± 60	∿ 2000
^{28}Na	- 1140 ± 80	∿ 350
^{29}Na	2650 ± 100	∿ 100
^{30}Na	8370 ± 200	∿ 20
^{31}Na	10600 ± 800	∿ 5
^{32}Na	16400 ± 1100	∿ 0,4

Tab.1. First determination of the ^{11}Li-mass together with results on Na-isotopes as reported at the Cargèse 76 conference [18].

isotopes. A first not sufficiently recognized step in this direction was also reported already at 76-Cargèse meeting, concerning the thermal neutron reaction (n^{th}, p) from a radioactive target of ^{84}Rb (see Fig. 11) produced by on-line mass separation. [21]. In our days heavy ion accelerators equipped with fragment separators and storage rings provide powerful tools for this kind of research. They will be complimented by radioactive beams produced by ISOL - techniques and accelerated to energies above the Coulomb barrier.

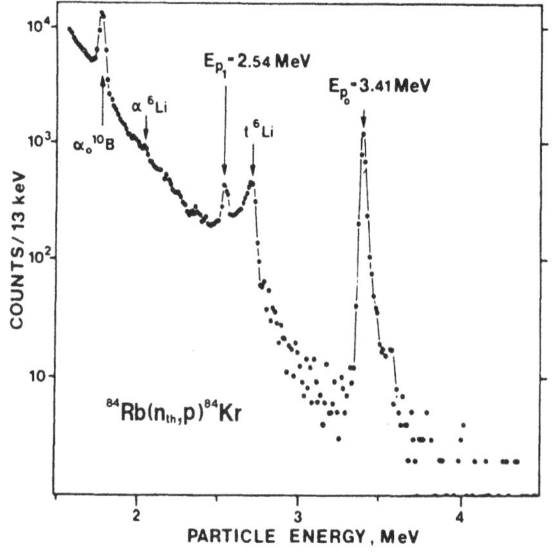

Fig.11. (n^{th},p)-reaction on a radioactive ^{84}Rb target as reported at the Cargèse 1976 conference [21].

We have finished our short walk along the boundaries between atomic and nuclear physics and I hope that the topics I selected meet the spirit and aims of this workshop. I am convinced that the scientific as well as social conditions for its success are well set.

References

[1] Proc. Int. Conf. On Nuclei Far From Stability, Cargèse, Corsica, (France), May 1976, CERN-Report 76-13, CERN, Geneva, 1976

[2] reviewed e.g. by F. Scheck, Phys. Rep. 44: 187 (1978)

[3] J. Zylicz, contribution to the ISOLDE workshop, Leysin, Switzerland, May 1991, unpublished

[4] A.S. Reiner and L. Wilets, Nucl. Phys. 36: 457 (1962)

[5] B. Hoffmann, G. Baur and J. Speth, Z.Phys. 315: 57 (1984)

[6] I. Ahmad, J.E. Gindler, R.R. Betts, R.R. Chasman and A.M. Friedman, Phys.Rev.Lett. 49: 1758 (1982)

[7] H.J. Kluge, in ref. [1], p 177 ff

[8] J.L. Wood et al. in ref. [1], p 364 ff

[9] G. Huber, R. Klapisch, C. Thibault, T.H. Duong, P. Juncar, S. Liberman, J. Pinard, J.L. Vialle and P. Jaquinot, ref. (1), p 188 ff

[10] reviewed recently by E.W. Otten in Treatise on Heavy Ion Science, Vol.8, ed. D.A. Bromley, Plenum Publ. Corp., New York, 1989, p 517 ff

[11] E. Arnold, W. Borchers, M. Carré, H.T. Duong, P. Juncar, J. Lerme, S. Liberman, W. Neu, R. Neugart, E.W. Otten, M. Pellerin, J. Pinard, G. Ulm, J.L. Vialle and K. Wendt, Phys.Rev.Lett. 59: 771 (1987) and ref.s therin

[12] H.T. Duong, J. Pinard, S. Liberman, G. Savard, J.K.P. Lee, J.E. Crawford, G. Thekkadath, F.L. Blanc, P. Kilcher, J. Obert, J. Oms, J.C. Putaux, B. Roussière and J. Sauvage, Phys.Lett. B217: 401 (1989)

[13] S. Liberman, J. Pinard, H.T. Duong, P. Juncar, P. Pillet, J.L. Vialle, P. Jaquinot, F. Touchard, S. Büttgenbach, C. Thibault, M. de Saint Simon, R. Klapisch A. Pesneile and G. Huber, Phys.Rev. A22: 2732 (1980)

[14] G. Huber, private commun.

[15] J. Eberz, U. Dinger, G. Huber, H. Lochmann, R. Menges, R. Neugart, R. Kirchner, O. Klepper, T. Kühl, D. Marx, G. Ulm and K. Wendt, Nucl.Phys. A464: 9 (1987)

[16] C. Wieman, contribution to the ISOLDE workshop, Leysin, Switzerland, May 1991, unpublished

[17] C. Monroe, W. Swann, H. Robinson and C. Wieman, Phys.Rev. Lett. 65: 1571 (1990)

[18] C. Thibault in ref. [1], p 93

[19] P.G. Hansen and B. Jonson, Euro-Phys.Lett. 4: 409 (1987)

[20] I. Tanihata, H. Hamagaki, O. Hashimoto, Y. Shida, N. Yoshikawa, K. Sugimoto, O. Yamakawa, T. Kobayashi and N. Takahashi, Phys.Rev.Lett. 55: 2676 (1985)

[21] E. Hagberg, G. Andersson, M. Asghar, A. Emsallem and B. Jonson in ref. (1), p 39

Commemorating Sylvain Liberman

At this point I would like to commemorate an outstanding physisist in the ORSAY collaboration. It is our late friend Sylvain Liberman who dedicated a good fraction of his scientific life to the laser spectroscopy of unstable isotopes, not only as director of the Lab. Aimé Cotton where he did everything to push and further develop this research, but also as an active collaborator on the spot. In later years when the LIOP method had exhausted its potential after having measured not less than about 100 alkali isotopes, the lab continued under his leadership this line of research by adapting other methods for the purpose of on line laser spectroscopy. I mention an NMR-variant of collinear laser spectroscopy applied to Ra-isotopes [11] and in particular resonance ionisation spectroscopy. The latter opened the door to the spectroscopy of refractory elements via laser desorption. Au and Pt were the first objectives of this research [12]. It forms the basis of the present COMPLIS collaboration at ISOLDE between groups from Orsay, McGill and Mainz which contributes to this conference.

A particular love of Sylvain was the spectroscopy of Fr, as interesting as demanding in many respects. Not only the hfs and IS, but also the gross structure of its atomic spectrum were completely unknown before; not a single spectral line was recorded in the tables. But many theoretical papers had already treated this interesting, highly relativistic case of a single valence electron spectrum. Their predictions differed, up to 10 % of the wavelength. Nevertheless his group mastered the terrible problem to find the extremely narrow atomic resonance lines by scanning a laser over this large interval which is like searching a needle in a hay stack [13]. When the series of Fr-experiments ended some years ago, not only the atomic spectrum of this element was established up to the Rydberg series, also the spins, moments and charge radii of 16 of its isotopes were measured, extracted from the h f s and IS.

Besides his engagement in the field of unstable isotopes, Sylvain was involved in a broad experimental programme in atomic physics and laser spectroscopy, including very subtle physics cases, like Rydberg states in external fields which is one of the pathways to quantum chaos, or subradiance which is complimentary to the better known superradiance, and others.

Sylvain Liberman died suddenly in August 1988 at the age of 54 by a heart attack. Those who have listened to his talks will remember his very personal sense and style of humour. Those who were closer to him are missing a good friend.

Sylvain Liberman

1934 1988

PANORAMIC NEW VISTAS OF NUCLEI IN AND NEAR THEIR GROUND STATES

J. H. Hamilton

Physics Dept., Vanderbilt University, Nashville, TN 37235 and Centre de Recherches Nucleaires, Universite Louis Pasteur, Strasbourg, France

INTRODUCTION

A panoramic overview of the new insights being gained from nuclei in and near their ground states as if taken with a fish-eye camera, to see the landscape in a broad prospective is given. The extension of our knowledge of the properties of nuclei in and near their ground states still farther off stability are continuing to reveal challenging surprises and unexpected diversities of nuclear shapes and structure that could not be seen by studying nuclei in and near the valley of beta stability. These include the discovery of new magic numbers which stabilize the nucleus for a spherical shape and the continuation of the importance of some and disappearance of other well-established spherical magic numbers along with the establishment of new magic numbers now associated with deformed shapes. Superdeformation, nuclei with 3:2 axis ratios and $\beta \sim 0.4$, is well established in the ground states and excited states of nuclei in the islands of superdeformation centered on the new deformed double magic numbers Z=N=38 and Z=38, N=60,62. Superdeformation, which has attracted so much attention at very high spins, is not only seen in nuclear ground states in these islands but in excited states beginning at spins as low as 4^+ and 6^+, for example in ^{72}Se and ^{194}Pb, respectively. The importance of the reinforcement of the protons and neutrons when they have shell gaps for the same spherical or deformed shape in stabilizing that shape is clearly seen. Many different varieties of nuclear shape coexistence are observed throughout the periodic table through the competition of spherical and deformed shapes associated with shell gaps in the single particle energies formed by intruder orbitals at different deformation. Many very large E0 admixtures are observed in $\Delta I=0$ transitions between competing shapes in the mass 180 and a few in the 120 region. In some cases the E0 radiation surprisingly totally dominates the $\Delta I = 0$ transition intensities. New shapes and collective modes of excitation have been discovered even in stable and near-stable nuclei. These include mixed symmetry states (scissor mode of excitation) where the protons and neutrons oscillate out of phase with each other like a scissor opening and closing and bands of parity doublets associated with stable octupole deformation.

Nuclear orientation facilities with low temperature helium refrigerators directly coupled to on-line isotope separators are opening

Nuclear Shapes and Nuclear Structure at Low Excitation Energies
Edited by M. Vergnes *et al.*, Plenum Press, New York, 1992

up the measurements of many previously inaccessible properties such as magnetic dipole and electric quadrupole moments, spins, parities, multipole mixing ratios in nuclei far off stability as well as new tests of fundamental interactions including the weak and the electromagnetic interactions. Very high rigidity fragment separators have been used to extend our knowledge of the proton drip line up to Z = 37, while establishing the existence of several astrophysical interesting new isotopes. Multinucleon transfer reactions and various types of induced and spontaneous fission are being used to extend our knowledge of very neutron-rich nuclei to heavier nuclei including new nuclei which are important in the synthesis of heavy elements. Experiments with exotic radioactive beams like ^{11}Li and ^{11}Be have shown such very neutron-rich nuclei have long tails of low density neutron matter. These nuclei offer unique new opportunities to probe the properties of low density neutron matter and the n-n interaction.

Highlights have been selected to illustrate the wealth and breadth of new physics being gained from studies of nuclei in and near their ground states and the exciting new future directions. New insights which challenge our understandings are being discovered all across the periodic table from the lightest to the heaviest elements on both the proton- and neutron-rich sides of stability. Many of the exciting discoveries have been made possible by the development of powerful new detection devices which can provide A and Z identification and new heavy-ion accelerators, for example at GANIL, MSU, GSI, LASL and RIKEN. Advances in instrumentation underway such as a new generation, high rigidity, high efficiency recoil mass spectrometer at HHIRF, large detector arrays such as GAMMASPHERE, EUROGAM and EUROBALL, and radioactive beam facilities will offer exciting opportunities to explore new nuclear terrain in and near the ground states under even more exotic conditions out to the proton and neutron drip lines at increasingly higher Z.

ESTABLISHING THE PROTON AND NEUTRON DRIP LINES

In their review Detraz and Vieira[1] described the extension of the proton drip line up to cobalt Z = 27 with gaps at iron (Z = 26) and chromium (Z = 24) and the neutron drip line up to neon Z = 10, which is farther off stability. Very recently Mohar et al.[2] in the first experiment with the newly-commissioned A1200 beam analysis device at the MSU National Superconducting Cyclotron Laboratory extended the proton drip line to Z = 37 as illustrated in Fig. 1. The six new isotopes, with circles around their mass numbers were identified from the fragmentation of a 65 MeV/A ^{78}Kr beam on a ^{58}Ni target. Note ^{65}As and ^{69}Br are beyond the predicted limits of stability in the mass model of Jänecke and Masson.[3] Not a single event could be attributed to ^{73}Rb, while ^{74}Rb has several hundred counts in its peak.[2] Thus, it seems the proton drip line has been reached for As(Z = 33), Br(Z = 35), and Rb(Z = 37). Mohar et al.[2] discuss the implications of their studies for the rp-process which is considered to be the means by which energy is generated in high temperature, high proton-density environments. For example, if a neutron star accretes matter from a hydrogen-rich binary companion then a thermonuclear explosion occurs on the surface with proton synthesis thought to continue up to the mass 100 region, as illustrated by arrows in Fig. 1. The low production of ^{65}As, which suggests it is weakly bound, and the absence of ^{73}Rb are clear barriers to extending the rp-process above this point.

On the neutron-rich side, the GANIL collaboration continues to push out to very neutron-rich higher Z nuclei. Guillemand-Mueller[4] has

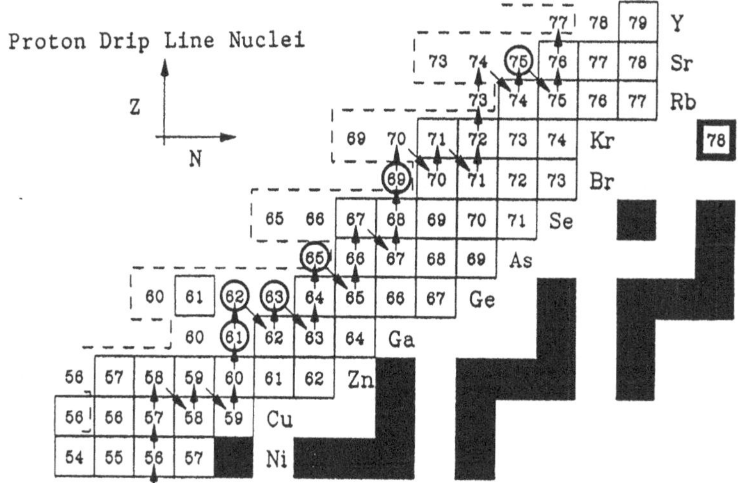

Fig. 1. Section of the chart of the nuclides in the region of interest. Stable nuclei are indicated by filled squares, and the projectile ^{78}Kr is specially noted. Open squares indicate nuclei that have been previously identified, and those circled were identified in ref. 2. Nuclei to the left of the dashed line are predicted to be unstable by the mass model of Jänecke and Masson.[3] The arrows indicate a possible flow for the rp-process.

summarized their work on the discovery of the new isotopes ^{35,36}Ng, ^{38,39}Al, ^{40,41,42}Si, ^{43,k44,45,46}P, ^{45,46,47,48}S, ^{46,47,48,49,51}Cl and ^{49,50,51}Ar. They also have begun to make measurements of half-lives as well as P_n values of these very neutron-rich isotopes to test the different theoretical predictions of these data. Indeed, as production yields continue to increase, measurements of the decay properties of these very neutron-rich nuclei will be a very fruitful field.

NEUTRON HALOS - DINEUTRONS IN ^{11}Li AND ^{11}Be

Since the discovery of ^{11}Li and its many different decay modes at ISOLDE, this nucleus has been of particular interest. Most of the recent interest has centered on the question of the existence of what has been called a neutron halo where ^{11}Li can be considered as a ^{9}Li-core with a two-neutron halo outside the core, i.e. a low-density neutron distribution tail which extends far out from the nuclear core.[5] Hansen and Jonson[5] also revived the old dineutron idea of Migdal[6] as an explanation of ^{11}Li. While the dineutron is unbound by about 70 keV, it may be bound in the field of the nucleus.[6]

Tanihata[7] and Bimbot[7b] summarized recent work with exotic, very neutron-rich beams. The presence of neutron halos are indicated for ^{11}Li, ^{11,14}Be, and ^{17}B. One evidence is the large increase in interaction radii for these three nuclei at the neutron drip line, as shown in Fig. 2. At GANIL the recoil separator, LISE, has recently been used to study[8] the cross-sections and forward angular distribution of fast neutrons for the reactions of radioactive beams of ^{11}Li, ^{9}Li on targets of Be, Ni and Au. The ^{11}Li angular distributions for the three targets are nearly identical with a very narrow half-angle of $(2.9 \pm 0.4)°$. They conclude[8] that all evidence is consistent with the existence of a neutron halo with a radius of about 12 fm.

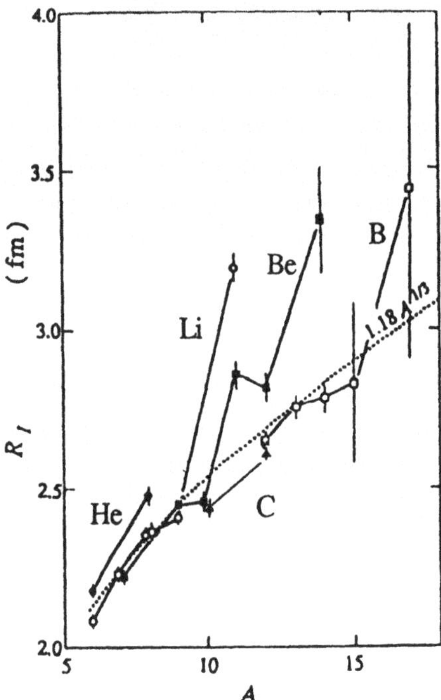

Fig. 2. Interaction radii of light p-shell nuclei. Neutron drip line nuclei show extremely large radii.

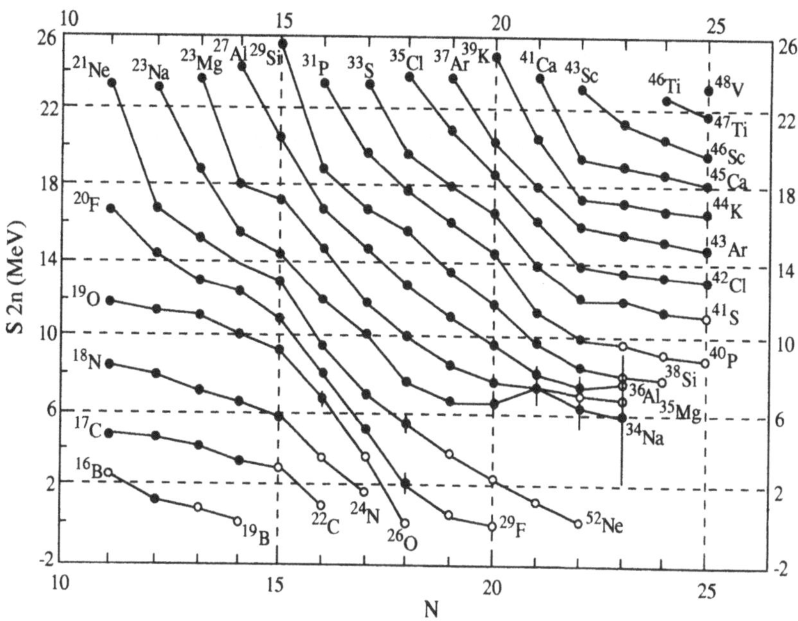

Fig. 3a. S_{2n} values in the region of the new "magic" N=15 and deformed ^{31}Na. Values estimated from systematic trends through smooth extrapolations are represented by open circles.10

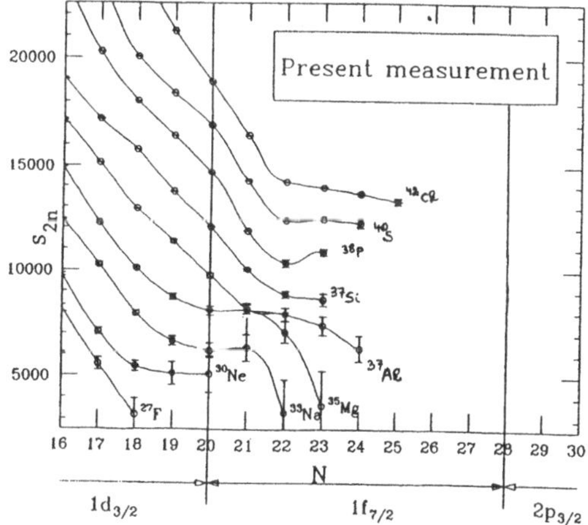

Fig. 3b. S_{2n} variation as a function of the neutron number for isotopes with Z=10 to Z=17 (ref. 13).

More recently, Seth and Parker[9] have reported studies of the missing mass spectra for the reactions leading to weakly unbound, very neutron-rich products $^{6}Li(\pi^{-}, \pi^{+})^{6}H$, $^{9}Be(\pi^{-}, p)^{8}He$, and $^{6}Li(\pi^{-}, p)^{5}H$. The best fits to their data are obtained when a dineutron is involved as one of the products in the breakup. Thus, their data suggest the presence of a dineutron as a cluster during the breakup of these neutron-rich systems. They suggest that these weakly unbound very neutron-rich nuclei may be analogous to the weakly bound nuclei ^{11}Li, ^{11}Be, and ^{17}Be. By analogy, these nuclei may be described as a core-plus dineutron. In any case, here we have another example of how nuclei under extreme conditions far from stability behave in ways not seen in nuclei near stability where the mass and charge distributions are nearly equal. These nuclei clearly offer exciting opportunities to study the n-n interaction and n-core interaction and to use their neutron "halos" as a laboratory to study pure neutron matter at low density.

MASS MEASUREMENTS IN LIGHT NUCLEI

Mass measurements are an important source of information about the structure of nuclei as well as being the basis for the development of improved mass formulas. The two-neutron separation energies (S_{2n}) provide a clear way of demonstrating the presence of spherical magic numbers and sudden changes in structure, as recently emphasized by Audi.[10] For example, smooth changes with change in N and Z are expected. However, it is well established that there is a sharp decrease in S_{2n} values when passing the $N = 82$ closed shell compared to the smooth slow decreases in S_{2n} for $N < 82$ nuclei. On the other hand, there is a well-established sudden increase in S_{2n} with N between $N = 88$ and 90 in Pm, Sm, and Eu nuclei associated with the sudden change from a near-spherical to a well-deformed shape. To illustrate the new physics coming out of mass measurements, look at Fig. 3a (refs. 10, 11). There one sees a clear bump in the S_{2n} which begins at the spherical magic number 20. This bump is interpreted as the onset of large deformation in the $N > 20$ Na nuclei and the disappearance of the magicity of 20 for a spherical shape.[12] Note in Fig. 3a the open circles are the extrapolated points based on the systematics of the behavior of the lighter mass isotopes. One can understand why sodium would show a large effect and $^{33}Al_{20}$ shows very little effect by studying the single particle orbitals in this region. The odd proton in sodium is in the $d_{5/2}$-3/2[211] Nilsson level which is independent of deformation. The $f_{7/2}$-1/2[330] neutron-orbital strongly down-slopes to "intrude" below the $d_{3/2}$-3/2[202] level at large prolate deformation. The binding energy is increased as a result of the deformed intruder state dropping below the energy of the normal spherical state. In $^{32,33}Mg$ the bump is only half that seen in $^{31,32}Na$ and good agreement with shell model calculations is seen for $^{33-35}Al_{20-22}$. In aluminum, the odd proton is in the up-sloping $d_{5/2}$-5/2[202] Nilsson level whose energy increases rapidly with increasing prolate deformation. The gain in energy for the down-sloping $f_{7/2}$-1/2[330] orbital in sodium is countered in $^{33-35}Al$ by the increased energy required to put the proton in the $d_{5/2}$ -5/2[202] orbital at large deformation.[1]

To see how real data change our expectations, compare Fig. 3a (from 1989) with Fig. 3b (1990) as presented by Schutz.[13] Now they have extended their mass measurements to include new members of each of these nuclei to carefully study the changes across $N = 20$. Note the quite different behavior of $^{29,30}Ne$ over expected values in Fig. 3a. For P, S and Cl one clearly sees the sudden decrease in S_{2n} above $N = 20$ as expected in crossing a spherical closed shell. However, now Ne, Na, Mg and Al all show anomalies suggestive of sudden changes to larger deformation with the point at which the change begins increasing in N as

Z increases. So the sharp increase in deformation occurs at N = 20 in the Ne chain, N = 21 in Na and Mg, N =22 in Al and N = 23 for P. Thus, new surprises continue. On the other hand, beta decay studies[14] of $^{34}Al_{20}$ to $^{34}Si_{20}$ revealed a quite new surprise for N = 20. The level schemes of the N = 20 isotones from ^{32}Mg to ^{40}Ca are shown in Fig. 4. The low energy of the first 2^+ state in ^{32}Mg indicates its deformed nature. The level scheme for ^{34}Si established in on-line separator work at ISOLDE[14] clears up discrepancies from earlier reaction work. The 2_1^+ energy is over three times higher than that in ^{32}Mg. Surprisingly, the levels of $^{34}_{14}Si_{20}$ look remarkably like those of spherical double magic $^{40}_{20}Ca_{20}$. Calculations[14,15] show the $(sd)^{-2}(fp)^2$ intruder states, which give ^{32}Mg a deformed ground state, are pushed up in ^{34}Si whose ground state is the more spherical sd configuration. Two-particle-two-hole intruder states are predicted between 3-4.5 MeV with the 2^+ level lowest, and the first non-intruder state is about 5.5 MeV dominated by the sd 2^+ excited state. However, just the sd model space is sufficient to reproduce the ^{36}S levels. The low-lying levels of ^{40}Ca also involve deformed intruder states. Remember, ^{40}Ca was the second case where shape coexistence was invoked[16] after Morinaga[17] introduced shape coexistence to explain the first few levels of double magic ^{16}O. The strong similarities of ^{34}Si and ^{40}Ca and the fact the mass formula based on α energies indicate Z = 14, N = 20 has a stronger shell closure than Z = N = 20 led Baumann et al.[14] to suggest that $^{34}_{14}Si_{20}$ is also a spherical double magic nucleus -- with Z = 14 a new spherical magic number.

Finally, in Fig. 3a, Audi[10] noted that in several nuclei there is also a sudden decrease in S_{2n} above N = 15 based on GANIL and LASL work. This could be evidence for a spherical magic character for N = 15. If so, this would be the first odd-integer spherical-magic number.

SUPERDEFORMATIONS AND DEFORMED SHELL GAPS IN MASS 70 AND 100 REGIONS

Superdeformation in nuclei has been and continues to be the focus of much attention in nuclear structure physics. Superdeformed prolate nuclei fall into two groups with simple ratios of their long to short axis: Group 1 with 3:2 axis ratios and deformations, $\beta_2 \sim 0.4$-0.45, and Group 2 with 2:1 axis ratios with $\beta_2 \sim 0.6$. In Table 1 the regions where superdeformed states are observed are given. Although most of the attention has been given to superdeformation at high spin following the discovery of a high spin superdeformed band in ^{152}Dy (ref. 18), I remind you that superdeformation was first discovered in nuclei in and near their ground states as shown in Table 1.

The opening up of this field began with Strutinsky's[19] interpretation of the fission isomers in the actinide elements as a shell

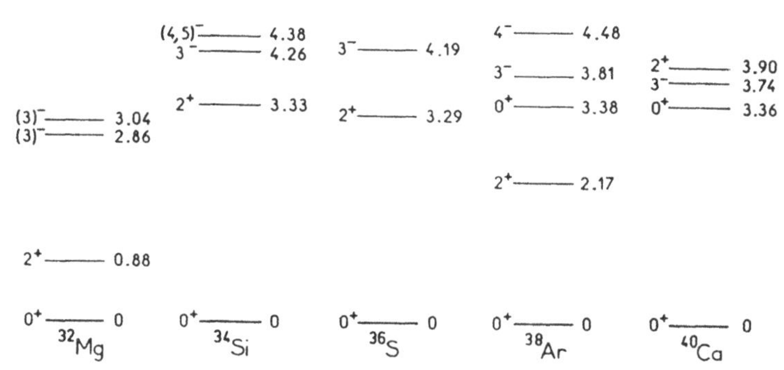

Fig. 4. Lowest excited states of N = 20 isotones (ref. 14).

structure effect which produced a second minimum in the potential at $\beta_2 \sim 0.6$. Bjørnholm and Lynn[20] have reviewed this field. It was many years before more direct evidence for these superdeformed minima were obtained. Recently Schirmer et al.[21], established the excitation energy of 2.75 ± 0.01 MeV for the 0^+ state in the superdeformed minimum for ^{236m}U by γ-decay to the ground state with the Heidelberg-Darmstadt Crystal Ball in the reaction $^{235}U(d,p)$.

Historically, the next observation of superdeformation was for an excited band in ^{72}Se, which crosses the near-spherical ground state band at low (2-4) spin[22]. However, the superdeformed character of this band was not appreciated until after our work on $^{74,76}Kr$. The next reference to this phenomena was the ISOLDE discovery of superdeformation in the ^{100}Sr ground state.[23] At that time, this was interpreted as arising from the neutron shell gap at $\beta \sim 0.4$ for N = 60. Next came the discovery of superdeformation in the ground states of $^{74,76}Kr$ (ref. 24). In these nuclei there is strong mixing between the superdeformed ground states and the excited near-spherical band which coexist at low spin. If one scales the 2^+-0^+ unperturbed energies in ^{74}Kr and ^{100}Sr by $A^{5/3}$ for comparison with the previously known most deformed ground state, in ^{240}Pu $E(2_1^+) = 43$ keV), their 2_1 energies of 28 and 30 keV, respectively, are much lower.

From an analysis of the level structures of their neighboring nuclei, these new superdeformed nuclei were found to occur in islands of superdeformation which were produced by the protons and neutrons having shell gaps at the same $\beta_2 \sim 0.4$ to reinforce each other.[25,26] The islands were proposed[25,26] to be centered on the shell gaps at $\beta_2 \sim 0.4$ for Z = N = 38 and Z = 38, N = 60,62 (see Fig. 5). Subsequent work confirmed ^{76}Sr as the center of this island.[27] Independently, the calculations of Möller and Nix[28] predicted the island of superdeformation centered around Z = N = 38.

Table 1. Regions where superdeformation have been discovered. The underlying physics involves the reinforcement of proton and neutron shell gaps for the same superdeformation to stabilize the superdeformation. The reinforcing shell gaps responsible for the superdeformation are shown.

Mass Region	Discovery Year	States	Reinforcing Deformed Shell Gaps
Nuclei With 3:2 Axis Ratios, $\beta_2 \sim 0.4$			
A ~ 100	1979	Ground States	Z=38, N=60,62
A ~ 76	1981	Ground States	Z=38, N=38
A ~ 130	1987	High Spin States	Z=58, N=72,74
A ~ 190	1989	High Spin States	Z=80, N=112-116
Nuclei With 2:1 Axis Ratios, $\beta_2 \sim 0.6$			
Actinides	1967	Fission Isomers	Z=94, N=144,148
A ~ 150	1986	High Spin States	Z=64,66,N=84,86

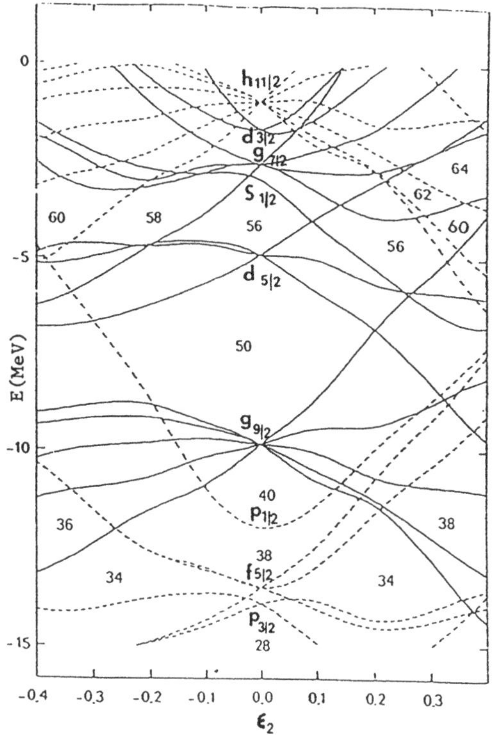

Fig. 5. Single particle levels as a function of deformation.

Recently we extended the levels of $^{72}_{34}Se_{38}$ to 26^+ (ref. 29), Fig. 6, where now one sees the very regular spacings characteristic of superdeformation. In the mass 130, 150 and 190 regions much effort has been given to finding the weakly (1-5%) populated superdeformed bands from which no connection to any known yrast state has been observed. By contrast, in ^{72}Se the superdeformed band is connected to the near-spherical ground band and is yrast by 4^+ above which it totally dominates the high spin state population in heavy ion reactions.[29] In summary, $^{76}_{38}Sr_{38}$ and $^{98,100}_{38}Sr_{60,62}$, are new ground state, double magic nuclei but now for deformed shapes. The underlying physics of superdeformation in all regions, as noted in Table 1, is the reinforcement of the protons and neutrons having shell gaps at the same superdeformation, $\beta_2 \sim 0.4$ and $\beta_2 \sim 0.6$ (refs. 25, 26). The reinforcing shell gaps responsible for each region are given in Table 1. Now $^{152}_{66}Dy_{86}$ and $^{192}_{80}Hg_{112}$ likewise have been called deformed double magic, but at high spins in those cases.

The ground states quickly become near spherical, as one moves away from these new deformed magic numbers by only four protons. Moreover, which spherical and deformed shell gap seen in Fig. 5 is important switches for N and Z of 38 and 40 depending on what number the partner nucleon has.[25,26] So, $^{90,96}_{40}Zr_{50,56}$ and $^{68}_{28}Ni_{40}$ are spherical double magic nuclei and $^{88}_{38}Sr_{50}$ is essentially double magic from the reinforcement of the Z = 38, 40 spherical gaps by the strong spherical shell gap at N = 50 and even the weaker subshell gap at N = 56 for Z = 40. The reinforcement of two 38 gaps yield larger deformation than the two 40 gaps. Very recently the N = Z line has been extended to $^{84}_{42}Mo$ (ref. 30). Only a single γ ray was seen in the recoil mass-Z - γ spectrum at 443.8 keV. This result shows that the Z = N = 42 shell gaps at larger deformation are already being over-ridden by the very strong Z = N = 50

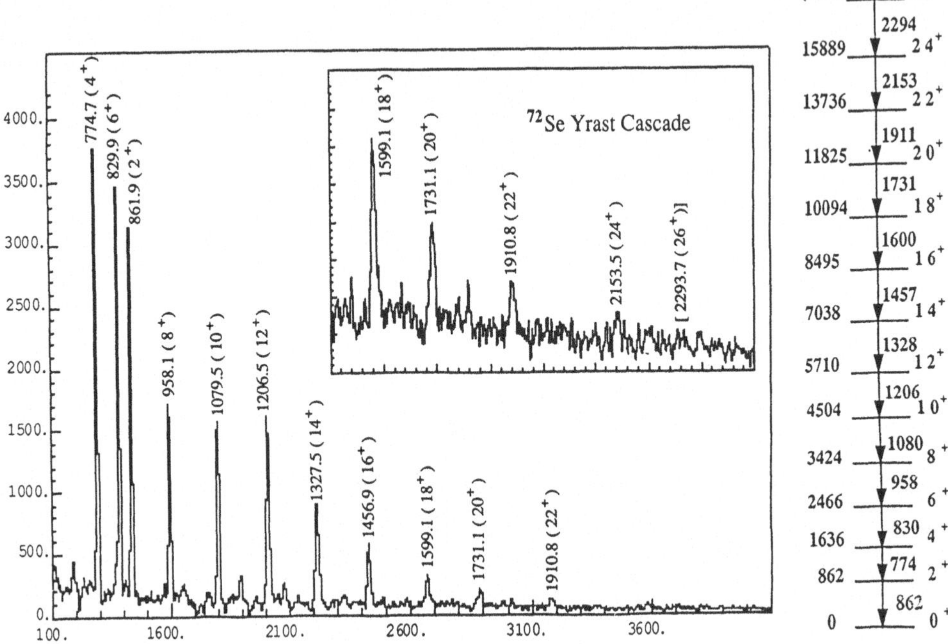

Fig. 6. Several coincidence spectra for ^{72}Se and levels observed.[29]

spherical shell gaps. This again shows how there can be switching back and forth between shell gaps at different deformation in determining the shape of a nuclear ground state. Also, Schuessler et al.[31] have extended measurements of the nuclear charge radii of the Kr isotope across the N = 50 shell closure. The charge radii are found to increase even more rapidly toward large deformation above N = 50 than below.

One of the important questions to be answered in this region is why the Z = N = 36 shell gaps at similarly large oblate deformation, $\beta_2 \sim -0.4$, is not clearly seen in $^{72}_{36}$Kr$_{36}$. Our recent work[32] has shown that the side band reported[33] in ^{72}Kr is not correct as placed and that there is a much greater anomaly in the behavior of the 2-0 transition (and moment of inertia) compared to the higher spin states in ^{72}Kr than in $^{74-80}$Kr (see Fig. 7). Following the same extrapolation used earlier,[24] our data point to a clear switch in ground state properties for ^{72}Kr compared to $^{74-76}$Kr and allow the possibility that the ^{72}Kr ground state has $\beta_2 \sim -0.4$.

A final note on the reinforcing and switching of shell gaps in nuclei comes from recent calculations of exotic cluster-decays for ^{78}Sr and ^{80}Zr by Gupta et al.[34] One of the exciting discoveries of the last decade was the new form of radioactivity, heavy cluster decays such as ^{14}C and heavier, as predicted by Sandulescu, Poenaru, and Greiner.[35] Gupta et al.[34] calculated the probabilities for cluster decays of $^{78}_{38}$Sr$_{40}$ and $^{80}_{40}$Zr$_{40}$. To illustrate the results, the emission of ^{24}Mg and ^{28}Si clusters have decay constants of (2.6×10^{-113}, 3×10^{-117}) and (2×10^{-54}; 1.8×10^{-56}) for ^{78}Sr and ^{80}Zr, respectively. So, ^{78}Sr is many orders of magnitude more stable. These calculations provide further evidence for the reinforcing effect and switch in importance from the spherical 40 to the superdeformed 38 shell gap.

The GSI separator group has studied by multinucleon transfer reactions, the proton-rich nuclei in the mass 60-70 region to extend our

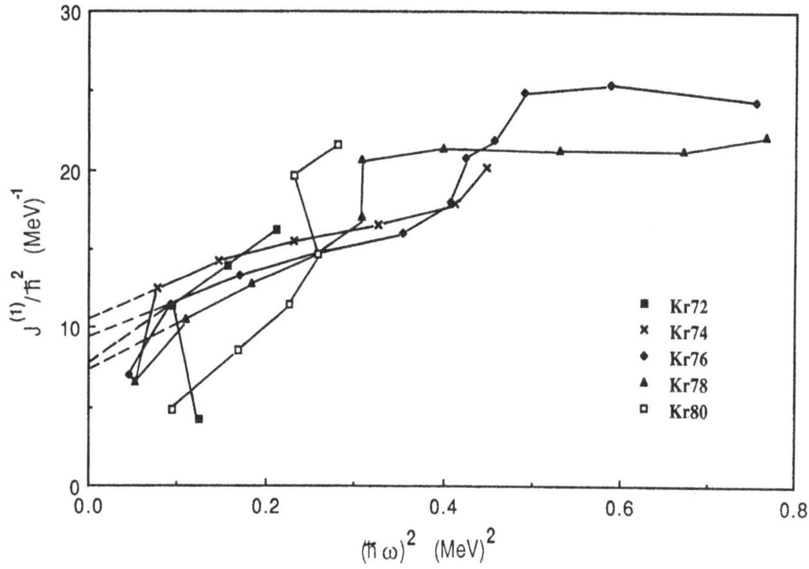

Fig. 7. Moments of inertia vs. $(h\omega)^2$ for $^{72-80}$Kr (ref. 32).

knowledge of these isotopes, including spherical-double-magic ^{68}Ni. These nuclei are of interest in calculations of the r-process in nuclear synthesis.[36] Two remaining major goals in probing whether spherical-double-magic nuclei persist very far off stability are $^{78}_{28}Ni_{50}$ and $^{100}_{50}Sn_{50}$. Recently, the new isotopes ^{71,72,73,74}Ni, produced in the very asymmetric thermal fission of ^{235}U and ^{239}Pu, were identified and their half-lives measured at the I.L.L. high flux reactor and Lohengrin spectrometer.[37] These data represent a major step toward ^{78}Ni. It is interesting that while the half-lives of the ^{66,67}Co and ^{69}Ni are significantly shorter than predicted by different theories, those of the new, heavier nickel isotope agree with the calculations.

Hill et al.[38] have sought to approach ^{78}Ni from above by studying the higher Z, N = 50 isotones. At the TRISTAN separtor, they have recently studied the decay of ^{83}Ge to $^{83}_{33}As_{50}$. In addition, they report studies of ^{74,76}Cu to $^{74,76}_{30}Zn_{44,46}$. They assumed ^{83}As could be considered as double-magic ^{78}Ni plus 5 protons in the $f_{5/2}$, $p_{3/2}$, $p_{1/2}$, and $g_{9/2}$ orbitals. They found reasonable agreement between the observed levels and calculations in a spherical shell model. There was no need to introduce any intruder orbitals to describe the low-lying levels.

The systematics[38] of the even A-zinc isotopes for A = 60 to 76 are shown in Fig. 8. The surprising feature of these levels is that for N = 30 to 38, the 2_1^+ energy is essentially constant around 1 MeV. Then it falls sharply to be constant around 600 keV for N = 42 to 46. This sharp drop to an approximately constant value of 0.6 MeV cannot yet be explained in terms of simple shell model calculations. It is particularly surprising for $^{76}_{30}Zn_{46}$ which is only four neutrons from the strong shell gap at N = 50. It will be very interesting to see what the 2_1^+ energy is in ^{78}Zn, only two neutrons from N = 50.

E0 TRANSITIONS BETWEEN SHAPE COEXISTING STRUCTURES IN MASS 120 REGION

Mantica et al.[39] have studied with the UNISOR separator the levels

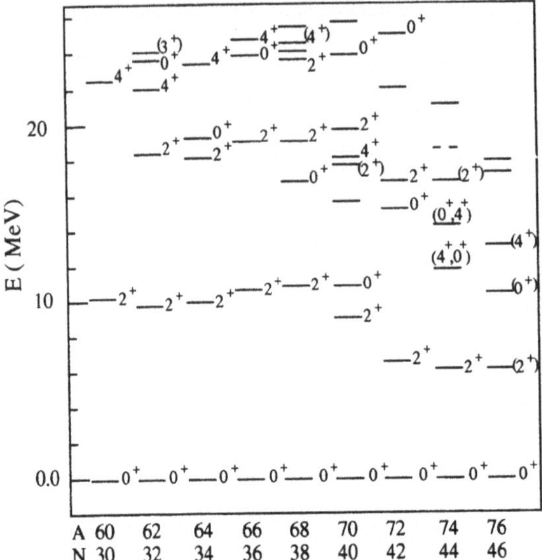

Fig. 8. Systematics for even-A Zn isotopes from A = 60 to 76.

of ^{120}Xe and compared its structure to the even-even Xe isotopes below N = 82 and the N = 66 mid-shell isotones above and below the Z = 50 closed shell with an emphasis on the role of intruder orbitals and shape coexistence. They note that the odd-mass nuclides in this region have low energy "intruder" $9/2^+$ states that have been attributed to the promotion of a proton from below the Z = 50 shell gap with the promoted proton paired with the "normal" $g_{7/2}$ or $d_{5/2}$ configuration to yield a $g_{9/2}{}^{-1}$ hole state. The mean square charge radii of the Xe nuclides show a gradual increase in deformation in going from N = 82 to mid-shell at N = 66 with β_2 = 0.3 for ^{120}Xe, using an axially-symmetric shape. The 2_1^+ energies in the even-even Xe nuclei decrease smoothly from both sides toward a minimum in mid-shell N = 66 ^{120}Xe.

The promotion of two protons across the Z = 50 proton shell closure can yield a relatively low energy 0^+ excited state in even-even nuclei, especially around N = mid-shell. This intruder state should have a quite different radius compared to the normal proton configuration. While γ decay is forbidden between a 0^+ intruder and 0^+ ground states, E0 internal conversion, with strength proportional to the difference in their nuclear radii, can be strong. Such E0 transitions also may occur in any ΔI = 0 transition between bands built on the two different shapes. No evidence for enhanced E0 strength in $^{124-126,128}$Xe has been found.[40,41,42] Careful measurements of the internal conversion coefficients in the UNISOR work revealed six pure E0(0^+ - 0^+) transitions and four ΔI = 0, E0 enhanced transitions in ^{120}Xe in sharp contrast with $^{124-130}$Xe. The levels of ^{120}Xe are shown in Fig. 9 where ΔI=0 transitions with enhanced E0 strength, and 0-0 transitions are shown by heavier lines. These levels are best described by a deformed ground band and γ vibrational band with 2_γ^+ at 876 keV with the same radius and a near-spherical excited band with one- and two-phonon-vibrational states built on it. The strong E0 transitions occur between these bands with different radii but not between the γ and ground band, as is well-known in rare earth nuclei. The levels of $^{120}Xe_{60}$

are just reversed from what is found in spherical closed shell Z = 50, ${}^{116}_{50}Sn_{66}$ (ref. 43) where the ground band is spherical, and a low-lying excited band is well deformed (Fig. 10). Similar structures to ${}^{120}Xe$, though somewhat distorted, are found in ${}^{118,122}Xe$. The N = 66 isotones (Fig. 10) show a remarkable symmetry around N = 50 where the deformed band is very high in energy (1757 keV) dropping to 1135 keV in ${}^{114}_{48}Cd$ and, 951 in ${}^{118}_{52}Te$, then becoming the ground state of ${}^{112}_{46}Pd$ and ${}^{120}_{54}Xe$ with the onset of well-deformed structures for Z = 56, 58 and 44, 42.

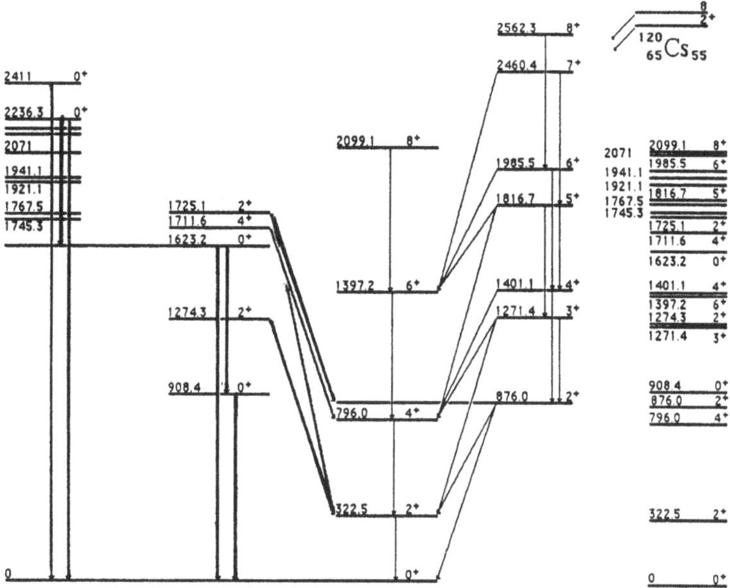

Fig. 9. ${}^{120}Xe$ levels with only selected γ-ray transitions are shown. Heavy lines indicate the presence of E0 radiation.

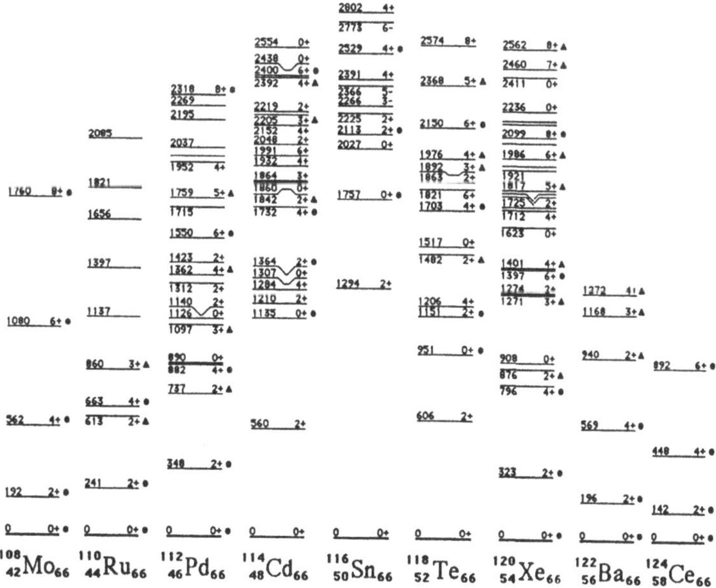

Fig. 10. Systematics of the N=66 isotones near the Z=50 closed shell. The dots indicate deformed states.

The region bounded by $78 \leq Z \leq 82$, $104 \leq N \leq 114$ was one of the first regions where shape coexistence was established[44] and continues as the most fertile ground for the studies of multiple coexistence of different shapes, intruder orbitals, electric monopole transitions, superdeformation and other phenomena. Here, I now can only select a few highlights. The even-even $^{184-188}$Hg isotopes were the first classic examples of nuclear shape coexistence where full bands of levels were seen above and below the crossing of their near-spherical ground states and well-deformed shapes. As seen in Fig. 11, in-beam studies by our group and the Canberra group traced the minimum and subsequent rise at lower N in the band head of the deformed minimum in ^{180,182}Hg. Although not predicted in some theoretical calculations, recent UNISOR work[45] located the deformed band in ^{190}Hg which has risen quite high in energy and has considerably smaller deformation. These data, coupled with the superdeformation discovered at moderately higher spins in ^{190}Hg demonstrate the complexity of competing shapes relatively near the ground state in this region. Recent calculations reproduce very nicely the prolate second minimum band and its smaller deformation in ^{190}Hg (ref.46).

One of the signatures of the shape coexisting structures in this region is the enhanced E0 transitions in $\Delta I = 0$ and $0^+ - 0^+$ transitions in this region. Zganjar et al.[47] recently reviewed the UNISOR work in which a remarkable richness of E0 transitions has been observed in this region, including for the first time in odd-A nuclei. Most surprising in the discovery[47] that some $I^\pi \rightarrow I^\pi$ $(I \neq 0)$ transitions have no observable γ intensity and are essentially pure E0. Earlier, four different shape coexisting structures were observed in ^{185,187}Au connected by enhanced E0 transitions. These four different structures are formed by coupling hole states to the two shape coexisting structures in ^{186,188}Hg and particle states to shape coexisting structures in ^{184,186}Pt (ref. 48). Recently they have identified very converted transitions (VCTs) ($\alpha_K >> \alpha_K$ M1 or E2) in the following nuclei: two VCTs in ^{195}Pb (ref. 49), ten VCTs in ^{185}Pt (ref. 50), and three VCTs in ^{193}Pt (ref. 51). Even more astonishing, they find[50] several transitions in ^{185}Pt and one in ^{194}Pt for which only conversion electrons are observed -- no gamma rays are seen! This is an unprecedented phenomenon in nuclei for $I^\pi \rightarrow I^\pi$ transitions with $I \neq 0$. However, a study of ^{181}Pt (ref. 52) has not yielded any VCTs. This will be shown to be consistent with a band-crossing hypothesis. The entire collection of low-energy, low-lying, odd-mass VCTs are shown in Fig. 12. The figure also displays the lowest known excited 0^+ states across the mass surface. The $176 \leq A \leq 187$ region obviously exhibits a distinct correlation between low-lying excited 0^+ and low-lying VCTs in odd-mass nuclei. The VCTs are interpreted as resulting from electric monopole transitions between shape coexisting structures which mix.

In the Pt isotopes the effect is dramatic because of the crossing[47] of the coexisting band heads near A = 186 where the $h_{9/2}$ intruder is shown to penetrate and fall below the normal ground state for A = 187 to 177. Mixing between the bands formed on these states would result in an abundance of E0 transitions in ^{187}Pt and ^{185}Pt at the crossing, diminishing for ^{183}Pt, and reappearing at the very light end of this isotope chain, probably near ^{178}Pt. The ten VCTs observed in ^{185}Pt, three in ^{183}Pt, and none in ^{181}Pt are consistent with this picture.

The lifetimes of the 0^+ states in ^{190,192,194}Pb have been measured by α-e-t coincidences.[53] For a deformation of 0.17-0.18 for the 0_2^+ excited shape coexisting states, the mixing of the spherical ground states and the deformed excited states was found to be quite small,

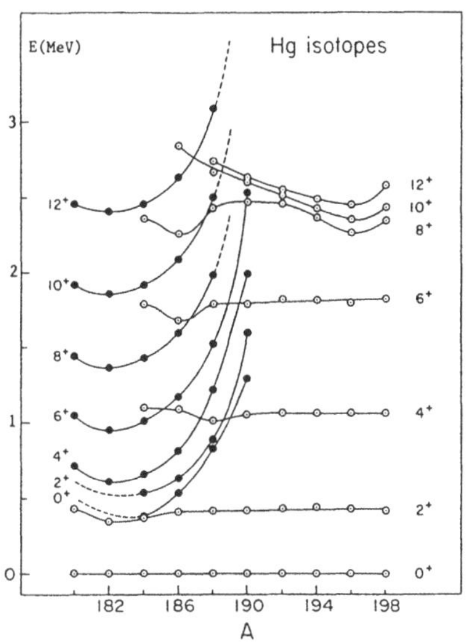

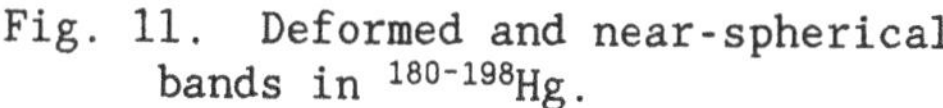
Fig. 11. Deformed and near-spherical bands in $^{180\text{-}198}$Hg.

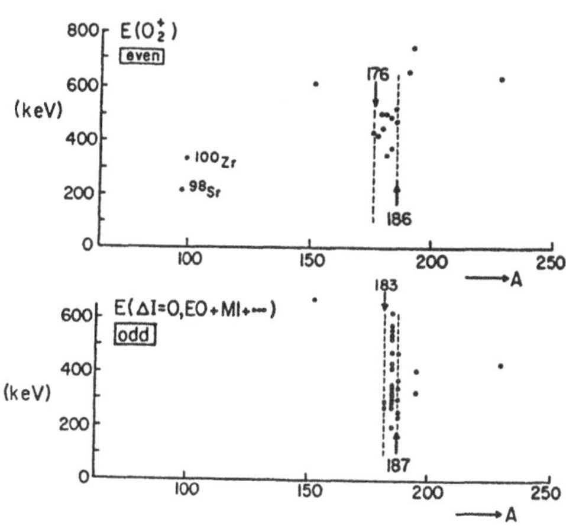

Fig. 12. The energy of 0_2^+ states in e-e nuclei below 800 keV (upper part) and the energy of transitions with large E0 components in odd-A nuclei.

b^2=0.003, 0.005 and ≥0.002 for ^{194}Pb to ^{190}Pb, respectively, where $|0^+\rangle_D$ = b|sph> - a|def>. These very small mixings are too small to explain the variations in the mean square charge radii which have been measured down to ^{192}Pb (ref. 54).

Finally, our recent work on ^{186}Hg brings to light several surprises to call our attention to the fact that when we look at greater detail nuclei in and near their ground states, they may be still further nuggets of gold to be mined. We collected 90 × 10^6 triple coincidence events in an in-beam study of ^{186}Hg with the close packed Ge ball at HHIRF. The level structure is shown in Fig. 13 (ref. 55). The first surprise is the large number of new bands which fall so near the ground state labeled 4-8 which were not seen in earlier work.[56] We have no explanation of the origin of so many bands so near the ground state of this even-even nucleus. Possibly some are related to a third minimum in the potential, such as a superdeformed minimum. The second point is the band labeled 3 is established in our work to be a negative-parity band based on the dipole character of the depopulating transitions. The earlier work without DCO ratios called this a positive-parity band because it was then yrast and positive-parity yrast bands are strongly populated in heavy ion reactions. This change in band structure underscores the dangers of even a very reasonable systematic extrapolation, and the importance of solid data to pin down interpretations--a point made earlier in the neon mass measurements. Finally, the last surprises are the two bands 6 and 7 built on a high K^π = 8^- isomer. This isomer is established in the N = 106 isotones from ^{176}Yb to ^{184}Pt, all of which have prolate-deformed ground states. To see this isomer in a near-spherical nucleus is surprising. I am reminded that in 1976 when I first presented our evidence for shape coexistence in ^{186}Hg in Dubna,[44] Jerry Brown did not believe there was shape coexistence there with two, so well-defined minimum in the potential so separate bands could be built in the second minimum. Now, we see the second minimum is so important that we find a high K isomer for the excited well-deformed prolate shape.

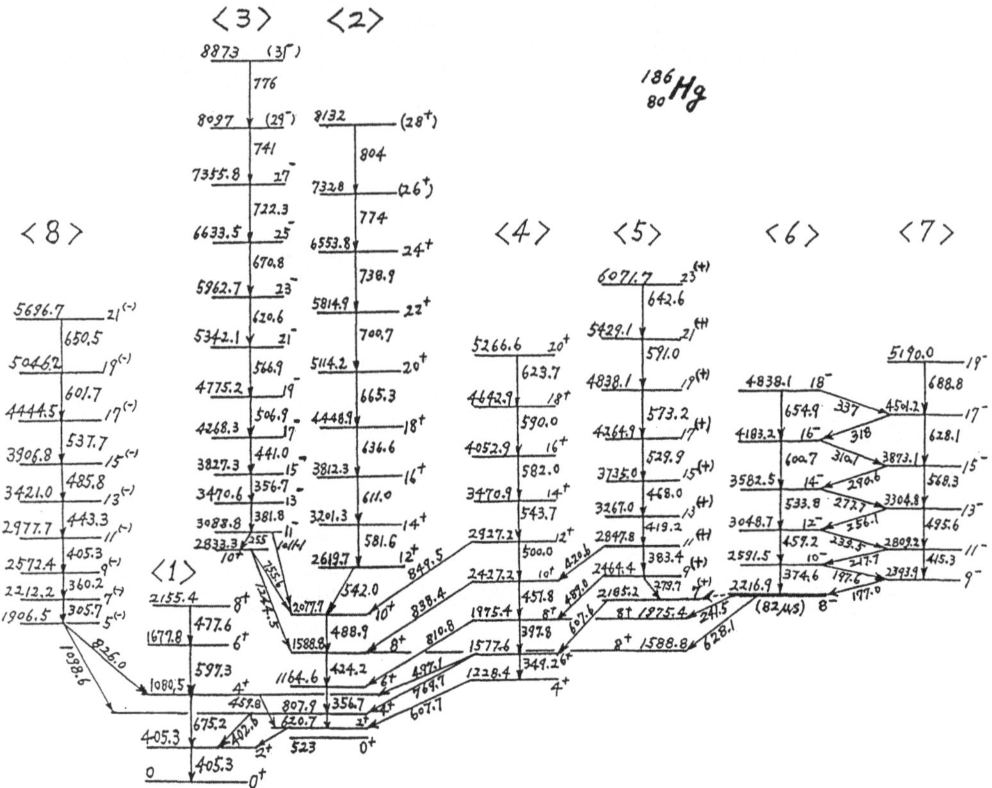

Fig. 13. Unexpected low energy bands in ^{186}Hg (ref. 55).

MIXED SYMMETRY STATES -- THE SCISSOR MODE

Mixed symmetry states occur in nuclei when the collective motions of the protons and neutrons are not in phase. Their out-of-phase motion is analogous to the opening and closing of a pair of scissors and has been called the scissor mode. These states are of particular interest from the standpoint of nuclear models because they require a separate treatment of neutrons and protons. In vibrational and γ-soft nuclei the neutron and proton vibrations can couple in phase to produce the 2_1^+ level and couple antisymmetrically to produce another 2_{ms}^+ state which is the lowest mixed symmetry state, typically around 2 MeV. In rotational nuclei a 1^+ level at about 3 MeV is the lowest MS level. The MS states are characterized by strongly enhanced M1 transitions compared to single-particle strength. In the 2^+ MS states, the 2_{ms}^+ - 2_1^+ transition will have very small E2/M1 mixing ratio since E2 is also retarded there. Such modes of out-of-phase neutron and proton motion were suggested many years ago by Faessler[57] and Maruhn-Rezwani et al.[58], but received little attention until the work of Richter and collaborators. Richter[59] has summarized their discovery of the so-called M1 scissor mode for 1^+ states around 3 MeV with enhanced M1 strength in well-deformed rare earth nuclei from ee′ scattering studies.

Mixed symmetry 2^+ states also have been identified in several nuclei including ^{56}Fe (ref. 60), ^{54}Cr (ref. 61), and N = 84 ^{140}Ba, ^{142}Ce and ^{144}Nd (ref. 62). Recently Robinson et al.[63] have analyzed the data for N = 30 ^{52}Ti, ^{54}Cr and ^{56}F, and the results are shown in Fig. 14. Note the surprising difference in the separation and mixing of the MS and second symmetric 2^+ states from nucleus to nucleus. In ^{54}Cr the MS strength is

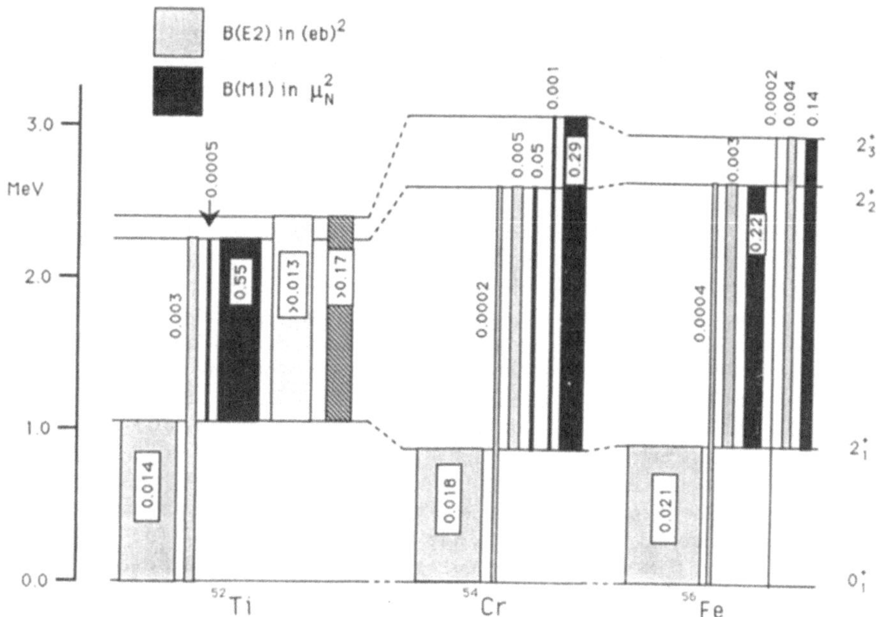

Fig. 14. Experimentally determined 2^+ levels and absolute transition strengths for the N = 30 isotones from ref. 62.

concentrated in the 2_3^+ level while in ^{56}Fe the 2_2^+ and 2_3^+ levels have strong mixing of the MS and symmetric states. In ^{52}Ti it appears that there is no mixing of the 2_2^+ and 2_3^+ states despite the fact they are the closest in energy of the three nuclei. They assumed in IBM-2 that only the 2-d boson symmetric and 1-d boson mixed symmetry states are close enough to show significant mixing and were able to nicely reproduce these different mixings and symmetries.[63] Areas where MS states may be easily identifiable have been discussed recently.[64] This analysis has lead to searches for such states in the Pt and Ge nuclei at UNISOR by using on-line nuclear orientation to measure δ(E2/M1) values for the 2_2^+ and 2_3^+ state decays to 2_1^+.

ON-LINE NUCLEAR ORIENTATION: NUCLEAR MOMENTS AND FUNDAMENTAL INTERACTIONS

On-Line Nuclear Orientation (OLNO) facilities in which a low temperature He refrigerator is coupled on line to an isotope separator which in turn is on line to an accelerator are in operation at Leuven (LISOL), Daresbury (DOLIS), Oak Ridge (UNISOR), and CERN (ISOLDE). The 1988 On-Line Nuclear Orientation and Hyperfine Interactions Conference Proceedings[65] document the rapidly developing importance of this technique to provide new and, in many cases, previously inaccessible information on the spins, parities and magnetic moments of nuclear levels, multipole mixing ratios of their connecting transitions, α and β correlations and other tests of fundamental interactions as well as solid-state properties of matter. Such data are providing significant new insights into the shapes and structures in and near the ground states of nuclei far from stability to test and extend current nuclear models.

OLNO studies of the $^{182-186}$Au and ^{185,187,189}Pt decays at ISOLDE[66] illustrate the power of the technique to measure nuclear moments. Both nuclear orientation (NO) and nuclear magnetic resonance on oriented nuclei (NMR/ON) studies were carried out. The g-factors and magnetic moments of ^{185,187}Pt were measured by NMR/ON. Strong electric quadrupole alignments were observed for ^{185,187,189}Pt in Zn. The ratios of the quadrupole moments for ^{189}Pt/^{191}Pt, ^{187}Pt/^{189}Pt and ^{185}Pt/^{189}Pt agree with the laser work[67] both in magnitude and in the sign change of the quadrupole moment in ^{185}Pt.

Another recent example of the power of OLNO is the study of the deformation of the light Br isotopes in an Oxford-Daresbury-Bonn-ISOLDE collaboration[68]. The strong competition among the oblate shell gaps at N = Z = 34,36, the spherical gaps at 38, 40, and prolate gaps at N = Z = 38 leads to shape coexisting structures as discussed above. The OLNO measurements of magnetic moments of ^{77,75,74m}Br combined with earlier values for ^{79,76}Br yield unambiguous assignments of the Nilsson orbitals involved and the deformation in these nuclei. There is a similar shift in ground state deformation between $^{79}_{35}$Br$_{44}$ and $^{77,75}_{35}$Br$_{42,40}$ (ref. 68) just as seen in the Kr isotopes[24], however, ^{74m}Br unexpectedly has only a small oblate deformation. There may be shape coexistence in ^{74}Br with the low spin ground state having large prolate deformation. This work also demonstrates the power of OLNO studies to measure ground state spins of nuclei far off stability which are essential to establish the spins of extensive band structures obtained by in-beam work where the ground state spin was unknown. We could not assign spins to bands in ^{73,75}Br (ref. 69) observed by in-beam studies because the ground state spins were unknown. This new work now provides the ground state spin of ^{75}Br.

Finally, consider two recent exciting examples of using both on- and off-line nuclear orientation to probe the weak and the electromagnetic interactions. Severijn et al.[70] have reported the first OLNO study of the β-decay anisotropies of mirror nuclei ^{15}O and ^{17}F with the Leuven KOOL facility. OLNO measurements with β-rays significantly enlarge the number of nuclei which now can be studied. More important, they open up measurements of the longitudinal polarization of betas emitted by polarized nuclei. In such measurements, the presence of right-handed currents in nuclear beta decay can be tested, in some cases, with up to an-order-of-magnitude enhancement in sensitivity compared to current measurements with unpolarized nuclei.[71] Studies of ^{15}O and ^{17}F are being pursued at Leuven.[72] Rikovska et al.[73] are investigating Time reversal symmetry violation which is expected at a level of 10^{-3} in the electromagnetic interaction. Since no T violation has been found in particle physics where it was predicted from CP violation in the decay of neutral kaons, it is of great interest to search for it in nuclear processes. They have estimated they can reach an accuracy of 10^{-4} in the 605 keV, M1 + E2 transition in ^{192}Ir decay to significantly improve the reported value[74] of $4(5) \times 10^{-3}$ in ^{192}Ir.

OCTUPOLE DEFORMATION

Nuclei with permanent octupole deformation do not have reflection symmetry (see the classic paper of Leander et al.[75]). Reflection asymmetry gives rise to a doubling of states with opposite parity in odd-A nuclei, as first observed in the actinide elements is illustrated for ^{223}Th shown in Fig. 15 (ref. 76). The sequences of rotational states in ^{223}Th can be understood as the coupling of an octupole deformed core with an unpaired neutron. Octupole deformation and octupole instability are the subject of much experimental and theoretical effort.[77] In even-even nuclei one observes even- and odd-parity rotational bands, as shown in

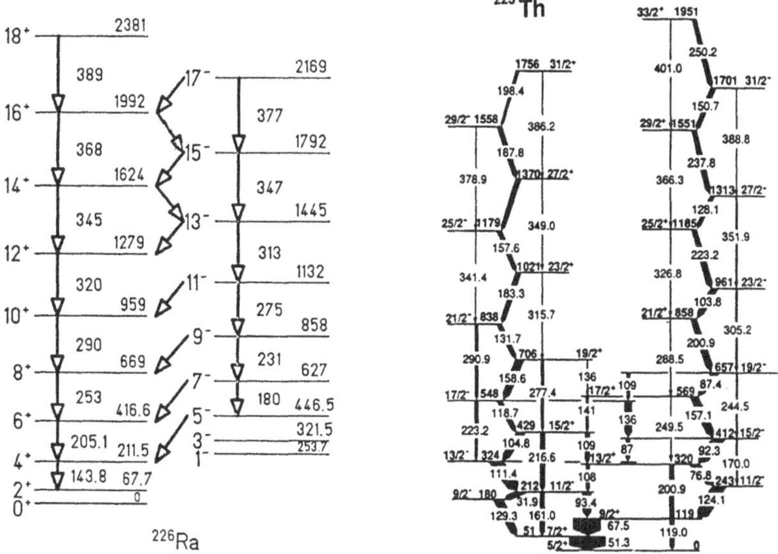

Fig. 15. Partial level schemes of ^{223}Th and ^{226} Ra (ref. 76).

Fig. 15. The GSI Coulomb excitation group have Coulomb excited the radioactive target ^{226}Ra($t_{1/2}$ = 1600 yr) with a ^{208}Pb beam.[78] They extracted electric dipole, quadrupole and, for the first time, octupole transition moments up to spin 18. Their extracted β_2, β_3 and β_4 deformations for ^{226}Ra are in excellent agreement with theoretical values[75] for a stable octupole deformation. Similar octupole deformations have been observed in other regions of the periodic table, for example in ^{144,146}Ba (ref. 79) which exhibits structures like those seen[78] in ^{226}Ba. Effects of octupole instability in super-deformed rotational configurations also are expected.[77]

Both theoretical and experimental investigations of octupole deformations have focused on the Ra-Th (Z ≈ 88, N ≈ 134) and Ba-Sm (Z ≈ 56, N ≈ 88) nuclei. There exists some experimental evidence that indicates an instability with respect to the octupole degree of freedom for N ≈ 94 nuclei. In ^{162}Er and ^{164}Yb, the lowest negative-parity band is seen to be systematically depressed about 200 keV relative to the next lowest negative-parity band, and there is a lack of signature splitting in the lowest negative-parity decay sequence in the neighboring odd-N isotopes. To clarify the possibility of strong interaction between the octupole and aligned quasineutron configurations observed in the N = 94 erbium and ytterbium, we extracted the B(E1)/B(E2) branching ratios from our ^{164}Yb lifetime data.[80] As a result of the strong octupole band character, one might expect enhanced E1 transitions from the (-, 1)-band to the yrast band. The E1/E2 branching ratios were extracted for ^{162,166}Yb and ^{164}Yb and, indeed, the branching ratios in ^{164}Yb are enhanced by more than a factor of 4 with respect to the neighboring isotopes. Furthermore, the absolute B(E1) values of ^{164}Yb are of the order of 10^{-3} W.u. comparable to those observed in the Ba-Sm nuclei. The enhanced octupole strength for ^{164}Yb probably is the result of the proximity of the second pair of octupole-driving orbits; i.e. the $\Omega = 3/2$ components of the $\nu i_{13/2}$ and $\nu f_{7/2}$ states near the Fermi level for N = 94. However, the octupole instability in ^{164}Yb is less pronounced than for Ba-Sm nuclei [$\pi(h_{11/2} \otimes d_{5/2})$ and $\nu(i_{13/2} \otimes f_{7/2})$] and Ra-Th nuclei [$\pi(i_{13/2} \otimes f_{7/2})$ and $\nu(j_{15/2} \otimes g_{9/2})$], where at least one set of the octupole driving orbitals both have $\Omega = 1/2$. These systematics indicate that the larger spacial localization of the $\Omega = 1/2$, high-j configurations probably is important in stabilizing the octupole shape. Nevertheless, ^{164}Yb provides an example of enhanced octupole deformation arising from the less pronounced pair of octupole-driving orbitals, i.e. $\nu(i_{13/2} \otimes f_{7/2})$ with $\Omega = 3/2$.

ACKNOWLEDGEMENTS

I wish to thank my co-workers for their permission to use our results prior to publication and the many people who provided preprints. The work at Vanderbilt and UNISOR was supported by the U.S. Dept. of Energy under Grants DE-FG05-88ER40407 and DE-AC05-76OR00033. I particularly thank Prof. A. Gallmann for his kind hospitality in Strasbourg.

REFERENCES

1. C. Detraz and D. J. Vieira, Ann. Rev. Nucl. Part. Sci. 39:402 (1989).
2. M.F. Mohar et al., Phys. Rev. Lett. 66:1571 (1991).
3. J. Jänecke and P. J. Masson, At. and Nucl. Data Tables 39:265 (1988).
4. D. Guillemand-Mueller, in "Int. School-Seminar on Heavy Ion Physics," Y. Oganessian, ed., Dubna Press, USSR (1990), p. 376.
5. P.G. Hansen and B. Jonson, Europhys. Lett. 4:409 (1987)
6. A.B. Migdal, Sov. J. Nucl. Phys. 13:238 (1973).
7. I. Tanihata, "Nuclear Structure in the Nineties," N. R. Johnson, ed., North-Holland, Amsterdam; Nucl. Phys. A520:411c (1990).
7b. R. Bimbot, in "Heavy Ion Physics, Today and Toorrow, World Scientific (in press.
8. R. Anne et al., Phys. Lett. B250:19 (1990).
9. K.K. Seth and B. Parker, Phys. Rev. Lett. 66:2448 (1991).
10. G. Audi, in "Workshop on Nuclear Structure of Light Nuclei Far From Stability: Experiment and Theory", (Obernai, France, Nov. 1989).
11. A.G. Artukh et al., in "Int. School on Heavy Ion Physics," Dubna, USSR, 1989, p. 385.
12. C. Thibault et al., Phys. Rev. C12:644 (1975).
13. Y. Schutz, in "Nuclear Structure in the Nineties," N. R. Johnson, ed., North-Holland, Amsterdam; Nucl. Phys. A520:401c (1990).
14. P. Baumann et al., Phys. Lett. B228:458 (1989).
15. A. Poves and J. Retamosa, Phys. Lett. B184:311 (1987).
16. W. Gerace and A. Green, Nucl. Phys. A93:110 (1967);A123:241 (1969).
17. H. Morinaga, Phys. Rev. 101:254 (1956).
18. P. Twin et al., Phys. Rev. Lett. 57:811 (1986).
19. V.M. Strutinsky, Ark. Fys. 36:629 (1967); Nucl. Phys. A95:420 (1967).
20. S. Bjørhnolm and J. E. Lynn, Rev. Mod. Phys. 52:725 (1982).
21. J. Schirmer et al., Phys. Rev. Lett. 63:2196 (1989).
22. J.H. Hamilton et al., Phys. Rev. Lett. 32:239 (1974); 36:340 (1976).
23. R.E. Azuma et al., Phys. Lett. 86B:5 (1979).
24. J.H. Hamilton et al., in "Proc. IVth Int. Conf. on Nuclei Far From Stability," CERN 81-09 (1981), p. 391; R. B. Piercey et al., Phys. Rev. Lett. 47:1514 (1981).
25. J.H. Hamilton, in "Proc. Int. Symp. on Nuclear Collectivity," A. Arima and T. Marumori, eds., INS, Tokyo (1981), p. 87.
26. J.H. Hamilton, in "Progress in Particle and Nuclear Physics, A. Faessler, ed., Pergamon Press, NY (1985), p. 107.
27. C. J. Lister et al., AIP Proc. 164 Nuclei Far From Stability, I. Towner, ed., AIP, NY (1988), p. 354.
28. P. Möller and J. R. Nix, Nucl. Phys. A301:199 (1981); At. Data Nucl. Data Tables 267:165 (1981).
29. L. Chaturvedi et al., Phys. Rev., in press.
30. W. Gelletly et al., Phys. Lett. 253B:289 (1991).
31. H.A. Schuessler et al., Phys. Rev. Lett. 65:1332 (1990).
32. H. Dejbakhsh et al., Phys. Lett. 249:195 (1990).

33. B.J. Varley et al., Phys. Lett. B194:1463 (1987).
34. R.K. Gupta, W. Scheid and W. Greiner, preprint in press.
35. A. Sandulescu et al., Sov. J. Part. Nucl. 11:528 (1980).
36. U. Bosch et al., Phys. Lett. 164B:22 (1985).
37. M. Bernas et al. A. Phys. A:Atomic Nuclei 336:41(1990).
38. J.C. Hill et al., in "Exotic Nuclear Spectroscopy," W. C. McHarris, ed., Plenum Press, NY (1990), p. 111.
39. P.F. Mantica et al., in "Exotic Nuclear Spectroscopy," W. C. McHarris, ed., Plenum Press, NY (1990), p. 495.
40. W.B. Walters et al., Hyperfine Int. 43:343 (1988).
41. P.F. Mantica et al., Bull. Am. Phys. Soc. 34:1236 (1989).
42. S. Ohya et al., INS Tokyo Prog. Rept., 40 (1980).
43. J. Bron et al., Nucl. Phys. A318:335 (1979).
44. J.H. Hamilton et al., Phys. Rev. Lett. 35:562 (1975); "Proc. Int. Conf. on Selected Topics in Nuclear Structure, Vol. 1, V. G. Soloviev et al., eds., JINR, Dubna (1976), p. 303.
45. M.O. Kortelahti et al., Phys. Rev. C43:488 (1991).
46. J.Y. Zhang and J. H. Hamilton, Phys. Lett., in press.
47. E.F. Zganjar and J. L. Wood, in "Nuclear Structure in the Nineties," N. R. Johnson, ed., North-Holland, Nucl. Phys. A520:427c (1990).
48. C.D. Papanicolopulos et al., Z. Phys. A330:371 (1988).
49. J.C. Griffin, Ph.D., GA Tech., 1987; and work in progress.
50. J. Schwarzenberg et al., Bull. Am. Phys. Soc. 33:1603 (1988).
51. J. Schwarzenberg et al., Bull. Am. Phys. Soc. 31:1825 (1989).
52. J. Schwarzenberg et al., work in progrss.
53. P. Dendoover et al., Phys. Lett. B226:27 (1989).
54. U. Dinger et al., Z. Phys. A328:253 (1987).
55. W.C. Ma et al., Vanderbilt University, private communication.
56. R.V.F. Janssens et al., Phys. Lett. 131B:35 (1983).
57. A. Faessler, Nucl. Phys. 85:635 (1966).
58. V. Maruhn-Rezwani et al., Phys. Lett. 57B:109 (1975).
59. A. Richter, in "Interactions and Structures in Nuclei," R. J. Blin-Stoyle and W. D. Hamilton, eds., Adam Hilger: Bristol
60. S. H. A. Eid et al., Phys. Lett. 166B:267 (1986); S. P. Collins et al., J. Phys. G Nucl. Part Phys. 51:321 (1989).
61. K. P. Lieb et al., Phys. Lett. 215B:50 (1988).
62. W. D. Hamilton et al., Phys. Rev. Lett. 53:2469 (1984).
63. S. J. Robinson et al., private communication.
64. W. D. Hamilton, J. Phys. G Nucl. Phys. 16:741 (1990).
65. T. L. Shaw and N. J. Stone, Hyperfine Int. 43 (1988); and Atomic and Nucl. Data Tables, to be published.
66. Report of Research at NICOLE, 4/89, private comm. from N.J. Stone.
67. B. Roussiere et al., Hyperfine Int. 43:473 (1988); Phys. Lett. B217:401 (1989).
68. A. G. Griffiths et al., private communication form N. J. Stone.
69. S. Wen et al., J. Phys. G. Lett. 11:L173 (1985).
70. N. Severijn et al., Phys. Rev. Lett. 63:1050 (1989).
71. P. A. Quin and T. A. Girard, Phys. Letts., to be published.
72. J. Deutsch et al., to be published.
73. J. Rikovska et al., Int. Nucl. Physics Conf., Brazil, Vol. 1 (1989), p. 102.
74. M. J. Holmes et al., Nucl. Phys. A199:401 (1973).
75. G. A. Leander et al., Nucl. Phys., A388:452 (1982).
76. M. Dahlinger et al., Nucl. Phys. A484:337 (1988).
77. W. Nazarewicz,"Nuclear Structure in the Nineties," N. R. Johnson, ed., North-Holland, Amsterdam; Nucl. Phys. A520:333c (1990).
78. H. J. Wollersheim, GSI-90-27 preprint (1990).
79. W. R. Phillips et al., Phys. Rev. Lett. 57:3257 (1986).
80. H. Xie et al., in "Proc. Int. Conf. on Nuclear Structure in the Nineties," Vol. 1:47 (1990); and Ph.D. Thesis Vanderbilt Univ.

NUCLEAR SHAPES AND NUCLEAR STRUCTURE AT LOW EXCITATION ENERGIES :

THEORETICAL ASPECTS

K. Heyde

Institute for Theoretical Physics and Institute for Nuclear Physics
Proeftuinstraat 86, B-9000 Gent, Belgium

1. INTRODUCTION

In describing nuclear structure and observables that give information on the nuclear shape at low excitation energy, one has to outline the different approximation methods with which to describe the interacting A-nucleon system as close as possible. In the present discussion we shall describe three approximation schemes that contain most of the theoretical efforts carried out up to now. In a first approximation we start from a spherical Hartree-Fock field to which the effects of residual interactions amongst a number of valence nucleons are added. This method presents its limitations when the number of valence nucleons is increasing so that the residual interaction energy becomes a substantial part of the total nuclear binding energy. We shall discuss various "signals" indicating a

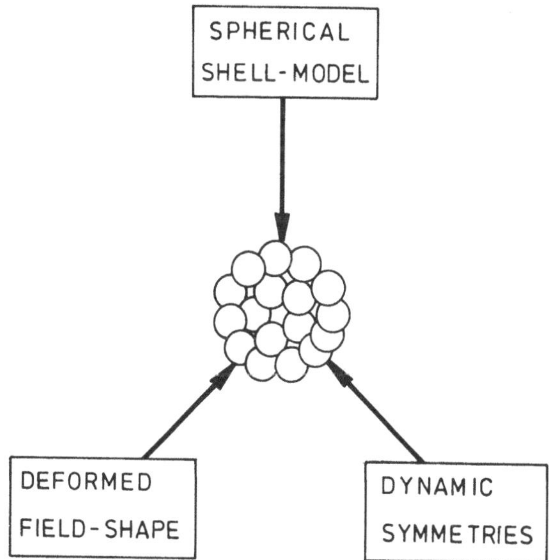

Fig.1. Schematic presentation of three large classes of approximation schemes to nuclear structure : (i) a spherical shell-model approach, (ii) incorporating nuclear deformation and shape variables, explicitely and, (iii) using (dynamical) symmetries to classify nuclear structure.

Nuclear Shapes and Nuclear Structure at Low Excitation Energies
Edited by M. Vergnes *et al.*, Plenum Press, New York, 1992

breakdown of this spherical shell-model approach. In a second approximation, the concept of shape is introduced from the very beginning and the importance of various multipole field producing interactions in accentuated. Here, the importance of the proton-neutron interaction is illustrated from a more microscopic approach. We present some state-of-the-art studies in this field. In a third approximation, the importance of symmetries in outlining basic low-lying nuclear structure properties is discussed. In this respect, the role played by various group-theoretical methods (such as the interacting boson model (IBM)) will be presented.

In assessing which avenue to use for describing a certain class of nuclei according to the above three approximation methods, one should study first which zero-order picture comes closest to the particular mass region under study. This will strongly depend on the number of valence nucleons outside of the closed shells and the ratio of the residual interaction energy $<H_{res}>$ to the energy related to the zero-order Hamiltonian $<H_0>$ (see Fig.1.).

We also indicate the various links and relations that exist between the different major approximation schemes, outlining in some regions their equivalence and in other regions their limitations and possible complementarity.

2. SPHERICAL SHELL-MODEL APPROACH TO NUCLEAR STRUCTURE

For nuclear interacting A-nucleon system, using self-consistent Hartree-Fock studies leads to the possibility of a natural separation between nucleons, contained into a rather stable spherical closed-shell configuration and a remaining small number of extra valence nucleons. This separation is at the basis of the tractability of the nuclear shell-model itself and leads to a solvable Schrodinger equation if the model space available for the extra valence nucleons, to be distributed over the single-particle orbitals, remains within certain limits. The present large-scale shell-model techniques, though, have led to a consistent description of e.g. the full sd shell-model space spanning nuclei from ^{16}O up to ^{40}Ca[1] (see Fig.2). We shall point out later various difficulties that can arise in treating such very large-scale shell-model calculations up to its technical limits.

In the shell-model approach, one should try, in particular, to use a single nucleon-nucleon interaction in order to determine both the average field and the remaining residual interaction in a consistent way. This is a rather ambitious task but a number of groups have tried such an approach using density-dependent nucleon-nucleon interactions (Skyrme forces)[2-5]. We shall illustrate the

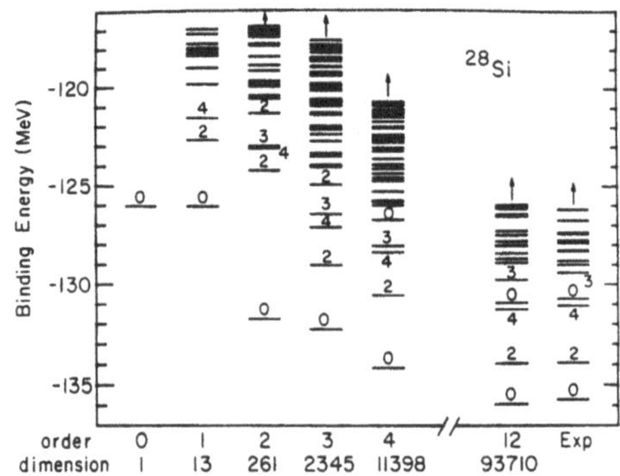

Fig.2. Binding energies of levels in ^{28}Si as a function of the number of excitations from the $(1d_{5/2})$ orbital (closed subshell $(1d_{5/2})^{12}(2s_{1/2})^0(1d_{3/2})^0$ configuration) (taken from Brown and Wildenthal[1]).

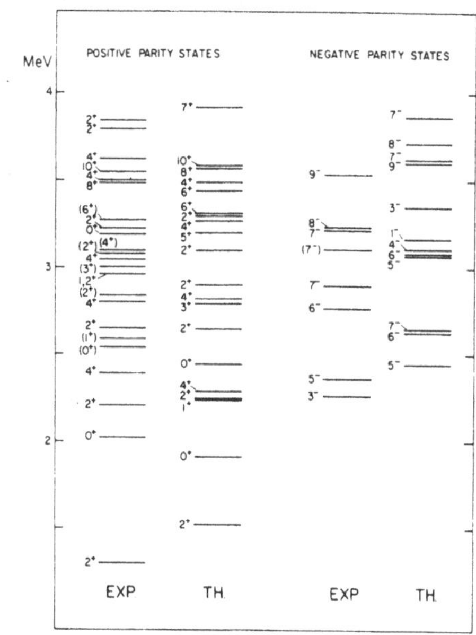

Fig.3. Experimental and calculated low-lying positive and negative parity states for ^{116}Sn. Levels belonging to the "intruder" band, based on a proton 2p-2h excitation across the Z=50 closed shell, are not included in the figure.

above procedure, starting from the extended Skyrme force constructed by Waroquier et al.[2], pointing out that both global (ground-state) properties such as nuclear binding energy, mass- and charge radii as well as detailed (nuclear energy spectra) properties can be correctly described. Most of the calculations using the above self-consistent shell-model studies, were performed with a rather small number of valence nucleons present, with a nucleus like ^{116}Sn presenting some of the most complex situations that could still be treated (see Fig.3). Of course, the possibility exists to use the self-consistent two-body matrix elements as the input to the large-scale shell-model studies. Instead, we like to point out the presence of some "signals" when comparing the shell-model calculations with the actual observed energy spectra and related observables.

Near, or even at the closed shells, extra states do show up even at rather low excitation energy that are outside of the standard chosen shell-model valence spaces[6,7]. It seems that instabilities of the spherical, closed core do show up. The coexisting, low-lying 0^+ states in ^{16}O, ^{40}Ca are some of the key examples in that respect[8,9]. In these situations, the spherical shell-gaps at the magic numbers are still large enough to prevent certain multi-particle multi-hole states to modify the ground-state structure. Zamick has shown the possibility for 4p-4h, 8p-8h and even 12p-12h excitations to remain rather close to the ground state because of the very large residual proton-neutron energy gain, compared to the very high-lying unperturbed picture[10]. In various other mass regions, either such low-lying "intruder" excitations show up in even-even, odd-mass and odd-odd nuclei[11] or, even worse, problems occur when comparing calculated and experimental ground-state binding energies (see Fig.4). A typical example is presented by the neutron-rich Na, Mg nuclei at the N=20 shell closure. It was more recently pointed out by various authors[12-14] that pair excitations across the N=20 neutron closed shell i.e. $(sd)^n \rightarrow (sd)^{n-2}\,(fp)^2$ can accomodate the extra binding energy (Fig.5).

Anyhow, it has become clear that even by extending the shell-model spaces and eventually reproducing the nuclear observables, the best zero-order picture no longer is a spherical shell-model (inert) core with a given number of interacting valence nucleons outside. Here, in a rather natural way,

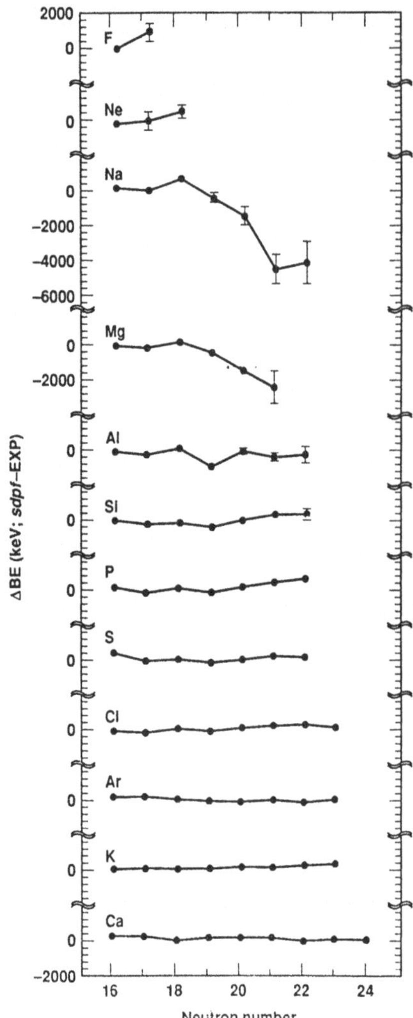

Fig.4. Difference between measured and calculated[13] binding energies for sd and sdpf nuclei with neutron number N≥16. (taken from Warburton et al.[13]).

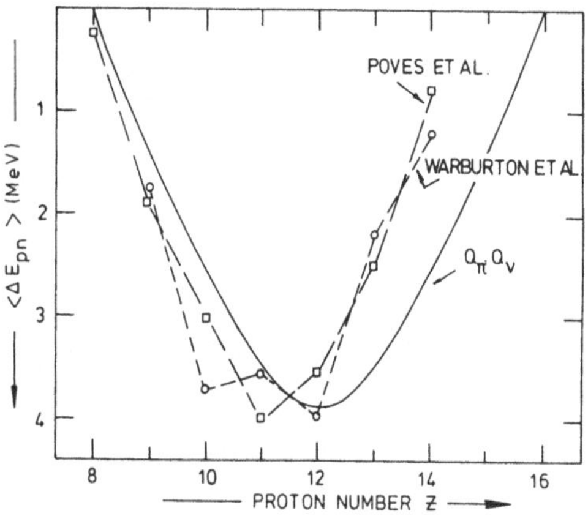

Fig.5. Proton-neutron correlation energy using various shell-model calculations[13,14] compared with the quadrupole binding energy withing the shell-model 0^+ intruder picture of Heyde et al.[11,12] (N=20).

the idea of "deformation" of the spherical field by low multipole components in the nucleon-nucleon interaction comes in.

3. NUCLEAR DEFORMATION

As forced upon us when using a spherical shell-model picture as long as possible, in certain regions of the nuclear mass table, the residual interaction energy can become a substantial part of the ground-state zero-order unperturbed energy or

$$\sum_i \varepsilon_i \approx \langle \psi_0 | H_{res} | \psi_0 \rangle \quad . \tag{1}$$

This signals the break-down of a separation in a closed, spherical core outside of which a number of valence nucleons feel the residual N-N interaction (see Fig. 6).

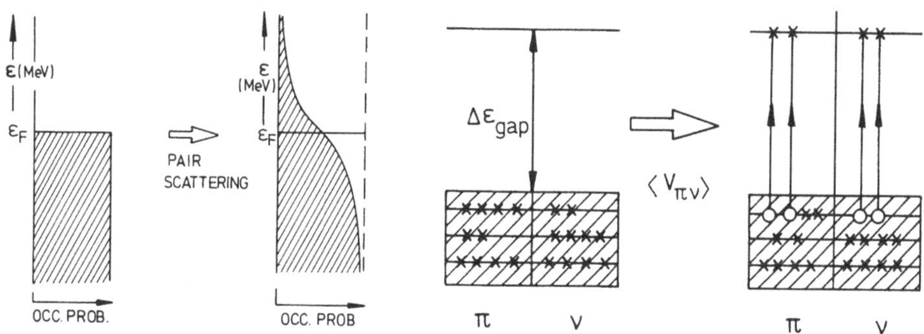

Fig.6. Examples (drawn in a schematic way) of instabilities of the zero-order spherical shell-model nucleon ground state distributions vs. (i) the pairing correlations and, (ii) proton-neutron residual interactions in creating coexisting excitations.

It has been pointed out that it is mainly that part coming from the proton-neutron interaction that is responsible for the onset of deformation and for the occurrence of intruder configurations at low excitation energy[6,7]. In such a regime, the pairing- and field producing (low multiple components) interactions act in a combined way to generate a new zero-order picture i.e. the picture of quasi-particle excitations acting as the elementary modes of excitation in a deformed, average field. So, the ground-state energy can be obtained as[15]

$$E_{HF+BCS} = Tr(T\rho) + \frac{1}{2} Tr(\Gamma\rho) + E_{pair} \quad , \tag{2}$$

where ρ denotes the one-body nuclear density, T describes the kinetic energy and Γ indicates the one-body average field, related with the two-body interaction V by the expression

$$\Gamma = Tr(V\rho) \quad . \tag{3}$$

Here, Tr(...) is the trace of matrices representing the operators T,ρ and Γ expressed in a suitable basis and E_{pair} describes the pair correlation energy.

A more interesting decomposition of the ground-state energy can be obtained by using a multipole decomposition and even separating out the proton-proton (pp), neutron-neutron (nn) and proton-neutron (pn) parts. Then, one can write

$$E_{HF+BCS} = E_0^{nn} + E_0^{pp} + E_0^{pn} + E_2^{nn} + E_2^{pp} + E_2^{pn} + \ldots \quad , \tag{4}$$

where $E_K^{\rho\rho'}$ denotes the K-multipole contribution from the (ρ,ρ') nucleon interaction. Applications using the above method to the Ge and Mo nuclei were performed by Dobaczewski et al.[16,17], using a Skyrme force as nucleon-nucleon interaction, with the main results

i) the (p,n) quadrupole part E_2^{pn} is much larger than E_2^{nn} and E_2^{pp} (see Fig.7),

ii) in plotting the difference δE = E − E(Q = 0) (with Q expressing the quadrupole moment), the monopole and quadrupole parts make up for the larger part of E (see Fig.8). A quite striking result from the above HF+BCS calculations is the cancellation effect between the monopole and the total quadrupole deformation energy. These same ingredients also showed up in the spherical shell-model description where the spherical shell-gap, the pairing- and monopole binding energy effects stabilize a spherical shape whereas it is the proton-neutron force mainly that causes "deformation" effects via the very low intruder energy or even changes the ground-state structure[11].

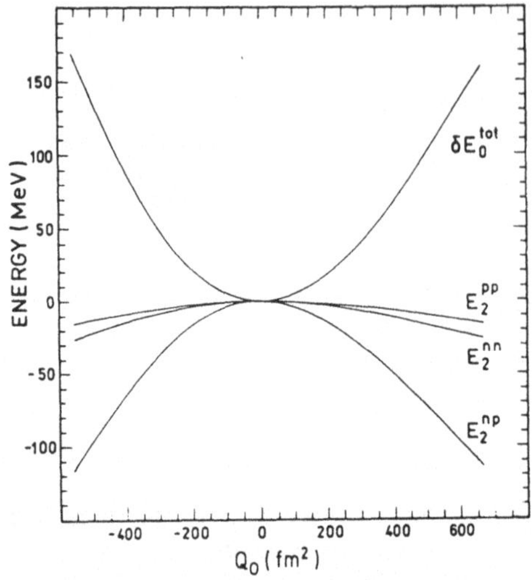

Fig.7. The total monopole, δE_0^{tot} and the quadrupole E_2^{nn}, E_2^{pp}, E_2^{pn} energies for ^{72}Ge as a function of the quadrupole moment Q_0 (taken from Dobaczewski et al.[17]).

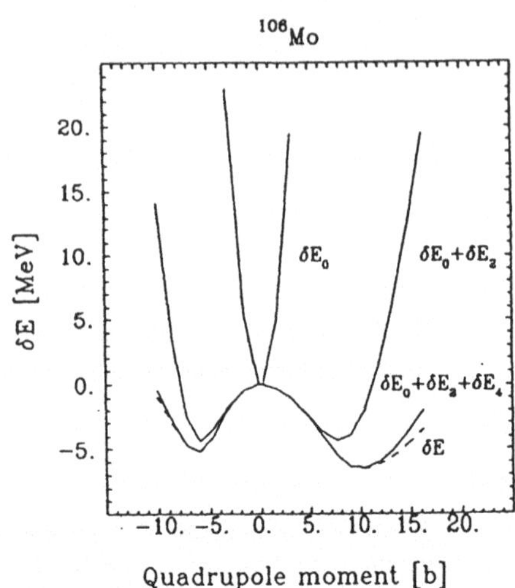

Fig.8. Deformation energy $\delta E = E\text{-}E(Q_0=0)$ for ^{106}Mo (dashed line), its monopole part δE_0 alone and corrections by the quadrupole (δE_2) and the hexadecapole (δE_4) binding energies (taken from Dobaczewski[16]).

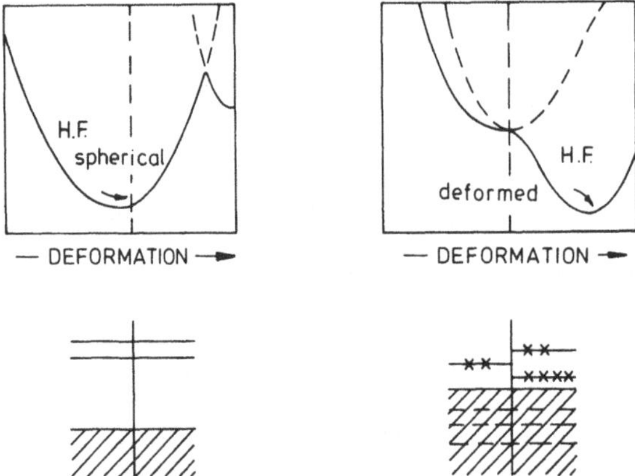

Fig.9. Balancing of spherical shape stabilizing force components (spherical shell gaps at magic numbers, pairing force components) vs. the deformation driving proton-neutron force components. On the left-hand side, a spherical ground-state and coexisting, excited deformed shape occur. On the right-hand side, deformation has set in to stabilize the nucleus in a permanently deformed shape.

Dobaczewski et al.[16,17] even pointed out that a close relation seems to exist between the value of the total deformation energy

$$\delta E^{eq.} \equiv E(Q_0^{eq.}) - E(Q_0{=}0) \quad , \tag{5}$$

calculated at the prolate minimum and the quantity N_pN_n (with N_p and N_n the number of valence protons and neutrons, respectively) for the Ge and Mo nuclei[18].

So, it becomes clear that by using fully microscopic HF + BCS calculations, starting from Skyrme forces (SkIII), one can gain deeper insight in the role of quadrupole vs. monopole components to the binding energy but also of the relative importance of the nn, pp and pn forces. Moreover, the HF + BCS calculations shed light on the appearance of low-lying intruder excitations near closed shells when simplifying the nuclear residual interaction to pairing and monopole variations of the average field and a quadrupole proton-neutron interaction[11]. We resume the above outcome in the schematic figure 9, where the balancing effects of a large shell-gap enforced by pairing vs. the deformation tendency of the quadrupole p-n force is depicted. As another example, we illustrate the changing nuclear structure along the N = Z line going from ^{16}O, ^{40}Ca towards the very neutron deficient N = Z nuclei near ^{84}Mo, ^{88}Ru, ...[19]. The sequence of low-lying coexisting deformed mp-mh 0^+ excitations in ^{16}O, ^{40}Ca, ^{56}Ni and deformed N = Z nuclei in the ground-state like ^{76}Sr, ^{80}Zr, ^{84}Mo is an interesting example of the above spherical-deformation competition which is awaiting the study of ^{100}Sn and even heavier N = Z nuclei with the advent of Radioactive Beam facilities (Fig.10).

In order to go now beyond the results obtained from HF + BCS ground-state energy deformation curves, obtained as adiabatic curves and approximating the static part of the nuclear system, methods to go beyond calculating the potential energy have been developed. These methods range from configuration constrained Hartree-Fock (CCHF) calculations including configuration mixing and Generator-Coordinate method (GCM) studies where various collective parameters (Q_2, Q_3, Δ, ...) are considered as dynamic collective variables to a microscopic determination of both the static and kinetic collective energy contributions with the subsequent study of the collective Bohr-Mottelson like Hamiltonian. These various methods are discussed in a recent review article by Aberg, Flocard and Nazarewicz[20] where references to the various original articles are given. In the present workshop, some of these methods will be discussed in more detail. It needs no comment that these studies very rapidly give rise to increasing numerical and technical problems (angular momentum projection,

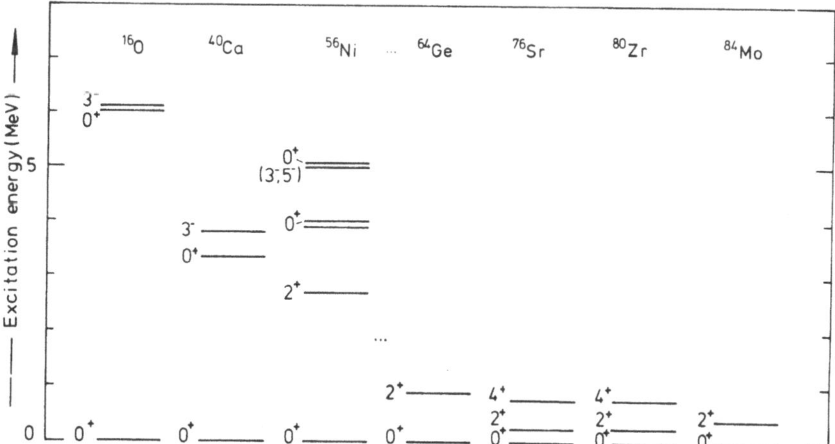

Fig.10. Low-lying excited states for some selected nuclei along the Z = N line. For ^{16}O, ^{40}Ca, ^{56}Ni, excited coexisting 0^+ states occur. In the heavier Z = N nuclei, the subshell effects at 38, 40 cannot prevent these nuclei to become deformed already in the ground-state band.

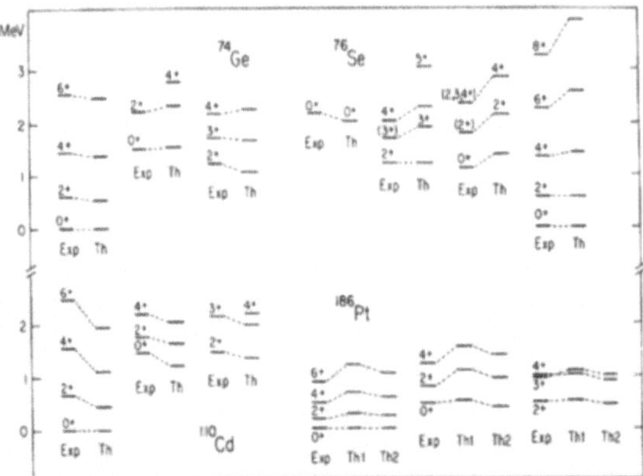

Fig.11. Comparison of experimental (Exp) and calculated (Th) spectra for four transitional nuclei. The two calculated sets in ^{186}Pt correspond to slighthly different pairing strengths (taken from I.Deloncle et al.[21]).

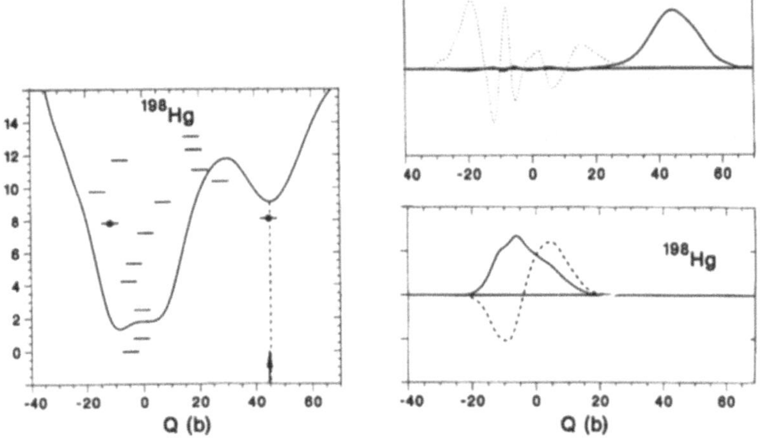

Fig.12. Hartree-Fock + BCS (HF + BCS) deformation energy curve (in MeV) as a function of the mass quadrupole moment Q (in barn). The horizontal line segments give the region where, in GCM, the mean values for the quadrupole moment for the corresponding state are localized. The solid dots give the same information but now for the decoupled states. The dashed vertical line points to the superdeformed minimum and the arrow to the GCM superdeformed state (left-hand part). In the right-hand part, the GCM collective wave functions for ^{198}Hg. In the lower part : (solid-line) ground-state and (dashed-line) first excited state and upper-part : (solid line) superdeformed and its nearest neighbor in energy (dotted line) (taken from Bonche et al.[22]).

particle number projection, ...). Still, we present a few examples (Figs.11,12) to illustrate the state-of-the-art using such methods in order to comprehend nuclear structure using the language of nuclear deformation as a zero-order picture but building the collective ingredients as far as possible starting from an underlying microscopic picture i.e. starting from a nucleon-nucleon two-body interaction as only input in order to determine the main degrees of freedom of the atomic nucleus[21,22)].

The various, microscopic calculations by K.W.Schmid et al[23)] in Tübingen center along similar ideas (MONSTER, VAMPIR, ...) using a self-consistent description of nuclear structure by starting from a HFB theory with spin and particle number projection. Using a variational set of calculations in determining the optimal basis for describing ground-state and excited states, the essential degrees of freedom in order to describe nuclear low-lying excitations, coexistence, oblate-prolate transitions are quite well taken care of. Applications relating to the study of shape coexistence at high spin in the ^{68,72}Ge nuclei were carried out recently[24)].

Concluding on this second general approach to describe low-lying nuclear structure properties by incorporating the concept of a deformed average one-body-field, it becomes clear that in order to describe the full dynamics of the nucleus where proton and neutron number are far removed from the spherical, closed-shell configurations, once again one is running into a very complex numerical problem. At this point, we like to pick-up a third, very elegant approach to study nuclear, collective motion mainly by starting from the underlying symmetries that might be present in the nuclear interacting A-nucleon system.

4. NUCLEAR SYMMETRIES

Before accentuating the importance of the concept of symmetries in describing low-lying nuclear physics properties, let us first go back to get a better feeling for the way symmetries and dynamical symmetries have been a constant guidance in describing various features of the nuclear many-body system.

1932 : The concept of isospin symmetry, describing the charge independende of the nuclear forces by means of the isospin concept with the SU(2) group as the underlying mathematical group[25)]. This is the simplest of all dynamical symmetries and expresses the invariance of the Hamiltonian against the exchange of any two nucleon coordinates.

1937 : Spin and isospin are combined by Wigner into the SU(4) supermultiplet scheme with SU(4) as the group structure[26)]. This concept has been extensively used in the description of light α-like nuclei (A = 4 x n).

1948 : The spherical symmetry of the nuclear mean field and the realization of its major importance for describing the nucleon motion in the nucleus has been put forward by Mayer[27)], Haxel, Jensen and Suess[28)].

1958 : Elliott remarked that in some cases, the average nuclear potential could be depicted by a deformed, harmonic oscillator containing the SU(3) dynamical symmetry[29,30)]. This work opened a first possible connection between the macroscopic collective motion and its microscopic description.

1943 : The nucleon residual interaction amongst identical nucleons is particularly strong in $J^\pi = 0^+$ coupled pair states. This "pairing" property is a corner stone in accounting for the nuclear structure of many spherical nuclei near closed shells in particular. Pairing is at the origin of seniority, itself related to the quasi-spin classification and group, as used first by Racah in describing the properties of many-electron configurations in atomic physics[31)].

1951-53 : The nuclear deformed field is a typical example of the concept of spontaneous symmetry breaking. The restoration of the rotational symmetry, present in the Hamiltonian, leads to the formation of nuclear rotational spectra. These properties were discussed before in a more phenomenological way by Bohr and Mottelson[32-34)].

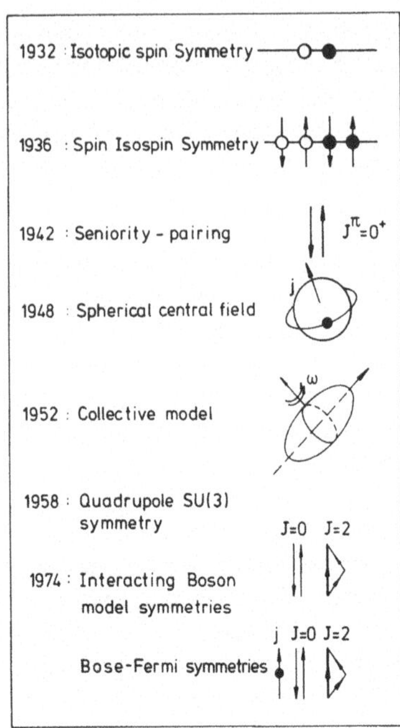

Fig.13. Pictorial representation of the most important nuclear symmetries developed over the years.

1974 : The introduction of dynamical symmetries in order to describe nuclear collective motion starting from a many-boson system with only s(L=0) and d(L=2) bosons is introduced by Arima and Iachello[35]. The relation to the nuclear shell model and its underlying shell-structure has been studied extensively[36]. These boson models have been giving rise to a new momentum in nuclear physics research. These symmetries are depicted schematically in Fig.13.

Here, we discuss in some more detail the dynamical symmetry properties related to classifying nucleon pairs, bound by the short-range attractive nucleon force mainly in 0^+ and 2^+ coupled pairs[35]. In a process of mapping the shell-model Hamiltonian, acting in a shell-model truncated space with 0^+ (S)-pairs and 2^+ (D)-pairs only, onto an exact boson model Hamiltonian acting in the corresponding sd-model space, an interacting boson model approximation to the nuclear shell-model is obtained[36]. This space, spanned by the s^+, d_μ^+ ($-2 \leq \mu \leq 2$) bosons spans a 6 dimensional space for which the transformation properties are given by the U(6) group. In Fig.14, the various acts of truncating and the final mapping are illustrated.

It has been shown over many years (since 1974) that this IBM-approach is a highly powerfull and elegant tool to describe nuclear, collective quadrupole motion accentuating the U(6) group structure and there is no need discussing the various well-known results [35,37] at the present workshop nor the various extensions to incorporate odd-mass nuclei (IBFM), odd-odd nuclei (IBFFM) and the more global systematic studies including ideas such as boson-fermion symmetries[38]. Explicit discussions can be found in two recent monographs[35,38]. I would like though to take some time to discuss a set of new proton-neutron "collective" modes obtained in coupling the proton and neutron IBM degrees of freedom into the IBM-2[39-41]. Within this extended model, it is possible to construct states that are not fully symmetric under the interchange of proton and neutron bosons. Such states, that are typical for all model descriptions where proton and neutron excitations are coupled, are called non-symmetric or mixed-symmetry states within the IBM-2. Such states present themselves as

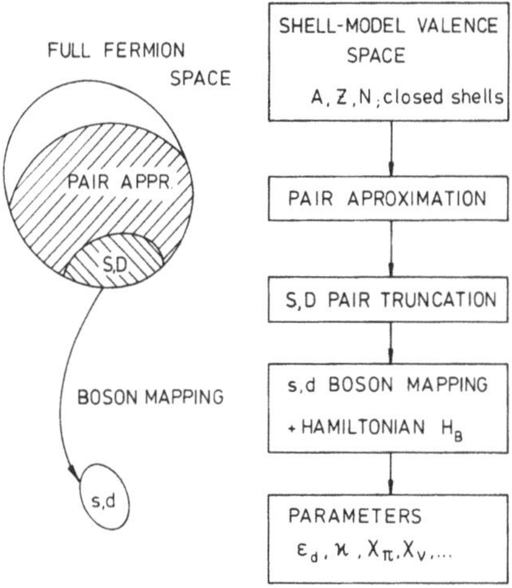

Fig.14. Schematic representation of the approximation underlying the interacting boson model when starting from the complete shell-model valence space.

i) Isovector-type of oscillatory motion of protons vs. neutrons with 2^+ vibrational-like excitations as the lowest-lying levels of such type (see Fig.15),

ii) rotational scissor-like excitations in deformed nuclei that give rise to 1^+ excitations as the lowest-lying levels of such type (see Fig.15).

The above type of excitations were also discussed within purely geometric models before the IBM-2 studies came along[42-44]. It was with the (e,e') inelastic scattering studies carried out by Richter et al[45] that the study of such 1^+, 2^+,...excitations fully started in a systematic way throughout most of the nuclear mass region. Detailed experimental work using different reaction probes such as (e,e')(γ,γ')(p,p') pointed towards important fragmentation going beyond a purely strongly concentrated type of low-lying collective excitation[46]. In studying summed B(M1) (or B(E2) in vibrational-like nuclei) values, however, a large part of the data can still be correlated using the IBM-2.

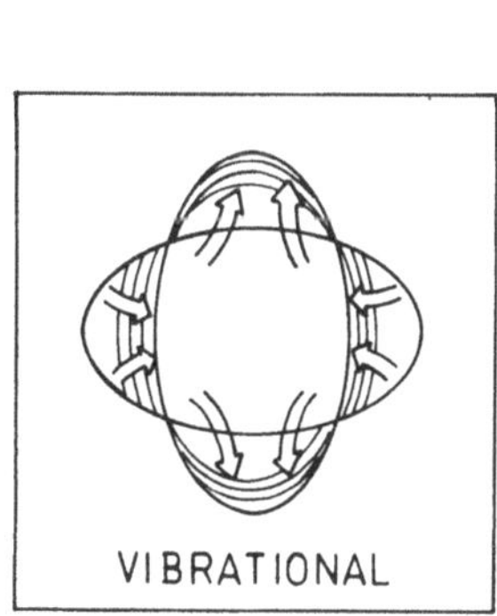

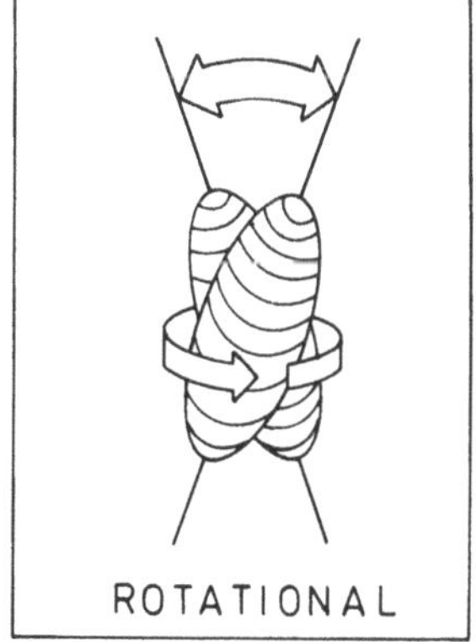

Fig.15. Pictorial representation of the lowest mixed-symmetry IBM-2 like excitations (i) vibrational, isovector 2^+ excitations and, (ii) rotational 1^+ scissor-like states.

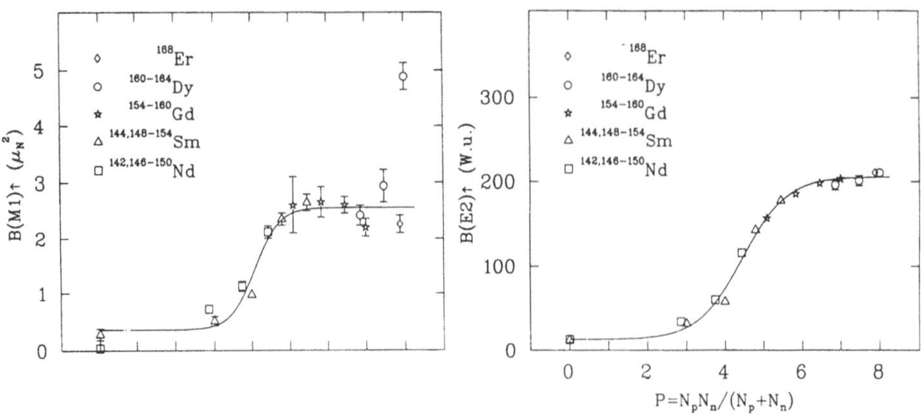

Fig.16. A comparison between the functional forms of the summed B(M1) strenght, up to $E_x \leq 4$ MeV (left part) and of the $B(E2;0_1^+ \rightarrow 2_1^+)$ values (right part) (taken from Rangacharyulu et al.[47]).

Recently, a close correlation between the summed $B(M1;0_1^+ \rightarrow 1_f^+)$ strength (with $E_x(1_f^+) \lesssim 4$ MeV) and the $B(E2;0_1^+ \rightarrow 2_1^+)$ E2 transition in transitional and deformed nuclei was observed[47,48], pointing towards a deeper connection between M1 and E2 observables (see Fig. 16). Making use of the dynamical symmetry exhibited by simple IBM-2 Hamiltonians, describing transitional and deformed nuclei, the linear energy-weighted M1 sum rule could be evaluated with the result

$$\sum_f B(M1;0_1^+ \rightarrow 1_f^+)\, E_x(1_f^+)$$

$$= \frac{\sqrt{3}}{2} < 0_1^+ \mid [[\hat{H}, T(M1)], T(M1)]^{(0)} \mid 0_1^+ >$$

$$= \frac{9}{2\pi} \left\{ \begin{matrix} -\kappa_{\pi\nu}/2 \\ -\kappa \end{matrix} \right\} (g_\pi - g_\nu)^2 < 0_1^+ \mid \hat{Q}_\pi \cdot \hat{Q}_\nu \mid 0_1^+ > \quad . \tag{6}$$

with $\hat{Q}_\rho$ ($\rho \equiv \pi,\nu$) the quadrupole boson operator, $\kappa_{\pi\nu}(\kappa)$ the quadrupole strength of the $\kappa_{\pi\nu}\hat{Q}_\pi \cdot \hat{Q}_\nu$ (or $\kappa(\hat{Q}_\pi + \hat{Q}_\nu) \cdot (\hat{Q}_\pi + \hat{Q}_\nu)$) proton-neutron interaction and $g_\rho(\rho \equiv \pi,\nu)$ the boson M1 gyromagnetic factors. By inserting a full set of 2_1^+ states in the extreme right-hand side of eq.6, the relation

$$\sum_f B(M1\,;0_1^+ \rightarrow 1_f^+) E_x(1_f^+) = \sum_f B(E2\,;0_1^+ \rightarrow 2_f^+) \quad , \tag{7}$$

can be obtained. Incorporating now the following reasonable approximations in eq. 7, i.e.

i) In the B(E2) sum rule, most of the isoscalar E2 strenght resides in the 2_1^+ level,

ii) The effective charge to be used in the E2 operator is taken to be a constant value of ≈ 0.12 eb,

iii) In the M1 sum rule, in particular when using the IBM-2 approach, most strength is concentrated in a scissor-like 1^+ state,

one can simplify eq. 7 into the simpler form

$$\sum_f B(M1\,;0_1^+ \rightarrow 1_f^+)\, \bar{E}_x(1^+\,;coll.) \cong B(E2\,;0_1^+ \rightarrow 2_1^+) \quad . \tag{8}$$

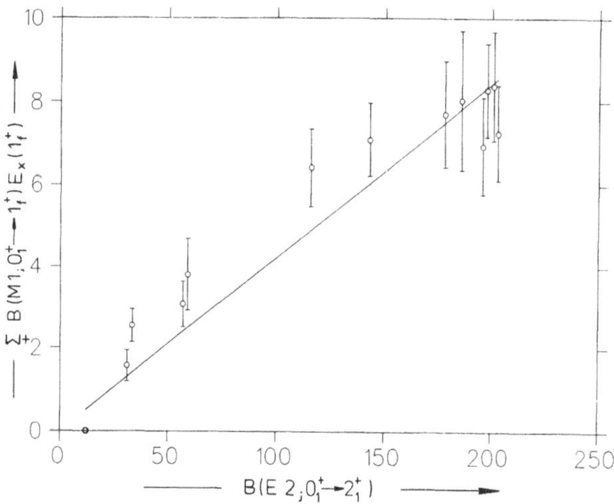

Fig.17. Comparison of the linear, energy weighted B(M1) sum rule (with $E_x \leq 4$MeV) with the B(E2;$0_1^+ \to 2_1^+$) value. The data correspond to the nuclei displayed in fig.16[49].

The data in the Nd, Sm, Gd, Dy, Er, Yb region indeed show such a relation to hold to a rather good degree[49](see Fig.17).

This somewhat unexpected result shows the powerful method contained in group-theoretical descriptions of nuclear collective motion and points out that summed M1 strength is proportional to the ground-state expectation value of the proton-neutron quadrupole interaction, and thus to the B(E2;$0_1^+ \to 2_1^+$) value.

Nuclear symmetry considerations go, of course, much beyong the U(6) sd-boson model of quadrupole collectivity. Recently, the importance of the pseudo-spin scheme to classify nuclear single-particle motion for heavy nuclei has been demonstrated. In particular, the general group structure for the many-particle pseudo LS shell model and the pseudo SU(3) scheme contain promising results relating to high-spin physics and symmetries connected with superdeformed shapes[50].

Finally, a microscopic approach to understand the nuclear collective model as a submodel of the nuclear shell model has been worked out, mainly by Rowe et al[51]. Using the symplectic Sp(3,R) group structure one can go a long way to accomplish this goal and, I am sure, D.Rowe will discuss this particular approach in some more detail.

5. CONCLUSION

Starting from the nuclear shell-model approach as the fundamental starting point in order to describe the nuclear observables, we have indicated three major ways of attack to the description of the elementary modes of excitation. The extended shell-model studies using large-scale shell-model diagonalizations within a spherical basis, the approach starting from a deformed basis incorporating the low multipoles from the nucleon-nucleon force in a non-spherical average field and the powerfull methods concentrating on (dynamical) symmetries of the A-nucleon system can give rise to ways for handling certain aspects of the nucleus in an easy way. These are the regions of the nuclear mass table where specific spherical shell-model aspects, specific deformation effects or specific symmetries dominate the observed phenomena and thus give a clear insight in the physics via the zero-order Hamiltonian. We have also discussed the "signals" that delineate the limits of easy applicability of the above approaches without running into enormeous complex numerical studies. Sometimes a given approximation can overtake another one that fails (onset of deformation near N=20 for Mg, Na nuclei). It is in these overlapping regions that it is the most difficult to come into grips with a transparent picture of what precisely determines the physics before another approximation takes over to determine the nuclear structure.

ACKNOWLEDGEMENTS

The author is most grateful to M.Waroquier, J.Ryckebusch, D.Van Neck for discussions on the self-consistent shell-model calculations ; to C.De Coster for discussions on magnetic M1 and collective scissor excitations. He is indebted to A.Richter, W.Ziegler and H.Wörtche for the use of material expressing a clear relation between M1 and E2 electromagnetic properties. He also likes to thank the IIKW and NFWO for financial support. Part of this research was supported by a research grant NATO RG/0452/86.

REFERENCES

1. B.A.Brown and B.H.Wildenthal ; Ann.Rev.Nucl.Part.Sci. 38 (1988), 29.
2. M.Waroquier, K.Heyde and G.Wenes, Nucl.Phys. A404 (1983), 269, 298.
3. D.Vautherin and D.M.Brink, Phys.Rev. C5 (1972), 626.
4. D.Gogny in : Nuclear Self-consistent fields, ed. G.Ripka, M.Porneuf (North-Holland, Amsterdam, 1975), p 333.
5. P.Quentin and H.Flocard, Ann.Rev.Nucl.Part.Sci. 28 (1978), 523.
6. K.Heyde, P.Van Isacker, M.Waroquier, J.L.Wood and R.A.Meyer, Phys.Repts. 102 (1983), 291.
7. J.L.Wood, K.Heyde, W.Nazarewicz, M.Huyse and P.Van Duppen, Phys.Repts, to be publ.
8. G.E.Brown and A.M.Green, Nucl.Phys. 75 (1966), 401 ; ibid 85 (1966), 87.
9. W.J.Gerace and A.M.Green, Nucl.Phys. A93 (1967), 110 ; ibid A113 (1968), 641 ; ibid A123 (1969), 241.
10. D.C.Zheng, D.Berdichevsky and L.Zamick, Phys.Rev.Lett. 60 (1988), 2262 ; Phys.Rev. C38 (1988), 437 ; ibid C42 (1990), 1004.
11. K.Heyde et al., Nucl.Phys. A466 (1987) 189 ; ibid A484 (1988), 275.
12. K.Heyde and J.L.Wood, J.Phys. G17 (1991), 135.
13. E.Warburton et al., Phys.Rev. C41 (1990), 447.
14. A.Poves and J.Retamosa, Phys.Lett. 184B (1987), 311.
15. P.Ring and P.Schuck, The Nuclear Many-body Problem, (Springer, Berlin-Heidelberg, 1980).
16. J.Dobaczewski in : Contemporary Topics in Nuclear Structure Physics (World Scientific, Singapore, 1988), p 227.
17. J.Dobaczewski, W.Nazarewicz, J.Skalski and T.Werner, Phys.Rev.Lett. 60 (1988) 2254.
18. R.F.Casten, Phys.Lett. 152B (1985), 145 ; Nucl.Phys. A443 (1985) 1.
19. K.Heyde and P.H.Heenen in : The Arenas Project (1990), p36.
20. S.Åberg, H.Flocard and W.Nazarewicz, Ann.Rev.Nucl.Part.Sci. 40 (1990), 439.
21. I.Deloncle, J.Libert, L.Bennour and P.Quentin, Phys.Lett. 233B (1989), 16.
22. P.Bonche et al., Nucl.Phys. A519 (1990), 509.
23. K.W.Schmid, N.Kyotoku, F.Grümmer and A.Faessler, Ann.Phys. (N.Y.) 190 (1989), 182.
24. A.Petrovici, K.W.Schmid, F.Grümmer and A.Faessler, Nucl.Phys. A504 (1989), 277 ; ibid A517 (1990), 108.
25. W.Heisenberg, Z.Phys. 77 (1932), 1.
26. E.P.Wigner, Phys.Rev. 51 (1937), 106.
27. M.G.Mayer, Phys.Rev. 75 (1949), 1969.

28. O.Haxel, J.H.D.Jensen and H.E.Suess, Phys.Rev. 75 (1949) 1766.
29. J.P.Elliott, Proc.Roy.Soc. A245 (1958), 128 ; 562.
30. J.P.Elliott and M.Harvey, Proc.Roy.Soc. A272 (1963), 557.
31. G.Racah, Phys.Rev. 63 (1943), 367.
32. A.Bohr, Phys.Rev. 81 (1951), 134.
33. A.Bohr, Mat.Fys.Medd.Dan.Vid.Selsk. 26, n° 14 (1951).
34. A.Bohr and B.Mottelson, Mat.Fys.Medd.Dan.Vid.Selks. 27, n°16 (1953).
35. F.Iachello and A.Arima, The Interacting Boson Model (Cambridge Univ. Press, 1988).
36. T.Otsuka, A.Arima and F.Iachello, Nucl.Phys. A309 (1978), 1.
37. R.F.Casten and D.D.Warner, Revs.Mod.Phys. 60 (1988), 389.
38. F.Iachello and P.Van Isacker, The Interacting Boson-Fermion Model (Cambridge Univ. Press), in press.
39. F.Iachello, Phys.Rev.Lett. 53 (1984), 1427.
40. P.Van Isacker, K.Heyde, J.Jolie and A.Sevrin, Ann.Phys. (N.Y.) 171 (1986), 253.
41. O.Scholten et al., Nucl.Phys. A438 (1985), 41.
42. N.Lo Iudice and F.Palumbo, Phys.Rev.Lett. 41 (1978), 1532 ; Nucl.Phys. A326 (1979), 193.
43. A.Faessler, Nucl.Phys. 85 (1966), 653.
44. W.Greiner, Nucl.Phys. 80 (1966), 417.
45. D.Bohle et al., Phys.Lett. 137B (1984), 27.
46. A.Richter, in The Nuclear shell model (eds. T.S.H.Lee and R.B.Wiringa) Nucl.Phys. A507 (1990), 99C.
47. C.Rangacharyulu et al., Phys.Rev. C43 (1991), 949.
48. K.Heyde and C.De Coster, to be publ.
49. H.Wörtche and C.De Coster, priv.comm.
50. J.P.Draayer and H.A.Naqvi, to be publ.
51. D.J.Rowe, Rep.Progr.Phys. 48 (1985), 1419.

ATOMIC PHYSICS AND NUCLEAR PROPERTIES

Jacques Bauche

Laboratoire Aimé Cotton
CNRS II
Bâtiment 505
91405 ORSAY, France

INTRODUCTION

The three topics which I propose to review in the present talk are, in my opinion, the main traces of nuclear shapes and nuclear structure in atomic spectra: (i) the electric-quadrupolar hyperfine structure, (ii) the magnetic-dipolar hyperfine-structure anomaly, and (iii) the field isotope shift.

This will leave aside several interesting others, among which: (i) the parity-violation effects, which will be addressed in this workshop by Ann-Marie Mårtensson-Pendrill, and (ii) the electronic and the muonic X-ray transitions, which are traditionally studied by nuclear, not atomic, physicists. Furthermore, I will not deal with the sophisticated experimental methods developed in the field, which Neugart will review here. However, I acknowledge that, historically, the progress in the theory of nuclear effects in atomic spectra has been thoroughly dependent on the experimental refinements.

As an introduction to the very heart of the review, it is interesting to write the simplest expansions of the hamiltonians of these nuclear effects, namely, of hyperfine structure (hfs) and isotope shift (IS).

Hyperfine Structure

$$H_{hfs} = H_m + H_Q$$

$$H_m = (2\,\beta\,\beta_n\,\mu_n/I)\ \Sigma_i \left([\,\mathbf{N}_i^{(1)}/r_i^3 + (8\pi/3)\ \delta(\mathbf{r}_i)\ \mathbf{s}_i^{(1)}\,] \cdot \mathbf{I}^{(1)} \right)$$
(if $I \geq 1/2$)

with $$\mathbf{N}^{(1)} = \boldsymbol{\ell}^{(1)} - (10)^{1/2}\ \{\mathbf{s}^{(1)}\ \mathbf{C}^{(2)}\}^{(1)}$$

$$H_Q = (3/2)^{1/2}\ [e^2\,Q\ /\ I(2I-1)]\ \Sigma_i \left((\mathbf{C}_i^{(2)}/\ r_i^3\ ) \cdot \{\mathbf{I}^{(1)}\mathbf{I}^{(1)}\}^{(2)} \right)$$
(if $I \geq 1$)

In the magnetic-dipolar hamiltonian H_m and the electric-quadrupolar hamiltonian H_Q, β and β_n are the Bohr and nuclear magnetons, respectively, μ_n is the nuclear magnetic-dipolar moment expressed in nuclear magnetons, I is the nuclear spin, Q is the nuclear electric-quadrupolar moment, and the sums run over all the electrons of the atom.

Isotope Shift

$$H_{IS} = H_{mass} + H_{vol}$$

$$H_{mass} = -\mathbf{P}^2_{nucl} / 2M_{nucl} = -(\Sigma_i \mathbf{p}_i)^2 / 2M_{nucl}$$

$$H_{vol} = \Sigma_i \delta(\mathbf{r}_i) \cdot k <r^2>_{nucl}$$

In the mass-shift hamiltonian H_{mass} and the field-shift hamiltonian H_{vol}, $\mathbf{P}_{nucl}$ and $\mathbf{p}_i$ are the momenta of the nucleus and of electron i, respectively, M_{nucl} is the nuclear mass, $<r^2>_{nucl}$ is the mean-square radius of the nuclear-charge distribution, k is a constant which will be addressed later, and the sums run over all the electrons of the atom.

It is noteworthy that, in the above expansions, both the hfs and IS hamiltonians are the sums of two parts, and that all four parts are products of an electronic and a nuclear term. Moreover, the nuclear term is a simple constant, or, in the quadrupolar part, a trivial tensor operator acting on the nuclear spin. Thus, in this presentation, the remaining problem is purely electronic.

When hfs is present, i.e., if $I \neq 0$, the total angular momentum of the atom is no more $\mathbf{J}$, but the sum $\mathbf{F} = \mathbf{I} + \mathbf{J}$, and all levels with $J \neq 0$ split into a number of F *sublevels*. For a reason given in the following, I have restricted the hfs hamiltonian to the first two terms of its multipolar expansion.

ELECTRIC-QUADRUPOLAR HYPERFINE STRUCTURE

Hyperfine-Structure Radial Parameters

The most successful theoretical scheme in Atomic Physics is the central-field model, invented by Slater in 1928. It is analogous to the nuclear shell model. But it works much better, because in the atom the nucleus is the president of the electron crowd, or cloud, whereas inside the nucleus there is no president!

From the expansion of the hfs hamiltonian, one deduces that, for any state of an electronic configuration, the hfs expectation value is a linear combination of electronic radial integrals, denoted

$$< r^{-3}>_{n\ell} \quad (\text{if } \ell \neq 0) \qquad \text{or} \qquad |\psi(0)|^2_{ns} \quad (\text{if } \ell = 0),$$

one for each electronic subshell which is open, i.e., neither complete nor empty.

These integrals are also called the radial parameters, because their values, which are difficult to determine *ab initio* accurately, are often freely optimised, in phenomenological interpretations. In this way, using his very accurate experimental measurements in the oxygen atom, Harvey[1] deduced in 1965 that the mixing with highly-excited configurations can be accounted for by associating one "effective" radial parameter to each of the angular operators of the hfs hamiltonian. For the magnetic hfs, these parameters read

$$\langle r^{-3}{}_{\ell}\rangle_{n\ell}\ ,\ \langle r^{-3}{}_{sC}\rangle_{n\ell}\ (\text{if } \ell \neq 0) \quad \text{and} \quad \langle r^{-3}{}_{s}\rangle_{n\ell}\ (\text{for any value of } \ell).$$

Among these parameters, the third one, for an ns electron, is equivalent to $|\ \psi(0)\ |^2_{ns}$. For the electric hfs, only

$$\langle r^{-3}{}_{Q}\rangle_{n\ell} \quad (\text{if } \ell \neq 0)$$

is needed. It differs from the simple (first-order) $\langle r^{-3}\rangle_{n\ell}$ value because the electrostatic field of the nuclear electric-quadrupolar moment Q polarises, i.e., it distorts, the complete electronic subshells.

The most striking difference between the first and the second lists of magnetic parameters is that, in the latter, an $\langle r^{-3}{}_{s}\rangle$ parameter can be assigned to any subshell, whatever its value. This is a consequence of the phenomenon of *spin-core polarisation*, discovered experimentally by Abragam et al.[2] in 1955. Physically, it can be described in two ways. Taking as a simple example the $1s^22p$ electronic configuration of the lithium atom, one can say equivalently that an $\langle r^{-3}{}_{s}\rangle_{2p}$ parameter occurs: (i) because the 2p electron polarises the complete $1s^2$ subshell in such a way that the probability amplitudes of the "spin-up" and "spin-down" 1s electrons at r = 0 do not cancel any more in the total hfs expectation value, and (ii) because the $1s^22p$ configuration mixes with high singly-excited 1sns2p configurations, as an effect of the interelectronic electrostatic interaction.

The relativity phenomenon also brings in corrections, the so-called Casimir corrections, which have been analysed anew by Sandars and Beck[3] in 1965. The latter authors have shown that each magnetic-hfs parameter is simply multiplied by a factor depending on the atomic number Z, but that for the electric hfs one must introduce three parameters, denoted

$$\langle r^{-3}{}_{02}\rangle_{n\ell}\ ,\quad \langle r^{-3}{}_{11}\rangle_{n\ell}\quad ,\text{and}\quad \langle r^{-3}{}_{13}\rangle_{n\ell}\ ,$$

the former being equivalent to $\langle r^{-3}{}_{Q}\rangle_{n\ell}$.

Of course, one must also take into account the mixing with close configurations, which has not such simple effects as that between far configurations, and the mixing between close atomic levels which is induced by the hfs interaction itself. The latter mixing results in sizeable second-order corrections. Therefore, it is difficult to put into evidence the genuine effects of a possible octupolar magnetic moment of the nucleus, because the expansion of the second-order effects of an electric-quadrupolar operator contains an octupolar (and a hexadecapolar) part.

Determination of the Electric-Quadrupolar Moment

For extracting Q from the experimental results, one must obtain *ab initio* an accurate value of an $\langle r^{-3}{}_{02}\rangle$ quantity. This is difficult. Methods of various accuracies have been devised. I will only quote five of them.

The Hartree-Fock method (e.g., see Ref. 4), is available for the evaluation of any atomic quantity.

The simplest method for evaluating the effects of far-configuration mixing on the electric hfs was worked out by Sternheimer[5] in 1951, on the basis of the non-relativistic Hartree-Fock atomic results. It consists in solving one inhomogeneous radial differential equation for each occupied $n\ell$ subshell. Many such corrections have been computed by Sternheimer himself until the eighties. One speaks of "shielding" ("antishielding") if the correction yields a larger (smaller) value for Q.

In the relativistic multiconfigurational Hartree-Fock method, also called the multiconfigurational Dirac-Fock method (MCDF),[6] the radial functions of the $n\ell$ subshells are determined in a self-consistent manner, and a large number of configurations can be entered in the calculation for expanding the wavefunction. One introduces in priority the singly-excited configurations which are liable to bring in the largest corrections. But it is difficult to warrant better than a 5% accuracy.

The many-body perturbation theory (MBPT) is another all-purpose method, which was developed systematically by Kelly (e.g., see Ref. 7) on the basis of the Bethe-Goldstone equations. It uses Feynmann diagrams for listing the various orders and terms of the perturbation expansion. It is a powerful method, which has also been beautifully applied to hfs by the Göteborg group.

Last, but not least, accurate methods have been developed by the Russian atomic physicists (e.g., see Ref. 8), and deserve consideration.

In conclusion, it is estimated that, through much work, a 1% accuracy for the Q value can presently be reached in the light elements ($Z < 30$), but maybe not in the heavy ones. I can add that deducing the value of the intrinsic quadrupole moment Q_0 from that of Q is a nuclear, not an atomic, problem.

MAGNETIC-DIPOLAR HYPERFINE-STRUCTURE ANOMALY

In the same way as the electric-quadrupolar hamiltonian is a multiple of Q, the magnetic-dipolar hamiltonian H_m is a multiple of μ_n, the nuclear magnetic moment. More precisely, the expectation value $\Delta W_m(J,F)$ of H_m in some hyperfine F sublevel of a J level is expected to be proportional to μ_n/I: when one goes from isotope 1 to isotope 2, the ratio $\Delta W_2 \; I_2/\mu_2$ ought to be equal to $\Delta W_1 \; I_1/\mu_1$. In these ratios, ΔW_1 and ΔW_2 are measured magnetic-hfs values, and μ_1 and μ_2 are the values of the magnetic moments, which can often be measured directly, e.g., by the Nuclear Magnetic Resonance method, or by the Electron Paramagnetic Resonance method in magnetic fields $\mathbf{B}$ large enough for the direct interaction between $\mathbf{B}$ and μ to overcome the hfs interaction.

Now, it generally appears that high-precision measurements do not check the equality of the above ratios. Thus, the quantity

$$\Delta_{12} = \frac{\Delta W_1 \; I_1}{\mu_1} \; \frac{\mu_2}{\Delta W_2 \; I_2} - 1$$

with isotope 2 heavier than isotope 1, is called the hyperfine-structure anomaly for the relevant sublevel and isotopes.

Why does Δ_{12} differ from zero ?

First, the fact that any nucleus possesses a spatial extension is the reason why the electrostatic potential experienced by the electrons which enter it is not that of a point charge located at r = 0. The corresponding change in the electronic wavefunction can be considered a crossed-second-order effect of the hfs and IS operators. It is appreciable only for s electrons, for which it may reduce the computed value of $|\psi(0)|^2$ by as much as 20% in very heavy atoms[9]. The reduction factor is denoted $(1-\delta)$ and is called the Crawford-Schawlow correction. The differential change between two isotopes is much smaller.

Secondly, the magnetic-dipolar interaction between the nucleus and the electrons is not exactly described by the H_m operator given in the Introduction. When an electron enters the nucleus, it does not experience the field of a point magnetic momentum, but of a spatial distribution of magnetism. Therefore, the product $\mu_n\delta(\mathbf{r})$ in the H_m expansion, called the Fermi contact term, must be changed. The corresponding effect on the hfs value[9] is called the Bohr-Weisskopf correction. This factor is denoted $(1-\varepsilon)$.

In contrast, the direct magnetic-dipolar interaction between a nucleus and a uniform magnetic field does evidently not depend on the charge and magnetism spatial distributions of the nucleus. This is the origin of the hfs anomaly. In a given spectrum, this anomaly has its largest value in the configurations containing an open s subshell, e.g., in the ground levels of the alkalis. But it has the same value in the spin-core-polarisation contributions, which are due to some s electrons, as was seen above. Bordarier et al.[10] have shown how to relate the hfs anomaly of an atomic level to that of a single s electron, denoted $\Delta_{12}(s)$, which just depends on the chosen pair of isotopes. For instance, they have determined the value $\Delta_{12}(s) = -0.65\%$ for ^{151}Eu and ^{153}Eu in Eu, Eu^{+}, and Eu^{++}.

The order of magnitude of this value is typical. Therefore, a 10^{-4} relative accuracy ought to be obtained in the hfs and magnetic-moment measurements, for the deduced value of $\Delta_{12}(s)$ to be significant. Then, for testing the nuclear-magnetism distributions of various nuclear-structure models, it will be necessary to compute accurately the shape of the electronic radial wavefunction inside the nucleus.

FIELD ISOTOPE SHIFT

Mass and field isotope shifts

Atomic isotope shift is the fact that any line linking two atomic levels is split into as many close components as isotopes, if the hyperfine-structure phenomenon is neglected. It is well-known that it is the sum of the mass and field isotope shifts. From the measured values of the shifts of well-chosen lines, shifts can be derived for levels, in the same way as level energies are derived from line energies, through Ritz's Combination Principle.

The mass isotope shift exists because no nucleus has an infinite mass, i.e., it is not exactly the centre of gravity the atom, but it moves a little around the genuine centre of gravity, just as the Sun in the solar system. It has been recalled in the Introduction that the corresponding hamiltonian, for an isotope of mass number A, is a multiple of 1/A. For a given mass difference between two isotopes, say $\Delta A = 2$, the mass isotope shift between two isotopic components tends to be larger in lighter elements, because the relative change of mass is larger.

The field isotope shift exists because no nucleus is a point, i.e., each nucleus has a volume, roughly proportional to the cubic root of its mass number. Its computation by means of the hamiltonian H_{vol} given in the Introduction looks simple, but the complicated aspects are contained in the k constant. Indeed, that hamiltonian is just an "effective" one for use in electronic calculations. But it is clear that the field isotope shift depends on the specific shape of the s electronic wavefunctions inside the nucleus, not only on their values at $r = 0$. Moreover, this wavefunction must be of the relativistic type. Everything being included, the magnitude of the field shift in the "optical" lines appears to be roughly proportional to the fourth power of Z.

Thus, the mass and field isotope shifts are the predominant effects in the light and heavy elements, respectively. Nevertheless, appreciable mass-shift contributions have been detected about thirty years ago in samarium and heavier elements, whereas more and more accurate experiments carried out on light elements have brought evidence of field mass shift[11].

Separation of the mass and field shifts

In the lowest (and excellent) first-order approximation, one deduces from the hamiltonian H_{IS} that the shift between the components of the isotopes i and j in the line α is the sum

$$\Delta\sigma^{\alpha}_{ij} = K^{\alpha} A_{ij} + E^{\alpha} C_{ij}$$

of two terms, each term being the product of an electronic quantity (only depending on the line) and a nuclear quantity (only depending on the isotopes). In this equation,

$$A_{ij} = (A_j - A_i)/(A_i A_j) = \Delta A_{ij}/(A_i A_j)$$

The C_{ij} quantity is shown to be essentially proportional to the difference of the mean-square radii of the two isotopes:

$$C_{ij} = u\ (\langle r^2\rangle_j - \langle r^2\rangle_i)$$

where u is a constant. Only C_{ij} is of interest in Nuclear Physics.

In a series of isotopes of a given element, if a pair of isotopes 1 and 2 with $A_2 - A_1 = 2$ is chosen for the reference, the ratios C_{ij}/C_{12} are the *relative isotope shifts* (RIS). For example, in tin, the RIS of the six consecutive pairs of even stable isotopes, with A = 112 through 124, are approximately those listed in Table 1.

In general, in a given spectrum, for a fixed value of ΔA_{ij}, A_{ij} is nearly constant, but not C_{ij}. Keeping this difference in mind, one concludes that the *modified shifts*

Table 1. Relative isotope shifts in tin (Z=50)[12]

$C_{112-114}$	$C_{114-116}$	$C_{116-118}$	$C_{118-120}$	$C_{120-122}$	$C_{122-124}$
1	0.93	0.72	0.72	0.32	0.39

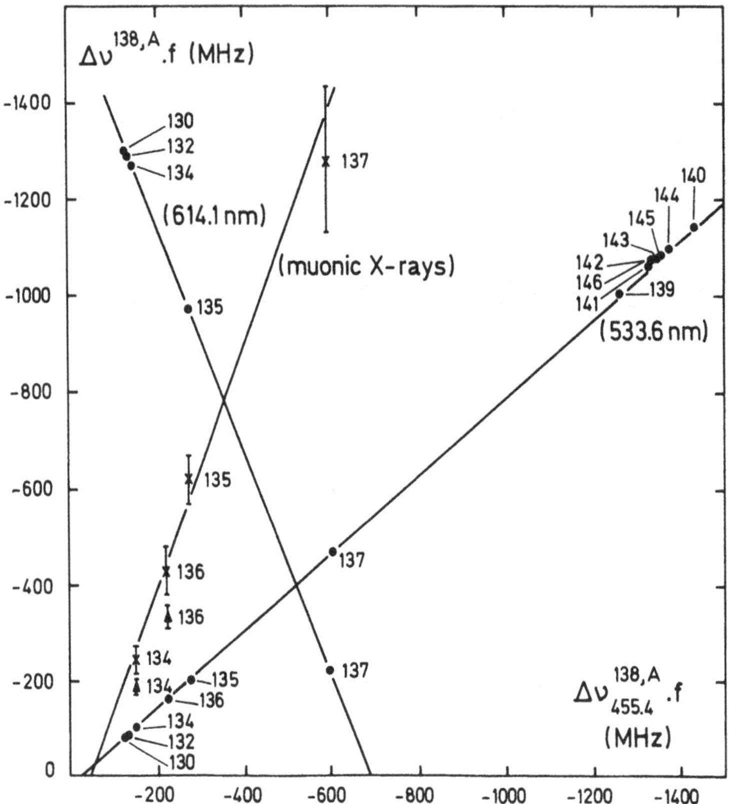

Fig. 1. King plot for the isotope shifts of 4 lines in the spectrum of barium[16]. The abscissa of each plotted point is the modified shift of the isotopic pair ^{138}Ba-^{A}Ba in the line λ=455.4 nm. Its ordinate is the modified shift in the line λ=614.1 nm or λ=533.6 nm, or in the muonic 1s-2p line.

$$\Delta\sigma^{\alpha}_{ij} A_i A_j / \Delta A_{ij} = K^{\alpha} + E^{\alpha} C_{ij} A_i A_j / \Delta A_{ij}$$

of any line α are in a linear relation with those of any line β, when several i,j pairs can be measured. This relation is clearly apparent in a *King plot*[13] like that presented in Fig. 1. It is an accurate way of checking a set of experimental measurements. The slopes of the straight lines, in Fig.1, are pure field-shift quantities; their intercepts are pure mass-shift quantities. But they are all just electronic quantities...

Determination of C_{ij}

Thus, from an isotope shift value measured in line α, C_{ij} can be deduced only after the mass shift part $K^{\alpha} A_{ij}$ and the electronic factor E^{α} are evaluated. Both these evaluations are problems of accurate ab-initio calculations in Atomic Physics. However, these problems are more difficult than those encountered in the determination of Q: (i) the expectation values of the mass and field shifts of an atomic level are sums which involve *all* its occupied subshells, (ii) the relative change of each of these sums from the upper to the lower level of a line is *very small*, and (iii) for refined evaluations involving configuration mixing, both shifts are difficult cases: the mass-shift operator contains a two-electron part, and the field-shift operator contains the "pathological" operator $\delta(\mathbf{r})$, whose accurate calculation requires electronic excitations to very high continuum energies.

Nevertheless, accurate calculations of K^{α} and E^{α} have already been performed for light elements, and recent results[14] obtained in caesium and thallium are promising.

Better situations could be, in principle, those of the X-ray and of the muonic-atom atomic transitions, where the mass shift and the E^{α} factor can be easily calculated. Unfortunately, the experimental accuracy is poor, because the linewidths are large. Another difficulty with the muon is that it does not experience the same nuclear potential as the electrons do, because it polarises the nucleus, a phenomenon hard to compute accurately. Nevertheless, many mean-square radii have been determined in the past twenty years by this method (e.g., see Ref.15).

In 1987, Otten[16] could conclude that the relative accuracy, in the determination of the mean-square radii, can be 10 to 30%, depending on the complexity of the spectrum studied, with much better values for their ratios. The best cases relate to the lines where the predominant part of the mass shift is the trivial "normal" mass shift[9], and where one can use safely the semi-classical Fermi-Segrè-Goudsmit formula[9] for evaluating E^{α}. Best among the best are the ns-np transitions in very heavy elements, where the mass shift is nearly negligible. Impressive results have already been obtained for long series of isotopes, including neutron-deficient and neutron-rich nuclei, especially in the alkalis, and in gold, mercury, and lead. The cases of caesium and mercury are presented in Figs.2 and 3.

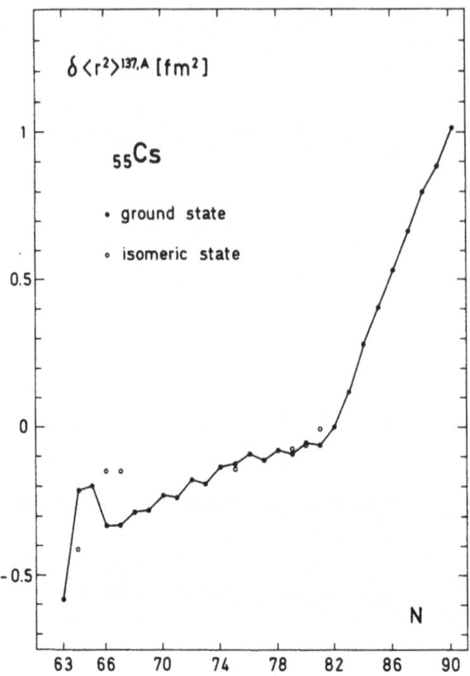

Fig.2. Root-mean-square nuclear radii of caesium isotopes[17]. Each ordinate of a black point is the value of the root-mean-square radius of the charge of the Cs nucleus with neutron number N minus that of the ^{137}Cs nucleus. The blank points correspond to the isomeric states of some nuclei.

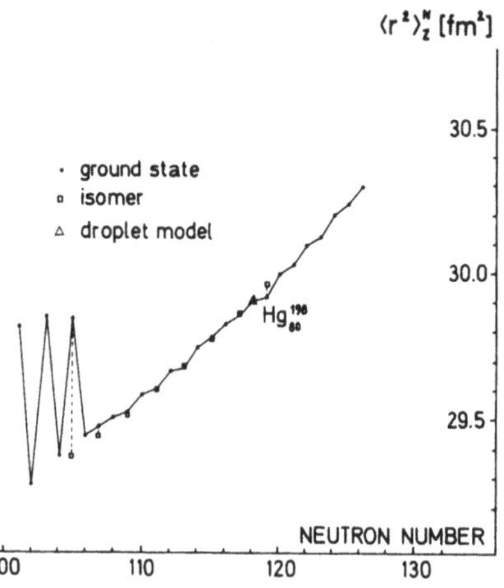

Fig. 3. Root-mean-square nuclear radii of mercury isotopes[16]. Each ordinate of a black point is the value in $(fm)^2$ of the mean-square radius of the charge of the mercury nucleus with N neutrons. The blank squares correspond to the isomeric states of a few nuclei. The triangle indicates the theoretical value for ^{198}Hg derived, for calibration, by means of the droplet model.

REFERENCES

1. J. S. M. Harvey, Hyperfine structure in the ground multiplets of ^{17}O and ^{19}F, Proc. Roy. Soc. A 285:581 (1965).

2. A. Abragam, J. Horowitz, and M. H. L. Pryce, On the hyperfine structure of paramagnetic resonance: the s-electron effect, Proc. Roy. Soc. A 230:169 (1955).

3. P. G. H. Sandars and J. Beck, Relativistic effects in many electron hyperfine structure, Proc. Roy. Soc. A 289:97 (1965).

4. C. Froese Fischer, "The Hartree-Fock method for atoms," John Wiley and Sons, New York (1977).

5. R. Sternheimer, On nuclear quadrupole moments, Phys. Rev. 84:244 (1951).

6. J.-P. Desclaux, A multiconfigurational relativistic Dirac-Fock program, Comput. Phys. Commun. 9:31 (1975).

7. H. P. Kelly, Hyperfine structure of oxygen calculated by Many-Body Theory, Phys. Rev. 180:55 (1969).

8. V. A. Dzuba, V. V. Flambaum, A. Ya. Kraftmakher, and O. P. Sushkov, Summation of the high orders of perturbation theory in the correlation correction to the hyperfine structure and to the amplitudes of E1-transitions in caesium atom, Phys. Lett. A142:373 (1989).

9. H. G. Kuhn, "Atomic spectra," Longmans, London (1969).

10 Y. Bordarier, B. R. Judd, and M. Klapisch, Proc. Roy. Soc.A 289:81 (1965).

11. G. Huber, F. Touchard, S. Büttgenbach, C. Thibault, R. Klapisch, H. T. Duong, S. Liberman, J. Pinard, J.-L. Vialle, P. Juncar, and P. Jacquinot, Spins, magnetic moments and isotope shift of $^{21-31}$Na by high resolution laser spectroscopy of the atomic D_1 line, Phys. Rev. C 18:2342 (1978).

12. H. Kopfermann, "Nuclear moments," Academic Press, New York (1958).

13. W. H. King, "Isotope shifts in atomic spectra," Plenum, New York (1984).

14. A. C. Hartley and A.-M. Mårtensson-Pendrill, J. Phys. B: At. Mol. Phys. 24:1193 (1991).

15. K. Heilig and A. Steudel, Changes in mean-square nuclear charge radii from optical isotope shifts, At. Data Nucl. Data Tables 14:613 (1974).

16. E. W. Otten, Nuclear radii and moments of unstable nuclei, in: "Treatise on heavy ion physics," D. A. Bromley, ed., Plenum, New York (1987).

17. C. Thibault, F. Touchard, S. Büttgenbach, R. Klapisch, M. de Saint-Simon, H. T. Duong, P. Jacquinot, P. Juncar, S. Liberman, P. Pillet, J. Pinard, J.-L. Vialle, A. Pesnelle, and G. Huber, Hyperfine structure and isotope shift of the D_2 line of $^{118-145}$Cs and some of their isomers, Nucl. Phys. A 367:1 (1981).

II - ATOMIC METHODS IN NUCLEAR SPECTROSCOPY

PROGRESS IN ATOMIC PHYSICS EXPERIMENTS ON NUCLEAR PROPERTIES

Rainer Neugart

Institut für Physik, Universität Mainz
D - 6500 Mainz, Germany

Introduction

The measurement of nuclear properties by atomic physics methods has influenced a great deal of our present understanding of the nuclear structure. This started from the discovery of the nuclear spin and magnetic moment, the observation of isotope shifts related to nuclear radii, and the resolution of quadrupole interaction effects in the hyperfine structures. The invention and improvement of many spectroscopic techniques has led to a comprehensive knowledge of the electromagnetic ground-state properties of all stable and many radioactive nuclides. On the other hand, the atomic and thus the nuclear masses of stable isotopes were determined very precisely by electromagnetic mass spectrometry. These masses formed the basis of indirect mass determinations for many radioactive isotopes whose decay or formation energies were measured.

The continuous interest in atomic physics methods for the investigation of nuclear properties arises from the fact that a large (and still increasing) number of isotopes far from stability have been discovered and made accessible to detailed studies. At the same time, laser spectroscopy brought about the revolution of experimental techniques for high resolution studies of the atomic energy levels, going along with an enormous gain in sensitivity. Low-energy ion beams of unstable nuclides from in-line isotope separators like ISOLDE turned out to be the ideal objects for the application of laser spectroscopy methods.

The new approach to direct atomic mass measurements has somewhat different reasons. Although the masses of many known nuclei have been determined from reaction and decay data, there is the general problem of the errors increasing with the distance from the directly measured masses. Therefore, a reliable evaluation of atomic masses requires at least a sufficient number of directly measured reference data extending into the regions far from stability. Such mass measurements have been started very successfully at ISOLDE. They are based on the measurement of the cyclotron frequency of ions captured in a Penning trap. These experiments are discussed in the contribution of Bollen et al. (these proceedings).

This paper concentrates on the status and possible further developments of laser methods for the measurement of nuclear spins, moments and radii. The problem of extracting

Nuclear Shapes and Nuclear Structure at Low Excitation Energies
Edited by M. Vergnes *et al.*, Plenum Press, New York, 1992

these nuclear properties from the measured atomic hyperfine structures and isotope shifts is discussed in the paper by J. Bauche (these proceedings).

From Fluorescence Spectroscopy to Ion or Atom Detection

The typical setup for a laser experiment on a sample of free atoms consists in an atomic beam crossed by a laser beam and a detection of fluorescence photons from the atoms excited by the laser light (see, e.g., ref. [11]). This scheme eliminates Doppler broadening by the atomic beam collimation and a right angle between the directions of both beams. Many experiments have used it, but for unstable nuclides which are available in very small intensities, it has the disadvantage of wasting most of the material by the formation of a collimated beam. One way out of this problem was found in compensating the losses by very efficient detection schemes. This concept was very successful for the alkali elements, where the optical pumping by a laser beam was combined with magnetic state selection and single-atom counting after surface ionization and mass separation [1]. More recent experiments along these lines use the laser ionization as an extremely background-free detection of optical transitions (see the contributions by G.D. Alkhazov et al. and J.E. Crawford et al.)

The opposite approach was chosen for experiments using fast beams of ions or atoms. First of all, such beams are available for many unstable isotopes from on-line mass separators, where the individual products of rather unspecific nuclear reactions are selected. Then it was realized [2] that in such beams the distribution of velocity components along the beam direction is so narrow due to the electrostatic acceleration that high-resolution spectroscopy just requires the collinear superposition of the ion beam with a laser beam. Neutral atoms with essentially the same phase space properties as the original ion beam are readily obtained by charge exchange with an alkali vapour at low pressure [3]. The major advantage for weak beams of unstable isotopes is thus the efficiency of the optical excitation in which high resolution is achieved without selecting a special velocity class of atoms.

The detection of optical resonance can be performed as usual by collecting the fluorescence light emitted along the beam. This is the standard setup of collinear laser spectroscopy shown in Figure 1a. It is very convenient that in a fast beam the Doppler-shifted optical frequency in the rest frame of the atoms can be changed by an electrical potential at the charge exchange cell. Typically for a 60 keV beam a $\pm 10\%$ change in the beam energy gives a tuning range of more than 50 GHz which is largely sufficient for most hyperfine structure and isotope shift measurements.

Although the fluorescence detection is universal, it is far from ideal from the sensitivity point of view. The main reasons for this are the relatively low photon collection and detection efficencies and the background from stray-light. Typically one needs beams of at least 10^6 atoms per second for an experiment, unless the particular atomic level scheme permits multiple excitation without optical pumping losses.

Several approaches towards much more efficient detection techniques have recently become quite successful. They are all based on the counting of fast atoms or ions, or their radioactive decay. In a first type of experiments, the optical pumping from the initial into a different final state is used to label the ions or atoms that have interacted with the laser light. The state selectivity of charge-changing collision processes is then used to distinguish

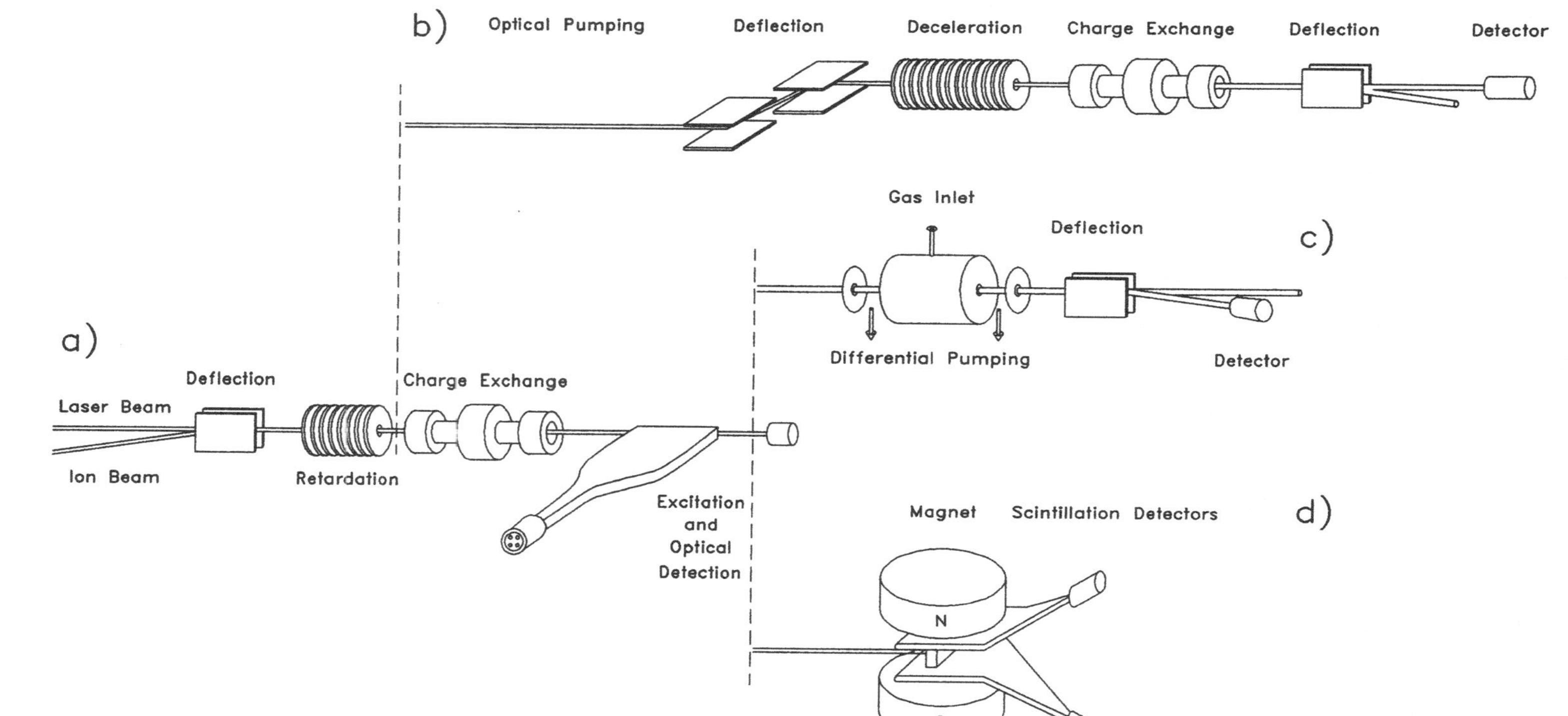

Figure 1. Schematic view of the apparatus for collinear laser spectroscopy with a) fluorescence detection, b) state-selective neutralization, c) state-selective collisional ionization and d) β-asymmetry detection.

between both classes of ions or atoms: The neutral and charged components of the beam are separated in an electrostatic deflector and transmitted to a secondary electron multiplier.

For noble gases, atoms produced in the first excited metastable $np^5\,(n+1)s$, $J=2$ state are further excited by the laser light and decay into the strongly bound ground state for which the collisional ionization cross-section is much smaller than for the metastable state. The experimental setup is as shown in Figure 1c.

On the other hand, the alkaline earth ions have accessible resonance transitions in the blue or the near UV spectral regions and metastable d states lie below the first excited p states. In pumping the atoms to these metastable states by the $s-p$ resonance excitation and the subsequent decay, one can use the difference in the charge-exchange neutralization cross-sections of both states. This leads to another modified setup which is shown in Figure 1b. An efficient discrimination and the suppression of background from the alkali isobars can usually be achieved at relatively low beam energies of about $5\,keV$. A deceleration lens is installed therefore in front of the neutralization cell which is put on a high positive potential.

A third alternative, shown in Figure 1d, is based on the polarization of the beam by optical pumping and the detection of β-ray asymmetries after implantation into a crystal. This method is well adapted to the short half-lives and high β-decay energies of light unstable nuclides and has been applied very successfully to the lithium isotopes - in particular ^{11}Li.

All these high-sensitivity schemes have in favourable cases permitted measurements with beam intensities as low as $100\,atoms/s$. A few examples of recent measurements are given in the following.

Recent Results

Fluorescence Spectroscopy in the Rare Earth Region

A typical region for conventional fluorescence measurements is that of the rare earth elements which all have rather similar chemical properties. The products from spallation reactions in a ^{181}Ta target cover a wide range of proton and neutron numbers. Therefore, the beams from the mass separator are a mixture of isobars. It is the great advantage of fluorescence detection to be selective for the chosen spectral line of only one element, although the isobaric contamination gives an additional background from collisional excitation and radioactivity. From such experiments and from resonance ionization spectroscopy on thermal atomic beams (see the contribution by G.D. Alkhazov et al.), the systematics of nuclear moments and radii between $Z = 62\,(Sm)$ and $Z = 70\,(Yb)$ is rather complete, except for the lightest and haeaviest short-lived isotopes.

As an extension of this systematics, the lutetium isotopes were recently investigated in the range of $^{161-179}Lu$. The results are particularly interesting in the neighbourhood of the stable ^{175}Lu, where isomers (ground states) of very high spin coexist with low spin ground states (isomers). For a number of these, earlier measurements of the magnetic moments were not sufficiently accurate or even contradictory. In addition to reliable values of all magnetic moments, the measurements provide quadrupole moments and the isotope and isomer shifts.

Figure 2a shows the changes in the mean square charge radii $\delta\langle r^2\rangle$ of all ground states and isomers with respect to the stable ^{175}Lu, together with the droplet model predictions [4, 5]. This curve can be compared with the development of quadrupole deformation. In

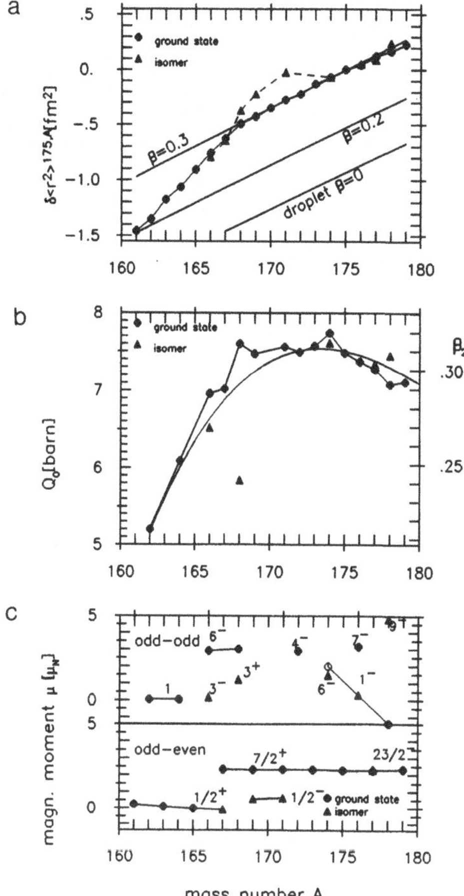

Figure 2. Results of the *Lu* experiment: a) mean square charge radii in comparison with the droplet model, b) quadrupole moments with predicition of [6] and c) spins and magnetic moments.

Figure 2b are plotted the intrinsic quadrupole moments as extracted from the measured spectroscopic quadrupole moments via the strong-coupling projection formula,

$$Q_0 = \frac{(I+1)(2I+3)}{I(2I-1)} Q_s. \tag{1}$$

This is justified, because all nuclei considered here are assumed to be rather well deformed and the quadrupole moments are large. It becomes obvious that apart from a regular increase with the neutron number the shape of the $\delta <r^2>$ curve is determined by changes in the β_2 deformation of the nuclear charge distribution.

There is only one remarkable exception from the smooth trend of the radii, and this can be understood by looking into the systematics of the spins and magnetic moments. Figure 2c shows the magnetic moments separately for the odd-even and the odd-odd nuclei, with the values for nuclei with the same spin connected by full lines. One realizes that the spin $I = 1/2$ states which appear in all nuclei between $N = 90$ and 100 have different origin for ^{169m}Lu and ^{171m}Lu than for the sequence $^{161-167m}Lu$. The $1/2^+$ states which become ground states in the lighter isotopes arise from the proton configuration [411]1/2, whereas the isomers ^{169m}Lu and ^{171m}Lu are $1/2^-$ arising from the $h_{9/2}$ intruder orbital [541]1/2. This explains the exceptionally large isomer shifts. All other isomer shifts around the stable ^{175}Lu are small, indicating that there is no difference in deformation between the high-spin and low-spin states of the same nucleus.

For the magnetic moments plotted in the lower part of Figure 2 it is remarkable to see how close they are to the Nilsson model predictions given by Ekström et al. [6]. For some long-lived isotopes they agree very well with the recent nuclear orientation results [7].

Krypton and Strontium

Measurements on the strontium ($Z = 38$) isotopes have been reported by several groups. The most recent and extended investigation [8, 9] includes the isotopes $^{78-100}Sr$, ranging from ^{88}Sr with the closed neutron shell $N = 50$ to the neutron-rich ($N \approx 60$) and neutron-deficient ($N \approx 40$) regions of strong deformation. For the shortest-lived isotopes very far from stability, this experiment took advantage of the ultrasensitive method of ion spectroscopy with detection by state selective charge-exchange neutralization (see Figure 1b). A similar experiment on the krypton ($Z = 36$) isotopes would be interesting mainly for comparisons in the $N \approx 40$ and $N \approx 60$ regions.

After earlier investigations of the noble gases radon and xenon [10], krypton had been a promising candidate for the exploitation of the collisional ionization detection scheme (Figure 1c). Here the $60\,keV$ beam of neutral atoms passes through a gas target of about $10^{15}\,molecules/cm^2$, and the ions created in collisions are deflected onto a secondary electron detector. Optical excitation from the metastable level and decay to the ground state goes along with a drop in the ionization rate because of the difference in the cross-sections. It is obvious that this technique requires very pure beams of the weakly produced isotopes. Krypton is probably the lightest noble gas for which the conventional ISOLDE techniques give sufficiently low isobaric contamination.

The measurements on krypton now include the full sequence of $^{72-96}Kr$, with the exception of ^{73}Kr. Figure 3 shows the changes in the mean square nuclear charge radii evaluated

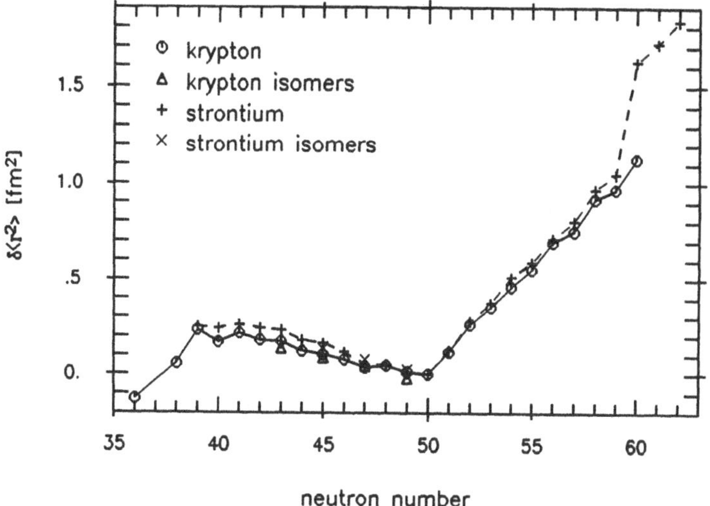

Figure 3. Changes in mean square charge radii for krypton and strontium isotopes around the magic neutron number $N = 50$ including the deformation regions $N \leq 40$ and $N \geq 60$.

from the isotope shifts in comparison with the strontium results. Outstanding features of the krypton curve are the absence of a deformation jump for the neutron number $N = 60$ and the effects of shape instabilities around $N = 40$. It should be mentioned that very interesting details could be added to these results by measurements on ^{73}Kr, ^{76}Sr and $^{101-102}Sr$.

Calcium

Already the strontium isotopes far from stability were investigated [8, 9] by the non-optical detection of laser-ion interaction. The principle of ground state depopulation by optical pumping into the metastable d states has been shown in Section 2. The d-state population is detected by selective neutralization in the charge exchange with sodium and counting of the atoms on a secondary electron multiplier. A problem for this technique was the large rubidium contamination of the strontium beams. For suppressing the neutralization of rubidium ions, the beam was decelerated to 5 keV in front of the charge exchange cell.

An important step towards further experiments using this concept is the recent application to calcium beams which give a value for the isotope shift of ^{50}Ca in the transition $4s\,^2S_{1/2} - 4p\,^2P_{1/2}$. This is the first extension of the famous results on the radii in the Ca chain of isotopes (refs. [11, 12] and references therein) beyond one of the magic neutron numbers $N = 20$ and $N = 28$. The new value is included in the $\delta <r^2>$ curve shown in Figure 4. For the previously known $f_{7/2}$-shell calcium nuclei, Barranco and Broglia [13] calculated $\delta <r^2>$ in very good agreement with the experimental values by taking into account all collective deformation contributions from excitations to low- lying $0^+, 2^+, 4^+, 3^-$ and 5^- levels. On the other hand, Talmi [14] proposed a shell-model based theory which reproduces

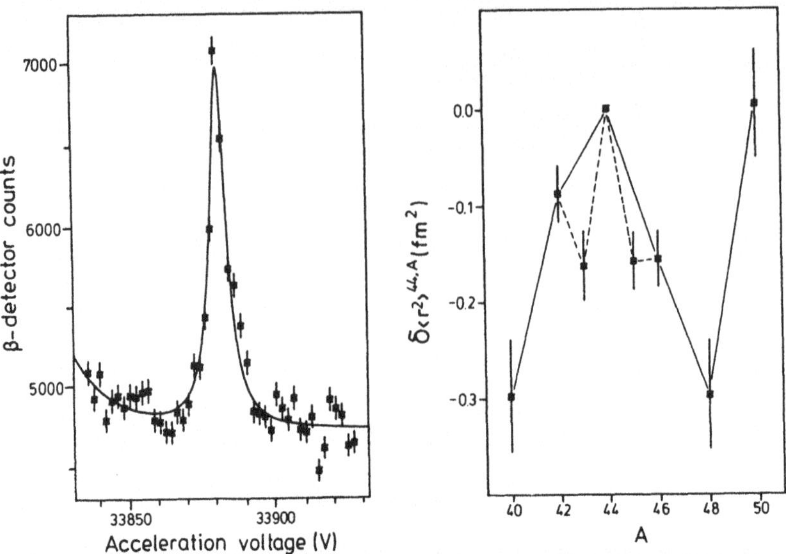

Figure 4. Left: β-radiation detected resonance signal of ^{50}Ca in the transition $4s\,^2S_{1/2} - 4p\,^2P_{1/2}$. Right: Changes in mean square charge radii for $^{40-50}Ca$.

well the parabolic trend and the odd-even staggering by describing the ground states as a mixture of the lowest- seniority states with low-lying 2^+ core-excited states. It will be very interesting to examine the new result of a rather steep increase of $< r^2 >$ beyond $N = 28$ in the light of these theories. In any case, it indicates a rather strong coupling between the $p_{3/2}$ neutrons and the core protons.

The Ca^+ resonance wavelength of 397 nm in the near UV was reached by using the near-UV laser dye Exalite 400. The relatively short half-life of 13.9 s for ^{50}Ca offered the possibility of selectively detecting the calcium atoms on a huge contamination from stable titanium by counting the β - decay electrons. This is a perfect scheme also for the other isotopes below $N = 20$ and above $N = 28$, for which experiments are planned in the future.

Lithium

In the lighter elements optical hyperfine structure and isotope shift measurements become more and more difficult, because both effects decrease and partly vanish in the natural linewidths. Nevertheless, there are possibilities to combine the described optical techniques with well-known methods of rf spectroscopy like the nuclear magnetic resonance (NMR). This has been demonstrated in the experiments on the very light neutron-rich nuclei ^9Li and ^{11}Li.

The observation of an exceptionally large matter radius of ^{11}Li in reactions of high-energy fragment beams [15], together with laser spectroscopy measurements of the spin and the magnetic moment [16], has shown that weakly bound nuclear systems at the neutron drip line have a spacially extended spherical structure which is difficult to describe in the usual mean field approach. The observations are explained consistently by the neutron halo model of Hansen and Jonson [17]. This concept also involves a narrow momentum distribution of the halo neutrons and large cross-sections for the Coulomb break-up which both have been found experimentally [18, 19].

A crucial experimental test of possible distortions of the spherical shape of the system by the two halo neutrons, or the final exclusion of a previously postulated deformed structure of ^{11}Li would be provided by a measurement of the quadrupole moment.

It had been shown for the example of ^{9}Li that the quadrupole moment can be determined from the splitting of the NMR signal of nuclei implanted into a non-cubic crystal [20]. The 60 keV beam of neutral atoms was polarized by optical pumping with circularly polarized laser light of the resonance wavelength 670.7 nm. The nuclear polarization is conserved during implantation and detected by the angular asymmetry in the β - decay.

The difficulty of using the same principle for ^{11}Li arises from the low production rate (corresponding to count rates of $100 - 200\,sec^{-1}$), the relatively low β - asymmetry of a few percent and the small reduction of this asymmetry in each of the resonances which are split by the quadrupole interaction. To overcome these difficulties, one had to introduce several improvements of the experimental technique.

The decisive trick consists in a simultaneous irradiation of the sample with all three equidistant transition frequencies, of which the central one is independent of the quadrupole moment and depends only on the known g_I - factor. Furthermore, the Zeeman-pumping was made more efficient by the simultaneous excitation of both hyperfine structure components of the ground state. Finally, the modification of the magnetic field configuration (homogeneous longitudinal field in the pumping region and rotating field in front of the magnet) increased the asymmetry by a factor of 2.

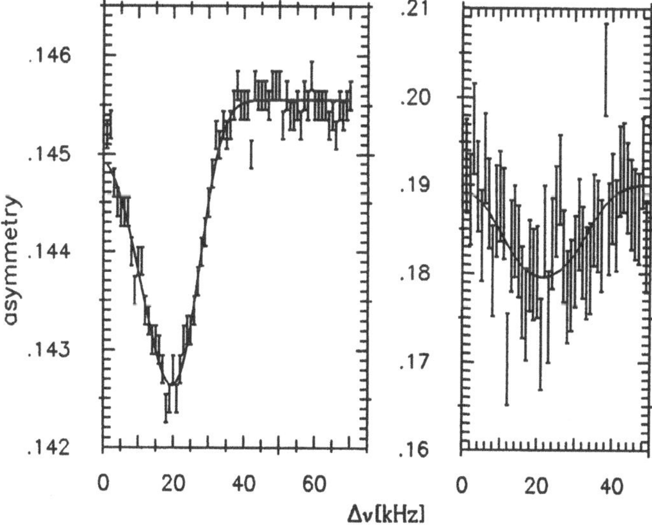

Figure 5. Quadrupole splitting of NMR signals of ^{9}Li and ^{11}Li in $LiNbO_3$, measured by simultaneous irradiation of three equidistant frequencies.

With these improvements it was possible to measure the quadrupole interaction of ^{11}Li in $LiNbO_3$. The resonance curve, in comparison with the one of ^9Li, is shown in Figure 5. Both curves were recorded by fixing the central frequency to the value corresponding to the known g_I - factor (about $1.6\,MHz$) and symmetrically sweeping the two additional frequencies at a distance of $\pm\delta\nu$ between 0 and $50\,kHz$. It is obvious from Figure 5 that the quadrupole moments of both isotopes are very nearly the same. A preliminary analysis of the curves in a rate equation approximation yields the ratio $Q(^{11}Li)/Q(^9Li) = 1.14(16)$. This result is in perfect consistency with the assumption of a ^9Li core being nearly unchanged in the presence of the two halo neutrons.

Conclusion

After many experiments have yielded systematic data on the spins, moments and radii of unstable nuclei, one should ask what problems remain to be solved by the future work in this field.

The lutetium results shown as a first example are typical for the kind of information extracted from the measurements in long sequences of isotopes. They usually include the unambiguous determination of the spins of all observed nuclear ground states and isomers. This is already an important ingredient of the nuclear structure analysis based on existing or future data from nuclear spectroscopy. In fact, the discussion of octupole-deformed nuclear shapes was very much based just on the observed spin sequence in the heavier radium isotopes, $^{221-229}Ra$ (ref. [21]). In the lutetium case, about half the previous spin assignments were unsafe and the spins of the lightest isotopes were essentially unknown.

In addition to the spins, the magnetic moments give much more detailed information about the wavefunction of the unpaired nucleon in the mean field of the core and on the core polarization effects. In the present case they were predicted very well by the Nilsson model [6], of course with an effective g_s - factor. This situation is often found in the regions of well-deformed nuclei, which means that surprises reflect interesting phenomena.

The quadrupole moments and the mean square radii are shown to be explained very consistently as arising from a nearly pure collective quadrupole deformation effect. Again, this is expected in a region, where the rigid-rotor model describes well the low-lying states of the nuclei. Still, there are exceptions from this consistency like the quadrupole moments of $^{168m}Lu(I = 3)$ and $^{176m}Lu(I = 1)$. This, as well as the surprisingly low difference in deformation between the ground states and isomers of most nuclei has to be analyzed in more detail.

There are, however, regions of nuclei, where the detailed behaviour of the radii is much more difficult to understand. This can be illustrated by the krypton-strontium example. Here the droplet model description reproduces very well the differences in $<r^2>$ between the spherical nuclei at $N = 50$ and the strongly deformed ones in the $N \approx 60$ and $N \approx 40$ regions [8]. Still, there are difficulties to explain the neutron shell effect at $N = 50$. This major neutron shell closure gives a very pronounced V-shaped kink in the curve of $\delta <r^2>$. In the usual deformed droplet model interpretation one assumes a "spherical" part according to a smooth volume increase with the neutron number and a deformation contribution. Apart from higher order terms and corrections this corresponds to the geometrical formula

$$<r^2> = <r^2>_{sph} + \frac{5}{4\pi}<r^2>_{sph} < \beta^2 > \quad (2)$$

Thus the change in slope is ascribed to the increase of $< \beta^2 >$ from the shell closure towards lower and higher neutron numbers. A problem of quantitative interpretations is the fact that $< \beta^2 >$ according to the isotope shift curve (including, of course, static and dynamic deformation) should increase strongly in the direct neighbourhood of the shell closure. This is in contrast to the theoretical analyses [22] as well as the observations on $< \beta^2 >$ from $B(E2)$ values [23]. Both of them give rather shallow minima at the shell closure and predict too small radii for the transitional nuclei. It is not obvious how the inclusion of higher-order deformation effects should resolve the appreciable discrepancy. On the other hand, also microscopic Hartree-Fock calculations usually fail to reproduce the kink at major shell closures.

The situation for a deeper understanding of these observations is quite favourable in calcium. Here large-scale microscopic calculations can be performed in the Hartree-Fock approach or in the shell model using the full configuration space. As it has been possible to reproduce rather well the the parabolic curve of $\delta <r^2>$ between ^{40}Ca and ^{48}Ca, it will be a crucial test of the theory to extend the calculations beyond $N = 28$, where the new result on ^{50}Ca is available.

The hope for a detailed theoretical explanation of all the effects observed in the nuclear moments and radii will in the future put special weight on the investigation of lighter nuclear systems. This report should have demonstrated that the experimental progress towards this aim is going on.

References

[1] G. Huber, F. Touchard, S. Büttgenbach, C. Thibault, R. Klapisch, H.T. Duong, S. Liberman, J. Pinard, J.L. Vialle, P. Juncar and P. Jacquinot, Phys. Rev. **C 18** (1978) 2342

[2] S.L. Kaufman, Optics Commun. **17** (1976) 309

[3] K.-R. Anton, S.L. Kaufman, W. Klempt, G. Moruzzi, R. Neugart, E.W. Otten and B. Schinzler, Phys. Rev. Lett. **40** (1978) 642

[4] W.D. Myers and K.-H. Schmidt, Nucl. Phys. **A 410** (1983) 61

[5] D. Berdichevsky and F. Tondeur, Z. Phys. **A 322** (1985) 141

[6] C. Ekström, Physica Scripta **13** (1976) 217

[7] E. Hagn, private communication and
B. Hinfurtner, E. Hagn, E. Zech, R. Eder, J. Kern and ISOLDE Collaboration, Phys. Lett. **B 263** (1991) 29

[8] F. Buchinger, E.B. Ramsay, E. Arnold, W. Neu, R. Neugart, K. Wendt, R.E. Silverans, P. Lievens, L. Vermeeren, D. Berdichevsky, R. Fleming, D.W.L. Sprung and G. Ulm, Phys. Rev. **C 41** (1990) 2883

[9] P. Lievens, R.E. Silverans, L. Vermeeren, W. Borchers, W. Neu, R. Neugart, K. Wendt, F. Buchinger, E. Arnold and ISOLDE Collaboration, Phys. Lett. **B 256** (1991) 141

[10] W. Bochers, E. Arnold, W. Neu, R. Neugart, K. Wendt, G. Ulm and ISOLDE Collaboration, Phys. Lett. **B 216** (1989) 7

[11] A. Andl, K. Bekk, S. Göring, A. Hanser, G. Nowicki, H. Rebel, G. Schatz and R.C. Thompson, Phys. Rev. **C 26** (1982) 2194

[12] C.W.P. Palmer, P.E.G. Baird, S.A. Blundell, J.R. Brandenberger, C.J. Foot, D.N. Stacey and G.K. Woodgate, J. Phys. **B 17** (1984) 2197

[13] F. Barranco and R.A. Broglia, Phys. Lett. **B 151** (1985) 90

[14] I. Talmi, Nucl. Phys. **A 423** (1984) 189

[15] I. Tanihata, H. Hamagaki, O. Hashimoto, Y. Shida, N. Yoshikawa, K. Sugimoto, O. Yamakawa, T. Kobayashi and N. Takahashi, Phys. Rev. Lett. **55** (1985) 2676

[16] E. Arnold, J. Bonn, R. Gegenwart, W. Neu, R. Neugart, E.W. Otten, G. Ulm, K. Wendt and ISOLDE Collaboration, Phys. Lett. **B 197** (1987) 311

[17] P.G. Hansen and B. Jonson, Europhys. Lett. **4** (1987) 409

[18] T. Kobayashi, O. Yamakawa, K. Omata, K. Sugimoto, T. Shimoda, N. Takahashi and I. Tanihata, Phys. Rev. Lett. **60** (1988) 2599

[19] R. Anne, S.E. Arnell, R. Bimbot, H. Emling, D. Guillemaud- Mueller, P.G. Hansen, L. Johannsen, B. Jonson, M. Lewitowicz, S. Mattsson, A.C. Mueller, R. Neugart, G. Nyman, F. Pougheon, A. Richter, K. Riisager, M.G. Saint-Laurent, G. Schrieder, O. Sorlin and K. Wilhelmsen, Phys. Lett. **B 250** (1990) 19

[20] E. Arnold, J. Bonn, W. Neu, R. Neugart, E.W. Otten and ISOLDE Collaboration, Z. Phys. **A 331** (1988) 295

[21] S.A. Ahmad, W. Klempt, R. Neugart, E.W. Otten, P.G. Reinhard, G. Ulm, K. Wendt and ISOLDE Collaboration, Nucl. Phys. **483** (1988) 244

[22] W.D. Myers and P. Rozmej, Nucl. Phys. **A 470** (1987) 107

[23] S. Raman, C.H. Malarkey, W.T. Milner, C.W. Nestor, Jr. and P.H. Stelson, At. Data Nucl. Data Tables **36** (1987) 1

RESONANCE IONIZATION SPECTROSCOPY OF RARE-EARTH ELEMENTS AT IRIS FACILITY

G.D. Alkhazov, A.E. Barzakh, V.N. Panteleyev,
E.P. Sudentas, V.N. Fedoseyev, V.S. Letokhov,
V.I. Mishin and S.K. Sekatsky

Leningrad Nuclear Physics Institute
USSR Academy of Sciences
Institute of Spectroscopy
USSR Academy of Sciences, Moscow

INTRODUCTION

The investigation of nuclei far from stability with the aid of atomic spectroscopy is a rather rapidly developing branch of nuclear physics. Information on the isotopic changes of ms charge radii, spins and electromagnetic moments of radioactive nuclei deduced from the experimental data on the isotope shifts (IS) and hyperfine structures (hfs) of optical lines gives us important new insight into nuclear structure. Different laser spectroscopy techniques have been applied for the IS and hfs measurements during the past decade. Among these methods resonance ionization spectroscopy (RIS) has proven to be one of the most efficient ones. By now, more than 100 ground and long lived isomeric states have been studied by RIS at IRIS mass-separator. Here we shall describe briefly our experimental set-up, with the emphasis on a new direction of further development of our installation, and present the latest experimental data for Gd, Tb and Eu isotopes as well as some theoretical considerations concerned with the interpretation of our previous results [1].

EXPERIMENTAL SET-UP

The principle of the experimental method is the following. The atoms under study are excited sequentially from level to

Nuclear Shapes and Nuclear Structure at Low Excitation Energies
Edited by M. Vergnes *et al.*, Plenum Press, New York, 1992

level into autoionization states by laser radiation of several frequencies. The photoions arise from the decay of these states. The laser radiation frequencies for the second and the third excitation steps are kept in resonance with the atomic transitions chosen, the linewidths of these radiations being sufficiently broad to ensure overlapping both the IS and hfs of the corresponding atomic lines. For the first excitation step the narrow-band laser is used. Photoions appear when the scanned frequency of this laser coincides with the first-step atomic transition frequency.

The isotopes in question were produced in tantalum (or uranium) targets by 1 GeV protons from the LNPI synchrocyclotron. The target was heated,and reaction products diffused to the ion source of the mass-separator. In the conventional RIS-scheme (see ref.1),radioactive atoms from the target were ionized on the hot surface of a tungsten ionizer, mass-separated, implanted into a heated tantalum tube at the exit of the mass-separator and evaporated. The appearing atomic beam was intersected at a right angle by laser beams, which excited and ionized the atoms. This version of RIS ensures a rather high efficiency, which permitted to carry out experiments with mass-separated ion fluxes down to 10^4-10^5 particles per second. The results, presented in the next section, have been obtained with this approach.

To increase the sensitivity we have tested a new version of RIS based on the laser ion source of the IRIS mass-separator [2](RIS-LIS method). In this version dye laser beams were directed immediately into the ion source - a hot niobium tube, where a resonance photoionization of radioactive isotopes under study took place. The ions arisen in the tube of the ion source were extracted by electric field, mass-separated and directed to the registration system. The efficiency of this laser ion source can reach some dozens of percent . To avoid the background from the surface ionized atoms, selective registration of the isotopes under investigation was applied. In the case of light Yb isotopes (A=154-156) it proved to be convenient to detect α-particles from the decay of these isotopes. Thus, optical spectra represented in this case the number of α-particles of the given energy versus the scanned laser frequency. As it follows from the results of the 1-st experiments, a very high efficiency of RIS-LIS enables us to carry out optical spectroscopic measurements for the isotopes with the yield in the target down to 1-10 nuclei per second, provided that α-branching for these isotopes is sufficiently high. It is worth mentioning that in conventional RIS-scheme a lower sensitivity is accompanied by a better resolution, so both versions of RIS supplement each other.

NEW RESULTS FOR RADIOACTIVE RARE-EARTH ISOTOPES

Neutron-rich Eu isotopes

The inverted odd-even staggering in the isotopic changes of $\delta\langle r^2\rangle$ (when occuring in the region of stable quadrupole deformation) is connected now [3,4] with the octupole deformation. A clear example of the inverted odd-even staggering was observed in the laser experiments at the IRIS facility for neutron-rich Eu isotopes (A=153-156) [5]. Sheline [3] interpreted these data as one of the evidences in favour of the existence of reflection-asymmetric shapes for these nuclei. In this connection there arises a question whether this

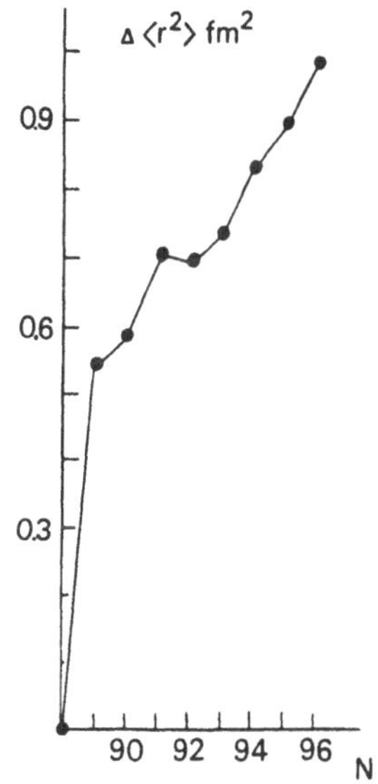

Fig.1. Isotopic changes in charge radii for Eu isotopes with N > 88.

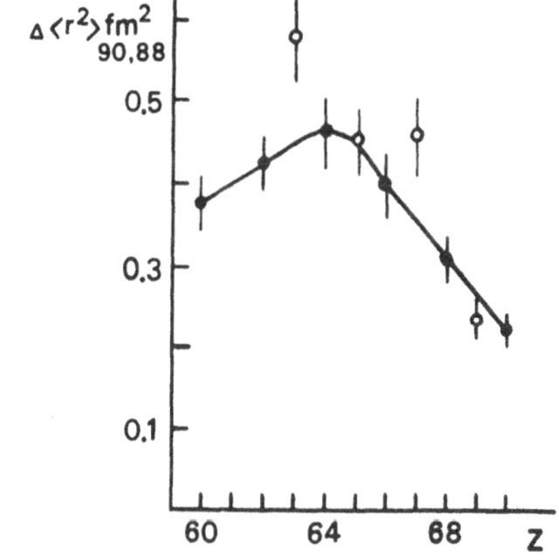

Fig.2. Value of $\delta\langle r^2\rangle_{90,88}$ as a function of Z for the even Z nuclei (full circles) and odd-Z nuclei (open circles). The continuous line is drawn through the points for the even - Z nuclei to guide the eye.

unusual behaviour of $\delta\langle r^2\rangle$ remains for the more neutron-rich Eu isotopes; or, in the other words, whether the region of stable octupole deformation comprises these nuclei too. To answer this question we have measured IS and hfs for $^{155-159}Eu$ in the 601.8 nm line of EuI. The restoration of the normal odd-even staggering for Eu isotopes with $N>92$ is obvious from fig.1. This means that the stable octupole deformation, attributed by Sheline [3] to the Eu nuclei around ^{154}Eu, does not display itself in the charge radii of heavier Eu isotopes. It is worth noting that the marked decrease of the $5/2^+$ ground state magnetic moment which was observed in our experiment for odd Eu isotopes when moving from A=155 to A=159, may also indicate the restoration of reflection-symmetric shapes for these nuclei.

Neutron-deficient Tb and Gd isotopes

Last decade a great deal of information was accumulated in the shape transition region near N=88-90. Here we complement these data by our new experimental results for Tb and Gd isotopes. In fig.2 the value of $\delta\langle r^2\rangle_{90,88}$ for the rare earth elements is plotted against Z. This value is directly connected with the corresponding change of nuclear deformation. It is clearly seen that the character of the transition between spherical and well-deformed shapes is dependent on the proton number Z: the nearer Z to the "magic" number Z=64, the sharper is the change of nuclear deformation at N=88-90. Although this feature is inherent to both the even-Z nuclei and the odd-Z ones, the transition from a gradual to an abrupt deformation change is sharper for the odd-Z nuclei. It is likely to be connected with the blocking effect caused by the odd proton. This conclusion is confirmed by our Hartree-Fock calculations: the deformation jump near N=88-90 is reproduced only by fixing the state blocked by the odd proton in accordance with the experimental spin and parity.

DISCUSSION

Theoretical analysis of the experimental data was performed within the framework of the Hartree-Fock (HF) method. For the strongly deformed nuclei ($N\geqslant 90$) the theoretical results on quadrupole moments coincide as a rule with the experimental ones within the limits of experimental errors. As to $\delta\langle r^2\rangle$, the HF theory overestimates $\delta\langle r^2\rangle$ values for these nuclei approximately by 20-30% . Theoretical and experimental values of $\delta\langle r^2\rangle$ in the average agree better for the nuclei with $N<82$ (fig.3). At the same time the $\delta\langle r^2\rangle$ predictions of the spherical HF method and the standard droplet model practically coincide for the nuclei with $N<82$ and differ by the similar quantity (20-30%) for the nuclei with $N>90$. This indicates that the disagreement between the HF theory and the experiment, found for the well-deformed nuclei, may lie in an incorrect evaluation of the "volume" contribution to $\delta\langle r^2\rangle$ in the theory. It is to be noted that the agreement between theory and experiment for the very neutron deficient nuclei with $N<82$ can be achieved only when the continuum states are taken into account in the pairing treatment (see fig.3).

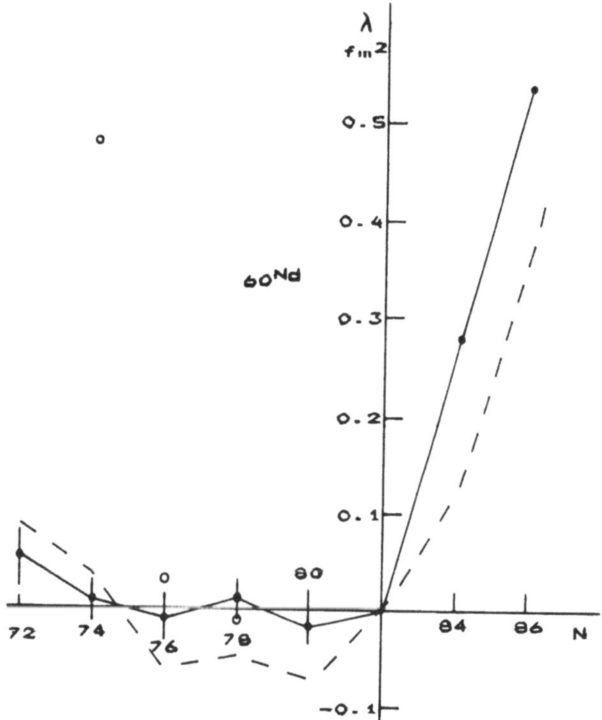

Fig.3. Comparison of experimental changes in ms charge radii for even-even Nd isotopes (full circles) connected by line with predictions from the HF method with SKM'force (dashed line). Open circles display the results of HF calculations without allowance for positive energy states in the BCS equations.

To conclude, RIS is a powerful and promising method for the nuclear shape investigation in long isotopic chains. New breakthrough in the sensitivity when using the laser ion source can pave the way to the proton drip line where unusual nuclear structure may be expected.

REFERENCES

1. G.D.Alkhazov et al. Investigation of Nuclear charge radii and electromagnetic moments of rare-earth elements by laser spectroscopy, in :"Proceedings of 5-th Conference on Nuclei far from stability", I.S.Towner, ed., American Institute of Physics, New Jork (1988).
2. G.D.Alkhazov, E.Ye.Berlovich and V.N.Panteleyev, A new highly efficient selective laser ion source, NIM, A 280 : 141(1989).
3. R.K.Sheline, Octupole deformation in the odd-odd nucleus ^{154}Eu, Phys.Lett.,B 219: 222 (1988).
4. R.K.Sheline, A.K.Jain and K.Jain, Possible octupole deformation in Cs and Ba nuclei from their differential radii, Phys.Rev., C 38: 2952(1988).
5. K.Dörshel et al., Investigation of europium isotopes with neutron numbers between 84 and 93 by collinear laser ion-beam spectroscopy, Z.Phys. A - Atoms and Nuclei, 317: 233 (1984).

ATOMIC SPECTROSCOPY AS A PROBE OF PARITY NON-CONSERVING EFFECTS IN THE NUCLEUS

Ann-Marie Mårtensson-Pendrill

Department of Physics
University of Göteborg and Chalmers University of Technology
S-412 96 Göteborg, Sweden

1. INTRODUCTION

Atomic spectroscopy is an old tool to study nuclear properties. This tool has been sharpened through the ever-increasing accuracy of experimental measurements, but also through the development of methods to perform accurate atomic theory calculations necessary to relate nuclear properties to experimental results. The accuracy of calculated hyperfine structures of low-angular-momentum states of alkali atoms is approaching the percent level,[1] and also for more complex systems, such as Tl, accuracies of a few percent are reached.[2] The improved accuracy makes possible a more reliable extraction of e.g. nuclear charge radii[3], as discussed elsewhere at this workshop.

Atomic physics is now also exploring interactions which violate mirror symmetry. The exchange of a neutral Z_0 boson leads to an electroweak electron-nucleon interaction, discussed in section 2, which is invariant under time reversal. This interaction is now well established and the experiments are on the verge of observing nuclear-spin dependent effects. Still hypothetical are the interactions that violate both parity (P) and time (T) reversal symmetry and the experiments establishing upper limits for atomic electric dipole moments (EDMs) can be used to set limits on several possible P and T violating effects, as discussed in Section 3. A brief review is given of the current experimental situation, with particular emphasis on effects that demand both atomic and nuclear theory for the interpretation

2 THE PARITY NON-CONSERVING WEAK INTERACTION

In the presence of weak interactions an atomic wavefunction no longer has pure parity, but receives a small admixture of opposite parity. This makes possible an electric dipole transition to a state of the same parity, where, without the PNC admixture, normally only magnetic dipole (and possibly electric quadrupole, etc) transitions would be allowed. The effect increases rapidly with nuclear charge, leading to observable effects for heavy atoms, as suggested in pioneering papers by Bouchiat and Bouchiat.[4] The most accurate PNC experiments[5,6] have been performed for the 6s→7s transition in ^{133}Cs and measure the ratio of the parity non-conserving electric dipole transition matrix element, E_{PNC}, to the vector Stark polarizability β, giving[5]

$$\mathrm{Im}(E_{PNC})/\beta = \begin{cases} -1.513(50)\,\mathrm{mV/cm}\ (F=3 \to F'=4) \\ -1.639(48)\,\mathrm{mV/cm}\ (F=4 \to F'=3) \end{cases} \tag{1}$$

Nuclear Shapes and Nuclear Structure at Low Excitation Energies
Edited by M. Vergnes *et al.*, Plenum Press, New York, 1992

The parity non-conserving electron-nucleus interaction mediated by the Z_0 boson can be written as[4]

$$h^{PNC} = \frac{G_F}{2\sqrt{2}} Q_W \rho_N(r) \gamma_5 - \frac{G_F}{\sqrt{2}} K_2 \rho_N(r) \frac{(\kappa_N - 1/2)}{I\,(I+1)} \alpha_{el} \cdot \mathbf{I} \qquad (2)$$

Here $G_F = G_\mu (\hbar c)^3 = 89.6$ eV $fm^3 \approx 2.22 \cdot 10^{-14}$a.u. is the weak interaction constant. The short-range character of the interaction is accounted for by $\rho_N(r)$, which is a normalized sum of the neutron and proton densities in the nucleus. The weak interaction in heavy atoms is dominated by the first ("nuclear-vector electron-axial" current) term which is independent of the nuclear spin. In this term, the change of parity is produced by the Dirac matrix $\gamma_5 = \begin{pmatrix} 0 & 1 \\ 1 & 0 \end{pmatrix}$, which interchanges the upper and lower component of a relativistic wavefunction. Combination of atomic theory with the experimental result in (1) makes it possible to extract values for the "weak charge", $Q_W = -2[\,(2Z+N)\,C_{1u} + (Z+2N)C_{1d}]$, of the nucleus, which is defined in terms of the vector coupling constants, C_{1u} and C_{1d}, of the up and down quarks to the neutral current. This part of the interaction is now well-established. The linear combination of C_{1u} and C_{1d} entering for heavy atoms is essentially orthogonal to that measured in high-energy experiments[7] and, even with a precision of a few percent, the atomic PNC experiments provide some of the most stringent limits to modifications of the standard model involving extra Z bosons[8]. The increasing experimental accuracy raises new demands on atomic theory. The most complete calculation so far[9] leads to the value $Q_W = -71.04(1.58)[0.88]$, where the first error bar refers to experiment and the second to the atomic calculation. Experimental[10] as well as theoretical[11] results have also been obtained for Tl, Pb and Bi.

In attempts to circumvent the need for atomic theory, experiments are also underway to study PNC in a sequence of isotopes of an element, such as Sm[12] and Cs[13], following the suggestion by Dzuba *et al.*[14] The motivation for these studies is that the electronic wavefunction is determined mainly by the nuclear charge (and to a small extent by its distribution) and can be expected to be essentially the same for all the isotopes of an element. The dependence on the atomic structure can then be factored out and the results would be proportional to Q_W. However, these experiments may shift the burden to nuclear theorists: The changes in *neutron* distribution may affect the result at the desired level of accuracy[15], but cannot be obtained directly from experiment, in contrast to changes in nuclear *charge* distribution, which may be deduced from optical isotope shift studies.

The second term in (2), which involves the nuclear spin through κ_N, may cause a difference between the results for the two transitions in (1). This part has a much smaller interaction constant $K_2 \approx -0.05$, than the nuclear-spin independent part. In addition, the enhancement is smaller (proportional to Z^2 rather than Z^3) since only unpaired nucleons contribute. Another term, identical in form, may arise through weak interactions within the nucleus which can lead to a helical current distribution, causing the nucleus to have an *anapole* moment[16,17], defined as $\mathbf{a} = -\pi \int dr\, r^2 \mathbf{j}(r) = a\mathbf{I}$. The anapole moment leads to a vector potential $\mathbf{A}(\mathbf{r}) = \mathbf{a}\,\delta(\mathbf{r})$ which gives rise to magnetic fields inside the system but vanishes outside. The ordinary electromagnetic interaction with an anapole moment can be treated together with the PNC interaction Eq (2) by replacing K_a with $K = K_a - K_2(\kappa_N - 1/2)/\kappa_N$. As pointed out by Flambaum *et al*[17] amplifying effects in the nucleus may make this interaction larger than $h_{(2)}^{(PNC)}$ and they estimate K to fall in the range 0.25 – 0.33. The spin-dependent contributions then enter at the percent level[9,18]. The experimental result[5] in (1) corresponds to K=0.72(39). The increased accuracy of experiments underway may lead to the detection of the hitherto unobserved anapole moment.

3. P AND T VIOLATING EFFECTS - ELECTRIC DIPOLE MOMENTS

An electric dipole moment (EDM) of an atom or elementary particle would violate not only parity but also time reversal symmetry.[19] Although CP violation (implying T violation through invocation of the CPT theorem) was observed[20] in the decay of long-lived K-mesons in 1964, it has not yet been observed for any other system in spite of intensive search and the cause of the T violation remains obscure. Several models to explain the T violation have been excluded by the experimental limits e.g. of the neutron EDM[21,22] which have fallen by approximately an order of magnitude every 5-6 years[23]. The current best limit[22] for the neutron EDM is $d_n= (-3\pm5)\cdot10^{-26}$ ecm. The lowest limit for any system has been obtained by Fortson and coworkers for the closed-shell ground states of Xe[24] and Hg[25]. For Xe they quote $D(^{129}Xe)=(0.3\pm1.1)\cdot10^{-26}$ecm.

An *atomic* EDM can arise from several sources[26-28]. In the limit of non-relativistic point particles no atomic electric dipole moment can be induced by an EDM of a charged particle.[26] However, through relativistic effects an *electron* EDM can, in fact, induce an atomic EDM as much as 100-1000 times larger in easily polarizable systems.[29] A limit, $d_e=(-3\pm8)\cdot10^{-27}$ ecm for the electron EDM can thus be deduced from the recent experimental result for Tl, $D(Tl)=(1.6\pm5.0)\cdot10^{-24}$ ecm, from the Berkeley group. A *nuclear* EDM, caused by an intrinsic nucleon EDM or by P and T violating interactions in the nucleus, will be shielded from an external electric field through rearrangement of the electronic wavefunction. nevertheless, an atomic EDM may result if the nuclear EDM distribution is different from the charge distribution, as discussed in more detail below. The prediction of the size of the nuclear EDM requires nuclear theory, and atomic as well as nuclear theory is needed to determine the size of the atomic EDM it can induce. In this way, atomic experiments can probe nucleon-nucleon or quark-quark interactions and lead to limits on quark-quark interactions constants which are comparable and in some cases lower than those deduced from limits on the neutron EDM.[28,31] In addition, the atomic experiments can set limits on electron-nucleon interactions,[27] which, of course, do not affect the neutron EDM. The experiments also restrict a possible T-odd P-even interaction[32] since this could combine with the known P-odd T-even interaction, discussed in Sect. 2, to give a net P and T violation.

In general, the search for an EDM focuses on the energy splitting between the different spin directions. A magnetic moment interacting with an external magnetic field leads to energy differences between the spin states. Similarly, an EDM, if it exists, would lead to a small energy shift in the presence of an external electric field and would lead to different splittings if the external electric field is parallel or anti-parallel to the magnetic field. The effects are enhanced with nuclear charge and the atomic experiments are performed for heavy atoms, such as Xe, Cs, Hg and Tl. Experiments[33] have also been performed for TlF, following the observation by Sandars in 1967[34] that "a polar molecule can be almost completely polarized in attainable laboratory electric fields" making it possible to obtain large effective electric fields at the nucleus, which "will no longer average to zero, but will be parallel to the external field".

Nuclear Electric Dipole Moments

The nuclear structure has only a small influence on the atomic EDM caused by electron-nucleon interactions. For P and T violating effects within the nucleus, on the other hand, details in the nuclear wavefunction are crucial. An EDM of an unpaired proton, e.g., would lead to an atomic EDM proportional to the difference in the second moments of the charge and dipole distribution,[26] i.e.,to $\langle r_q^2\rangle - \langle r_d^2\rangle$. However, larger effects may arise through P and T violating interactions between the nucleons.[35] The P and T odd interaction between a non-relativistic nucleon with spin σ and the core of neutrons and protons with normalized density ρ_{core} can be described by the Hamiltonian[36]

$$H_{PT} = (G_F/\sqrt{2})\,(\eta\,/2Mc)\,A\;\sigma\cdot\nabla\,\rho_{core}(\mathbf{r}) \qquad (3)$$

where the model-dependent interaction parameter η (sometimes denoted by ξ) has the value[36] $\eta = \eta_0\,(1.34\,N\text{-}Z)\,/\,A = -0.67\cdot10^{-8}\,(1.34\,N\text{-}Z)\,/\,A$ in the Kobayashi-Maskawa model.[37] This interaction then gives rise to opposite parity admixtures in the nuclear wavefunctions that can interact with an electric field. The resulting effective electron-nucleon interaction is of essentially the same form as that caused by an intrinsic nucleon EDM and they can be treated together by introducing[36] a "Schiff moment" defined by

$$\mathbf{Q} = \frac{1}{6}\left(\frac{3}{5}\int \delta\rho_c(\xi)\,\xi^3\,\mathbf{C}^1\,d^3\xi - \int \rho_c(\xi)\,\xi^2 d^3\xi \int \delta\rho_c(\xi)\,\xi\,\mathbf{C}^1\,d^3\xi\right)$$

where $\delta\rho_c(\xi)$ is the P and T odd correction to the charge distribution $Ze\rho_c(\xi)$. Flambaum *et al*[38] have calculated Schiff moments for a number of nuclei in terms of the interaction parameter η in (3). The Schiff moment leads to an effective electron-nucleus interaction[36,39],

$$H_Q = 6\mathbf{Q}\cdot\mathbf{T}^{el}$$

where $\mathbf{T}^{el} = \sum_i \mathbf{t}_i = \sum_i (-2\pi/3)\,(\nabla\,\delta(\mathbf{r}_i))$.

Magnetic Effects

In addition to the finite size effects discussed above, an atomic EDM may arise through magnetic interactions. Hinds and Sandars[40] in their study of P and T violating effects in the polar molecule TlF, have analysed the interaction of a nuclear EDM with a magnetic field from the electrons. The resulting EDM is somewhat smaller than the finite size effect, but is also less uncertain, since it is insensitive to the details of the nuclear wavefunction. For nuclei with $I>1/2$, larger effects may arise through an induced nuclear magnetic quadrupole moment (MQM). Sandars[27] gives the estimate $Z^2\alpha^2(m/M)d_N$ for the atomic EDM caused by an MQM, to be compared with the magnitudes $Z^2(r_N/a_0)^2 d_N$ to be expected from the finite size effects and $Z\alpha^2(m/M)d_N$ to be expected from the interaction with the magnetic field from the electrons.

4. RESULTS AND DISCUSSION

The experimental limits for the closed-shell atoms Xe[24] and Hg[25] and for the polar molecule TlF.[33] give the lowest limits for interactions involving nuclear spins as well as for nuclear Schiff moments. These results are shown in Table 1. (The lowest limits for the electon EDM and for a scalar electron-nucleon interaction H_S, are obtained from the experiments on Cs and Tl, as discussed e.g. in Ref. 41). The sensitivities to the tensor-pseudotensor interaction

$$H_T = \left(2G_F C_{T,N}/\sqrt{2}\right)\sigma_N\cdot\sum_e i\beta_e\alpha_e\rho_N(r_e)$$

have been calculated for Xe[42,43] and TlF.[40,44] Flambaum and Khriplovich[42] have shown than these results can be used to obtain sensitivities also to an interaction

$$H_3 = \left(iG_F C_{3,N}/\sqrt{2}\right)\,(\beta\gamma_5)_N\sum_e i\beta_e\rho_N(r_e)$$

and to nuclear Schiff moments. More details on the calculations can be found in Ref. 45. The limits on the proton EDM obtained from these moments are significantly weaker than the neutron EDM limit. However, the Schiff moments also limit quark-quark interaction constants (related to η in (3)). As pointed out by Khatsymovsky *et al*[31] these limits are comparable, and in some cases lower than, those obtained from the neutron EDM. In this way the study of P and T violating effects in atoms and molecules form an important complement to the study of the neutron EDM and can help to shed light on the still obscure source of T violation in nature. The calculations of sensitivities to these interaction constants pose an interesting challenge to nuclear, atomic and molecular theorists.

Table 1. Experimental limits of atomic (or molecular) electric dipole moments and derived limits for various P- and T- violating parameters.

	^{129}Xe	^{199}Hg	TlF[a]
D_{exp} [e cm][a]	$(0.3\pm1.1)\cdot10^{-26}$[b]	$(0.7\pm1.5)\cdot10^{-26}$[c]	$-(.13\pm.22)$[d]
C_T	$(0.6\pm2.1)\cdot10^{-6}$	$-(1.1\pm2.5)\cdot10^{-7}$	$-(1.5\pm2.6)\cdot10^{-7}$
C_3	$(2.5\pm9.1)\cdot10^{-4}$	$(4.0\pm8.6)\cdot10^{-5}$	$-(5\pm9)\cdot10^{-5}$
Q [e fm^3]	$(0.8\pm3.2)\cdot10^{-9}$	$(0.2\pm0.4)\cdot10^{-9}$	$(2.3\pm3.9)\cdot10^{-10}$
d_p [ecm]	$(0.6\pm2.4)\cdot10^{-21}$		$(-3.7\pm6.3)\cdot10^{-23}$
η	(0.05 ± 0.18)	(0.012 ± 0.026)	(-0.013 ± 0.021)

a) For TlF the interaction constant D/h for the effective interaction $H_{eff} = -D\sigma_{Tl}\cdot\hat{\lambda}_{Tl}$ is given in mHz

b) Vold *et al*, Ref 24 c) Lamoreaux *et al*, Ref. 25 d) Cho *et al*, Ref. 33,

REFERENCES

1 S A Blundell, W R Johnson and J Sapirstein, Phys. Rev. A**43** *3407* (1991)
V A Dzuba, V V Flambaum, A Ya Kraftmakher and O P Sushkov Phys. Lett. **142A** *373* (1989)
S Salomonson and A Ynnerman, Phys. Rev. A**43** *88* (1991)

2 A C Hartley and A-M Mårtensson-Pendrill, Z. Physik D**15** *309* (1990)

3 A-M Mårtensson-Pendrill, L Pendrill, S Salomonson, A Ynnerman and H Warston J. Phys. B **23** *1749* (1990)
A C Hartley and A-M Mårtensson-Pendrill, J. Phys. B **24** *1193* (1991)

4 M A Bouchiat and C Bouchiat, Phys. Lett. B**48** *111* (1974) J. Physique (Paris) **35** *899* (1974)

5 M C Noecker, B P Masterson and C E Wieman, Phys. Rev. Lett. **61** *310* (1988)

6 M A Bouchiat, J Guena and L Pottier, J de Physique **46** *1897* (1985)
M A Bouchiat, J Guena, L Pottier and L Hunter, J de Physique **47** *1175, 1709* (1986)

7 C Bouchiat, C A Piketty and D Pignon, Nucl. Phys. B**221** *68* (1983)

8 W J Marciano and A Sirlin, Phys. Rev. D**27** *552* (1983)
U Amaldi, A Böhm, L S Durkin, P Langacker, A K Mann, W J Marciano, A Sirlin and H H Williams, Phys. Rev. D**36** *1385* (1987)

9 S A Blundell, W R Johnson and J Sapirstein, Phys. Rev Lett. **65** *1411* (1990)

10 A recent review of the experimental situation was given by M-A Bouchiat, Atomic Physics **12** (1991), Proceedings of the 12th ICAP, Ann Arbor, Michigan, Aug.1990

11 References to the calculations can be found in the review by
S A Blundell, A C Hartley Z W Liu, A-M Mårtensson-Pendrill and J Sapirstein, in Proc. from the Workshop on Coupled-Cluster Theory at the Interface of Atomic Physics and Quantum Chemistry, Harvard, Aug. 1990
Theor. Chim. Acta **80** *257* (1991)

12 P E G Baird et al, private communication

13 C E Wieman, private communication

14 V A Dzuba, V V Flambaum and I B Khriplovich, Z. Physik D**1** *243* (1986)

15 E N Fortson, Y Pang and L Wilets, Phys. Rev. Lett. **65** *2857* (1990)

16 Ya B Zel'dovich, Zh. E.T.F. **33** *1531* (1957)
L M Barkov, I B Khriplovich and M S Zolotorev, Comments At. Mol. Phys. **8** *79* (1979)
I B Khriplovich, Comments At. Mol. Phys. **23** *189* (1989)

17 V V Flambaum, I B Khriplovich and O P Sushkov, Phys. Lett. **B146** *367* (1984)

18 P A Frantsuzov and I B Khriplovich, Z. Physik. D7 *297* (1988)
A Ya Kraftmakher, Phys. Lett A**132** *167* (1988)

19 E M Purcell and N F Ramsey, Phys. Rev. **78** *807* (1950)
L Landau, Zh.E.T.F. **32** *405* (1957) Sov. Phys. JETP **5** *336* (1957)

20 J H Christenson, J W Cronin, V L Fitch and R Turlay, Phys. Rev. Letts. **13** *138* (1964)

21 N F Ramsey, Rep. Prog. Phys. **45** *95-113* (1982)
I S Altarev *et al* Pizma Zh. E T F **44** *360* (1986), JETP Lett **44** *460* (1986)

22 K F Smith *et al* Phys. Lett. B**232** *191* (1990)

23 B Heckel, Atomic Physics **9** (Proc. 9th ICAP, Eds. R S van Dyck and E N Fortson, World Scientific, Singapore, 1984) *p285*

24 T G Vold, F J Raab, B Heckel, E N Fortson, Phys. Rev. Lett. **52** *229* (1984)

25 S K Lamoreaux, J P Jacobs, B R Heckel, F J Raab and E N Fortson, Phys. Rev. Lett. **59** *2275* (1987)

26 L I Schiff, Phys. Rev. **132** *2194* (1963)

27 P G H Sandars, Atomic Physics **9** (Proc. 9th ICAP, Eds. R S van Dyck and E N Fortson, World Scientific, Singapore, 1984) p*225*
P G H Sandars, Physica Scripta **36** *904* (1987)

28 I B Khriplovich, Atomic Physics **11** *113* (1989)

29 P G H Sandars, Phys. Lett. **14** *194* (1965), Phys. Lett. **22** *290* (1966)

30 K Abdullah, C Carlberg, E D Commins, H Gould and S B Ross Phys. Rev. Lett. **65** *2347* (1990)

31 V M Khatsymovsky, I B Khriplovich and A S Yelkhovsky, Annals of Physics **186** *1* (1988)

32 I B Khriplovich, Novosibirsk Institute of Nuclear Physics preprint, INP 90-69 (1990)
Nucl. Phys. B**352** *385* (1991)

33 D Cho, K Sangster and E A Hinds, Phys. Rev. Lett. **63** *2559* (1989) with references also to earlier work

34 P G H Sandars, Phys. Rev. Lett. **19** *1396* (1967)

35 W C Haxton and E M Henley, Phys. Rev. Lett. **51** *1937* (1983)
G Feinberg, Trans. N Y Acad. Sci. **38** *26* (1977)

36 O P Sushkov, V V Flambaum and I B Khriplovich, Zh. E. T. F. **87** *1521* (1984), Sov. Phys. JETP **60** *873* (1984)

37 M Kobayashi and T Maskawa, Prog. Theor. Phys. **49** *652* (1973)

38 V A Dzuba, V V Flambaum and P G Silvestrov, Phys. Lett. **154B** *93* (1985)
V V Flambaum, I B Khriplovich and O P Sushkov, Phys. Lett. **162B** *213* (1985)
V V Flambaum, I B Khriplovich and O P Sushkov, Nucl. Phys. A**449** *750* (1986)

39 P V Coveney and P G H Sandars, J. Phys. B**16** *3727* (1983) and P V Coveney, part II BA Thesis, Oxford, 1981

40 E A Hinds and P G H Sandars, Phys. Rev. A**21** *421* (1980)

41 E Lindroth and A-M Mårtensson-Pendrill, Europhys. Lett. **15** *155* (1991)

42 V V Flambaum and I B Khriplovich, Zh E T F **89** *1505*(1985) J E T P **62** *872* (1985)

43 A-M Mårtensson-Pendrill, Phys. Rev. Lett. **54** *1153* (1985)

44 P G H Sandars, Atomic Physics **4** (Proc. 4th ICAP, Ed. G zu Putlitz, E W Weber and A Winnacker, Plenum, 1975) p*71*

45 A-M Mårtensson-Pendrill, in Meth. Comp. Chemistry **6**, (Plenum. New York, Ed. S Wilson, 1992)

ISOTOPE SHIFTS AND HYPERFINE STRUCTURE IN POLONIUM ISOTOPES BY ATOMIC BEAM LASER SPECTROSCOPY

D. Kowalewska, K. Bekk, S. Göring, A. Hanser, W. Kälber, G. Meisel, and H. Rebel

Kernforschungszentrum Karlsruhe, Institut für Kernphysik
P.O.B. 3640, D-7500 Karlsruhe, Germany

INTRODUCTION

Atomic electrons can be used to probe nuclei by studying details of atomic spectra: the isotope shift of spectral lines contains information on the change of the mean square charge radius from one isotope to another one, while the hyperfine splitting allows to determine spins and the lower electromagnetic moments of nuclei.

The range of application of optical spectroscopy has been extended considerably by the introduction of tunable dye lasers. This is mainly due to the increase of sensitivity which enables one to obtain spectra from minute samples of, e.g., artificially produced isotopes. Thus systematic studies of the above mentioned gross properties for long chains of isotopes became possible.[1,2] As to the changes of nuclear charge radii of short lived isotopes, optical spectroscopy is the only source of information. We have chosen to investigate polonium isotopes since only little was known on them so far.[3-8]

EXPERIMENTAL DETAILS

It is required to have free atoms to obtain sufficiently unpertubed atomic spectra. The method we applied was laser induced fluorescence on collimated atomic beams.[9] Polonium was evaporated from graphite or molybdenum crucibles which were heated by electron bombardment to about 1000 K. The polonium beam was collimated to an angular spread of about $\pm 2°$ (sometimes less or more than that) leading to a residual Doppler width of about 50 MHz. For safety reasons the atomic beam apparatus was operated inside a glove box.

The optical transition induced was from the ground state $6p^4\ {}^3P_2$ to the excited state $6p^3\ 7s\ {}^5S_2$. The required light at a wavelength of 255.8 nm was obtained by doubling the frequency of a cw dye laser oscillating at 511.5 nm. The nonlinear medium used for frequency doubling was a beta barium borate (BBO) crystal;[10] to obtain a power of several milliwatts, the crystal was placed inside the laser resonator. The resonant absorption was detected by the strength of the spontaneously reemitted fluorescence at the same ultraviolet wavelength as was incident.

The polonium isotopes were prepared in different ways. The long lived ones ($^{208,209,210}Po$) were purchased dissolved in HCl or H_2SO_4. Po was spontaneously deposited onto silver foils which were dipped into the solution. The foil was put into the crucible and heated; the polonium then evaporated readily. The isotopes $^{206,207}Po$ were made from $^{209}Bi_2O_3$ targets by (p,5n) or (d,5n) reactions. The targets were dissolved in HCl after irradiation; the polonium was extracted from the solution as before. The lighter isotopes - $^{200,202,204,205}Po$ - were produced by irradiating metallic ^{204}Pb or ^{206}Pb targets with alpha particles. The Pb target thereafter was heated inside the crucible such that the Po was vaporized while most of the lead remained in the crucible.

The minimum sample size required (or the sensitivity) depends strongly on the nuclear spin of the isotope under investigation. Even mass isotopes with zero spin have only one component in their optical spectrum since they have no hyperfine splitting. Therefore it was possible to perform measurements for ^{200}Po with only 13 pg (or $4 \cdot 10^{10}$ Po atoms) in the ^{204}Pb target. In contrast to this, the odd isotopes have several hyperfine levels in the ground state that share the population. In addition, there are many hyperfine components rather than a single one to be recorded while some of the hyperfine transitions are weak and not yet saturated by the ultraviolet power available. Therefore the odd isotopes required considerably larger samples; as an example, 500 pg were needed for a survey record of ^{205}Po. If available, larger samples were used. Fig. 1a gives one of 25 scans obtained from a sample containing 13 pg of ^{200}Po and 70 pg of ^{202}Po. Fig. 1b is a survey spectrum for ^{207}Po; the sample in this case also contained some $^{206,208,209}Po$.

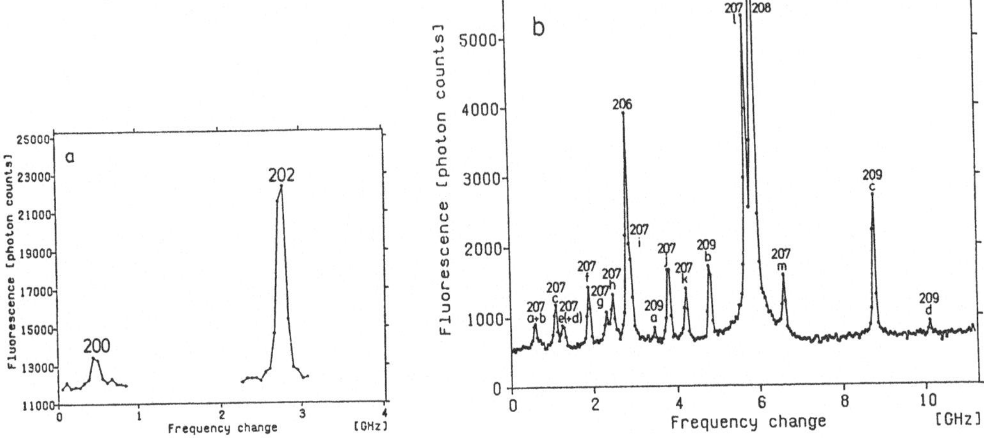

Figure 1. a: Record of the isotope shift between ^{200}Po and ^{202}Po.
b: Survey record of the hyperfine structure of ^{207}Po; the components of $^{206,208,209}Po$ are also present since these Po isotopes were produced along with ^{207}Po.

Table 1. Isotope shifts for Po in the transition $6p^4\,{}^3P_2 \rightarrow 6p^3 7s\,{}^5S_2$ ($\lambda = 255.8$ nm) and changes of the rms nuclear charge radii derived thereof. For the $\delta\langle r^2\rangle$ values a conversion factor $\delta v / \delta\langle r^2\rangle = 30.6$ GHz/fm^2 was used.

mass number A	nuclear spin I	isotope shift $v^A - v^{208}$[MHz]	rms radius change $\langle r^2\rangle^A - \langle r^2\rangle^{208}$ [fm^2]
210	0	+3464(8)	+ 0.113
209	1/2	+1295(7)	+ 0.042
208	0	0	0
207	5/2	−2306(6)	− 0.075
206	0	−3064(8)	− 0.100
205	5/2	−5457(16)	− 0.178
204	0	−6077(12)	− 0.199
202	0	−8852(12)	− 0.289
200	0	−11154(20)	− 0.365

RESULTS

Measurements were done for 9 polonium isotopes; Table 1 gives the isotope shift results. The isotope shift for heavy elements like polonium essentially is proportional to the change of the rms nuclear charge radius.[11,12] Fig. 2 is a compilation of the isotope shift data for polonium and some neighboring elements. They are plotted in such a way that the shift (or rms radius change $\delta\langle r^2\rangle$) is

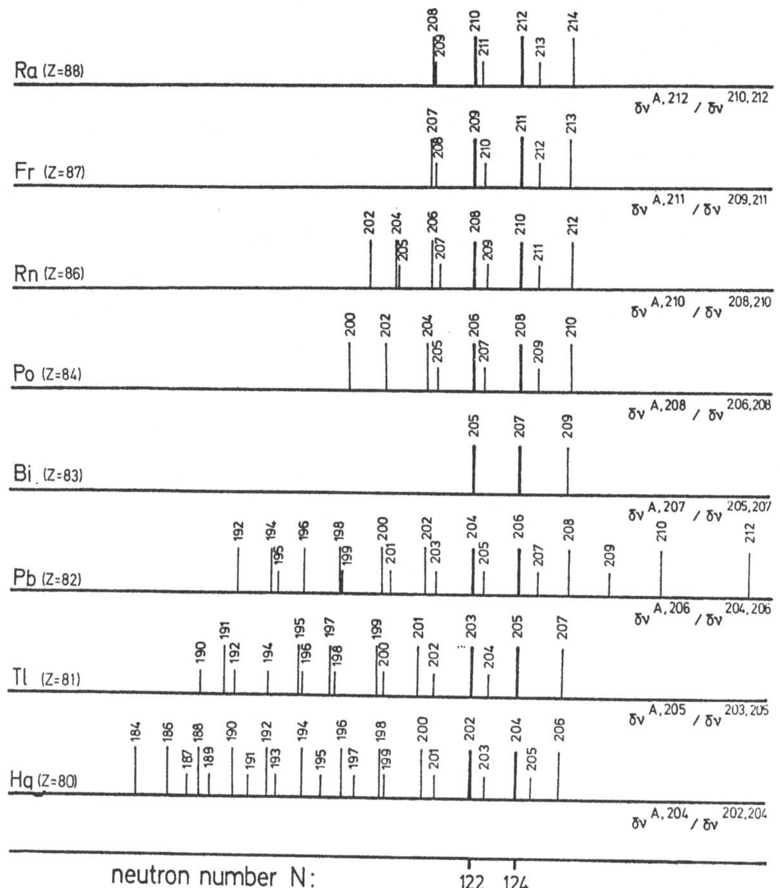

Figure 2. Isotope shift data for polonium and its neighboring elements. The data are normalized to $\nu(N=122)-\nu(N=124)$. The data were taken from [13](Ra), [14](Fr), [15](Rn), [16](Bi), [17,18](Pb), [19,20](Tl), and [21](Hg).

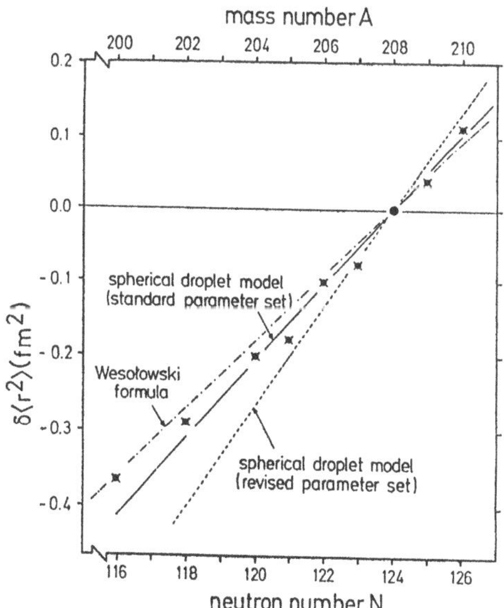

Figure 3. Plot of $\delta\langle r^2\rangle$ values for polonium. The experimental points were obtained using an electronic conversion factor of $\delta\nu / \delta\langle r^2\rangle = 30.6$ GHz/fm^2. The model curves correspond to the standard droplet model,[24,25] to a revised parameter set for the droplet model,[26] and to a semiempirical model by Wesolowski.[27]

made the same for the isotopes with neutron numbers $N=122$ and $N=124$. The plot shows that there are remarkable similarities as to the general trend with neutron number. This also applies to the so-called odd-even staggering, i.e. to the observation that the addition of one odd neutron results in a considerably smaller increase of $\langle r^2 \rangle$ than the addition of the next even one.

The conversion of $\delta\nu$ into $\delta\langle r^2 \rangle$ values poses some problems since an electron shell parameter $\Delta|\psi(0)|^2$ (the change of the electron density at the site of the nucleus upon optical excitation) is needed. Usually it is obtained by combining isotope shift results for stable isotopes with $\delta\langle r^2 \rangle$ values obtained for them in other ways. No such calibration data are available for polonium. We have taken a $\Delta|\psi(0)|^2$ value supplied by Fricke[22] which was obtained by a relativistic multi-configuration Dirac-Fock calculation,[23] leading to $\delta\nu = \delta\langle r^2 \rangle \cdot 30.6$ GHz/fm^2. The $\delta\langle r^2 \rangle$ results obtained in this way are given in the 4th column of Table 1. They are also plotted in Fig. 3 together with those from different models.

The transition from $\delta\nu$ to $\delta\langle r^2 \rangle$ might be incorrect to some degree but in such way that a correction factor has to be applied that is common to all data points. Therefore the data are rather precise as far as relative values or trends are concerned. Thus the upwards bending of the data curve with decreasing neutron number (Fig. 3) is an indication for the onset of an accelerated deformation; for a speculation on an independent observation of this effect related to this one see ref.[28]. The odd-even staggering clearly is very similar to that one for neighboring elements (Fig. 2). Its nature practically still is not understood since models require ad hoc assumptions that apply only locally; obviously its finer details such as size and sign of the staggering with neutron number are quite independent of the particular nucleus. It rather seems to be a common feature of the nuclei considered, favoring the view that it might be worthwhile to try to deduce it from the many unpopulated continuum states rather than the actually populated ones.[29]

REFERENCES

1. E. W. Otten, Nuclei far from stability, in: "Treatise on Heavy-Ion Science", vol. 8, D. A. Bromley, ed., Plenum Press, New York (1989).
2. H. Rebel, Laserspectroscopic studies of nuclear structure, in: "Recent advances in nuclear physics", M. Petrovichi and N. V. Zamfir, eds., World Scientific., Singapore - New Jersey - London - Hong Kong (1989).
3. G. W. Charles, D. J. Hunt, G. Pish, and D. L. Timma, J. Opt. Soc. Am. 45:869 (1955).
4. S. Axensten and C. M. Olsmats, Arkiv Fysik 19:461 (1961).
5. C. M. Olsmats, S. Axensten, and G. Liljegren, Arkiv Fysik 19:469 (1961).
6. G. W. Charles, J. Opt. Soc. Am. 56:1292 (1966).
7. P. Herzog, H. Walitzki, K. Freitag, H. Hildebrand, and K. Schlösser, Z. Physik A 311:351 (1983).
8. P. Raghavan, Atomic Data Nucl. Data Tables 42:189 (1989).
9. G. Meisel, K. Bekk, H. Rebel, and G. Schatz, Hyperfine Interact. 38:723 (1987).
10. D. Eimerl, L. Davies, S. Velsko, E. K. Graham, and A. Zalkin, J. Appl. Phys. 62:1968 (1987) and references therein.
11. P. Aufmuth, K. Heilig, and A. Steudel, Atomic Data Nucl. Data Tables 37 : 455 (1987).
12. W. H. King, Isotope shift in atomic spectra, Plenum Press, New York (1984).
13. K. Wendt, S. A. Ahmad, W. Klempt, R. Neugart, E.W. Otten, and H. H. Stroke, Z. Physik D 4:227 (1987).
14. A. Coc, C. Thibault, F. Touchard, H. T. Duong, P. Juncar, S. Liberman, P. Pinard, J. Lerme, J. L. Vialle, S. Büttgenbach, A. C. Müller, A. Pesnelle, and the ISOLDE Collaboration, Phys. Lett. B 163:66 (1985).
15. W. Borchers, R. Neugart, E. W. Otten, H. T. Duong, G. Ulm, K. Wendt, and the ISOLDE Collaboration, Hyperfine Interact. 34:25 (1987).
16. M. Barboza-Flores, O. Redi, and H.H. Stroke, Z. Physik A 321:85 (1985).

17. M. Anselment, W. Faubel, S. Göring, A. Hanser, G. Meisel, H. Rebel, and G. Schatz, Nucl. Phys. A 451:471 (1986).

18. U. Dinger, J. Eberz, G. Huber, R. Menges, S. Schröder, R. Kirchner, O. Klepper, T. Kühl, D. Marx, and G. D. Sprouse, GSI Report 88-1, p.17 (1988).

19. J. A. Bounds, C. R. Bingham, H. K. Carter, G. A. Leander, R. L. Mlekodaj, and E. H. Spejewski, Phys. Rev. C 36: 2560 (1987).

20. U. Dinger, J. Eberz, G. Huber, R. Menges, S. Schröder, R. Kirchner, O. Klepper, T. Kühl, D. Marx, and G. D. Sprouse, Hyperfine Interact. 59:77 (1990).

21. G. Ulm, S. K. Bhattacherjee, P. Dabkiewicz, G. Huber, H.-J. Kluge, T. Kühl, H. Lochmann, E.-W. Otten, K. Wendt, S. A. Ahmad, W. Klempt, R. Neugart, and the Isolde Collaboration, Z. Physik 325:247 (1986).

22. B. Fricke, private communication (1990).

23. G. Torbohm, B. Fricke, and A. Rosen, Phys. Rev. A 31:2038 (1985).

24. W. D. Myers and W. J. Swiatecki, Ann. Phys. 55:395 (1969).

25. W. D. Myers and K. H. Schmidt, Nucl. Phys. A 429:269 (1984).

26. D. Berdichevsky and F. Tondeur, Z. Physik A 322:141 (1985).

27. E. Wesolowski, J. Phys. G 11:909 (1985).

28. A. Maj, H. Grawe, H. Kluge, A. Kuhnert, K. H. Maier, R. A. Meyer, J. Recht, and N. Roy, Z. Physik A 324:123 (1986).

29. I. Rotter, private communication.

III - SINGLE PARTICLE ASPECTS

SINGLE-PARTICLE RESPONSE FUNCTION

COMPLEMENTARY ASPECTS OF (e,e'p) AND TRANSFER REACTIONS

S. Gales

Institut de Physique Nucléaire - Orsay
91406 Orsay, Cedex, France

ABSTRACT :

In real nuclei, the properties of single-particle states are strongly distorded as compared to a bare Hartree-Fock nucleus. These modifications have been revealed by means of high resolution (e,e'p) and one nucleon transfer reactions induced by intermediate energy probes. The depletion of the spectroscopic strengths for orbitals near the Fermi sea are discussed together with the assumptions made in the underlying reaction models used to deduce the single-particle response functions.
The shapes of the strength distributions deduced from one nucleon pick-up reactions and from (e,e'p) reactions are shown to agree quite well,whereas the absolute values of spectroscopic factors are shown to differ by a factor two. The various factors limiting the accuracy of the deduced spectroscopic strengths in both processes are investigated and remaining questions are discussed.

I - INTRODUCTION

It has been well established that a large body of data for low energy nuclear phenomena can be described rather accurately, if one assumes that the nucleons move independently in an average potential. The Independent Particle Model (IPM) can be justified by the empirical data on the existence of magic numbers reflecting the shell structure of nucleons in nuclei, the nucleon mean free path in nuclei, the well defined angular momentum and energy for a nucleon when it is added to (or removed from) the target nucleus in one nucleon transfer process.
The average potential is not local with respect to space variables, V(r,r'), in other words it is velocity dependent. The fluctuation of the average potential gives rise to collective modes. The coupling ot these collective modes to single-particle degrees of freedom drastically modifies the properties of the mean field which becomes energy dependent V(r,r',E). This new situation may be described by saying that the Independent Particle Model acquires its "dynamical content"[1,2].

Nuclear Shapes and Nuclear Structure at Low Excitation Energies
Edited by M. Vergnes *et al.*, Plenum Press, New York, 1992

The aim of this paper is to discuss the advantages and limitations of the two experimental approaches used to study the single-particle response function in nuclei. The advent of high-resolution data from (e,e'p) reactions[3] on heavy target nuclei, has revealed a large depletion of the proton single-particle spectroscopic strengths for orbitals just below the Fermi sea. The quantitative understanting of the reaction model of the knock-out process and the nuclear structure information employed to extract the spectroscopic factors and the occupancies of shell model orbitals are shown to lead to model dependent quantities.

On the other hand, the one nucleon transfer approach is also examined[4] and the deduced single-particle responses are compared to the ones deduced from (e,e'p) experiments.

The occupation numbers deduced from (e,e'p) and $(d,^{3}He)$ reactions for various subshells in ^{90}Zr and ^{208}Pb are compared to the predictions of nuclear microscopic models.

The limiting factors for the determination of absolute spectroscopic factors and occupation numbers are investigated and open questions are discussed in the last section of this paper.

II - SINGLE-PARTICLE RESPONSE - SPECTROSCOPIC FACTORS AND DAMPING PROCESS

Single-particle states, for the nucleus consisting of A ± 1 nucleons, are obtained by adding to (stripping) or removing from (pick-up) the A core one nucleon.

As an example, the ground-state of ^{209}Bi is obtained by the addition to the ^{208}Pb core of a $1h_{9/2}$ proton. The sequence of excited states in ^{209}Bi is according to the IPM : $2f_{7/2}$, $1i_{13/2}$, etc.

This expected shell structure is immediately visible in the spectrum of fig. 1 of the $^{208}Pb(\alpha,t)$ reaction at 80 MeV [5] which exhibits clearly only strong transitions up to 2 MeV excitation energy. On the basis of their angular distributions, these transitions are unambiguously interpreted as $1h_{9/2}$, $2f_{7/2}$ and $1i_{3/2}$ s-p states in agreement with IPM expectations.

A quantitative measure of the single-particle character of such transitions is given by the strength or the spectroscopic factor which is defined as follow :

$$S_f^{\alpha}(A \to A+1) = |\langle\Phi_f(A+1)|a_{\alpha}^{+}|\Phi_o(A)\rangle|^2 \qquad (1)$$

where $\Phi_{o,f}$ are respectively the ground and final state wave functions and a_{α}^{+} is the creation operator of a nucleon with quantum number α acting on the ground state $\Phi_o(A)$.

A similar expression can be derived in the case of one nucleon pick-up process $S_f(A \to A-1)$. In the framework of the IPM, S_f^{α} is equal to 1. This is only approximatively true for the nuclear levels observed in the spectrum of fig. 1. The empirical values of S_f^{α} deduced from the analysis of the transfer reaction cross-sections are smaller than one ($S_f^{9/2} = 0.7$, $S_f^{13/2} = 0.65$, etc).

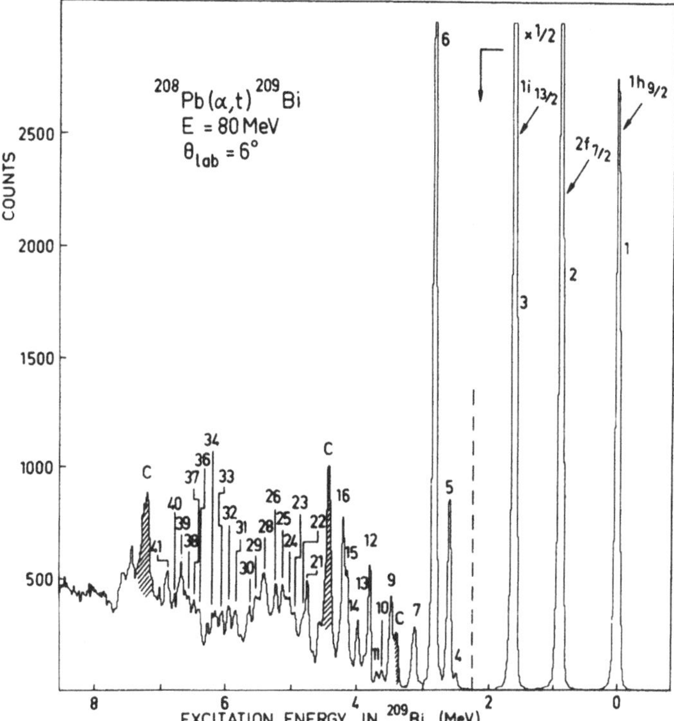

Fig. 1. Residual energy spectrum from the reaction $^{208}Pb(\alpha,t)^{209}Bi$ at E_α = 80 MeV. The few low-lying states correspond to the IPM orbits and are labelled according to these quantum members. (Taken from Ref.5)

This departure from the IPM predictions is even larger for the states located at higher excitation energy where a dense spectrum of states is populated. These states have been interpreted as fragments of the $1i_{13/2}$ and $2f_{5/2}$ orbitals.

In the case of a pick-up reaction, using the definition (1) of spectroscopic factors and summing all the strength belonging to the same orbital α in the A-1 spectrum one derives the well known sum rule.

$$\sum_f S_f^\alpha = \sum_f \langle \Phi_o(A)|a_\alpha^+|\Phi_f(A-1)\rangle^* \langle\Phi_f(A-1)|a_\alpha|\Phi_o(A)\rangle$$

$$= \langle \Phi_o(A)|a_\alpha^+ a_\alpha|\Phi_o(A)\rangle = n_\alpha(A) \tag{2}$$

for the number $n_\alpha(A)$ of α-nucleons in the target ground state.

This rigourous sum rule has been widely applied in the past ignoring two obvious difficulties which make its immediate application rather unadequate.

First it is clear that it holds if all the fragments of the full A-1 spectrum are identified which is in practice in contradiction with the experimental cut-off in excitation energy. This means that one always misses some strength at high excitation energy. It has to be considered as a lower limit to the occupation number.

Secondly the accuracy of absolute spectroscopic factors as it will be shown in the next section both for the (e,e'p) process and for the transfer reactions is too low to allow an accurate determination of $n_\alpha(A)$ through a sum rule.

Turning now to the mechanism responsible for the fragmentation of the s-p strength, it is believed to occur in two stages as shown in fig. 2. In the first stage, the bare single-particle state is spread over several doorway states. In the second step the doorway states themselves have their strengths damped over the many other degrees of freedom. The single-particle state acquires a spreading width $\Gamma^{\downarrow}$ which measures the energy interval where all the fragments belonging to the nlj orbital are concentrated. As it is illustrated in the bottom part of Fig. 2, depending on the level density of neighbouring complicated states, the shell structure of the nucleus will be strongly distorded until its complete disappearance.

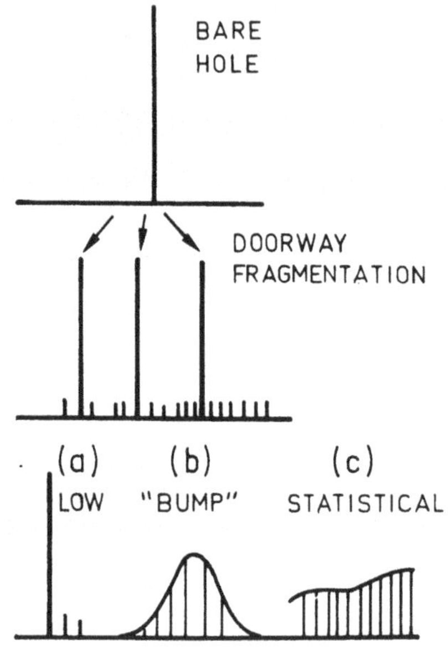

Fig. 2. The two stages of fragmentation of the single-particle state.

By analogy with the optical model for scattering states at positive energies, the spreading width $\Gamma^{\downarrow}$ can be explained by introducing an imaginary part to the average potential, W, being related to the spreading width $\Gamma^{\downarrow}$ by the relation

$$\Gamma^{\downarrow} = 2 \langle W \rangle \quad (3)$$

The empirical data on the spreading of valence or high-lying single-particle states is in qualitative agreement with this description. The origin, size and mechanisms responsible of the spreading are among the main motivations for the study of single-particle response functions.

III - EMPIRICAL SINGLE-PARTICLE STRENGTH DISTRIBUTIONS AND REACTION MODEL DEPENDENCE

Historically, the predictions of the IPM have first been tested using one nucleon pick-up reactions (p,2p)[6], (e,e'p)[7] and (p,d)[8]. However, in the early 70, both the available beams qualities as well as the experimental techniques led to a rather limited set of results on light and medium-weight nuclei (A > 60).

In the following section, we will describe how the nuclear structure information on single-particle degrees of freedom can be extracted from the experimental data coming from the investigation of the (e,e'p) process and from the alternative approach, the one-nucleon transfer process induced by hadronic probes. The links between nuclear structure aspects and reaction model dependence of the data analysis will be emphasized.

III.1 - The (e,e'p) knock out process

III.1.1 - Essentials of the reaction model

The diagram of the A(e,e'p)B reaction is shown schematically in fig. 3. An incident electron with momentum e_1 ejects a proton with momentum p from a target nucleus A, leaving the unobserved residual nucleus in a state B.

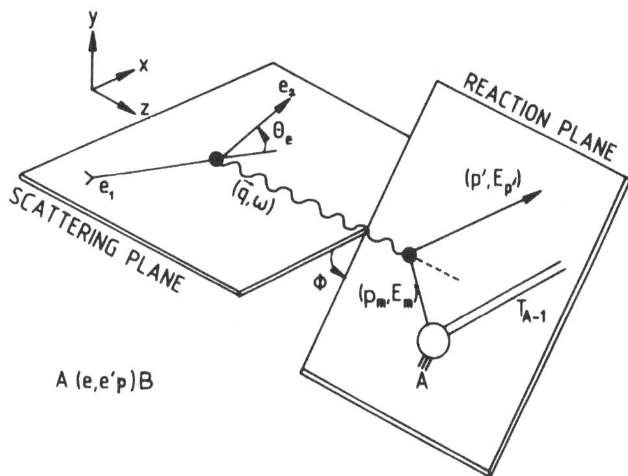

Fig. 3. Kinematics for the (e,e'p) reaction in the one-photon exchange approximation.

For parallel kinematics in which the proton is ejected in the direction parallel to the momentum transfer (i.e p//q) and in the quasi-free scattering limit (assuming the virtual photon to be absorbed by a single proton, Plane Wave Impulse Approximation), the reaction cross-section factorizes in a elementary electron-proton cross-section σ^*_{ep}, and the spectral function $S(E_m,p_m)$ which contains the nuclear structure information,

$$d^6\sigma \equiv d^6\sigma/de_2dp = K\, \sigma^*_{ep} S(E_m,p_m) \qquad (4)$$

Here K is a kinematical factor and $S(E_m,p_m)$ represents the joint probability to find a proton with momentum p_m and energy E_m in the nucleus. We note that the coupling of a virtual photon and a moving proton σ^*_{ep}, that is off the mass shell, must be extrapolated from the free electron-proton cross-section via a model.
In the IPM, the spectral function can be written as the sum of the momentum density distributions for all occupied orbitals nlj

$$S(E_m, p_m) = \sum_{nlj}^{occ} S_{nlj}(E_m)\, \rho_{nlj}(p_m) \tag{5}$$

where S_{nlj} is the spectroscopic factor for the nlj orbital at the mean energy E_m and $\rho_{nlj}(p_m)$ is in PWIA related to the Fourier transform of the nlj bound state wave function by the relation

$$\rho_{nlj}(p_m) = \left| \int \Phi_{nlj}(r)\, e^{ip_m r}\, dr \right|^2 \tag{6}$$

From the spectral function integrated over all momenta p_m, one gets the fraction of the sum-rule for each orbital nlj

$$n(\alpha) = \frac{1}{2J+1} \int S_{nlj}(E_m)\, dE_m \tag{7}$$

This factorization of the PWIA cross-section (4) is in practice not applicable. The final state interaction between the ejected proton and the final state B implies to use distorded waves, calculated by means of an optical model. This leads to the Distorded Wave Impulse Approximation (DWIA) and a new momentum density distribution $\rho^D_{nlj}(p_m)$.

Moreover, the distorsions in the Coulomb field of the incident and out going electrons have also to be taken into account (CDWIA).

The density distributions plotted in fig. 5 show clearly the effects of the corrections to the first order PWIA for the first two low-lying hole states in ^{208}Pb.

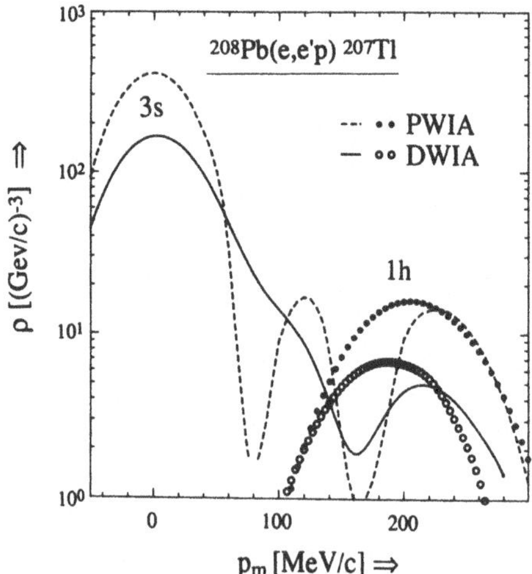

Fig. 4. Calculated density distributions for the 3s and 1h proton hole states in ^{208}Pb using the PWIA and the DWIA reaction models. (Taken from ref.9).

The absolute value and the width of the density distribution are strongly modified by the inclusion of the FSI and Coulomb distorsions. We note in addition that the shell model wave-functions $\Phi_{nlj}(r)$ are calculated by means of an average Woods-Saxon potential which depth is ajusted to reproduce the binding energy of the proton bound state and with a fixed geometry (r_o = 1.15 fm, a_o = 0.65 fm).

III.1.2 - Experimental method, results and data analysis

In order to illustrate the nuclear structure information recently obtained via the (e,e'p) reaction, we will describe the high-resolution data obtained at the NIKHEF-K linear accelerator which delivers 500 MeV electron beams with a 2 % duty cycle[3].

The scattered electron and the knocked-out proton are detected in coincidence by means of two large magnetic spectrographs with a momentum resolution of 5.10^{-5} and a momentum band width of 10 %. Protons hole states up to 25 MeV excitation energy have been studied on a number of target nuclei ranging from ^{3}He to ^{208}Pb (^{3}He, ^{4}He, ^{6}Li, ^{12}C, ^{40}Ca, ^{51}V, ^{90}Zr, ^{208}Pb). As an example, the response function measured for proton hole states near the Fermi level (E_x = 0 - 5 MeV) and for an average momentum of 220 MeV/c is shown in fig. 5 for the ^{208}Pb(e,e'p) reaction. It is worth pointing out the obtained nice energy resolving power allowing to measure very precisely over a large momentum range the density distributions of the hole states. The momentum distributions are displayed in fig. 6 for the first five proton hole states in ^{208}Pb.

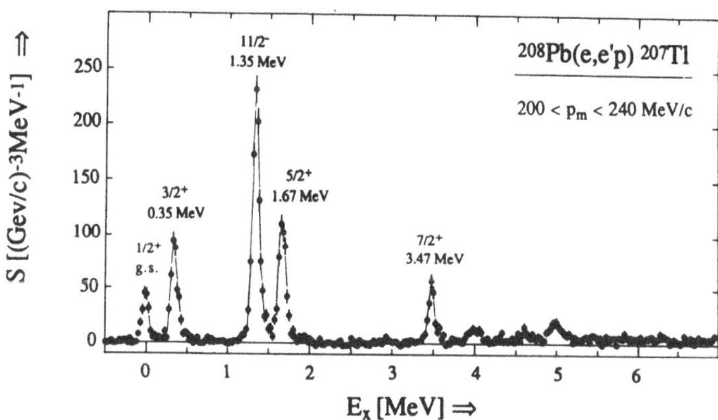

Fig. 5. Excitation energy spectrum of ^{207}Tl from the ^{208}Pb(e,e'p) reaction at 220 MeV/c. (Taken from Ref. 9).

The best fits to the experimental points (solid curves) are obtained through DWIA calculations. The shell model wave functions $\Phi_{nlj}(r)$ have been derived from a Woods-Saxon potential which depth has been ajusted to fit the experimental binding energies, but and this has to be emphasized, the radius of each orbital is ajusted to reproduce the experimental distribution.

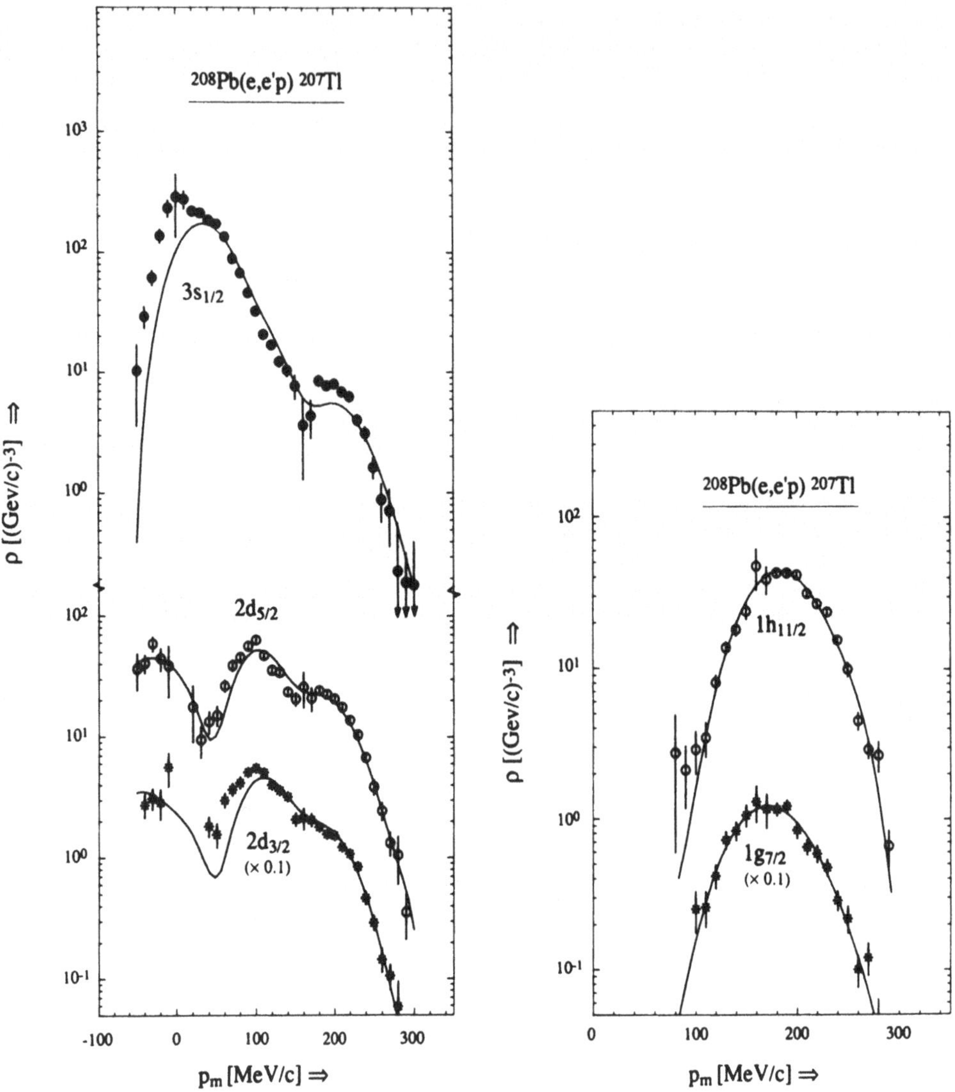

Fig. 6. Momentum distributions corresponding to knock-out from several orbits close to the Fermi surface in ^{208}Pb. The solid curves correspond to the best fits obtained within the DWIA reaction model (Taken from Ref. 9).

The normalization between the experimental points and the DWIA curves gives S_f^{α}, the spectroscopic factors for the $3s_{1/2}$, $2d_{3/2}$, $1h_{11/2}$ and $2d_{3/2}$ orbits. The deduced values are all of the order of 0.5 suggesting a strong quenching of single-particle strengths for these valence orbitals. They are much smaller than the ones obtained via one nucleon pick-up reactions[10)]

For low l orbitals, the fits to the experimental momentum distributions (see fig. 6) are not very good. The discrepancies have been attributed to distorsions not fully accounted for by the DWIA or CDWIA calculations. A remedy to the situation has been found by introducing a shift Δp_m in the recoil momentum as a free parameter. With that procedure the spectroscopic factors increase by 30 % ($S(3s_{1/2}) = 0.49$ Ref.9), $S(3s_{1/2}) = 0.65$ Ref.11).
The overall picture that emerges from the systematic of S_f^α from (e,e'p) studies at NIKHEF is displayed in fig. 7. It can be summarized as follows :

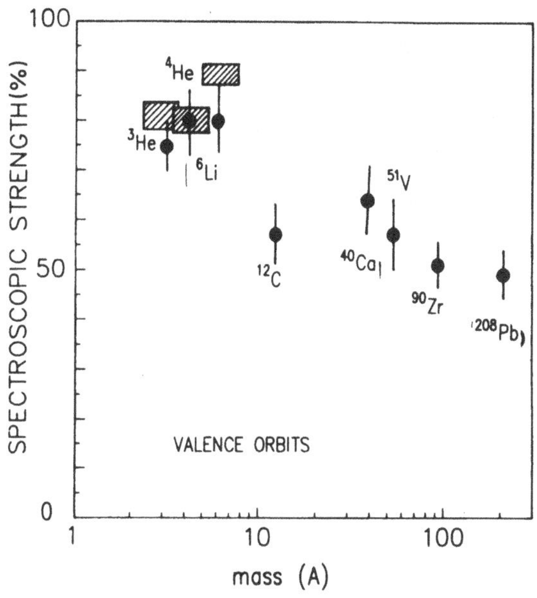

Fig. 7. Observed spectroscopic strength S_f^α (in percentage of the value of the IPM) for least bound orbits from (e,e'p) reactions. Theoretical results are indicated by hatched regions. (Taken from ref. 3).

For valence orbitals, the observed absolute spectroscopic strengths for A $\geq$ 12 nuclei is only about 50 % of the shell model limit.
In order to study, the response function for deeper orbitals, the momentum distributions ρ(pm) have to be analyzed with multipole decomposition of a bin slice, 0.5 MeV wide. The deduced strength functions for both low-lying and deeper proton holes in ^{208}Pb are shown in fig. 8.
The radius of the potential well for the deeper orbits has been taken equal to the one predicted by Hartree-Fock calculations using a density dependent force (DME 72). We note that the strength distributions follow closely the overall picture expected from the damping of s-p states which have been discussed in sec. 2, namely valence orbitals display narrow peaks where about 50 % of the strength is concentrated whereas for the deeper orbits, the spreading of the distribution increases with the excitation energy (see fig. 8).

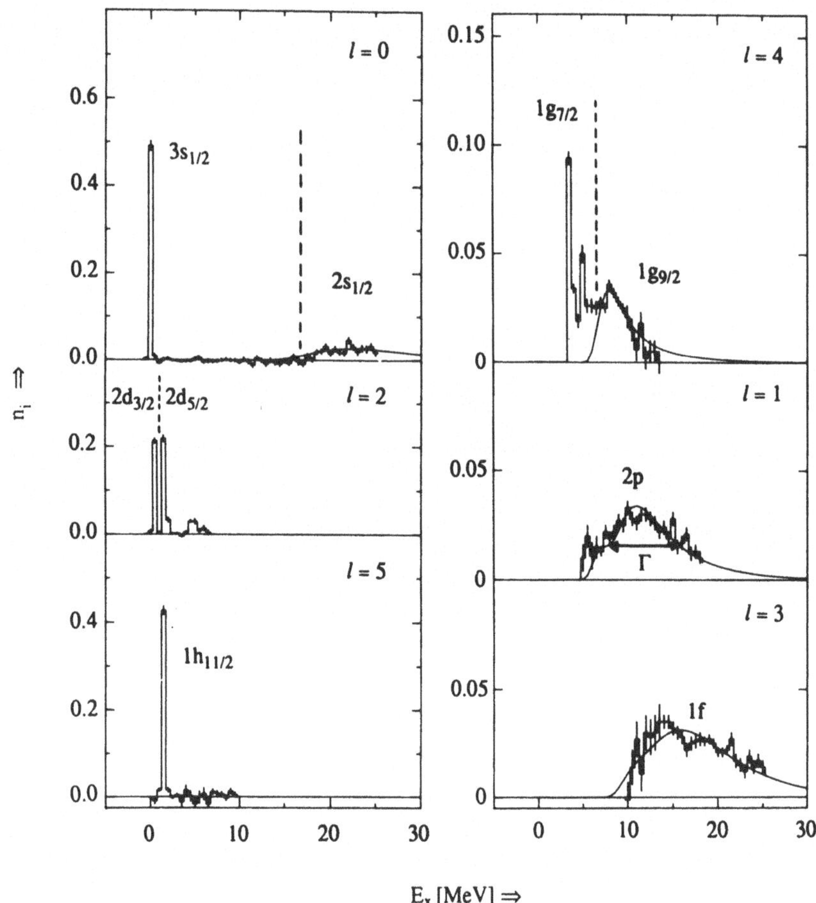

Fig. 8. Proton hole strength functions deduced from the analysis of the $^{208}Pb(e,e'p)^{207}Tl$ reaction for the low-lying (left) and deeper hole states (rigth). (Taken from Ref. 9).

To summarize the properties of the (e,e'p) reaction, we have shown that the proton hole response functions for both valence and deep orbitals can be measured.
The momentum distributions are characteristic of the multipolarity l of the transition but no J assignement can be made. The mean radius could be determined with a good precision.
The impulse approximation (one photon exchange) is well verified.
However, the deduced spectroscopic strengths remains largely dependent on the reaction model :

a) the distorsion due to the FSI between the struck proton and the residual nucleus are treated in a phenomenological way by means of an optical potential.

b) a model dependence is introduced by inserting the electron proton coupling in the nuclear medium σ^*_{ep} since only the on-shell matrix elements of the nuclear current are experimentally accessible.

c) the Coulomb distorsion is also treated in a approximate way for the electrons in the entrance and exit channels.

From the experimental point of the view, decisive progresses have been accomplished in particular in the study of heavy nuclei with closely spaced orbitals. However the very low cross-section makes difficult to extend this method to the neutron hole states via (e,e'n) reactions. Moreover, single-particle states located at positive energies are not accessible.

III.2 - <u>One nucleon transfer reactions</u>

III.2.2 - The Distorded Wave Born Approximation (DWBA) reaction model

In the case of a one nucleon transfer reaction induced by a hadronic probe A + a → b + B where b = a ± n and n is the transfered nucleon, the reaction model is based on the following assumptions.

a) One step process

In the framework of the DWBA theory of nuclear reactions, the transition amplitude from the initial state A to the final state B can be written

$$T_{BA} = \iint dr_{aA}\, dr_{bA}\, \chi_b^{(-)*}(k_b, r_{bB}) F(r_{aA}, r_{bB}) \chi_a^{(+)}(k_a, r_{aA}) \qquad (8)$$

where $\chi_a^{(+)}$, $\chi_b^{(-)*}$ are the distorded waves describing the elastic scattering in the entrance (aA) and exit (bB) channels. They are calculed by means of phenomenological optical potentials. F is the from factor and contains all the nuclear structure information, in particular the overlap between the wave functions in the final state B = A ± n and in the initial state A.
The experimental differential cross-section for the one nucleon transfer reaction is related to the amplitude T_{BA} by the relation.

$$\left(\frac{d\sigma}{d\Omega}\right)_{exp} = S(E)\, K|T_{AB}|^2 = S(E) \left(\frac{d\sigma}{d\Omega}\right)_{DW} \qquad (9)$$

where K is a constant representing the kinematics and spin statistics for a given transition and $\left(\frac{d\sigma}{d\Omega}\right)_{DW}$ is the calculated DWBA cross-section for the same transition.

According to (9) $\left(\frac{d\sigma}{d\Omega}\right)_{exp}$ is directly proportional to S(E) the single-particle strength contained in the final state nlj at energy E.
We note that the assumption of the one-step process is generally verified at sufficiently high incident energy.

Besides the optical parameters ambiguities, two other sources of uncertainties are present in the calculated $\left(\frac{d\sigma}{d\Omega}\right)_{DW}$ cross-sections.

b) Exact-Finite Range DWBA cross-sections (EFR-DWBA)

The computation of the transition amplitude T_{BA} implies the knowledge of the interaction V_{bn} between the ejectile b and the transferred nucleon n. This interaction is usually deduced from a phase-shift analysis of the nucleon-nucleon or nucleus-nucleon scattering at low energy. Under this condition, the EFR-DWBA cross-sections can be calculated exactly from eq(8) and no further normalization is needed to compute the σ_{DW} cross-sections.

c) The radial dependence of the form factor

A static Woods-Saxon potential with ad hoc geometry is used to compute the form factor of the transferred nucleon. The depth, like in DWIA analysis of (e,e'p), is ajusted to reproduce experimental binding energies.
However a strong dependence of σ_{DW} on the radius $R = r_o A^{1/3}$ of the well is observed. To be more quantitative, a reduction of the rms radius by 1 % leads to a 5-10 % increase of the spectroscopic factors. A recent analysis of data on pick-up reactions[12] utilized the information on the rms radius obtained from (e,e'p) reaction. Then, the deduced S_f^α are found to agree within 15-25 % with those from the DWIA analysis of (e,e'p). In contrast, if the smaller rms radius obtained from elastic magnetic scattering[13] is used, the spectroscopic factors are much larger (30-40 %). Clearly the (e,e'p) and the magnetic scattering are probing density distributions which appear to have different rms radius.
Keeping in mind the model dependence of the deduced strength functions, we will now discuss the use of one nucleon pick-up reactions which were employed to reach a detailed knowledge on the single-particle response functions in heavy nuclei.

III.2.3 - Experimental aspects and analysis of one nucleon pick-up reactions

Various factors such as the selectivity in angular momentum transfer, the characteristic patterns of the angular distributions, the use of polarized beams, the energy resolution in the exit channel, the explored energy range, the nature and the magnitude of the "true" underlying continuum, play key role in the localization and determination of the amount of strength for a given subshell observed in the transfer reaction spectra.
In the case of deep-neutron hole states in heavy nuclei which cannot be studied up to now by (e,e'n) reaction, the strong selectivity for large angular momentum is clearly visible in the energy spectrum from the reaction $^{208}Pb(^3He,\alpha)$ at 283 MeV[14], displayed in fig. 9.
At low excitation one observes the strong population of the 1 $i_{13/2}$ (l=6) valence hole state at 1.63 MeV. Between 4 and 20 MeV, a broad bump

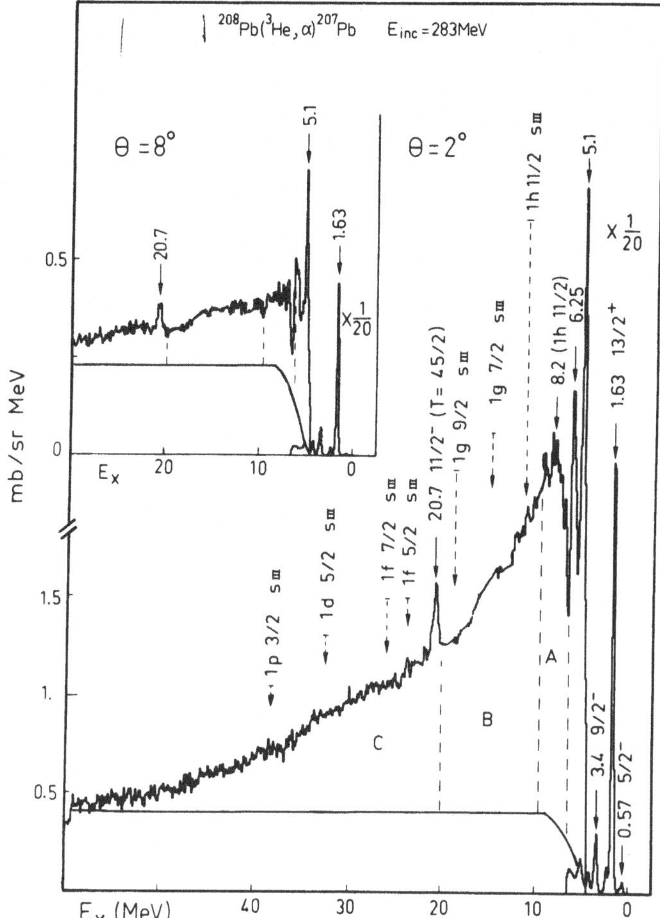

Fig. 9. Residual energy spectra from the reaction $^{208}Pb(^3He,\alpha)^{207}Pb$ Valence and deep neutron-hole states are indicated by solid arrows. Dashed arrows indicate the HF calculated positions of other subshells. The horizontal solid line represents the shape of the assumed background. (Taken from Ref. 14)

dominates the residual energy spectrum. It has been interpreted as neutron pick-up from the inner high-spin $1h_{11/2}$, $1g_{7/2}$ and $1g_{9/2}$ orbitals.

Another remarkable feature of the displayed spectrum is the smooth and continuous decrease of the cross-section beyond 20 MeV. Since an excitation energy range larger than 50 MeV has been explored, the total neutron hole strength of all orbits down to the bottom of the well may be involved. Note that beyond 50 MeV, the pick-up cross-section is quite low and this magnitude of the continuum (see horizontal line in fig. 9) has been extrapolated down to 5 MeV and subtracted from the experimental cross- sections. This ad hoc continuum estimate is a major source of uncertainty for the measured strength using the hadronic probe.

Among the major advantages of the transfer reaction, very high energy resolution can be achieved using magnetic spectrograph but, even more important, spin assignments of the transitions can be made using polarized beams.

First spin determinations of deeply-bound proton hole states have been reported in the study of $(\vec{d},{}^3He)$ reaction at 52 MeV on ^{90}Zr[15]. Shortly after, combining high energy resolution and asymmetry measurements, detailed knowledge of the overlapping 1g and 2p neutron-hole strength in the Sn isotopes has been obtained by means of the $^{112,116}Sn(\vec{d},t)$ reactions using 40 MeV polarized deuterons beams [16].

In fig. 10, we show the energy spectra from the reaction $^{112}Sn(\vec{d},t)^{111}Sn$. The good energy resolution allows us to study in detail the "fine structure" peaks observed in the energy range 3.0 to 7.0 MeV where a large fraction of the $1g_{9/2}$ and 2p inner hole strengths have been located[16-19]. The angular distributions $\sigma(\theta)$ and the analyzing power $Ay(\theta)$ data for each "peak" have been extracted and compared to DWBA predictions. In fig. 10, the angular momentum dependence of these experimental cross-sections is

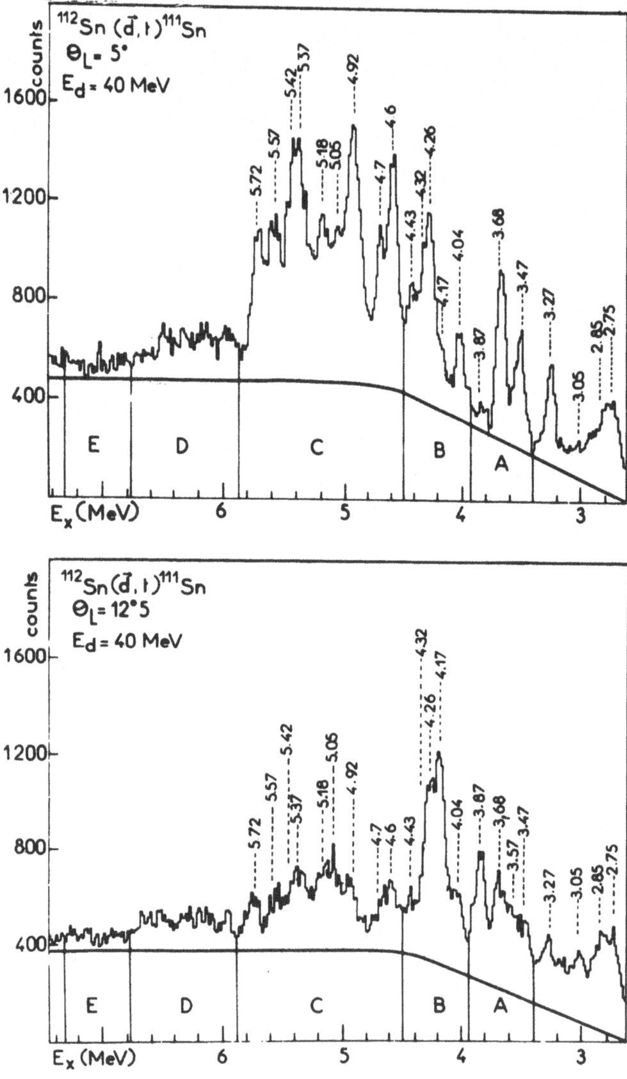

Fig. 10. Triton energy spectra from the $^{112}Sn(\vec{d},t)^{111}Sn$ reaction at 5° an 12°5 lab. angles. The horizontal line shows the assumed background. Above 3.5 MeV the spectra were divided into adjacent energy bins. (Taken from ref. 16).

clearly visible. At forward angles (θ = 5°), l = 1 components are strongly excited whereas at larger angles (θ = 12°5) one observes mainly the l = 4 contributions and the overlapping of the two strength distributions.
As an example of the analysis, in fig. 11, the angular dependence of $\sigma(\theta)$ and Ay (θ) for the 4.17 MeV peak are very well reproduced by a mixing of J^{π} = $9/2^{+}$ and J^{π}= $1/2^{-}$ states without ambiguity.
The extension of this detailed analysis to the whole region 3 - 7 MeV leads to the distributions displayed in fig. 12.

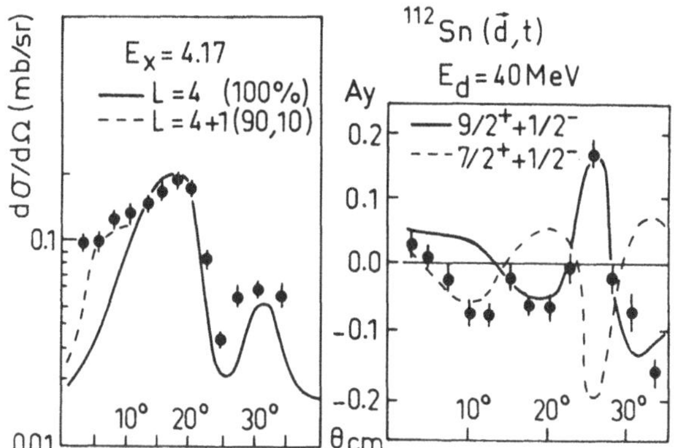

Fig. 11. Experimental differential cross-section and analyzing power for the "state" at 4.17 MeV (see fig. 10) from the $^{112}Sn(\vec{d},t)^{111}Sn$ reaction. The solid and dashed curves are the DWBA predictions for the indicated l and J values. (Taken from Ref. 16).

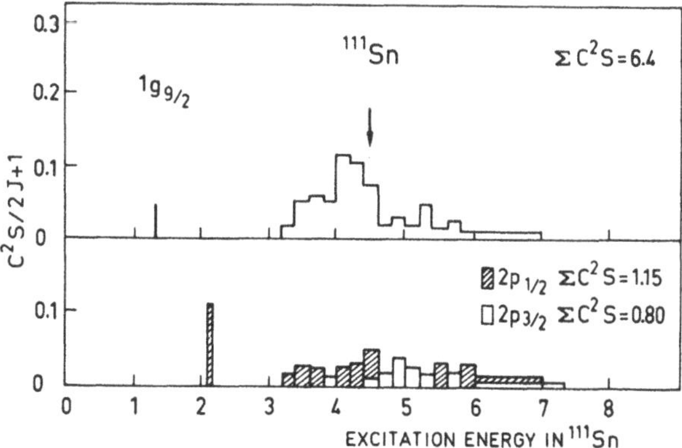

Fig. 12. Experimental $1g_{9/2}$, $2p_{1/2}$ and $2p_{3/2}$ inner-hole strength distributions extracted from the analysis of the $^{112}Sn(\vec{d},t)$ reaction at 40 MeV. For each subshell, the total amount of strength located in the energy range, $\sum S^{\alpha}_{f}$ is indicated. (Taken from Ref. 16).

In summary one neutron pick-up reaction has been widely used to study valence and inner hole states in medium-heavy nuclei.
The DWBA reaction model employed in the analysis of transfer reactions is well understood but suffers from the sensibility of the DWBA cross-sections to the radius of the Woods-Saxon well, from the optical model ambiguities and from the widely used zero-range approximation which neglects the energy dependence of the nucleon-nucleon and nucleus-nucleon interactions. This makes the deduced absolute spectroscopic factors strongly model dependent.
On the experimental aspects, a wide range of reactions, with in some cases polarized projectiles, and in general very good energy resolution are accessible to study the single-particle response. However a substantial continuum limits seriously our quantitative knowledge on the total strength and the spreading of inner-hole states.

IV - COMPARISON OF (e,e'p) AND (d,^{3}He) STRENGTH DISTRIBUTIONS

According to the reaction model description of the (e,e'p) and (d,^{3}He) processes the two approaches are probing the single-hole structure of the nucleus.
In fig. 13 is presented a comparison between the spectra obtained from the ^{90}Zr(d,^{3}He)^{89}Y reaction at 80 MeV with an energy resolution of 50 KeV[20) and the ^{90}Zr(e,e'p)^{89}Y reaction[21) for a mean recoil momentum of 190 MeV/c which favors $l = 3$ transfer.
These spectra illustrate in an remarkable way the similarity of the information contained in both types of direct reactions. Not only the one to one correspondance of the low-lying hole states (0-3 MeV) is impressive, but the strongly fragmented gross structure peak extending between 3 to 9 MeV, which is largely due to $1f_{7/2}$ knock-out, is very similar in both reactions.
A quantitative analysis of the (e,e'p) and (d,^{3}He) strength functions may be carried out as shown in fig. 14 where the $1g_{7/2}$ and $1g_{9/2}$ proton hole distributions in ^{207}Tl are compared.
In fig. 14, one notices the striking similarity of the strength distributions as emphasized in the raw spectra from the proton pick-up on ^{90}Zr (see fig. 13), but as far as "absolute" spectroscopic strengths are concerned, the hadron process gives S^f_α values a factor of two larger than the ones obtained from (e,e'p) reactions. Similar conclusions have been reached in the analysis of many medium and heavy weight nuclei when direct comparison between (e,e'p) and (d,^{3}He) results is possible. The results for one nucleon pick-up reactions and (e,e'p) lead to different values of spectroscopic factors both for low-lying and deeper orbitals. By summing all the fragments belonging to a given subshell up to about 20 MeV, when available, one obtains "absolute" occupation numbers. An illustration of the present empirical knowledge is presented in fig. 15 where occupation numbers from (d,^{3}He) and (e,e'p) studies on ^{90}Zr, ^{144}Sm, ^{208}Pb are plotted against binding energy and are compared to the predictions of microscopic models.
In the framework of the IPM, the subshells located below the Fermi sea are fully occupied, as indicated by the full line (HF) of fig. 15.
Using the dispersion-relation approach to mean field, Mahaux and Sartor[22) have developped a more realistic model for both negative and positive single-particle states. Their prediction for occupation probabilities in ^{208}Pb in shown by the dashed line of fig. 15.

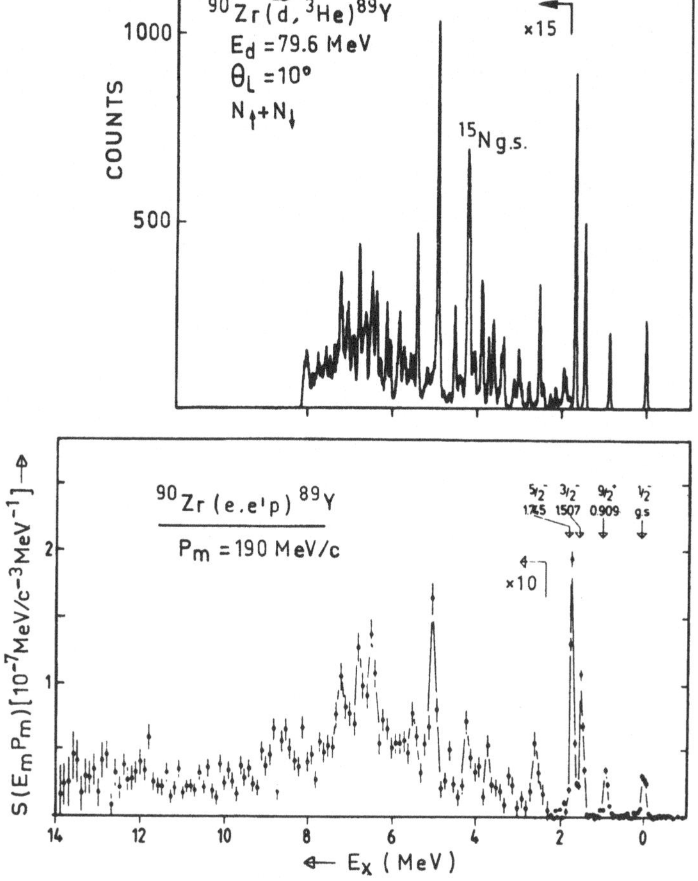

Fig. 13. Comparison of a $^{90}Zr(e,e'p)^{89}Y$ spectrum (Taken from Ref.21) and a $^{90}Zr(\vec{d},{}^3He)$ spectrum (Taken from Ref. 20).

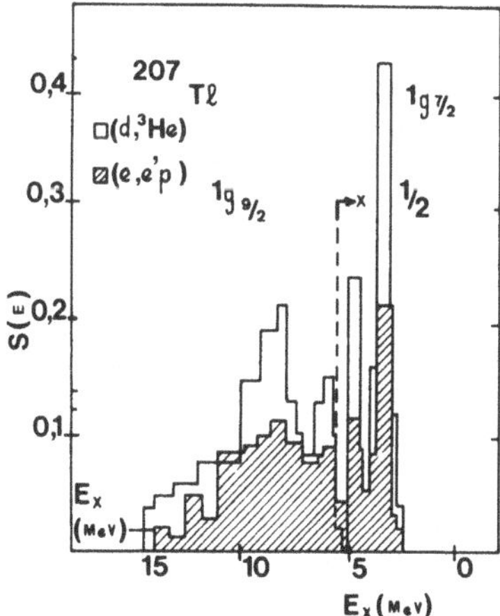

Fig. 14. $1g_{7/2}$ and $1g_{9/2}$ proton hole distributions in ^{207}Tl deduced from the analysis of the $^{208}Pb(d,{}^3He)$ and $^{208}Pb(e,e'p)$ reactions. (Taken from Ref. 10 and 9 respectively).

The prediction of another different approach developed by Pandharipande et al. at Urbana[23] is also considered in fig.15 (labelled PPW). This group has calculated with improved accuracy the ground state wave functions of interacting many-body systems which yields exact results for mass A = 4 and nuclear matter. For ^{208}Pb their predictions for occupation numbers are obtained by superimposing on the nuclear matter results the effets of surface vibrations. Note that the estimate of the Urbana group gives quite low occupation numbers (60-65 %) near to the Fermi sea.
The empirical values from (d,^{3}He) reactions are all close to one with large error bars due to reaction model uncertainties. The (e,e'p) values are very low and with an average value of n = 0.5 for valence subshells.
A possible renormalisation of (e,e'p) value of 30 to 40 % due either to the electron-proton coupling in nuclear medium and/or to uncertainties in the DWIA reaction model has certainly to be considered to make the empirical data compatible with the prediction of nuclear models.
In summary, both hadronic and electron knock-out processes are very efficient tools to map the hole response function in nuclei and give very similar strength distributions. Absolute value of spectroscopic factors with an accuracy of 10 % cannot be obtained in both approaches due to the reaction model dependence of these quantities. In addition since only a limited range of excitation energy can be covered, occupation numbers are even harder to deduce from the analysis of these processes.
However, both in the DWIA and DWBA reaction models, the relative spectroscopic factors and occupation numbers are quite comparable and can be obtained with a high degree of accuracy[24].

V - OPEN QUESTIONS

In this review of the present status of the proton pick-up and quasi-elastic proton knock-out reactions, the data confirm several aspects of IPM : the existence of shell structure is clearly manifest even for deeply bound orbitals. However, the same data provide strong evidence for a quenching of the s-p strength compared to the shell model predictions and therefore confirm the importance of coupling to collective modes. The reliability of the extracted spectral function depends on how accurately the reaction mechanisms are described.
For one nucleon pick-up process, EFR-DWBA calculations deriving the ejectile-nucleon interaction from phase shifts analysis and the use of HF or (e,e'p) rms radius should strongly reduce the scattering of the deduced experimental spectroscopic factors obtained at different energies and/or using different one nucleon pick-up reactions.
In the case of (e,e'p), the DWIA formalism provides a reliable framework. However, the overall normalization depends strongly on the assumption that the distorsions are well described in terms of an optical potential and this assumption requires further study.
Another source of uncertainty is the possible modification of the photon-proton form factor inside the nucleus. A powerful tool to study such effects is the separation of the longitudinal and transverse cross section by measurements at different kinematical conditions. For example, the agreement between the IPM and the inclusive quasi-elastic electron cross-section turned out to be completely fortuitous.
The same model failed to account for the separated longitudinal (Coulomb) and transverse (Magnetic) response functions $R_L(q,\omega)$ and $R_T(q,\omega)$ for a variety of nuclei[25]. Fig. 16, shows the longitudinal and transverse response functions for ^{40}Ca at a momentum transfer of 410 MeV/c [26].

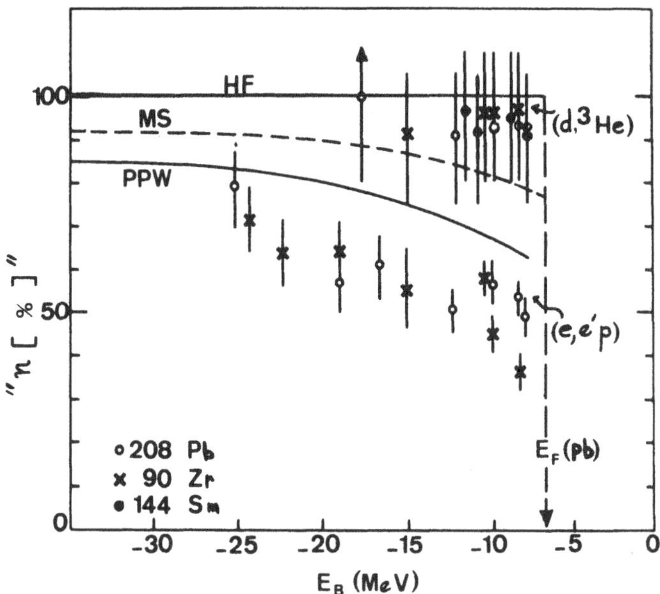

Fig. 15. Occupation numbers as a function of binding energy for IPM (full line), and microscopic models (M & S dashed line, PPW lower line). The experimental data is represented by empty, full circles and crosses from (d,^{3}He) and (e,e'p) analysis.

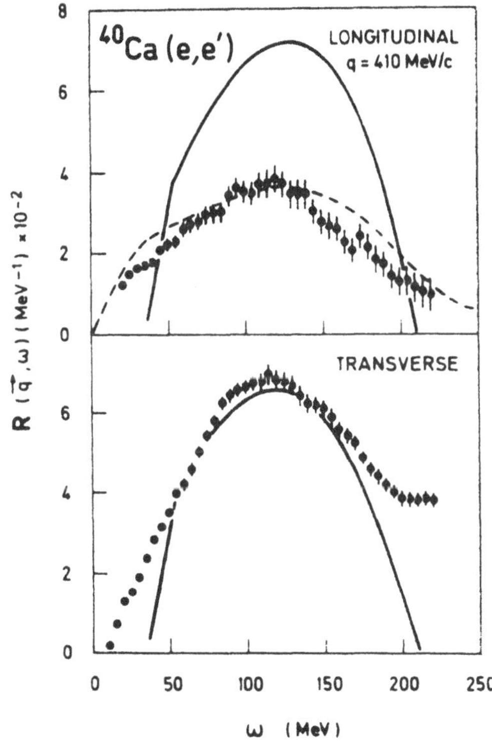

Fig. 16. ^{40}Ca longitudinal and transverse response functions at q = 410 MeV/c. The solid curve is the Fermi gas prediction. The dashed curve is a result of a calculation for nuclear matter with correlations (taken from Ref. 25).

The measured $R_L(q,\omega)$ shows 50 % less strength at the peak than predicted by the IPM. To explain this discrepancy both relativistic effects based on Dirac phenomenology or more exotic process, like the increase of the nucleon radius in the medium have been investigated without significant conclusions.

Many-body correlations can be very significant in influencing the longitudinal response as shown by the dashed curve of fig. 16 reproducing the calculations of the Urbana group. The good agreement with the experimental data is obtained because the effect of the correlations is to quench the amplitude of the quasi elastic peak and to shift a significant fraction of its strength at much higher excitation energy.

The transverse response would not be reproduced by calculations that neglect subnucleonic degrees of freedom. The agreement with the data shows that these effects become significant only at higher momentum transfer (q = 550 MeV/c).

An even stronger constraint of the possible medium effects in the electron-proton coupling may be obtained by a separation of R_L and R_M responses for scattering a bound proton. From the Saclay data[27] for the $^{40}Ca(e,e'p)$ reaction shown in fig. 17 a value of $R_{ep} = \sigma^L_{ep}/\sigma^T_{ep} = 0.70 \pm 0.05$, which is rather independent of the momentum transfer Q^2, has been deduced. Similar information deduced from several other experiments has corroborated a 20-30 % reduction of R_{ep} with respect to the value for free proton.

This anomaly may lead to an equivalent renormalization of the electron-proton coupling in nuclei which will increase the (e,e'p) spectroscopic factors accordingly. However if part of this anomaly can be explained by enhancement of the transverse photon-proton coupling, because of meson and charge exchange processes then the longitudinal response will be consistent with the free σ_{ep} and the reduced spectroscopic strength deduced from (e,e'p) data should be attributed to large correlation effects in the nuclear many body system.

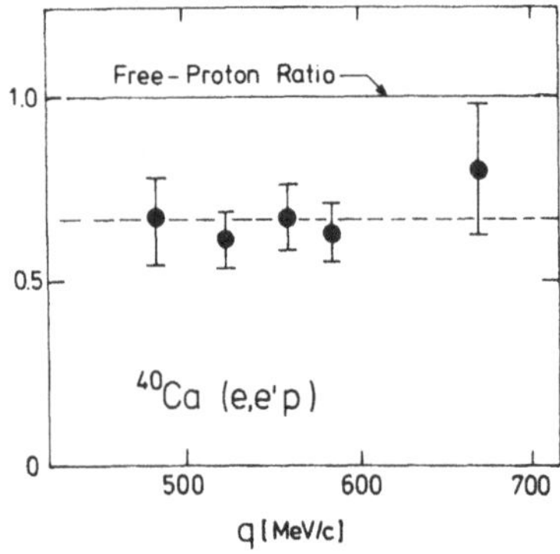

Fig. 17. Ratio of longitudinal and transverse electron-proton cross-section extracted from $^{40}Ca(e,e'p)$ data. (Taken from Ref. 3).

REFERENCES

1) G.E. Brown, J.H. Gunn and P. Gould, Nucl. Phys. 46 (1963) 598.

2) For a review see for example, C. Mahaux, P.F. Bortignon, R.A. Broglia and C.H. Dasso, Phys. Reports 20 (1985) 1.

3) See for example P.K.A. de Witts Huberts, J. Phys. G. Nucl. Part. Phys. 16 (1990) 507, A.E.L. Dieperink and PKA de witts Huberts, Ann. Rev. Nucl. Part. Sci. 40 (1990) 239.

4) S. Galès, Ch. Stoyanov and A.I. Vdovin, Phys. Reports 166 (1988) 127.

5) S. Galès et al. Phys. Rev. C31 (1985) 94.

6) G. Jacobs and Th. A.J. Maris, Rev. Mod. Phys. 45 (1973) 6.

7) J. Mougey, Nucl. Phys. A355 (1980) 35; A396 (1983) 390.

8) J. Källne et B. Fagerström, Proc. Fifth Int. Conf. on High energy Physics and Nuclear Structure, Uppsala, 1973, ed. G. Tibell (North-Holland, Amsterdam 1974) p. 369.

9) E.N.M. Quint, Ph.D. (Thesis), unpublished Amsterdam (1988).

10) H. Langevin-Joliot et al., J. Phys. G. : Nucl. Phys. 10 (1984) 1435.

11) J.P. Mc Dermott, Phys. Rev. Lett. 65 (1990) 199.

12) G.J. Kramer et al., Nucl. Phys. A477 (1988) 55.

13) S. Platchkov et al., Phys. Rev. C25 (1982) 2318.

14) H. Langevin-Joliot et al., Phys. Lett. B114 (1982) 103.

15) A. Stuirbrink et al., Z. Phys. A297 (1980) 207.

16) S. Galès et al., Nucl. Phys. A381 (1982) 40 and references herein.

17) E. Gerlic et al., Phys. Rev. C21 (1980) 124, Phys. Lett. B57 (1975) 338.

18) G.M. Crawley et al., Phys. Rev. C23 (1981) 307.

19) S.Y. Van der Werf et al. Phys. Rev. Lett. 33 (1974) 712, Nucl. Phys. A289 (1977) 14.

20) W.P. Jones et al., I.U.C.F. Annual Report (1984) p. 72.

21) J.W.A. den Herder et al., Phys. Lett. 161B (1985) 65.

22) C. Mahaux and R. Sartor, Advances in Nuclear Physics, 20 (1991) 1.

23) V.R. Pandharipande, C.N. Papanicolas and J. Wambach, Phys. Rev. Lett. 53 (1984) 1133.

24) G.J. Wagner, Prog. Part. Nucl. Phys. 24 (1990).

25) For a review see for example B. Frois and C. Papanicolas, Ann. Rev. Nucl. Part. Sci. 37 (1987) 133.

26) Z.E. Meziani et al., Phys. Rev. Lett. 52 (1984) 2130, 54 (1985) 1233.

27) A. Magnon et al., Phys. Lett. B222 (1989) 352.

INTRUDER STATES AND LOW ENERGY NUCLEAR SPECTROSCOPY

Ragnar Bengtsson

UNISOR
Oak Ridge Associated Universities
Oak Ridge, Tennessee 37831

1 Introduction

At moderate deformations and low spin an intruder subshell can be defined in an essentially unique way as the spherical subshell with $\ell = N$ and $j = \ell + 1/2$ which originates from the next higher main shell relative to the dominating shell in the vicinity of the Fermi surface, which then has the quantum number $N - 1$. Consequently, the intruder subshell has parity different from the majority of levels near the Fermi surface and the intruder states may be referred to as unique parity states. In many situations, the spin–orbit partners of the intruder subshells, i.e. the subshells with $\ell = N$ and $j = \ell - 1/2$, show features similar to those of the "true" intruder subshells and are often involved in a crucial way in nuclei exhibiting shape coexistence.

The crucial role intruder orbitals play is not limited to low spin and small deformations. In fact much of the physics at superdeformed shapes and at very high spin is a direct result of the presence of high–j intruder orbitals in the vicinity of the Fermi surface, including the introduction of intruder orbitals from still higher shells at extreme deformations.

Since the intruder subshell has a larger angular momentum (j) than any other subshell close to the Fermi surface it plays an essential role for building up angular momentum in the nucleus, and the alignment of intruder shell (quasi)particles is responsible for a large variety of bandcrossings observed in discrete spectroscopy along or near the yrast line.

Intruder shell orbitals are also associated with large quadrupole moments and the occupation of these orbitals may lead to significant changes of the deformation of the nucleus. Under favourable circumstances a nucleus may have totally different shapes depending on the occupation of the intruder orbitals, giving rise to a phenomenon which is usually referred to as shape coexistence .

In this paper, shape coexistence in several mass regions will be discussed with specifc attention focused on the relation between shape coexistence and the occupation of intruder shell levels. Shape effects associated both with spin aligned and non-aligned intruder orbitals will be investigated.

Nuclear Shapes and Nuclear Structure at Low Excitation Energies
Edited by M. Vergnes *et al.*, Plenum Press, New York, 1992

2 General background to shape coexistence

2.1 From stable deformation to shape coexistence

In nuclei exhibiting stable deformation, the shape of the nucleus is expected to be approximately the same for all configurations, at least at low spin. This situation is illustrated in fig. 1a. If, however, the nucleus is soft and in particular if there are configurations present which involve strongly deformation driving intruder orbitals, calculations show that the equilibrium shape of different configurations is not exactly the same, showing differences of the order 10 to 20 %. This is illustrated in fig. 1b which gives an example of moderatly configuration dependent equilibrium shapes. According to total routhian surface (TRS) calculations configuration dependent shape differences of this order seem to be very common in nuclei[1].

As the nucleus gets even softer the excitation of a particle to a strongly deformation driving orbital may displace the energy minimum to the point that the excited configuration, at some distance from its equilibrium deformation, may lie lowest in energy. In this case the lowest energy surface (defined adiabatically), will suddenly change structure at the place where the two configurations cross. This does not lead to a second energy minimum in the lowest potential energy surface, but the presence of a second configuration would immediately be revealed if the overlap between wavefunctions representing different shapes is calculated. Both configurations could be exploited in generator coordinate calculations by just using the lowest energy surface. Situations of this kind will be treated in more detail in section 4.

If the nucleus is very soft, the energy minimum of the excited configuration could be displaced so much away from the equilibrium deformation of the lowest configuration that the energy minimum of the excited configuration will also appear in the lowest (adiabatic) potential energy surface. In this case two energy minima coexist in the same energy surface. Since the shape of the nucleus at the two minima is different, we talk about shape coexistence.

2.2 Adiabatic versus diabatic configurations

We have so far not made a precise definition of a 'configuration'. In many applications it is useful to distinguish between **adiabatic** and **diabatic** configurations. A configuration can be defined by specifying which single-particle or quasiparticle orbitals are occupied. As the deformation is changed away from the original one the original configuration undergoes gradual changes as the occupied orbitals are interacting with unoccupied orbitals. Occupied and unoccupied orbitals are not allowed to cross unless, due to symmetry properties, the interaction between two crossing levels is exactly vanishing. In most high-spin calculations the only conserved quantum numbers are parity and signature and therefore only levels with different parities and/or signature will have sharp crossings. By not changing the occupation of the single-particle (or quasiparticle) levels at such crossings it is possible to trace the continuation of a configuration having a prescribed parity and signature and

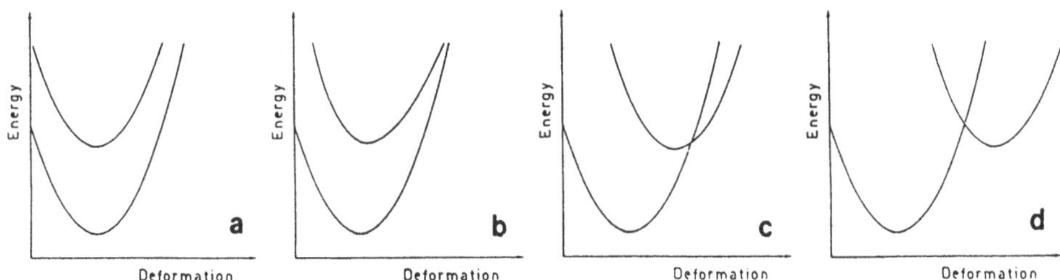

FIGURE 1. A schematic illustration of how two configurations which have the same deformation in a rigid nucleus gradually can turn into two shape coexisting configurations as the nucleus gets softer.

initial occupation of the single-particle levels throughout the deformation space. An initial configuration extended in this way is called an adiabatic configuration and its total energy can be displayed as adiabatic potential energy or total routhian surfaces.

In many applications, weak interactions between levels with the same parity and signature causes computational as well as conceptual problems. An adiabatic configuration defined through such a crossing will completely change its properties within the immediate vicinity of the crossing point. Often, it is more desirable to consider configurations, which have similar properties on both sides of a level crossing with a very weak interaction. In practice this

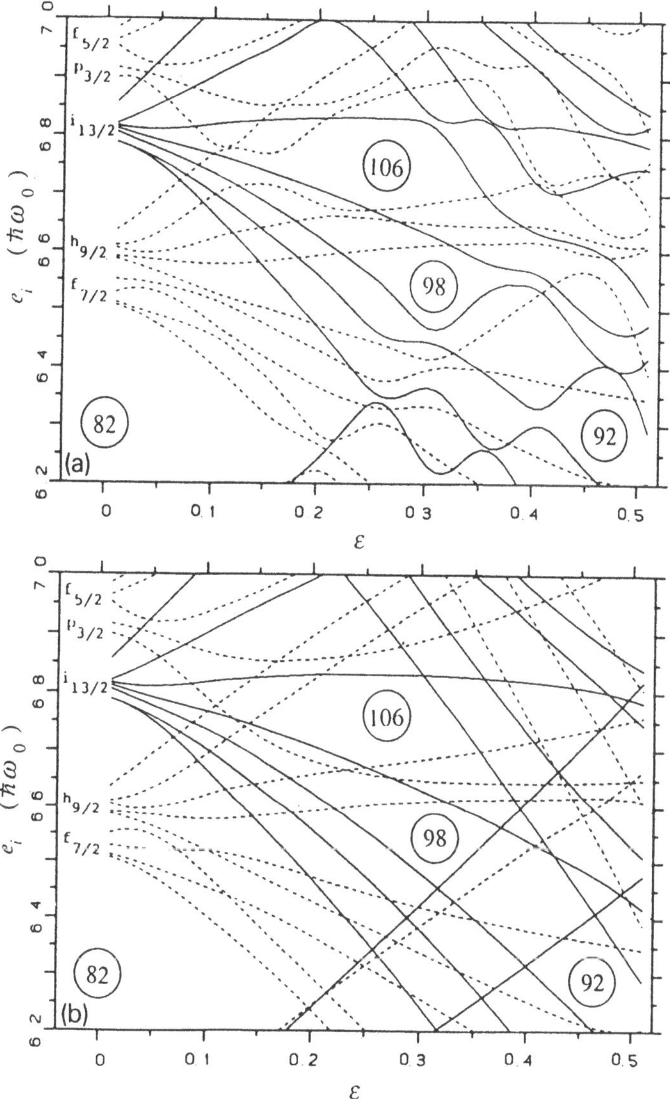

FIGURE 2. Single-neutron energies in the Nilsson potential as a function of the quadrupole-deformation parameter ϵ_2. The triaxiality and hexadecapole parameters are $\gamma = 20°$ and $\epsilon_4 = 0.02$, respectively. The parity is the only conserved symmetry and levels with positive parity are drawn as solid lines and those with negative parity as dashed lines. The upper part shows the adiabatic levels that result from the direct diagonalisation of the Nilsson hamiltonian. The lower part shows the corresponding diabatic levels. The figure is taken from ref.[2] in which a detailed description of the calculation of diabatic levels is given.

can be obtained by simply neglecting the interaction between the levels or by reconstructing non-interacting levels, see fig. 2. Crossings between such levels can then be treated in the same way as crossings between levels with different parity and/or signature. The technique for constructing diabatic levels has been highly developed in recent years and I refer to ref.[2] for details.

Clearly, with the definition of adiabatic configurations given above, sharp crossings of the energy curves as shown in fig. 1 will only (except for accidental degeneracies) occur between configurations belonging to different symmetry groups (defined e.g. in terms of parity and signature quantum numbers). With the introduction of diabatic configurations the number of sharply crossing energy curves (or energy surfaces) increases significantly. An adiabatic energy surface can then be decomposed into a number of suitably chosen diabatic energy surfaces. How this can be done in practice is exemplified in section 4.

3 The classical examples

3.1 The theoretical predictions

The most convincing and, by now, almost classical examples of shape coexistence at low spin can be found in the light mercury isotopes. Potential energy surfaces for these nuclei have two well defined minima, one at a small oblate shape, $\beta_2 \approx -0.12$, and another at larger prolate shape with $\beta_2 \approx 0.25$, cf. fig. 3. The oblate energy minimum is calculated to have a lower energy than the prolate minimum. Due to the difference in deformation a rotational band built on the prolate minimum must have a considerably larger moment of inertia than a rotational band built on the oblate minimum. This implies that although the ground state is oblate, the energy of the oblate band will increase rapidly as the angular momentum increases and it is expected to cross the prolate rotational band and become non yrast at a relatively moderate spin.

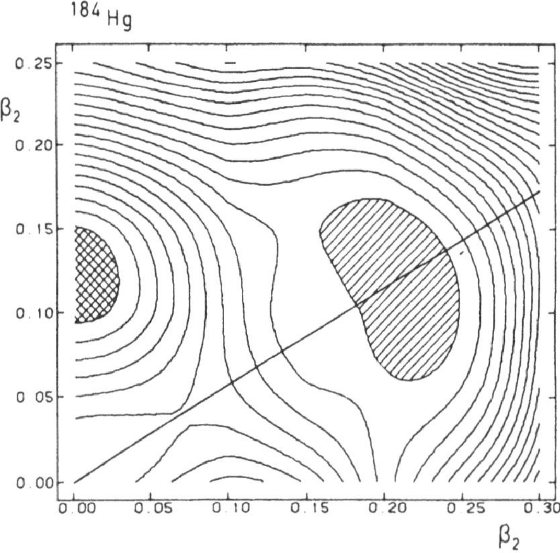

FIGURE 3. A spin zero potential energy surface for ^{184}Hg calculated with the Woods–Saxon potential. The figure clearly shows the coexistence of the oblate ground state (the cross-hatched area) with an excited prolate energy minimum. The energy difference between contour lines is 0.1 MeV.

3.2 The experimental observations

The experimental data have the theoretically expected features, showing rotational band sequences built on the two lowest 0^+ states. The two 0^+ states in the lighter isotopes are separated in energy by less than 0.5 MeV and the rotational bands built on the 0^+ bandheads are observed to cross each other at about spin 4. If the energy of the oblate band had continued to increase as indicated by the moment of inertia, it would, even at moderate spins, lie high above the yrast line and it would in practice be hard to observe. However, at the relevant oblate shapes the neutron Fermi surface is in the middle of the neutron $i_{13/2}$ intruder subshell and the subshell splitting is relatively small. This allows the low-K members of the subshell to mix easily to form states with a large aligned angular momentum as rotation sets on. This is a "cheap" way of gaining angular momentum and at spin 10 oblate shapes reappear at the yrast line with a configuration which is the neutron s-band. At prolate shape the neutron Fermi surface is also near the middle of the $i_{13/2}$ subshell, but due to the larger deformation the shell splitting is much greater. This is a less favourable situation for building up aligned angular momentum. Therefore, the $i_{13/2}$ neutrons align gradually and give only moderate contributions to the total angular momentum at prolate shape.

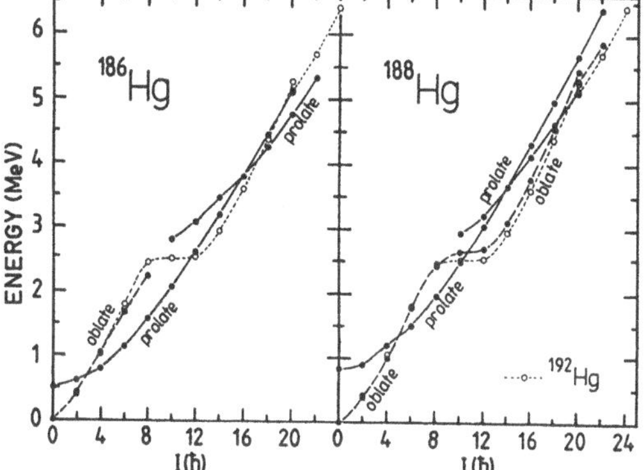

FIGURE 4. Experimentally observed coexisting oblate and prolate ratational bands in ^{186}Hg and ^{188}Hg. For comparison the dashed curve shows the yrast band of ^{192}Hg, which is known to have a slightly deformed oblate shape. The data are taken from refs.[3, 4, 5].

3.3 The interpretation

Neither the calculations nor the experimental observations reported above provide us with an immediate explanation for why coexisting shapes appear at low energy. For this purpose the microscopic structure of the nucleus has to be investigated. The single-particle energies for the neutron and proton systems are shown in figs. 5 and 6, respectively. For the relevant neutron numbers ($N = 100 - 108$) the neutron Fermi surface is in the middle of the $i_{13/2}$ intruder subshell and the intruder orbitals near the Fermi surface are therefore not strongly deformation driving. The density of single-particle levels is, however, much higher for spher-

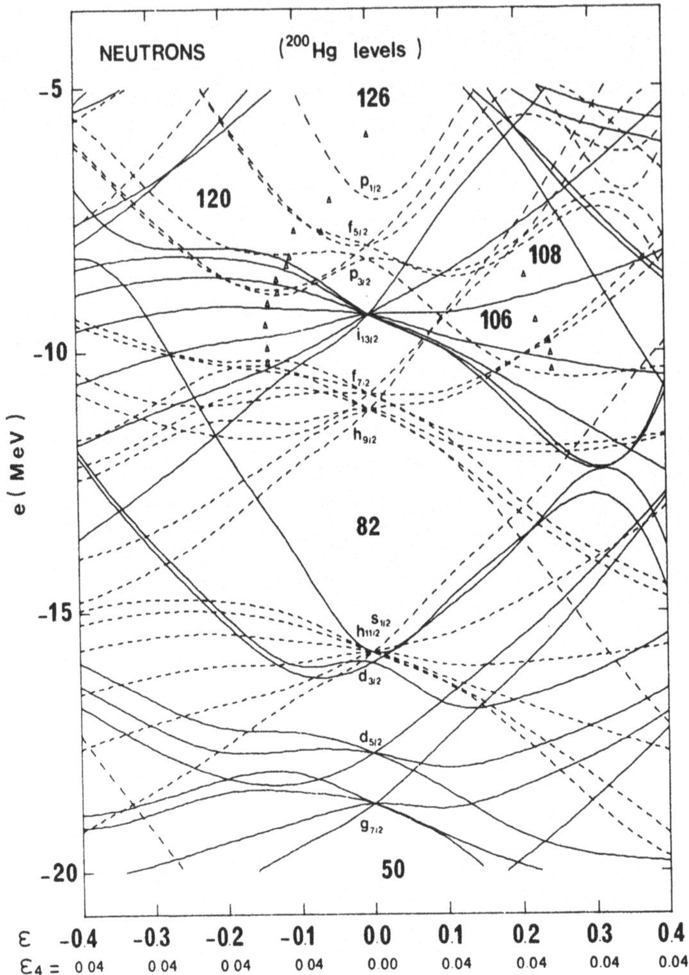

FIGURE 5. Single-neutron levels relevant for the Hg isotopes, calculated with the folded-Yukawa potential[6]. The triangles indicate the deformation of energy minima in the potential energy surface for the even–even Hg isotopes. The ground state always has an oblate deformation ($\epsilon < 0$). The prolate deformations ($\epsilon > 0$) indicated for $N \leq 108$ are the deformations of the coexisting excited energy minimum that appears in the potential energy surfaces of these light Hg isotopes, cf. fig. 3.

ical shape than for both prolate and oblate shape, with the lowest level density appearing for $0.15 < \beta_2 < 0.30$ and $-0.25 < \beta_2 < -0.10$, respectively. The neutrons are therefore generally prefering deformed shapes, which could be either oblate or prolate.

In the proton system the Fermi surface lies near the top of the $h_{11/2}$ subshell and therefore the deformation driving force of these intruder orbitals is much stronger. For mercury with $Z = 80$ the strongly upsloping levels near the Fermi surface for $\beta_2 > 0$ prevent the development of prolate shapes. On the other hand the large energy gap at $Z = 80$ for moderate ($\beta_2 \approx -0.15$) oblate shapes is enforced by the low density of neutron levels at similar oblate deformations. This is what establishes the weakly oblate shape of the ground state of the light mercury isotopes.

The proton single-particle levels also show that for $\beta_2 \geq 0.2$ strongly downsloping levels from the $h_{9/2}$ and $f_{7/2}$ subshells cross the upsloping levels at the Fermi surface and the proton contribution to the total energy decreases. Combined with the favourable neutron shell energy for $N \leq 108$ a local energy minimum develops at prolate shape in the light mercury isotopes. This minimum is identified as the one associated with the excited 0^+ state.

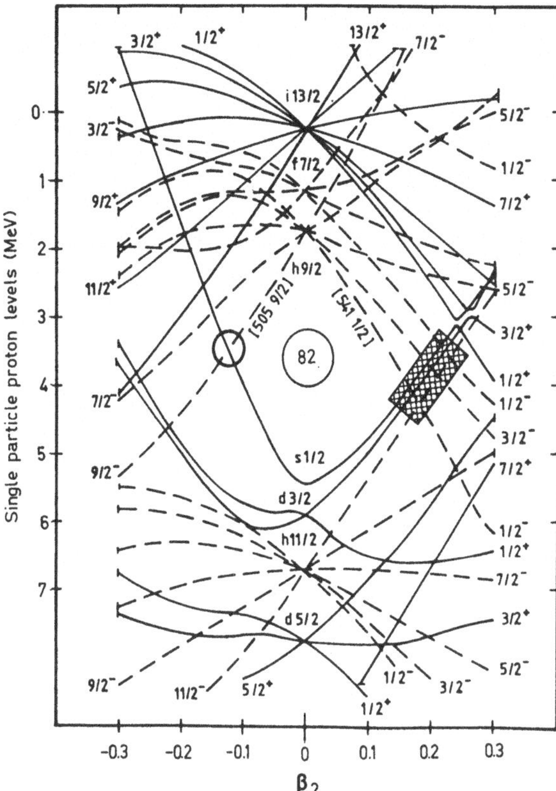

FIGURE 6. Single-proton levels relevant for the Hg isotopes, calculated with the Woods–Saxon potential[7]. The particular crossings of levels from above and below the spherical $Z = 82$ gap, which play a crusial role in the interpretation of shape coexistence in the Hg and Pb isotopes are indicated both on the prolate side (the cross-hatched region) and on the oblate side (encircled, important only in lead).

The proton configuration at the two minima is clearly different. On the oblate side the $h_{11/2}$ intruder subshell is fully occupied whereas the $h_{9/2}$ subshell is empty. On the prolate side particles are excited to the deformation driving low-K levels of the $h_{9/2}$ subshell from the upsloping orbitals (including the $K = 11/2$ level of the $h_{11/2}$ subshell) located between the spherical $Z = 82$ gap and the deformed $Z = 76$ gap.

4 The use of diabatic configurations

The difference between adiabatic and diabatic configurations and the advantages that can be achieved by using diabatic configurations is illustrated in fig. 7 for ^{186}Pb and ^{196}Pb. The lowest adiabatic configuration contains the spherical ground state expected for the lead isotopes. In ^{186} Pb it also contains an excited prolate minimum at $\beta_2 \approx 0.25$, which has a configuration similar to the one at the prolate minimum in the mercury isotopes. On the oblate side the adiabatic energy curve is perturbed at $\beta_2 \approx -0.15$, where it shows a clear shoulder both in ^{186}Pb and in ^{196}Pb, but no minimum is developed. The proton single-particle energy diagram in fig. 6 immediately reveals that the reason for the shoulder is the crossing between the $K = 9/2$ level from the $h_{9/2}$ subshell and the single level from the $s_{1/2}$ shell. However, because of the pairing interaction no levels can penetrate the Fermi surface. Therefore, the crossing between the two single particle levels does not result in a sharp crossing between two configurations in fig. 7, where the pairing correlations have been considered.

The energy of excited two–quasiparticle configurations can easily be calculated. At moderate deformations the lowest two–quasiparticle excitation is, on the oblate side, an excitation of a pair of quasiparticles to the $K = 9/2$ levels of the $h_{9/2}$ subshell and, on the prolate side, an excitation to the $K = 1/2$ levels of the same subshell. The total energy of these two configurations is also shown in fig. 7. These two excited adiabatic configurations have energy minima at $\beta_2 \approx -0.15$ and $\beta_2 \approx 0.1$ for the excitation to the $K = 9/2$ and $K = 1/2$ levels, respectively. At small deformations the excited quasiparticle pair has particle character, whereas at large deformations, beyond the single-particle level crossings it achieves a hole character. As a consequence, the adiabatic continuation of the ground state configuration, which at $\beta_2 = 0$ does not contain $h_{9/2}$ protons, will change its character at the single-particle level crossing points and, for $\beta_2 < -0.20$ and $\beta_2 > 0.25$, it will indeed contain occupied $h_{9/2}$ proton orbitals.

The structure of the even–even Pb isotopes can be made much more transparent by introducing diabatic configurations. Considering three such configurations with the characteristics that no protons, a pair of $K = 1/2$ protons and finally a pair of $K = 9/2$ protons are excited to the $h_{9/2}$ subshell, results in the dashed energy curves in fig. 7. These energy curves have sharp crossings and at some distance from the crossing points they join the original adiabatic configurations. By knowing the structure of the diabatic configurations one can immediately see how the original adiabatic energy curves can be decomposed into parts which correspond to specific diabatic configurations. What seemed to be just a faint shoulder at oblate shapes actually is a manifestation of a configuration which has a $K = 9/2$

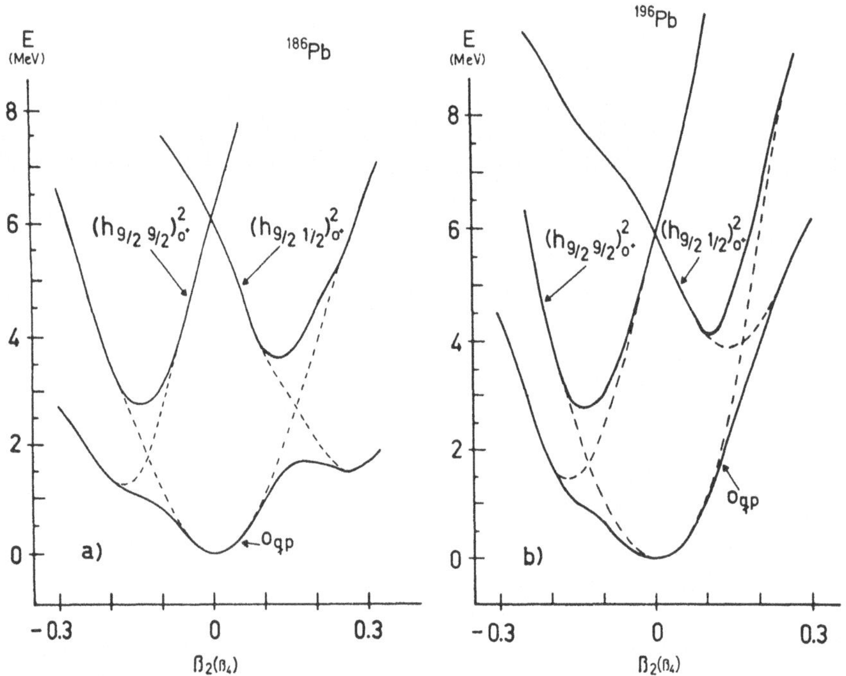

FIGURE 7. Adiabatic (solid lines) and diabatic (dashed lines) configurations in ^{186}Pb and ^{196}Pb. Only after the construction of the diabatic configurations does it become clear that the lowest adiabatic configuration completely changes its structure in the region where the energy curve has a shoulder near $\beta_2 = -0.15$. The excited energy minimum in the lowest adiabatic configuration (0qp) in ^{186}Pb has a similar structure as the prolate configurations in the Hg isotopes. Although this configuration never gives an energy minimum in nuclei with $N > 108$ it can still be found by using diabatic configurations as illustrated for ^{196}Pb. For additional details, see ref.[7].

proton pair in the $h_{9/2}$ subshell. In fact, this configuration can be identified with that of the lowest excited 0^+ state in the light Pb isotopes, on top of which a rotational band has been observed. There is overall good agreement between calculations and experimental data as described in ref.[7].

5 Shape coexistence related to the occupation of the proton $h_{9/2}$ subshell

In the previous two sections we have discussed the role that the $h_{9/2}$ protons are playing for creating an oblate–prolate shape coexistence in the light mercury isotopes and a spherical-deformed (oblate as well as prolate) shape coexistence in the lead isotopes. Many other nuclei in this mass region show more or less convincing signs of shape coexistence.

In the light platinum isotopes a similar kind of shape coexistence, as in the mercury isotopes, is observed. However, in the platinum isotopes the oblate 0^+ state lies about 0.5 MeV higher than the more deformed prolate 0^+ ground state. The rotational band built on the excited 0^+ therefore never reaches the yrast line and it cannot be traced to high spins like in the mercury isotopes.

A summary of of the quadrupole deformations calculated for shape coexisting low-lying 0^+ states in Pt, Hg and Pb is given in fig. 8. 'Experimental' β_2 values[8] derived from measured $B(E2)$ values are also included in this figure.

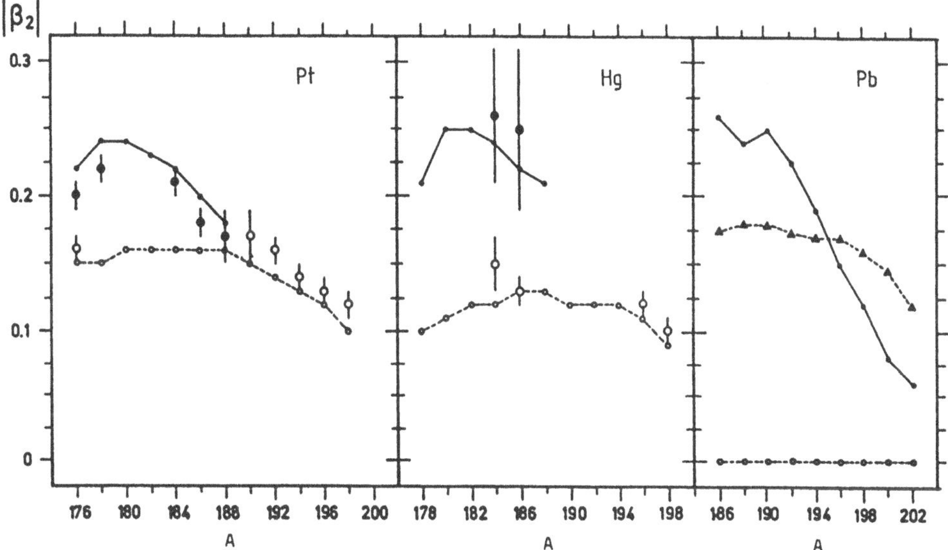

FIGURE 8. A summary of calculated quadrupole deformations for low-lying shape coexisting 0^+ states in even–even Pt, Hg and Pb isotopes[7, 8]. The solid lines show the deformation of the prolate states, which are the ground states of the light Pt isotopes, are observed as excited states in the light Hg isotopes and are predicted, but not yet observed, in the Pb isotopes. Like in lead (see fig. 7) the deformation of the prolate configuration can be calculated also in the heavier Hg anf Pt isotopes if diabatic configurations are used. The dashed lines with open circles show the deformation of the oblate or spherical states, which have no protons occupying $h_{9/2}$ orbitals, and which are the ground states of all the Pb, Hg and heavy Pt isotopes and which have been observed as excited 0^+ states in the lighter Pt isotopes. The dashed line with triangles shows the deformations of the oblate configuration in which a pair of protons is excited to the $K = 9/2$ level of the $h_{9/2}$ subshell. This configuration will have a low energy only in lead, where it also has been observed.

Both in the odd mercury isotopes and in some of the gold isotopes there are strong evidence for an oblate–prolate shape coexistence associated with the filling of the proton $h_{9/2}$ intruder subshell.

In lighter elements with proton numbers below 78, there is neither experimental nor theoretical indications of a prolate–oblate shape coexistence. Nontheless shape coexistence has been discussed for a number of light Re, Os and Ir isotopes and the coexisting states are supposedly distinguished by the extent to which $h_{9/2}$ proton orbitals are occupied[9, 10, 11, 12]. Like in the mercury isotopes the occupation of $h_{9/2}$ proton orbitals results in a large prolate deformation ($\beta_2 = 0.25 - 0.30$). In states with the $h_{9/2}$ orbitals unoccupied the deformation is smaller ($\beta_2 < 0.20$) but, unlike in the mercury isotopes, it now has a prolate shape.

When the coexisting states both are prolate, there is less structural difference between them and the decay between states with different shape is less hindered. Therefore, there is no direct observation of shape coexistence in the sense that states with the same spin and the same general configuration but different deformation have been observed. At a given spin only the lowest of two predicted coexiting states are observed, with the states of small deformation only appearing at low spin, see fig. 9. The structure of the observed rotational bands changes adiabatically as the spin increases from a small to a large deformation. For detailed descriptions, see ref.[9] .

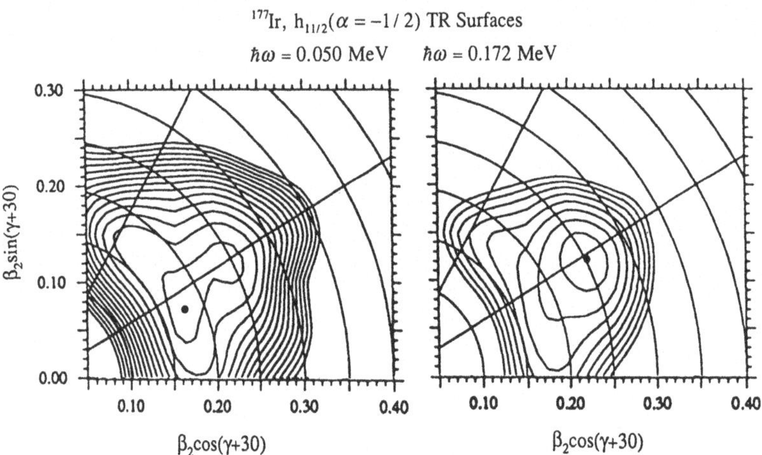

FIGURE 9. Examples of total routhian surfaces for the $h_{11/2}$ rotational band in ^{177}Ir, showing evidence for a shape coexistence between a less deformed prolate shape ($\beta_2 \approx 0.18$), which is energetically favoured at low rotational frequencies, and a more deformed prolate shape ($\beta_2 \approx 0.25$), which is energetically favoured at higher rotational frequencies. The surfaces are calculated using the Woods–Saxon potential and the details are given in ref.[9].

6 Shape transitions near $N = 90$

The transitional region from spherical to deformed shapes that appears for the rare earth nuclei at neutron numbers close to 90, is usually not considered as one of the classical examples of shape coexistence, although large changes in deformation takes place for these nuclei both as a function of neutron number and angular momentum. The reason is that easily recognizable coexisting rotational band structures at low spin have not been observed. However, at high spin largely different shapes do coexist, the best example being ^{152}Dy, where oblate single-particle like, prolate collective and superdeformed states all appear near the yrast line for spins in the range 25–35 as illustrated in fig. 10.

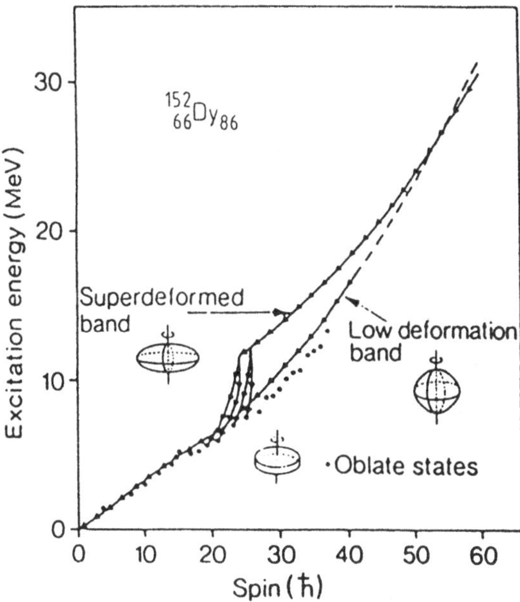

FIGURE 10. Coexisting oblate, prolate and superdeformed states in 152 Dy. The figure is from ref.[13].

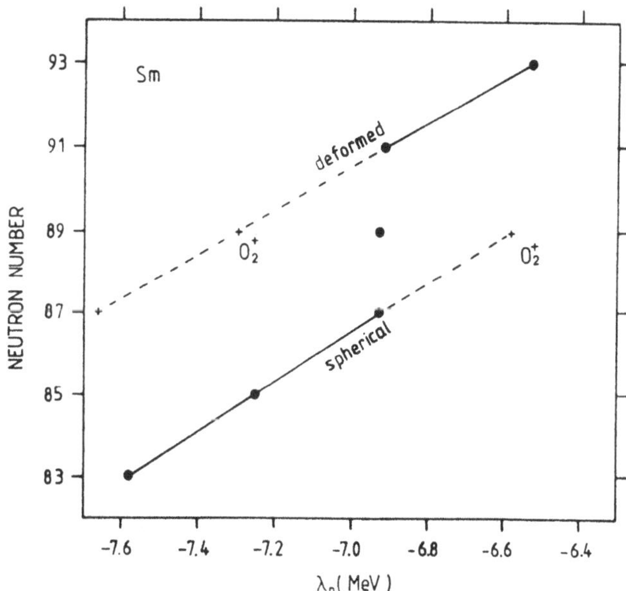

FIGURE 11. A so called gauge space plot for Sm, showing the neutron number as a function of neutron Fermi energy. The figure is constructed from experimental data. Further details are given in ref.[6].

At low spin the coexistence of 0^+ states of different shapes is very convincingly demonstrated in th even–even Sm isotopes with neutron numbers ranging from 86 to 90. In the lighter isotopes, with neutron numbers just above 82, the ground state is spherical or nearly spherical since 82 is a closed neutron shell. The heaviest Sm isotopes, with N higher than 90, already lie in the deformed region. Sudden deformation changes are easily recognized as an upbending or a backbending when the neutron (or proton) number is plotted as a function of the Fermi energy[6] and indeed fig. 11 shows that the Sm isotopes undergo a deformation change at $N \approx 90$. However, if instead of calculating the Fermi energy from the ground state masses, we calculate the Fermi energy by combining the 0^+ ground states in the isotopes with $N \geq 90$ with the first excited 0^+ states in the isotopes with $N < 90$, we get a perfectly smooth curve, indicating that all of these 0^+ states have a very similar structure, including a similar deformation. Since it is known that the ground state of the heavier isotopes is deformed, it can be concluded that the excited 0^+ states in the lighter isotopes also are deformed. Similarly, if the ground states of the lighter isotopes are combined with the excited 0^+ state in the $N = 90$ isotope the result is another smooth curve in the N versus λ plot, suggesting that the excited 0^+ state in the $N = 90$ Sm isotope has a more or less spherical shape. Thus spherical and deformed coexist in the even–even Sm isotopes from $N = 86$ to $N = 90$

7 Shape coexistence in the $A \approx 130$ mass region

The $A \approx 130$ mass region extends from the vicinity of the doubly magic, and therefore spherical, nucleus ^{132}Sn to the well deformed nuclei around ^{130}Ce. This is a particularly interesting mass region, since it is the heaviest region of nuclei in which the intruder subshell is the same for neutrons and protons, namely the $h_{11/2}$ subshell. The region is also interesting since deformation in the $A = 130$ nuclei is created by simultaneously removing neutrons from an originally full $h_{11/2}$ subshell and filling protons into an originally empty $h_{11/2}$ subshell. Both processes are efficient for creating a deformed nucleus, but whereas the $h_{11/2}$ protons outside the $Z = 50$ closed shell favour a prolate shape, the $h_{11/2}$ neutron holes favour an oblate shape. The preference for prolate and oblate shapes respectively for protons and neutrons leads to a close competition between prolate and oblate shapes, which becomes particularly striking once the (quasi)particles which occupy $h_{11/2}$ intruder orbitals start to align.

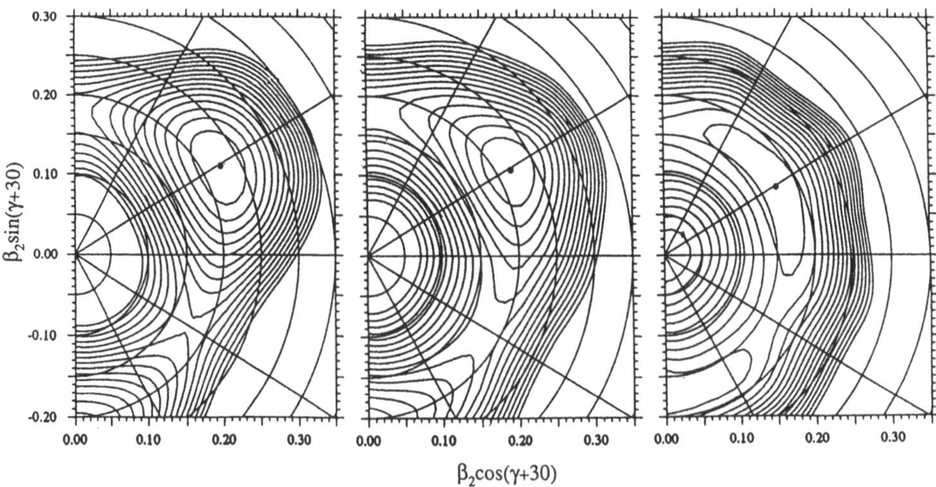

FIGURE 12. Total routhian surfaces for the ground band of ^{118}Xe, ^{122}Xe and ^{126}Xe at a rotational frequency of 60 keV.

Calculated total energy surfaces for the ground state rotational band in even–even nuclei in the $A = 130$ mass region do not show the same kind of strong evidence for shape coexistence as is the case for the light mercury isotopes. The energy surface for a number of nuclei is very γ–soft and the energy difference between oblate and prolate shapes is sometimes very small, but separate well defined minima do not develop, cf. fig. 12.

In several of the even–even Xe isotopes low-lying excited 0^+ states have been observed at energies below 1 MeV, which is much lower than is expected for a two–quasiparticle excitation across the paring gap. The low energies therefore tell us that these excited states cannot be of a two–quasiparticle nature.They could be shape coexisting states or they could be pushed down by collective correlations, being e.g. vibrational type states.

There is no theoretical support for the lowlying excited 0^+ states to be interpreted as evidence for an oblate–prolate shape coexistence. Unlike the situation in the mercury – lead region there is no correlation between the energy of the excited 0^+ states and the oblate–prolate energy difference. In fact, these two quantities depends on the neutron number in opposite ways: The energy of the excited 0^+ state is lowest at midshell, i.e. at $N \approx 66$ whereas the oblate–prolate energy difference is comparatively large at midshell and decreases steadily as the number of neutrons increases.

Even if the interpretation of the excited 0^+ states is unclear, there are clear evidence supporting a coexistence between oblate and prolate shapes to be found both in odd–Z and odd–N nuclei in this mass region. Also in the even–even nuclei, the observed s–bands can only be interpreted in a consistent way if the presence of both oblate and prolate shapes is assumed.

The most direct evidence for shape coexistence is found in the odd iodide isotopes where both strongly coupled and decoupled negative parity bands have been observed, cf. fig. 13. The only negative parity orbitals available close to the Fermi surface are those of the $h_{11/2}$ intruder subshell. In iodide $Z = 53$ and the Fermi surface is just below the intruder subshell at both moderate oblate and prolate shapes. An odd proton excited into this subshell will therefore occupy the $K = 11/2$ orbital if the nucleus has an oblate shape and the $K = 1/2$ orbital if the nucleus has a prolate shape. Since high K results in a strongly coupled band and low K in a decoupled band, it is reasonable to assume that the strongly coupled negative parity band is associated with an oblate shape and that the decoupled negative parity band is associated with a prolate shape.

Calculated total routhian surfaces for ^{121}I for the lowest negative parity (i.e. $h_{11/2}$) configuration have a prolate energy minimum, which through a valley in the γ–plane is connected with a low-energy region at oblate shape as shown in fig. 14. Although there is not a well-defined energy minimum at oblate shape, the configuration is very different compared to that at prolate shape (high K versus low K). With the same arguments as used for the lead isotopes in section 4, we conclude from the calculations that an oblate–prolate shape coexistence is to be expected. With the prolate shape having a lower energy than the oblate shape, the decoupled band should have a lower energy than the strongly coupled band. This is indeed observed experimentally[14].

In theoretical calculations, better examples of shape coexistence can be found in odd–N nuclei as illustrated in fig. 15. However, the neutron Fermi surface is usually lying closer to the middle of the $h_{11/2}$ intruder subshell and the coexisting states are often prolate–triaxial rather than prolate–oblate. The structural differences between coexisting states are therefore not as large as in the odd–Z isotopes. In particular, it is not possible to make a distinction between high–K and low–K states and decay between coexisting states should be less hindered. Direct experimental evidence for shape coexistence in negative parity one–quasiparticle bands in the odd–N isotopes is lacking. However, if three–quasiparticle bands are also considered, the evidence for coexisting shapes becomes much more clear.

As the number of neutrons increases, filling more and more levels of the $h_{11/2}$ subshell, one would expect that the signature splitting of the $h_{11/2}$ rotational bands in odd–N isotopes should gradually decrease, if the nucleus has a prolate deformation. This is true for the lighter isotopes of Ba and Ce with $N \leq 71$, but for larger neutron numbers the signature

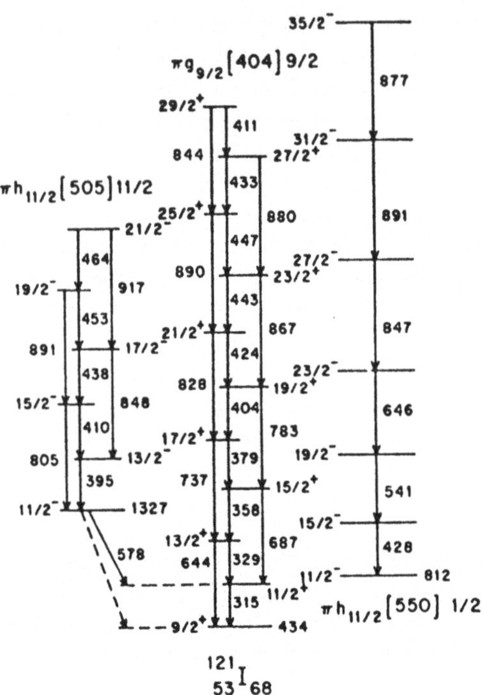

FIGURE 13. Partial level scheme for ^{121}I, showing the two negative parity bands $\pi h_{11/2}[550]$-1/2 and $\pi h_{11/2}[505]11/2$, which are interpreted as shape coexisting intruder configurations with prolate and oblate shapes, respectively. The bandhead energies are given relative to the $5/2^+$ ground state. The figure is from ref.[14].

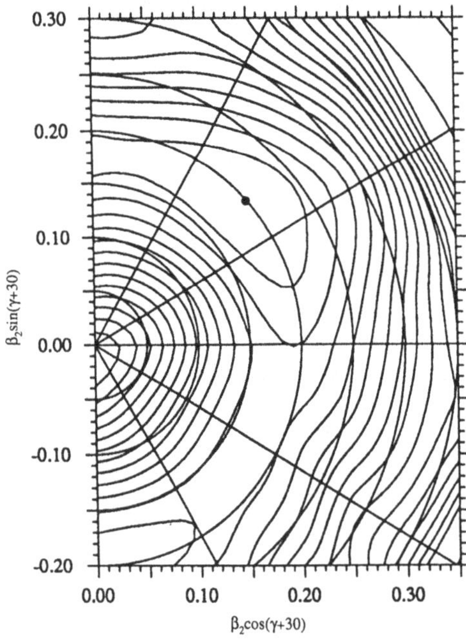

FIGURE 14. Total routhian surface for the favoured signature of the $h_{11/2}$ rotational band in ^{121}I at a rotational frequency of 60 keV.

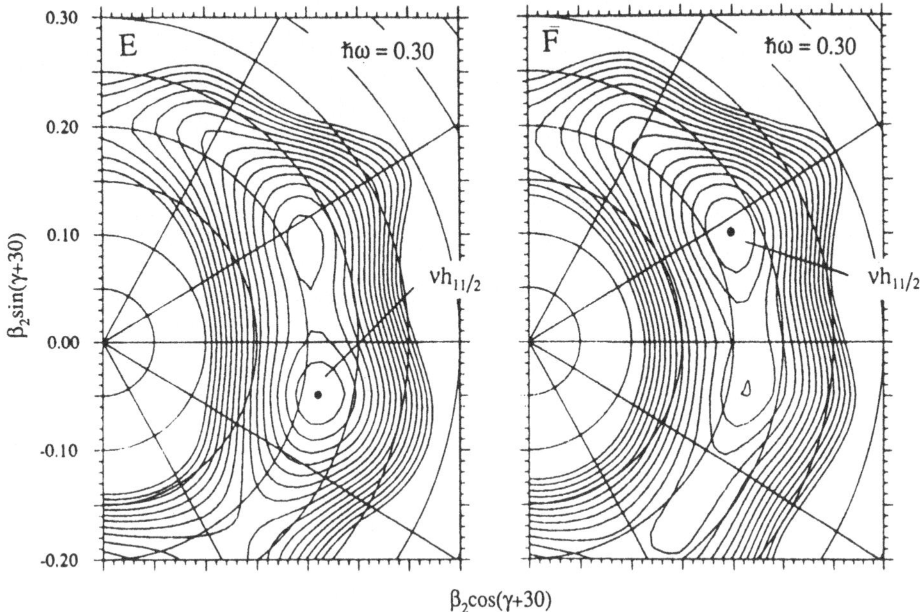

FIGURE 15. Total routhian surface for the favoured (E) and unfavoured (F) signatures of the $h_{11/2}$ rotational band in ^{127}Ba at a rotational frequency of 300 keV.

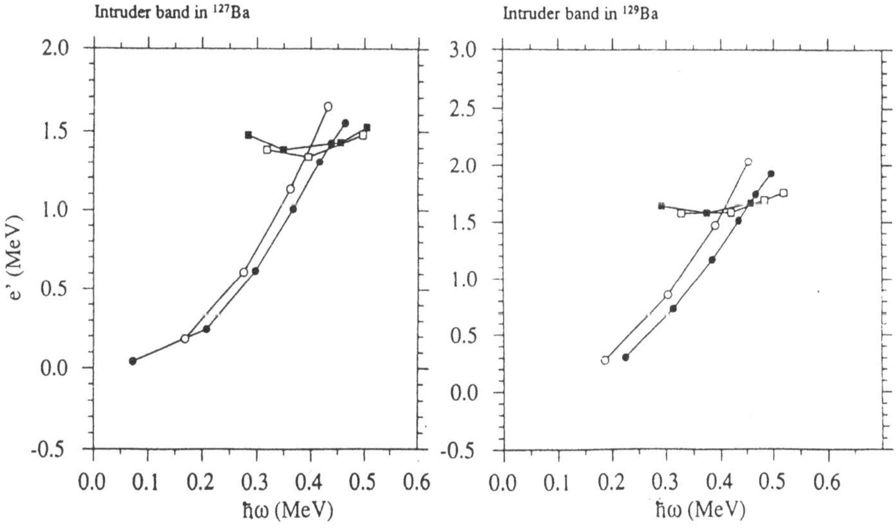

FIGURE 16. Experimental routhians for the $h_{11/2}$ rotational bands in ^{127}Ba and ^{129}Ba. Filled symbols are used for the favoured signature and open symbols for the unfavoured signature. Below the bandcrossing there is a clear signature splitting whereas above the bandcrossing the signature splitting is practically zero. An s–band reference has been used, making the routhians of the three–quasiparticle configuration above the bandcrossing approximately horisontal.

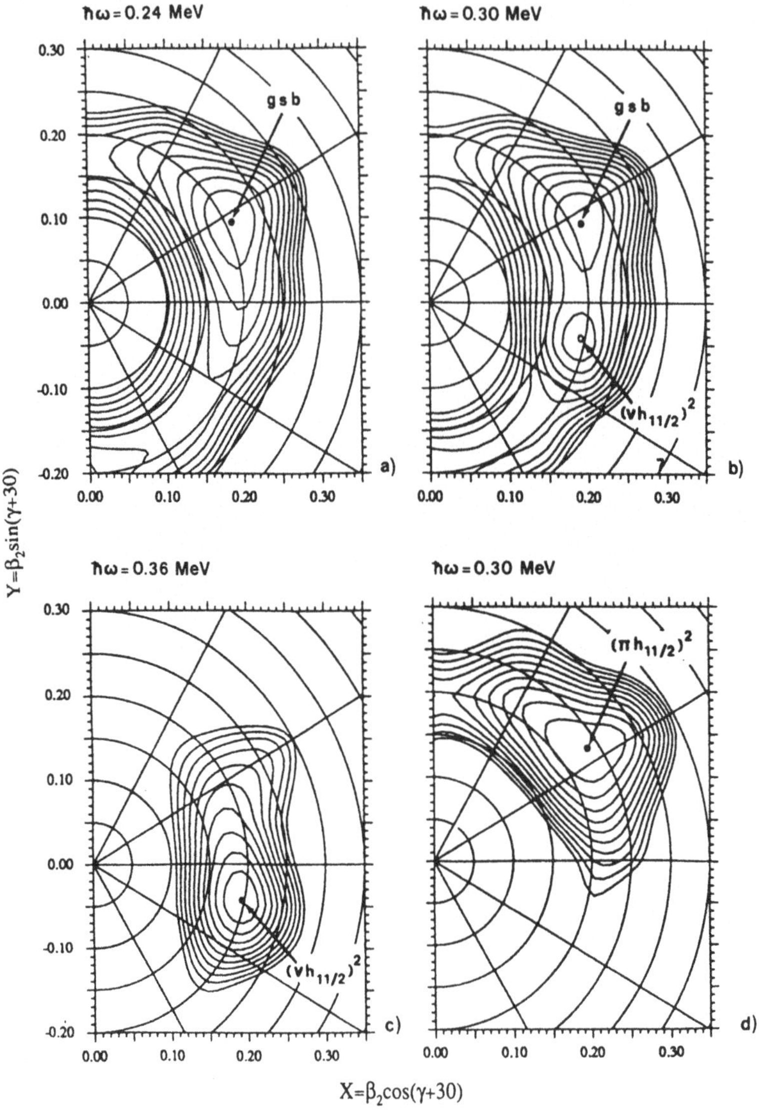

FIGURE 17. Examples of total routhian surfaces for ^{124}Xe. a), b) and c) show the vacuum configuration at the rotational frequencies 240, 300 and 360 keV while d) shows the lowest proton two–quasiparticle configuration at the rotational frequency 300 keV. The relevant configurations at the energy minima are indicated.

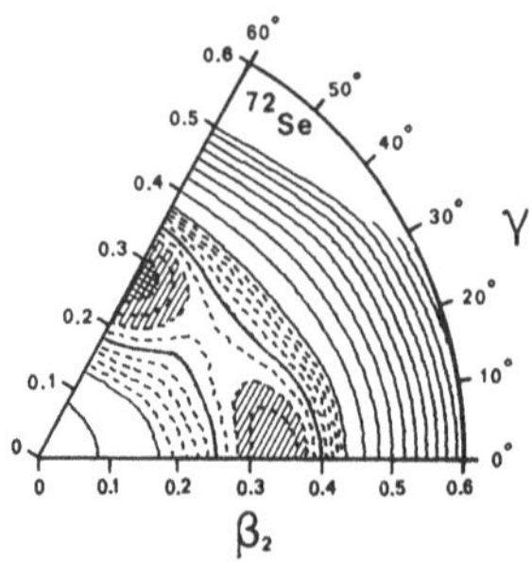

FIGURE 18. Potential energy surface for ^{72}Se at spin 0, showing coexisting oblate (lowest in energy) and prolate energy minima.

splitting increases again. The TRS calculations give a natural explanation to this observation, predicting a change to triaxial and finally oblate deformations in the heavier isotopes. At oblate shape the low-K levels lie in the top of the $h_{11/2}$ subshell. For $N > 70$ the Fermi surface is approaching the top of the subshell and at oblate shape $K = 1/2$ components can easily mix into the aligned $h_{11/2}$ orbital and create a signature splitting.

Another prediction of the theory is that above the first bandcrossing the deformation should change from oblate to prolate in the negative parity bands of the odd-N isotopes. This is because a pair of $h_{11/2}$ protons aligns and, with the proton Fermi surface in the bottom of the subshell, these protons are strongly prolate–driving. With the neutron Fermi surface in the upper half of the intruder subshell a deformation change from oblate to prolate should result in the disappearance of the signature splitting. The experimental data show that this actually happens (fig. 16), thus giving support to the thoeretically predicted deformation changes.

The shape coexistence discussed for the odd-N isotopes is of a special kind, since it is a coexistence between oblate one–quasiparticle bands and prolate three–quasiparticle bands. Whether there are shape coexisting prolate one–quasiparticle bands and oblate three–quasiparticle bands is not known.

Another kind of shape coexistence caused by the different deformation driving properties of protons in the bottom of the $h_{11/2}$ subshell and neutrons in the upper half of the same subshell appears in even–even nuclei in the $A = 130$ mass region. Two s–bands have been observed in several nuclei and have been interpreted as corresponding to the alignment of an $h_{11/2}$ quasineutron pair and an $h_{11/2}$ quasiproton pair. A careful analysis of band crossing systematics reveals that a reasonable explanation of the data can only be obtained if it is assumed that in the heavier isotopes (with $N \geq 70$) a prolate proton s–band coexists with an oblate neutron s–band. Such a shape coexistence is also predicted by TRS calculations. A few TRS are shown in fig. 17. A detailed discussion of the coexistence of proton and neutron s–bands with different deformations can be found in ref.[15].

At proton number 50 or just above, the $h_{11/2}$ proton levels are still too high above the Fermi surface for giving rise to shape coexisting states at low excitation energy and low spin. Still, deformed excited 0^+ states with collective rotational bands built on them are observed in a number of mid-shell Sn isotopes. The key deformation driving orbital in these isotopes seems to be the $K = 1/2$ level of the $g_{7/2}$ subshell, which seems to be as important for creating deformation near the $Z = 50$ gap as the $h_{9/2}$ subshell is for creating deformation near the $Z = 82$ gap. Furthermore, the deformation driving force is enhanced by holes created in the upsloping $K = 9/2$ levels of the $g_{9/2}$ subshell. This parallels the creation of holes in the $K = 11/2$ levels of the $h_{11/2}$ subshell (and the parallel [400 1/2] and [402 3/2] single-particle levels) below the $Z = 82$ gap.

8 Shape coexistence in the $A \approx 70$ mass region

In the $A \approx 70$ mass region both neutrons and protons are filling the $g_{9/2}$ intruder subshell. The experimentally investigated nuclei extend from $N = Z$ over into the neutron rich side. However, the number of neutrons is in general only slightly higher than the number of protons. Therefore, the same kind of conflicting situation between oblate and prolate shapes, which is characteristic for spin aligned configurations in the $A \approx 130$ mass region, does not exist in the $A \approx 70$ mass region. Still, shape coexistence appears to be a common phenomenon in the $A \approx 70$ mass region. It is a result of competing gaps at oblate and prolate shapes in the single-particle energy spectrum and the strong deformation driving force of the lowest–lying levels of the $g_{9/2}$ subshell, which can drive the deformation either prolate or oblate, whatever is favoured by the shell structure. For several combinations of N and Z, clear cut examples of shape coexistence appear[16]. In fact, the selenium isotopes ^{70}Se and ^{72}Se were among the very first nuclei in which shape coexistence at low spin was established. Cranking calculations based on the Woods–Saxon potential have been able to explain a variety of features related to coexisting shapes in this mass region[17]. It has also been made clear that coexisting shapes at low energy and low spin, as a rule, have different numbers of particles (neutrons as well as protons) occupying the $g_{9/2}$ intruder orbitals.

9 Summary

In a series of examples ranging from the $A \approx 180$ mass region down to the $A \approx 70$ mass region it has been shown that shape coexistence is a common phenomenon which often is closely related to the filling of either the lowest or the highest orbitals of an intruder subshell or as a combination of the two. Shape coexistence is usually not a coexistence between spherical and deformed shapes, as it was once often described. More common is apparently a coexistence between oblate and prolate shapes or between more or less deformed prolate states. Coexisting shapes can appear at any value of the angular momentum, and the superdeformed bands that have now been seen in several mass regions can be considered as another example of shape coexistence. Since no nucleus invesigated so far has been found to have a superdeformed ground state (although the groundstate deformation sometimes can be sizable), superdeformed states must always coexist with states with more normal shapes. Also in the configurations of superdeformed bands, the occupation of intruder orbitals from higher-lying oscillator shells plays a crucial role.

One may ask if it is enough that states with different shapes exist in the same nucleus in order to say that a nucleus exhibits shape coexistence, or should it be required that different shapes appear for the same value of the angular momentum, or should an even greater similarity between states of different shape be required in order to talk about shape coexistence? There may be some confusion on that point and it would be useful if one could agree to a common definition.

Whatever the definition should be, it now seems clear both from theoretical calculations and from experimental observations that there is not a single nucleus that has just one shape. On the contrary, any nucleus can appear in a variety of shapes, depending on its angular momentum, excitation energy or specific configuration. The well-known regular changes between spherical and deformed mass regions that is characteristic for nuclei in their ground states may have no or very little relevance at higher angular momentum or excitation energies, where totally new patterns of deformation changes may appear.

However, already at very low spins and excitation energies, there seem to exist in almost any mass region, states or families of states, which have a totally different deformation than the ground state. The shape of the nucleus therefore is a very complex and changeable property about which we have only a fragmentary knowledge as soon as we go beyond the ground state.

Acknowledgment

UNISOR is a consortium of universities, the State of Tennessee, Oak Ridge Associated Universities and Oak Ridge National Laboratory and is partially supported by them and by the U.S. Department of Energy under Contract No. DE-AC05-OR00033 with Oak Ridge Associated Universities.

References

[1] M.C. Carpenter, C.R. Bingham, L.H. Courtney, V.P. Janzen, A.J. Larabee, Z.-m. Liu, L.L. Riedinger, W. Schmitz, R. Bengtsson, T. Bengtsson, W. Nazarewicz, J.-y. Zhang, J.K. Johansson, D.G. Popescu, J.C. Waddington, C. Baktash, M.L. Halbert, N.R. Johnson, I.Y. Lee, Y.S. Schutz, J. Nyberg, A. Johnson, J. DeBuc, S. Monaro, S. Pilotte, K. Honkanen, D.G. Sarantites and D.R. Haenni, Alignment processes and shape variations in ^{184}Pt, Nucl. Phys. **A513** (1990) 125

[2] T. Bengtsson, Nucl. Phys. **A496** (1989) 56

[3] C. Bourgeois, L. Hildingsson, D. Jerrestam, N. Perrin, H. Sergolle, F. Hannachi, G. Bastin, M.G. Porquet, C. Schück, J.P. Thibaud, F.A. Beck, T. Byrski, J.C. Merdinger and J. Dudek, Proc. XXV Int. Winter Meeting on Nucl. Phys., Bormio (Italy) 1987, p. 447

[4] W.C. Ma, A.V. Ramayya, J.H. Hamilton, S.J. Robinson, J.D. Cole, E.F. Zganjar, E.H. Spejewski, R. Bengtsson, W. Nazarewicz and J.-y. Zhang, Phys. Lett. **167B** (1986) 277

[5] H. Hübel, A.P. Byrne, S. Ogaza, A.E. Stuchberry, G.D. Dracoulis and M. Guttormsen, Nucl. Phys. **A453** (1986) 316

[6] R. Bengtsson, P. Möller, J.R. Nix and J.-y. Zhang, Phys. Scripta **29** (1984) 402

[7] R. Bengtsson and W. Nazarewicz, Z. Phys. **A334** (1989) 269

[8] R. Bengtsson, T. Bengtsson, J. Dudek, G. Leander, W. Nazarewicz and J.-y. Zhang, Phys. Lett. **183B** (1987) 1

[9] R. Bengtsson, Nucl. Phys. **A520** (1990) 201c

[10] G.D. Dracoulis, R.A. Beck, A.E. Stuchbery, A.P. Byrne, A.M. Baxter and F. Riess, Nucl. Phys. **A486** (1988) 414

[11] G.D. Dracoulis, A.E. Stuchbery, A.P. Byrne, A.R. Poletti, S.J. Poletti, J. Gerl and R.A. Bark, J. Phys. **G12** (1986) L97

[12] R.A. Bark, G.D. Dracoulis, A.E. Stuchbery, A.P. Byrne, A.M. Baxter, F. Riess and P.K. Weng, Nucl. Phys. **A501** (1989) 157

[13] S. Åberg, H. Flocard and W. Nazarewicz, Annu. Rev. Nucl. Part. Sci. **40** (1990) 439

[14] Y. Liang, D.B. Fossan, J.R. Hughes, D.R. LaFosse, R. Ma, E.S. Paul, P. Vaska, M.P. Waring, N. Xu and J.-y. Zhang, Proc. Int. Conf. on High Spin Physics and Gamma-Soft Nuclei, Pittsburgh, PA, September 17–21, 1990, p. 308

[15] R. Wyss, A. Granderath, W. Lieberz, R. Bengtsson, P. von Brentano, A. Dewald, A. Gelberg, A. Gizon, J. Gizon, S. Harrisopulos, A. Johnson, W. Nazarewicz, J. Nyberg and K. Schiffer, Nucl. Phys. **A505** (1989) 337

[16] R. Bengtsson, Proc. Int. Workshop on Nuclear Structure of the Zirconium Region, Bad Honnef, April 24-28, 1988, p. 17

[17] R. Bengtsson and W. Nazarewicz, Proc. XIX Winter School, Zakopane, Poland, 1984 (Report IFY No 12681PL), p. 171

ROTATIONAL STRUCTURES AND RESIDUAL INTERACTIONS IN DOUBLY ODD NUCLEI

Andrés J. Kreiner

Physics Department
CNEA
Buenos Aires, Argentina

INTRODUCTION

The study of doubly odd nuclei has become a subfield of nuclear structure in its own right. A large body of data has been accumulated and we shall attempt in this work an approximation to a review. This however has to be qualified, otherwise the scope of the subject would be unmanageably extensive. We shall concentrate ourselves on near-yrast states of deformed and transitional nuclei and hence to studies carried out predominantly, though not exclusively, through the (HI, xnypzα) reaction. Some nice examples of how high-spin and low-spin work complement each other will also be given.

A basic information one expects to extract from studies of doubly odd nuclei concerns the residual interaction between the odd valence proton and neutron. However, in order to do so, one must first thoroughly understand the different ways in which two distinguishable nucleons couple to each other[1-7]. Deviations with respect to this "free" behavior will carry the information on residual forces. The exploration of the coupling schemes[1-7] has been a major theme in the last fifteen years and will occupy a good part of this review.

The study of doubly odd nuclei, in addition of posing specific questions, has accompanied the general trends and ideas in nuclear structure physics and made contributions from its particular point of view. Some of these subjects are: Signature splitting[2,5] (or staggering); Phase inversion[8,9] of the signature splitting and its claimed relation to triaxiality[10,11] or p-n forces;[12] Nature of ground to S band crossings and double blocking experiments;[13,14] Identical bands both in the normal[15] and in the superdeformed regime.[16,17] Pseudospin,[16,18] and Octupole deformations.[19-22] We shall touch only some of these topics as we move through the examples.

SURVEY OF STRUCTURES

Here we shall restrict the discussion to structures in which both valence particles occupy high-j orbits. The motivation is multiple. The yrast structures of odd-odd nuclei will most likely be built out of these configurations since they have largest spins for smallest values of the collective rotational angular momentum. They show the largest

Nuclear Shapes and Nuclear Structure at Low Excitation Energies
Edited by M. Vergnes *et al.*, Plenum Press, New York, 1992

deviations with respect to the strong coupling limit because their large intrinsic spins produce large couplings to the rotational motion through the Coriolis force hence leading to a rich variety of situations. Finally these structures are more easy to analyze and draw conclusions from because of their relative pureness in terms of configuration mixings.

The theoretical approach to this problem has been conducted so far in the framework of the rotational model because a) it has most frequently been used to describe high-j bands in odd mass nuclei and b) it provides a physically more transparent picture than others. A correct application of this model for these cases requires an exact diagonalization of the particle-plus-rotor Hamiltonian[23] within a configuration space of strongly Coriolis-related orbits (namely the high-j multiplets). The appropriate version for odd-odd nuclei is the two-quasiparticle-plus-rotor model[2] (TQPRM) with an energy operator which in its simplest form is:

$$H = A\vec{R}^2 + h_{qp}(p) + h_{qp}(n) \tag{1}$$

Where $\vec{I} = \vec{R} + \vec{j}_p + \vec{j}_n$ and A is the rotational constant $\hbar^2/2\Im$.

The state of lowest energy for a given value of the total angular momentum I (yrast state) results from the compromise between the rotational energy and the intrinsic (quasi-particle) energy. For nuclei where the rotational energy is fairly large we expect bandhead structures which minimize the rotational angular momentum R. This is particularly true for transitional regions (small deformation implies both small moments of inertia, i.e. large A, and small energetic splittings of the high-j multiplets thus facilitating the mixing between the different Ω components). We can envisage here two extreme situations.

a) If both proton and neutron Fermi levels (λ_p and λ_n) lie near the Ω=1/2 components of their respective high-j multiplets both particles will tend to decouple, aligning their angular momenta with the collective spin $\vec{R}$. Such a structure is called doubly decoupled[2,3,7] and its bandhead spin is expected to be $I_o = I_M = j_p + j_n$ (I_o and I_M are respectively the bandhead and the maximum spin obtainable from the intrinsic motion), the yrast band being a sequence of states connected by $\Delta I = 2$ stretched quadrupole transitions (i.e. a spin sequence $I = j_p + j_n, j_p + j_n + 2, j_p + j_n + 4, ...$).

b) If for instance $\lambda_p \simeq \epsilon_{\Omega=1/2}$ and the neutron lies high up in his j-shell (i.e. $\lambda_n \simeq \epsilon_{\Omega=j}$) we expect the bandhead spin to result from the perpendicular coupling, namely $I_o + \frac{1}{2} \simeq (j_p(j_{p+1}) + j_n(j_{n+1}))^{1/2}$. Here the yrast band is a sequence of (more or less perturbed) $\Delta I = 1$ states. This kind of structure is called semidecoupled[2,5] (we also use this name when λ_n is from approximately mid-shell to upper-shell).

Table 1 collects examples of these structures found through most of the periodic table.

SIGNATURE SPLITTING AND PHASE INVERSION

The level staggering or signature splitting[23] phenomenon in odd mass nuclei has been known for a long time. In an axially symmetric nucleus this feature is linked to the reflexion symmetry (in a plane perpendicular to the symmetry axis) which leads to a diagonal matrix element in $\Omega = 1/2$ bands called decoupling parameter,[23] **a**:

$$\mathbf{a} = - < \Omega = 1/2 | j_+ | \overline{\Omega = 1/2} > \tag{2}$$

For single-j bands this takes the value $\mathbf{a} = (-1)^{j-1/2}(j + 1/2)$ leading to positive values for $j = 1/2, 5/2, 9/2, 13/2, ...$ and negative values for $j = 3/2, 7/2, 11/2, ...$ Since the decoupling parameter enters into the energy expression as:

Table 1. Structures related to high-j (j_p and j_n) orbitals in doubly odd nuclei. Examples of I_o and I_M values as deduced from odd-A data and comparison with measurements.

Region	Odd Proton Data			Odd Neutron Data			Doubly Odd Nuclei				
	Parent Orbital	Charac. of Band	Nuclei	Parent Orbital	Charac. of Band	Nuclei	Nuclei (Refs.)	I_o^{π} [b] Predic.	I_o^{π} meas.	I_M^{π}	Charac. of Band
28<Z<50 28<N<50	$g_{9/2}$ partic.	$\Delta I=2$	^{75,77}Br	$g_{9/2}$ mixed	$\Delta I=1$ ($\Omega_o \simeq 5/2$)	^{77}Kr	^{74,76}Br (89,24-30)	- [c]	4^+ [d]	9^+	$\Delta I=1$
40<Z<50	$g_{9/2}$	$\Delta I=1$	Tc	$h_{11/2}$	$\Delta I=2$	Mo	^{98}Tc (31,32)	$\sim 7^-$	8^-	10^-	$\Delta I=1$
50<N<82	hole	$\Delta I=1$	Rh	partic.		Ru	^{100}Rh (32,33)	$\sim 7^-$	8^-	10^-	$\Delta I=1$
		$\Delta I=1$ ($\Omega_o \simeq 7/2, 9/2$)[a]	^{103,105}Ag			^{103,105}Pd	^{104,106}Ag (34,35)	7^-	7^-	10^-	$\Delta I=1$
		$\Delta I=1$ ($\Omega_o \simeq 9/2$)	^{107}In			^{107}Cd	^{108}In (36)	7^-	7^-	10^-	$\Delta I=1$
50<Z<82 50<N<82	$h_{11/2}$ partic.	$\Delta I=2$ $\Delta I=2$	$^{125-135}$La $^{135-137}$Pr	$h_{11/2}$ hole	$\Delta I=1$ ($\Omega_o \simeq 9/2$)	^{131}Ba	$^{130-132}$La (37,38)	$\sim 8^+$	8^+	11^+	$\Delta I=1$
					$\Delta I=1$	^{133}Ce	^{132}Pr (39)	$\sim 8^+$	8^+	11^+	$\Delta I=1$
					($\Omega_o \simeq 9/2$)		^{134}Pr (40)	$\sim 8^+$	7^+	11^+	$\Delta I=1$
50<Z<82 82≤N	$h_{11/2}$ mixed	very distorted	^{149,151}Eu - [d]	$i_{13/2}$ partic.	$\Delta I=2$	^{151}Gd	^{152}Eu (41)	d	9^-	12^-	$\Delta I=1$ large stagger.
Z≤82	$h_{9/2}$ partic.	$\Delta I=1$ ($\Omega_o \simeq 9/2$)	$^{195-199}$Tl	$i_{13/2}$ [f] hole	$\Delta I=2$	$^{195-199}$Hg	^{196,198}Tl (1,42)	8^-	8^-	11^-	$\Delta I=1$
N≤126	$h_{11/2}$	$\Delta I=2$	^{195}Au		$\Delta I=2$	^{195,197}Hg	^{196}Au (9)	12^-	12^- [e]	12^-	$\Delta I=2$
	hole						^{194}Au (43)	$\sim 12^-$	11^-	12^-	

[a] The symbol Ω_o denotes the most probable value of the projection of the intrinsic spin on the symmetry axis. [b] The symbol ~ in this column indicates an estimated uncertainty of one unit. [c] Because Ω_o for the neutron is distinctly different from j_n, the I_o value can not be predicted by a simple rule. [d] Since $\Omega_p > 1/2$ this case does not correspond to double decoupling which would give $I_o^{\pi} = 12^-$. [e] Only the 12^- state has been identified.[68] [f] For neutron deficient Tl and Au isotopes λ_n penetrates into the $i_{13/2}$ shell.

$$Aa(-1)^{I+1/2}(I+1/2) \tag{3}$$

it causes the levels with $I = 1/2, 5/2, 9/2, ...$ to be pushed down (up) for positive (negative) **a**. Hence positive signature (α=1/2) states (namely $I = 1/2, 5/2, 9/2, ...$) are favored (unfavored) for a positive (negative) decoupling parameter and viceversa for negative signature ($\alpha = -1/2$) states (namely $I = 3/2, 7/2, 11/2, ...$). This effect persists for bands with $\Omega > 1/2$ since it is transmitted to higher Ω orbits through non-diagonal (but signature independent) Coriolis matrix elements. In an odd-odd nucleus there is[3,7] an equivalent matrix element to that in (3):

$$(\mathbf{a}_p + (-1)^{(I+1)}\mathbf{a}_n)(I(I+1))^{1/2} \tag{4}$$

which in this case connects the $K = 0(=\frac{1}{2}_p - \frac{1}{2}_n)$ to the $K = 1(=\frac{1}{2}_p + \frac{1}{2}_n)$ state.

When both decoupling parameters have the same sign there is[3] a reinforcement of the coupling for odd spin states (since this is a nondiagonal effect the overall sign is irrelevant) and a destructive interference for even spin states and viceversa for decoupling parameters of opposite sign. Hence states with $\alpha = 1(I = 1, 3, 5, ...)$ are favored (unfavored) when both decoupling parameters have the same sign (opposite sign) and viceversa for $\alpha = 0(I = 0, 2, 4, ...)$ states.

This is the behavior we expect for sufficiently high spins (see below) unless another completely different ingredient enters the scene. We shall now discuss the first example of Table 1, namely the $\tilde{\pi}g_{9/2} \otimes \tilde{\nu}g_{9/2}$ structure of the Br region.

$\tilde{\pi}g_{9/2} \otimes \tilde{\nu}g_{9/2}$ Structure

This structure is well known in several odd-odd Br isotopes ($^{72-78}$Br, refs.8,9,24-30). Let us briefly recall the discussion[8] for the N=41 case, namely the Coriolis-distorted structure in ^{76}Br and its apparent relationship with the $5/2^+$ ground-state band[44] in the isotone ^{77}Kr.

The first two excited states in that band lie nearly at the same energy as the corresponding levels in the ^{77}Kr band (see fig.1), and while the higher levels in the doubly odd nucleus become depressed relative to those in ^{77}Kr, the phase of the level staggering remains the same (up to the 9^+ state, which was the last state known at the time when ref.8 was written; see below). This observation seemed to imply that the coupling of the aligned $g_{9/2}$ proton excitation to the ^{77}Kr ground-state band does not alter significantly the structure of this band's low lying spectrum. The description was based on the assumption that the intrinsic motion of the two valence quasiparticles involved in the 4^+ band takes place within the $\tilde{\pi}g_{9/2} \otimes \tilde{\nu}g_{9/2}$ configuration space; that is, both the proton and the neutron move in the five Nilsson states of $g_{9/2}$ parentage. This assumption was strongly suggested by the experimental information from neighboring odd-A nuclei. The $\Delta I = 1, 5/2^+$ ground-state band of ^{77}Kr corresponds to a predominantly $\Omega = 5/2$ neutron of $g_{9/2}$ parentage which feels the strong influence of the Coriolis interaction giving rise to the observed level staggering. On the other hand, a decoupled $\Delta I = 2$ band is known in ^{77}Br and ^{75}Br (refs.45,91). The different characters of the bands in these two nuclei are essentially determined by the position of the respective neutron and proton Fermi levels (in the latter case this lies nearest to the $\Omega = 1/2$ orbit).

The results for ^{76}Br are shown at the right-hand side of fig.1 (neutron and core parameters were chosen so as to reproduce the spectrum of ^{77}Kr shown at the left-hand side). They are seen to reproduce all qualitative characteristics present in the data namely: (a) The 4^+ state is the lowest and the calculated spin sequence of the yrast states is the correct one. (b) The phase of the staggering in both ^{77}Kr and ^{76}Br is the same, in agreement with experiment for $I \leq 13/2$ and $I \leq 9$, respectively, but a change of phase is predicted for the upper states of ^{76}Br. This represented an unforeseen result of the model which made the investigation of the states with $I \geq 10$ very attractive. Indeed a recent measurement[30] nicely confirmed that prediction for ^{76}Br (see states with $I = 10, 11, 12, 13$ in fig.1) which was already verified[25] for ^{74}Br. In order to better understand these results it is convenient to display the calculated behavior of the main physical quantities. In fig.2(a) we show the proton (dot-dashed line) and neutron (solid line) occupation probabilities of the different Ω components for some states of the 4^+ band. The dashed line indicates the same quantity for the odd neutron in ^{77}Kr. The neutron configuration in both ^{77}Kr and ^{76}Br remains essentially the same (mainly $\Omega_n = 5/2$) for the first three states.

On the other hand, the proton is not found to be totally decoupled, being, nevertheless, predominantly in a state of mixed $\Omega_p = \frac{1}{2}$ and $\frac{3}{2}$. Hence, from the point of view of the intrinsic energy, the states with $K = |\Omega_n \pm \Omega_p| = 1,2,3$, and 4 and $I = K$ lie lowest.

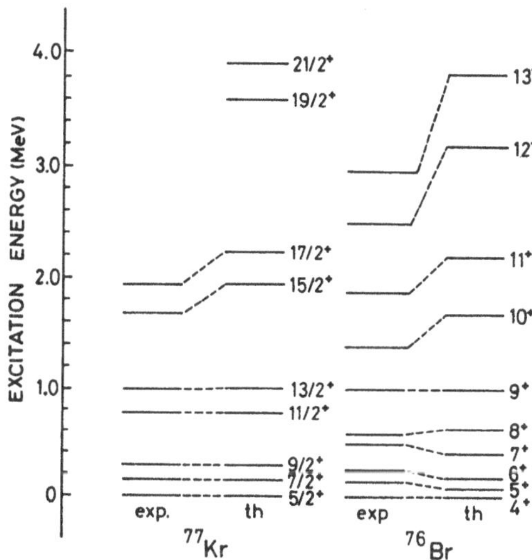

Fig.1. Experimental and calculated $\frac{5}{2}^+$ and 4^+ bands in ^{77}Kr and ^{76}Br.

However, the attractive Coriolis interaction is most effective for the 4+ state because the number of contributing $\Delta K = \pm 1$ matrix elements is largest. In Fig.2(c) the mean quasiparticle energies are displayed as functions of total angular momentum. It is seen that the proton is significantly more inert than the neutron. Its intrinsic energy remains

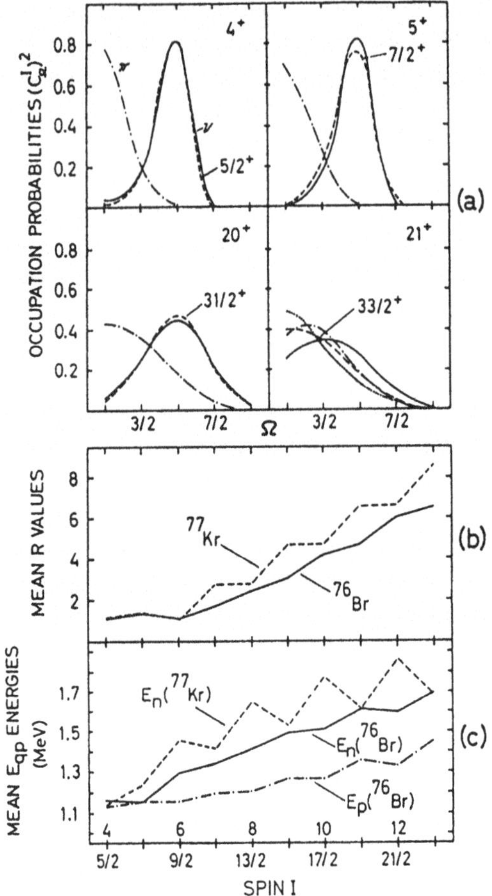

Fig.2. Occupation probabilities for the valence proton (dot-dashed line) and neutron (solid line) in ^{76}Br and neutron (dashed line) in ^{77}Kr are plotted in (a) for the first two high-spin states. The dotted line indicates the decoupled limit. In the lower half, mean values of the core angular momentum R in (b) and mean quasiparticle energies in (c) are shown as functions of spin I.

relatively constant at the beginning of the band, particularly for the first three states. The neutron in ^{76}Br is seen to exhibit a staggering of the same phase as that in ^{77}Kr for the first states, beyond which there is a transition region until a phase change sets in for $I^{\pi} > 9^{+}$.

The amount of collective angular momentum R for each state of ^{76}Br (solid line) is given in Fig.2(b) and it is compared with that of ^{77}Kr (dashed line). It becomes apparent from this plot that the first three states in ^{77}Kr and ^{76}Br have essentially the same R parentage. However, while the $I^{\pi} = \frac{11^{+}}{2}$ and $\frac{13^{+}}{2}$ states in ^{77}Kr are seen to carry approximately two additional units of core angular momentum and the sequence continues in similar steps for the upper states, the R values for ^{76}Br do not increase as fast (providing a natural explanation for the observed deviation between the ^{76}Br and ^{77}Kr spectra above the first three states) and a Coriolis-distorted rotational sequence is developed only from the 9^+ state onwards. This has to do with the fact that the maximum spin obtainable from the intrinsic motion for the doubly odd system is 9 while for the odd nucleus it is $\frac{9}{2}$. In such a situation the R values for the I=7,8, and 9 states represent the minimum-energy compromise between rotation and realignment of the intrinsic spin as means of increasing the total angular momentum. From this point of view the $I^\pi = 9^+$ state is a favored state (as is the $\frac{9}{2}^+$ state in ^{77}Kr) and ultimately a band resembling that of ^{77}Kr should develop on top of this state. In the present case this circumstance produces a change of 180° in the phase of the staggering above the 9^+ state. To complete the picture at high spins, Fig.2(a) shows the wave functions at $I = 20(\frac{31}{2})$ and $21(\frac{33}{2})$ where the identification of states in the odd and doubly odd systems has been made by requiring same core rotation asymptotically. Unfavored neutron distributions are very similar (I=20 and $I = \frac{31}{2}$), while favored distributions of both proton (I=20) and neutron (I=21 and $I = \frac{33}{2}$) approach the totally decoupled limit (shown in dotted line).

The state with $I = I_M = j_p + j_n$ plays a special role in the spectrum of a doubly-odd nucleus (somewhat similar to that of the $I = j$ state in an odd mass nucleus). It separates antialigned from aligned states in a low deformation situation. We see that eventually on top of the $I = I_M$ state a band has to develop showing a signature splitting which follows the rules discussed above. This picture may be somewhat compounded by the presence of a residual p-n interaction as will be discussed later on.

$\tilde{\pi}h_{11/2} \otimes \tilde{\nu}i_{13/2}$ Structure

This structure has been extensively studied for prolate deformations in nuclei of the first half of the rare-earth region, namely Eu (ref.41), Tb (ref.10), Ho (refs.46-49), Tm (refs.50-53) and recently Lu (ref.54). As shown in Table 1 this structure has $I_M^\pi = 12^- = 11/2^- + 13/2^+$ and the favored signature is $\alpha = 0(= -1/2_p + 1/2_n)$ which in fact turns out to be invariably true for all these structures. In all these bands even spin states become favored at sufficiently high spin. All these structures show a change of phase of the level staggering (or signature inversion) at a certain spin value. The splitting is anomalous (i.e. opposite to the one predicted by the rules given above) in the low spin regime.

This behavior has been taken[10] as a strong indication of triaxial deformation in the range $5° \leq \gamma \leq 15°$ since a theoretical analysis in the frame of the cranked shell model shows a crossing of the two signatures for the $h_{11/2}$ proton in the calculated Routhians as functions of rotational frequency. This interpretation is somewhat puzzling in view of the fact that non of the $h_{11/2}$ bands seen in neighboring odd isotopes show this feature, and it has been[11] questioned on the basis of calculations performed in the frame of the axially symmetric TQPRM and results similar to those discussed for the $\tilde{\pi}g_{9/2} \otimes \tilde{\nu}g_{9/2}$ structure.

$\tilde{\pi}h_{9/2} \otimes \tilde{\nu}i_{13/2}$ Structure

This structure was first discovered[1,2,42] and extensively discussed[55-57] in the neutron deficient doubly odd Tl isotopes in an oblate deformed situation and then followed into the prolate upper rare-earth region through Ir (refs.58-60), Re (refs.13,61-63), Ta (ref.64), Lu (ref.54) and Tm (ref.53) isotopes.

These bands in Tl represent the first examples of collective structures based on high-j excitations in doubly odd transitional nuclei and were successfully described as semidecoupled systems by using the TQPRM. They start with a set of transitions whose energies are small compared to energies in related bands of neighboring odd-mass nuclei, i.e., $\tilde{\pi}h_{9/2}$ and $\tilde{\nu}i_{13/2}$ bands in odd Tl and Hg nuclei, respectively. In contrast to high-j bands in odd-mass nuclei, where a single state is associated with vanishing collective angular momentum (the state with $I = j$), in doubly odd nuclei, where the intrinsic spin, J, is built out of two parts ($\vec{J} = \vec{J}_p + \vec{J}_n$), there is a multiplet associated with small core excitation (that with $I = J$ and $I \leq j_p + j_n$). The fact that the lowest-lying state within this multiplet has a spin value of around 8 is related to the positions of proton and neutron Fermi levels λ_p and λ_n, relative to the $\tilde{\pi}h_{9/2}$ and $\tilde{\nu}i_{13/2}$ Nilsson multiplets, respectively. The proton orbital is intruding from the next shell across the $Z = 82$ closure and since the deformation is oblate, the Nilsson state with the largest projection on the symmetry axis ($\Omega_p = \frac{9}{2}$) lies nearest to λ_p, giving rise to "normal" $\Delta I = 1$ bands in odd Tl isotopes. On the other hand, for $N \simeq 117$ the $\Omega_n = \frac{1}{2}$ member of the $\tilde{\nu}i_{13/2}$ orbital lies close to λ_n, resulting in decoupled bands in the odd Hg isotopes, in which $\vec{j}_n$ is approximately in a plane perpendicular to the symmetry axis. Hence, the energetically most favorable situation for the heavy doubly odd Tl isotopes is the perpendicular coupling of $\vec{j}_p$ and $\vec{j}_n$ which according to the relation $J(J+1) = \frac{9}{2}(\frac{9}{2}+1) + \frac{13}{2}(\frac{13}{2}+1)$ corresponds to $J \simeq 8$. Higher-spin states in the multiplet (in principle up to $I = \frac{9}{2} + \frac{13}{2} = 11$) are obtained by progressive alignment of $\vec{j}_p$ and $\vec{j}_n$ with respect to each other at the expense of intrinsic energy. In going to the neutron-deficient side, however, λ_n starts penetrating the $i_{13/2}$ subshell and neutron quasiparticle states with larger projections on the symmetry axis ($\Omega_n = \frac{3}{2}, \frac{5}{2}, ...$) become successively lower. The implication for the doubly odd system is that states with higher total angular momentum within the multiplet come down in energy. Eventually the state with $I = 9$ will cross below the one with $I = 8$ and subsequently $I = 10$ below $I = 9$, etc. This behavior[57] can be observed in Fig.3, where excitation energies of some yrast states for the $\tilde{\pi}h_{9/2} \otimes \tilde{\nu}i_{13/2}$ system (A band) have been calculated as functions of λ_n. For the observed structures in the Tl nuclei, which depopulate into long-lived 7^+ states, this means that the character of the isomeric out-of-band transition will change from $E1$ to $M2$, $E3$, etc., thus giving rise to increasingly longer lifetimes. It should be stressed that the picture outlined above corresponds to a negligible p-n force. The presence of a strong J-dependent force would cause deviations with respect to this behavior, allowing the extraction of the p-n interaction matrix elements $V_J =< (j_n j_p)J|V_{p-n}|(j_n j_p)J >$ (ref.12). These predictions have in fact been verified.[57]

The other special feature of these bands is that they show a pronounced level staggering very similar in magnitude and neutron number dependence to that observed in $\tilde{\pi}h_{9/2}$ (B) bands[12]. Concerning the origin of this staggering there are two completely different explanations. The first approach,[1,2] which utilizes a two noninteracting quasiparticle plus rotor model and which is in line with the discussion developed so far, finds this staggering to be associated with the signature dependence introduced by the Coriolis interaction [i.e., a$(-1)^I$ effect, where I is the total angular momentum]. This picture has the advantage of being able to account, on the same footing, for the staggering in both A and B bands, but requires the assumption of an unobserved low energy transition in order to obtain agreement between the calculated and experimental phase of the staggering. The other approach,[65-66] which is almost identical except for the inclusion of a residual proton-neutron (p-n) force, resorts to a completely unrelated mechanism for the explanation of the staggering namely a J dependence in the p-n interaction acting between the valence quasiparticles [i.e., a $(-1)^J$ effect, J being the total intrinsic angular momentum]. This treatment does not need to assume an unobserved transition but looses the very suggestive analogy which exists between A and B bands.

A careful analysis[12] of the influence of the residual p-n force was performed and might allow the reconciliation, at least for the heavier Tl isotopes, of the two different points of view: The staggering is due to the $(-1)^I$ dependence of the Coriolis force even in the presence of an appropriate p-n force while its inclusion renders unnecessary the assumption of an unobserved transition. However, as we shall see, the consistent use of this p-n force for lighter Tl nuclei leads to disagreement between theory and experiment.

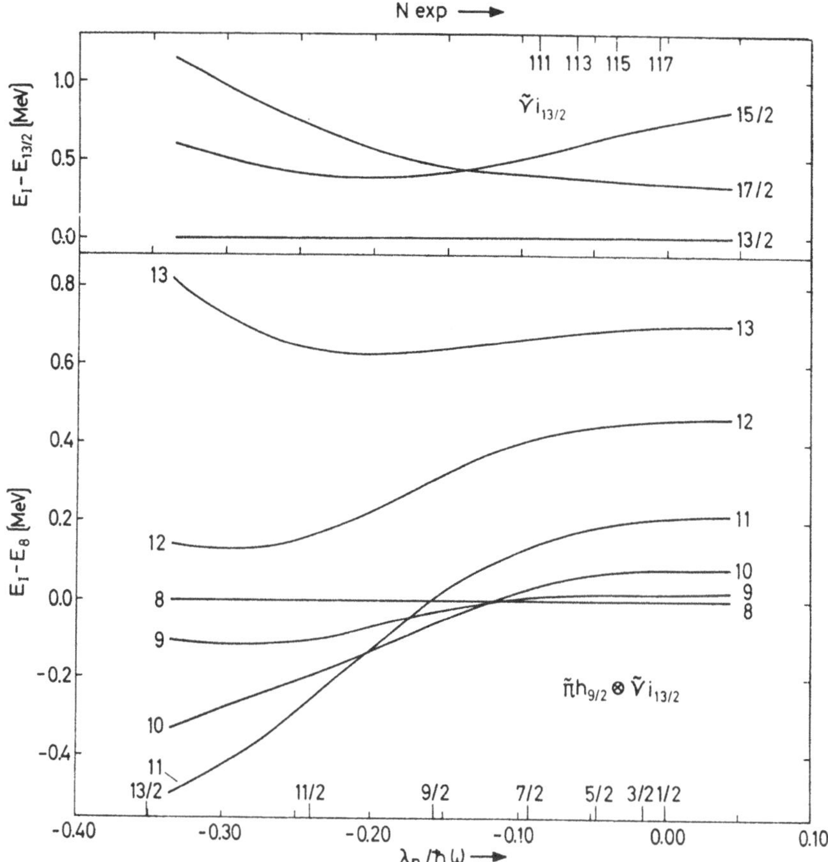

Fig.3. Calculated energies of some yrast states for the $\tilde{\pi}h_{9/2}\otimes\tilde{\nu}i_{13/2}$ systems as a function of the neutron Fermi level (λ_n in units of the oscillator constant $\hbar\omega = 41/A^{1/3}$). Positions of single-particle Nilsson energies of the $i_{13/2}$-parentage states are indicated on the λ_n axis. The distance between the first favored member $(\frac{17}{2}^+)$ of the $\tilde{\nu}i_{13/2}$ bands in odd Hg nuclei and the first unfavored one $(\frac{15}{2}^+)$, which depends very sensitively on λ_n, has been used to establish the correspondence between neutron number (N_{exp}) and position of λ_n by comparison with the experimental figures.

Figure 4 displays most of the available data concerning A and B bands. Spins in A bands are undetermined to some extent because the experiment can only ensure the value for the band head and the stretched dipole character of the observed transitions. Most likely, values for ΔI (see Fig.4) are 0 or 1 because of the existence of the two conflicting approaches just mentioned. The similarity of the level staggering in A and B bands has already been pointed out. In contrast to the constancy of level spacings in A [from the $(10+\Delta I)^-$ state onwards; see discussion below] and B bands in the mass range A=192-196 and A=191-197, respectively, a gradual reduction in the amount of staggering occurs in A bands for A=198 and 200 (ref.67) in exactly the same way as in B bands for A=199 and 201. This behavior is directly related to the constancy of the excitation energies of the ground state bands in the even-even Hg cores for A=192-196 and their subsequent decrease for A=198 and 200 (ref.68) which in terms of the rotational model implies a diminished Coriolis force and hence a less pronounced staggering. Such a behavior for the level staggering is in fact very suggestive of a particle-core coupling effect rather than the result of a residual p-n force.

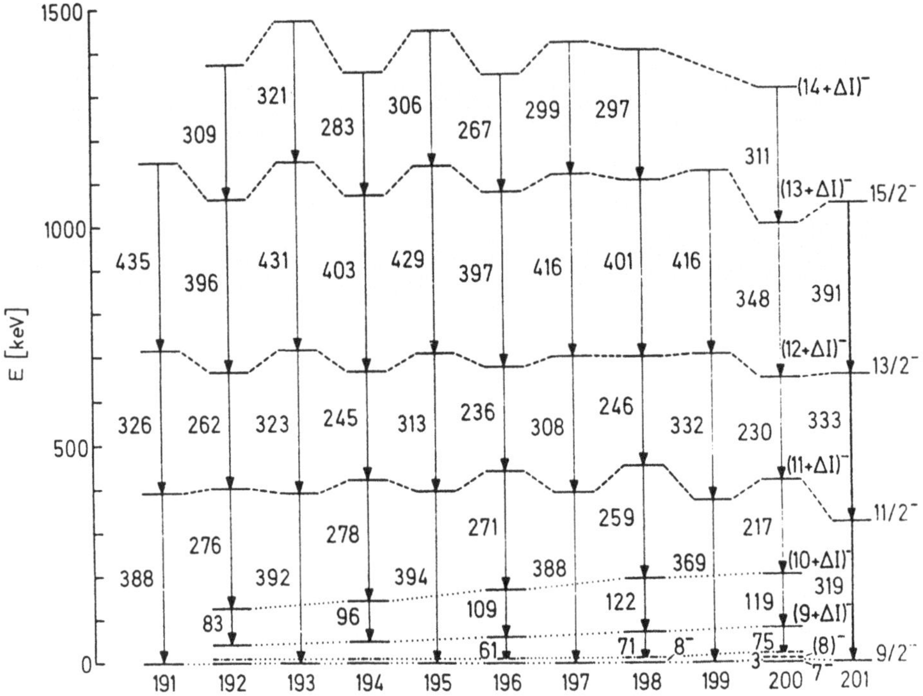

Fig. 4. Experimental $\tilde{\pi}h_{9/2}$ and $\tilde{\pi}h_{9/2} \otimes \tilde{\nu}i_{13/2}$ bands in Tl isotopes. The dashed-dotted level in doubly odd nuclei is drawn to indicate the possibility of an unobserved low energy transition which would correspond to $\Delta I = 1$ (see also text).

In order to draw the most reliable conclusions, essentially parameter-free calculations were performed for the Tl isotopes with A=190-200. Proton and neutron quasiparticle parameters were taken from neighboring odd and even-even nuclei: gap values from odd-even mass differences, quadrupole deformations from Hg-core $B(E2, 2^+ \rightarrow 0^+)$ values, and Fermi levels from fits to B and $\tilde{\nu}i_{13/2}$ (C) bands. In addition a procedure was used which incorporates the experimental core spectrum instead of utilizing a rigid rotor. With these choices no free parameters are left for A bands and several calculations were performed only differing in the p-n interaction used. Table 2 gives the resulting spectra for three different options. $C1$ corresponds to the noninteracting case, $C2$ to a pure particle-hole (p-h) force as obtained from ^{208}Bi (ref.69), and $C3$ corresponds to a mixed p-h and particle-particle (p-p) case to be discussed below. While $C1$ shows a bunching of the $I^\pi = 8^- - 11^-$ states and level staggering above the 11^- state, $C2$ leads to an 8^--10^- multiplet and staggering from there on. The differences between $C1$ and $C2$ will be illustrated using the (R, J) basis for ^{196}Tl. There are no appreciable differences between $C1$ and $C2$ for the 8^--10^- states. For the 11^- and higher states, however the $(R, J = 11)$ amplitude is drastically reduced in $C2$ as compared to $C1$. An examination of the p-n force as obtained[69] from the experimental $\tilde{\pi} h_{9/2} \otimes \nu i_{13/2}^{-1}$ multiplet in ^{208}Bi reveals the following features. The matrix elements $< (j_p j_n^{-1})J|V_{p-n}|(j_p j_n^{-1})J >= V_J^{-1}$ display small deviations ($\sigma \simeq 60$ keV) around a mean of about 150 keV for the J=5-10 states while the

Table 2. Theoretical transition energies for $\tilde{\pi} h_{9/2} \otimes \tilde{\nu} i_{13/2}$ bands in doubly odd Tl isotopes as functions of spin I and neutron number N for three different calculations.

N	119			117			115			113			111			109		
	ΔE_I^a			ΔE_I			ΔE_I			ΔE_I			ΔE_I			ΔE_I		
I	$C1^b$	$C2^b$	$C3^b$	$C1$	$C2$	$C3$	$C1$	$C2$	$C3$	$C1$	$C2$	$C3$	$C1$	$C2$	$C3$	$C1$	$C2$	$C3$
8																		
9	39	58	52	31	56	42	22	52	29	17	48	22	7	38	10	-8	26	-6
10	104	90	105	95	72	95	78	49	80	69	39	75	50	24	61	20	-5	34
11	170	235	220	168	263	225	163	274	206	156	278	195	142	277	174	119	283	148
12	242	206	229	260	203	242	267	198	253	274	205	266	283	223	283	296	242	297
13	259	369	345	269	412	355	270	418	321	274	423	303	282	422	282	295	424	261
14	346	263	293	381	259	319	383	246	343	393	252	369	399	268	402	406	279	421
15	308	408	373	308	444	381	295	434	336	303	460	324	319	475	299	329	477	273

[a] Transition energies are given in keV. $\Delta E_I = E_I - E_{I-1}$.

[b] $C1$ corresponds to the noninteracting case ($V_{p-n} = 0$), $C2$ to the pure particle-hole case, and $C3$ to the effective residual force between BCS quasiparticles.

value for J=11 lies at 886 keV, approximately 740 keV above the mean.(The particle-particle force would give, on the contrary, a lowest 11^- state. This is consistent with the short range character of the force since the 11^- state corresponds to the largest spatial overlap.) This figure is large when compared with the E_2^+ energy (426 keV) for ^{196}Hg. So, the behavior of the system with p-n force included is very similar to that described previously[2] with the only difference that the J=11 component is practically excluded from the spectrum of intrinsic excitations. Up to the 10^- state both proton and neutron align their spins to each other, from there on a further alignment [which implies a sizable admixture of the $(R=0, J=11)$ component] is energetically more costly and the system prefers to increase its total angular momentum at the expense of collective energy. (However for sufficiently high core angular momenta it may be cheaper for the system to go to the J=11 state, thus changing the phase of the staggering along the band. However, in the Tl case we may not reach this regime since other structural changes, like band crossings, may set in before). Also from the 10^- state onwards the neutron is far more inert (essentially decoupled) than the proton which moves alternatively towards and away from the symmetry axis exhibiting the same phenomenon as in B bands. This behavior, which in the strong coupling basis appears analytically as the signature dependence of the Coriolis interaction, is a direct consequence of the fact that the core ground band is restricted to even values of R.

There is a second feature of the A bands which is relevant for the discussion of the possible role played by the residual p-n force. As shown in Fig.4 the splitting of the $(8)^-$-$(10+\Delta I)^-$(8^--11^- for $C1$ and 8^--10^- for $C2$) multiplet (states connected by dots) diminishes towards the neutron deficient side. This behavior will be shown to be related to the position of the neutron Fermi surface λ_n relative to the $i_{13/2}$ Nilsson multiplet. As already discussed this posittion can be extracted from C bands in neighboring odd Hg nuclei. For λ_n at the top of the $i_{13/2}$ multiplet $[\lambda_n > \epsilon(\Omega_n = \frac{1}{2})]$ the 8^- state lies lowest (although the 7^- state is very close to it) because from the point of view of the intrinsic energy the proton tends to be maximally aligned with the symmetry axis $[\lambda_p \simeq \epsilon(\Omega_p = \frac{9}{2})]$ whereas the neutron has its spin pointing in a direction perpendicular to that axis. Higher spins within the multiplet are obtained mainly by successively tilting the spins of proton and neutron, each one towards the other at the cost of some intrinsic energy. As λ_n penetrates into the $i_{13/2}$ subshell, however, orientations of the neutron corresponding to higher Ω_n values come down in energy, the multiplet shrinks, and level crossings occur in qualitative agreement with the experiment.

Concerning the p-n interaction and while the use of V_J^{-1}, as obtained from ^{208}Bi, might have some justification for Tl isotopes with $A \geq 198$, it should undergo a major alteration as λ_n penetrates the $i_{13/2}$ subshell and the hole character of the neutron quasiparticle excitation changes. Within the BCS formalism the supposedly bare ^{208}Bi p-h interaction is modified to include both p-h and p-p vertices and the matrix elements of the p-n force in the strong coupling basis are obtained as

$$< KIM|V_{p-n}|K'IM >= \delta_{K,K'} < \alpha^{\dagger}_{\Omega_n}\alpha^{\dagger}_{\Omega_p}|V_{p-n}|\alpha^{\dagger}_{\Omega_{n'}}\alpha^{\dagger}_{\Omega_{p'}} >$$

$$= \delta_{K,K'} \sum_{J=|j_p-j_n|}^{j_p+j_n} (2J+1) \begin{pmatrix} j_n & j_p & J \\ \Omega_n & \Omega_p & -K \end{pmatrix} \begin{pmatrix} j_n & j_p & J \\ \Omega_{n'} & \Omega_{p'} & -K' \end{pmatrix} V_J^{eff},$$

with

$$V_J^{eff} = (u_{\Omega_n}u_{\Omega_p}u_{\Omega_{n'}}u_{\Omega_{p'}} + v_{\Omega_n}v_{\Omega_p}v_{\Omega_{n'}}v_{\Omega_{p'}})V_J + (u_{\Omega_n}v_{\Omega_p}u_{\Omega_{n'}}v_{\Omega_{p'}} + v_{\Omega_n}u_{\Omega_p}v_{\Omega_{n'}}u_{\Omega_{p'}})V_J^{-1} \quad (5)$$

by just transforming the two-body force operator into quasiparticle language. $K = \Omega_n + \Omega_p$ and $\alpha^{\dagger}_{\Omega_n(\Omega_p)}$ is a neutron (proton) quasiparticle creation operator. u's and v's are the usual BCS occupation amplitudes. V_J is the particle-particle matrix element obtained from V_J^{-1} applying the Pandya transformation[70]. V_J shows some fluctuations for J=5-10 (of opposite sign as V_J^{-1}) and is strongly attractive for J=11. As λ_n penetrates the $i_{13/2}$ shell, the effective force and, in particular, the strong repulsion for J=11 becomes smaller, comparable to the collective energies at some point and finally negligible due to the mixed p-h character of the neutron quasiparticle for λ_n in the middle of the shell. This behavior can be observed for $C3$ in Table 2. According to this calculation ^{196,194}Tl lie in an intermediate region where the system is still undecided between rotation or total alignment for the 11^-, 13^-, etc. states, while for ^{192,190}Tl one recovers the situation corresponding to $C1$. Thus, if the ^{208}Bi p-h force is taken and modified according to a prescription which follows from the standard Nilsson BCS treatment of rotational-like structures in transitional nuclei, the "phase" of the staggering inverts from A=200 to A=190. Such an inversion is not observed experimentally and moreover the attenuation of the level staggering towards the neutron rich side is just opposite to what would be expected if it were caused by the $(-1)^J$ oscillating component in the p-n force which should become stronger as the heavier Tl isotopes come closer to the pure p-h situation in ^{208}Bi.

It should be mentioned that recently[71] the suggestion has been made that $\Delta I = 0$ (see Fig.4), on the basis of α-decay measurements from isomers in doubly odd Bi isotopes into Tl. However this conclusion hinges on an $I^\pi = 10^-$ assignment for the α emitting isomers.

Summarizing, the present interpretation of a two-quasiparticle plus rotor calculation including the p-n residual force for the $\tilde{\pi}h_{9/2} \otimes \tilde{\nu}i_{13/2}$ system as obtained from ^{208}Bi might allow the explanation of the staggering phenomenon in A bands as due to the signature dependence in the Coriolis interaction without the assumption of an unobserved transition. However, the measured neutron number dependence of the level staggering does not show the behavior predicted if the p-n force is modified according to the position of the neutron Fermi surface. This circumstance casts some doubts on the adequacy of the ^{208}Bi or a similar residual force. Perhaps the fragmentation of the single particle states in the oblate mean field has sufficiently weakened the effective residual force to make its effect negligible. This may not necessarily be true for other systems like for example the $\tilde{\pi}g_{9/2} \otimes \tilde{\nu}g_{9/2}$. Here as one sweeps across the neutron $g_{9/2}$ shell (Br isotopes) one finds a 9^+ very low[27] for ^{72}Br (λ_n low in the shell) consistent with an attractive p-n particle-particle matrix element in the J=9 state while[26] for ^{78}Br the $I^\pi = 9^+$ state is pushed up in energy perhaps partially by a repulsive particle-hole matrix element in the J=9 state.

In general one may say that for particle-particle (or hole-hole) systems the p-n force will act coherently with the Coriolis interaction (the $I_M = j_p + j_n$ state will be favored) and also with the particle-core interaction (deformed field), while for particle-hole situations they will oppose each other. This last case may apply to ^{98}Tc (ref.31).

As the nuclear shape changes from oblate in Tl to prolate in Ir, Re, etc. on its way to the well deformed center of the rare earth region the semidecoupled $\tilde{\pi}h_{9/2} \otimes \tilde{\nu}i_{13/2}$ structure has been shown to persist. Interestingly, the qualitative features of this structure remain, although the roles of proton and neutron are reversed. In fact here, in a prolate deformed field, the proton orbit nearest to the Fermi surface is the Ω=1/2 (1/2$^-$ [541]) giving rise to decoupled bands in many odd Z nuclei of this region. On the other hand the neutron occupies orbits of $i_{13/2}$ parentage which at one end (heavy Os) have Ω=11/2 while at the lighter one (Yb,Er) have Ω=5/2. Here the neutron bands display a strong staggering (the larger the lower the position of the neutron's Fermi surface in the $i_{13/2}$ shell). This same staggering appears in the semidecoupled structures with the same amplitude and neutron number dependence as is documented in Fig.5 which shows the systematics for doubly odd Ir and Re and the associated $i_{13/2}$ bands in odd Os and W (remember that in the Tl case the staggering was associated with the proton while the neutron was decoupled). In a cranking model language one may say that the yrast structure results from the coupling of the favored α_p=1/2 signature trajectory of the h9/2 proton to the favored α_n=1/2 and unfavored α_n=-1/2 parts of the $i_{13/2}$ neutron structure. The neutron's signature splitting increases as the neutron Fermi level approaches the Ω_n=1/2 orbit of $i_{13/2}$ parentage. In this prolate region the ambiguity associated with the exact number of low energy M1 transitions in the initial multiplets of states is less because of more spectroscopic information and the staggering has the phase predicted by the "free" TQPRM. No phase inversions are known and the proton neutron force is expected to play a smaller role than in Tl on account of the larger deformation. As the neutron Fermi surface penetrates deeper and deeper into the $i_{13/2}$ shell one starts loosing the semidecoupled character entering a regime called doubly decoupled where unfavored states may cross above the favored ones. This is illustrated in Fig.6 and it is realized in Tm and Lu isotopes (refs.53 and 54).

There has been a tendency in the literature in recent years to associate staggering with triaxiality. In order to explore this question triaxial TQPRM calculations have been performed[72] on this structure. The result is clear: The staggering present at γ=0 tends to rapidly dissappear as one moves to positive values of γ (here prolate collective corresponds to γ=0 while γ=60° corresponds to oblate collective). The reason: the signature splitting of the semidecoupled band reflects the same splitting of the neutron when coupled to an aligned (and hence inert) proton. As soon as γ departs to positive values the decoupled character of the proton changes (unfavored states come down) and this suppresses the staggering. This overcomes the "enhancement" of the staggering experienced by the $i_{13/2}$ neutron as γ increases.

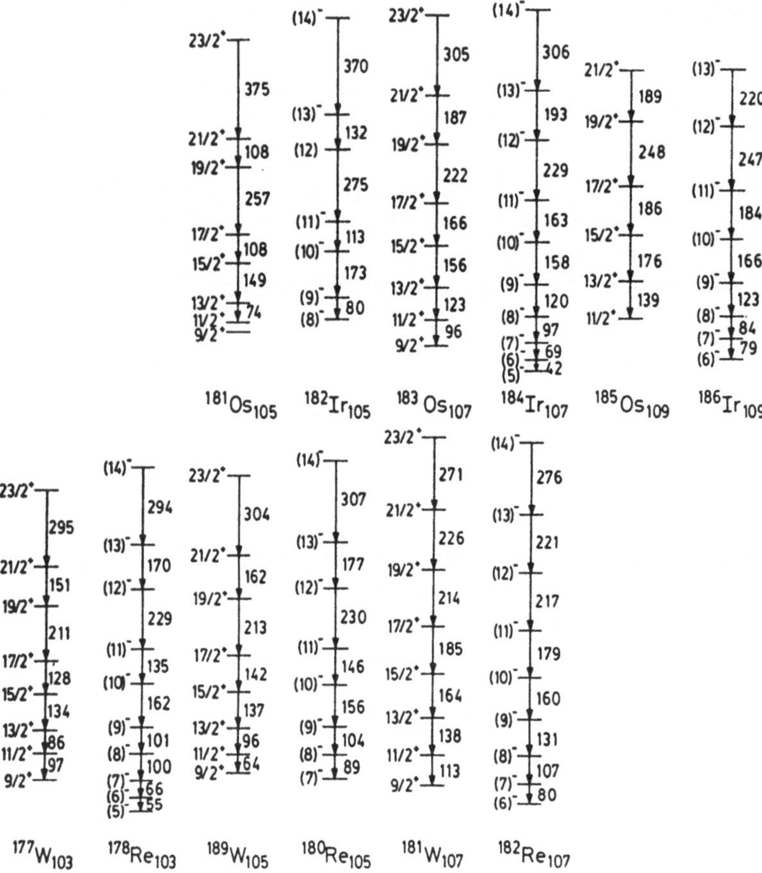

Fig.5. Systematics of $\tilde{\nu}i_{13/2}$ and $\tilde{\pi}h_{9/2} \otimes \tilde{\nu}i_{13/2}$ bands.

Before closing the discussion on $\tilde{\pi}h_{9/2} \otimes \tilde{\nu}i_{13/2}$ bands, it is worth mentioning their relevance in connection with the problem of the nature of the first (g-s) crossing in the heavy rare earth region.[13,14,64] The first crossing is consistently delayed both in $\tilde{\pi}h_{9/2}$ and $\tilde{\nu}i_{13/2}$ bands. The $\tilde{\pi}h_{9/2} \otimes \tilde{\nu}i_{13/2}$ bands allow the performance of so called double blocking experiments[13,14,64] since both critical, $\tilde{\pi}h_{9/2}$ and $\tilde{\nu}i_{13/2}$, orbits are blocked, showing the largest delay in crossing frequency among all the known bands in this mass region. The alternative deblocking of either one of these orbits brings the crossing frequency down. A systematic study of all shifts in crossing frequency shows that the roles of $\tilde{\pi}h_{9/2}$ and $\tilde{\nu}i_{13/2}$ are largely equivalent suggesting that both pairs may simultaneously participate in the structure of the S band in this mass region.

There are two categories of doubly odd nuclei from which information on p-n residual forces may be extracted in a relatively well defined way.

The first one comprises spherical (near doubly magic) nuclei and their study leads to coupling rules which are mainly known under the names of Nordhein, Schwartz, and Brennan and Bernstein (refs.73-75). Schiffer and True[69], for instance, have made a comprehensive review and analysis of $\vec{j}_p + \vec{j}_n = \vec{J}$ multiplets extracting valuable information on the residual force and use has already been made of it here.

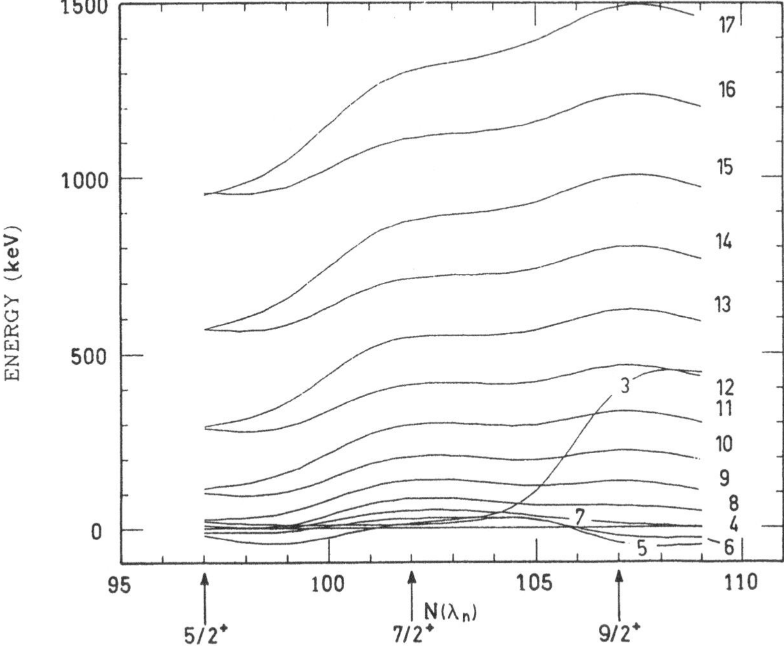

Fig.6. Calculated energies of some yrast states (relative to $I = 4$) for the $\tilde{\pi} h_{9/2} \otimes \tilde{\nu} i_{13/2}$ bands as a function of the neutron Fermi level.

The second category refers to nuclei which have a permanent (and large) deformation. Their spectrum is a series of bands built on intrinsic states which are labeled by quantum numbers which result from the combination of the quantum numbers of the associated proton and neutron orbits in a mean field, say a Nilsson potential. In this case the information about residual forces is contained mainly in the relative location of the bandheads. Also here some rules have been deduced, are associated mainly with the names of Gallagher and Moszkowski[76] and essentially state that the intrinsic spins tend to couple paralell to each other. This matter has been the subject of several investigations.[77-79]

However, most of the structures discussed here do not fall into the above categories. As we have seen they involve high-j excitations, where the asymptotic Nilsson representation may be inadequate, and imply strong couplings of several intrinsic states through the rotational motion. Here the information which is extracted may be more appropriately characterized as the expectation value of the residual force in the lowest state of a given structure. In order to extract this information the excitation energy of the different band structures above the ground state has to be known which is not very often the case for these complex nuclei, where isomerism, high level density and large internal conversion may complicate the spectroscopy. In these cases the complementation of high and low spin work may be very valuable. Here we report data from a collaborative effort where the low spin part comes[80,81] for Orsay-ISOCELE.

A possible approach to obtain the expectation values V_{p-n} of the residual interaction in an (N+1,Z+1) nucleus is to consider the binding energies, B, of the different band structures in the odd-odd nucleus, in the (N+1,Z) and (N,Z+1) odd neighbors and the associated core (N,Z). In this context the expression would be[7]:

$$V_{p-n} = B(N+1,Z) + B(N,Z+1) - B(N+1,Z+1) - B(Z,N) \quad (6)$$

Table 3 gives information for some of the structures detected in doubly odd $^{182-186}Ir$ along with necessary data of the associated bands in odd mass Os and Ir isotopes.

Table 3. Band Structures in odd Ir, Os and doubly odd $^{182-186}Ir$ isotopes and residual interaction values V_{pn}.

Odd Z Nucleus		Odd N Nucleus		Odd-Odd		
Config.	$E^*(keV)$	Config.	$E^*(keV)$	Config	$E^*(keV)$	V_{pn}(keV)
$^{181}Ir^{g)}$		$^{181}Os^{a)}$		$^{182}Ir^{b)}$		
$h_{9/2}$	0	$1/2^-[521]$	$\simeq 0$	$\tilde{\pi}h_{9/2} \otimes 1/2^-[521]$	$\simeq 0$	$\simeq$ -510
		$7/2^-[514]$	$\simeq 0$	$\tilde{\pi}h_{9/2} \otimes 7/2^-[514]$	45	$\simeq$ -465
		$9/2^+[624]\ (i_{13/2})$	108	$\tilde{\pi}h_{9/2} \otimes \tilde{\nu}i_{13/2}$	127	$\simeq$ -491
$^{183}Ir^{g)}$		$^{183}Os^{e)}$		$^{184}Ir^{c)}$		
$h_{9/2}$	0	$9/2^+[624]\ (i_{13/2})$	0	$\tilde{\pi}h_{9/2} \otimes \tilde{\nu}i_{13/2}$	0	-310
		$1/2^-[521]$	171	$\tilde{\pi}h_{9/2} \otimes \tilde{\nu}1/2^-[521]$	226	-255
$^{185}Ir^{g)}$		$^{185}Os^{f)}$		$^{186}Ir^{d)}$		
$h_{9/2}$	0	$1/2^-[510]$	0	$\tilde{\pi}h_{9/2} \otimes \tilde{\nu}1/2^-[510]$	0	-312
		$\tilde{\nu}i_{13/2}$	276	$\tilde{\pi}h_{9/2} \otimes \tilde{\nu}i_{13/2}$	66	-522

$^{a)}$ The relative position of the $1/2^-[521]$ and $7/2^-[514]$ is unknown but small. The $i_{13/2}$ structure lies 108 keV above the $7/2^-[514]$ orbit.[85]

$^{b)}$ Ref.60; the lowest state so far detected (at $\simeq$0 keV) is not necessarily the ground state.

$^{c)}$ Refs.59,80; $^{d)}$ Refs.58,81; $^{e)}$ Ref.86; $^{f)}$ Ref.87; $^{g)}$ Refs.85,88-90

Other, more sophisticated, averaging prescriptions than the one given in eq.(6) may be used, but it is not clear how to separate single-quasiparticle quantities from collective effects (like changes in the mean fields) in the binding energies so that the absolute values for V_{p-n} may have large uncertainties. On the other hand the differences in V_{p-n} values may be more meaningful and these are relatively small suggesting that the p-n forces play a minor role in determining their structure. Table 3 gives information mainly on the semidecoupled $\tilde{\pi}h_{9/2} \otimes \tilde{\nu}i_{13/2}$ and the doubly decoupled[7] $\tilde{\pi}h_{9/2} \otimes \tilde{\nu}(1/2^-[521] or 1/2^-[510])$ structures. The latter one is known in a number of Ir (refs.3,82), Re (ref.83) and Ta (refs.64,84)isotopes and involves the decoupled $h9/2$ proton and very particular neutron states which can be interpreted as pseudo-spin aligned states.[18] Evidence is accumulating which suggests that the p-n force is playing a minor role in this structure and in particular in connection with its signature splitting.

SUMMARY AND CONCLUSIONS

A survey of rotational structures over a good part of the periodic table has been made and analized within the frame of the rotational model. Particular attention has been devoted to the signature splitting problem and it has been shown that the phase of the staggering predicted by an axially symmetric model coincides with the data at high spin.

In some situations the noninteracting two-quasiparticle plus rotor model leads to phase inversion (anomalous splitting) as a consequence of competition between intrinsic and collective energy, thus not necessarily requiring departures from axial symmetry.

It is also shown under which circumstances the inclusion of the residual proton-neutron interaction may enhance the normal or the anomalous signature splitting.

Finally some comments are made about bandhead energies and residual interactions for doubly odd Ir isotopes studied both through low and high-spin techniques.

p-n interactions may influence some details in these structures but the main features are shaped by the couplings of the particles to the mean field and to the rotation.

REFERENCES

1. A.J.Kreiner, M.Fenzl, S.Lunardi, and M.A.J.Mariscotti, Nucl.Phys.A282, 243 (1977).
2. A.J.Kreiner, Z.fur Physik A288, 373 (1978).
3. A.J.Kreiner, D.DiGregorio, A.J.Fendrik, J.Davidson, and M.Davidson, Phys.Rev.C29, R1572 (1984).
4. A.J.Kreiner et al., Phys.Rev.C36, 2309 (1987), C37, 1338E (1988).
5. A.J.Kreiner, in "Contemporary Topics in Nuclear Structure". eds., R.F.Casten et al., (World Scientific 1988), p.521-540.
6. A.J.Kreiner in "XII Workshop on Nuclear Physics" (World Scientific, 1990) pp.137-155.
7. A.J.Kreiner in, "Exotic Nuclear Spectroscopy", ed. Wm.C.McHarris (Plenum Press, 1990) Chap.26.
8. A.J.Kreiner and M.A.J.Mariscotti, Phys.Rev.Lett. 43, 1150 (1979).
9. A.J.Kreiner, and M.A.J.Mariscotti, J.of Phys.G.Lett. 6, 13 (1980).
10. R.Bengtsson, J.A.Pinston, D.Barneoud, E.Monnand, and F.Schussler, Nucl.Phys.A389, 158 (1982).
11. I.Hamamoto, Phys.Lett.B235, 221 (1990).
12. A.J.Kreiner, Phys.Rev.C22, 131 (1980).
13. A.J.Kreiner et al., Phys.Rev.C40, R487 (1989).
14. A.J.Kreiner in "Nuclear Structure in the Nineties", Nucl.Phys. A520, 225c-240c (1990).
15. A.J.Kreiner, Phys.Rev.C38, R2486 (1988).
16. A.J.Kreiner and A.O.Macchiavelli, Phys.Rev.C42, R1822 (1990).
17. A.J.Kreiner in, "Future Directions in Nuclear Physics with 4π γ Detection Systems of the New Generation", American Inst.of Phys. (1991) in press.

18. A.J.Kreiner, Phys.Lett.B (1991) Submitted.
19. M.E.Debray et al., Phys.Rev.C39, R1193 (1989).
20. M.E.Debray et al., Phys.Rev.C41, R1895 (1989).
21. N.Schulz et al., Z.Phys.A335, 107 (1990).
22. N.Schulz et al., Z.Phys.A, (1991), Accepted.
23. A.Bohr and B.R.Mottelson, Nucl.Struc., Vol.2 (Benjamin, New York, 1975).
24. A.J.Kreiner, G.García Bermúdez, M.A.J.Mariscotti, and P.Thieberger, Phys.Lett.B83, 31 (1979).
25. G.García Bermúdez et al., Phys.Rev.C23, 2024 (1981).
26. M.Behar et al., Nucl.Phys.A376, 131 (1982).
27. G.García Bermúdez, C.Baktash, A.J.Kreiner, and M.A.J.Mariscotti, Phys.Rev.C25, 1396 (1982).
28. W.D.Schmidt-Ott, A.J.Hautojärvi, and V.S.Schrewe, Z.Phys.A289, 121 (1978).
29. J.Döring, G.Winter, L.Funke, P.Kemnitz, and E.Will, Z.Phys.A305,365 (1982).
30. D.F.Winchell, J.X.Saladin, M.S.Kaplan, and H.Takai, Phys.Rev.C41,1264(1991).
31. A.M.Bizzeti-Sona, P.Blasi, A.A.Stefanini, and A.J.Kreiner, Phys. Rev.C36, 2330 (1987).
32. A.M.Bizzeti-Sona, P.Blasi, A.A.Stefanini, G.Galeazzi, and A.J.Kreiner, Europhys.Lett., 3(2), 163 (1987).
33. R.Duffait et al., Nucl.Phys.A454, 143 (1986).
34. R.Popli, F.A.Rickey, L.E.Samuelson, and P.C.Simms, Phys.Rev.C23, 1085 (1981).
35. J.Tréherne, et al., Phys.Rev.C27, 166 (1983).
36. N.Elias et al., Nucl.Phys.A351, 142 (1981).
37. E.S.Paul et al., Phys.Rev.C36, 1853 (1987).
38. J.R.B.Oliveira et al.,Phys.Rev.C39,2250(1989).
39. S.Shi et al., Phys.Rev.C37, 1478 (1988).
40. C.W.Beausang et al., Phys.Rev.C36,1810 (1987).
41. J.A.Pinston, R.Bengtson, E.Monnand, F.Schussler, and D.Barneoud, Nucl.Phys.A361, 464 (1981).
42. A.J.Kreiner, M.Fenzl, and W.Kutschera, Nucl.Phys.A308, 147 (1978).
43. A.Neskakis et al., Phys.Lett.B80, 194 (1979); Nucl.Phys.A390, 53 (1982).
44. E.Nolte and P.Vogt, Z.Phys.A275, 33 (1975).
45. M.A.Deleplanque, C.Gershell, N.Perrin, B.Alder, and M.Ishihara, J.Phys.Lett.35, 237 (1974).
46. J.R.Leigh, F.S.Stephens, and R.Diamond, Phys.Lett.B33, 410 (1970).
47. J.A.Pinston et al., Phys.Lett.B137, 47 (1984).
48. N.Rizk and J.Boutet, J.de Physique 37, 197 (1976).
49. G.Lovhoiden, Phys.Scripta 25, 459 (1982).
50. S.Drissi et al., Z.Phys.A302, 361 (1981).
51. S.Drissi et al., Nucl.Phys.A451, 313 (1986).
52. S.Drissi et al., Nucl.Phys.A466, 385 (1987).
53. S.André et al., Z.Phys.A333, 247 (1989).
54. D.Hojman, A.J.Kreiner, M.Davidson, J.Davidson, and M.Debray, Phys.Rev.C, (1991), submitted.
55. A.J.Kreiner, M.Fenzl, U.Heim, and W.Kutschera, Phys.Rev.C20, 2205 (1979).
56. A.J.Kreiner et al., Phys.Rev.C21, 933 (1980).
57. A.J.Kreiner, C.Baktash, G.García Bermúdez, and M.A.J.Mariscotti, Phys.Rev.Lett.47, 1709 (1981).
58. A.J.Kreiner, D.DiGregorio, A.J.Fendrik, J.Davidson, and M.Davidson, Nucl.Phys.A432, 451 (1985).
59. A.J.Kreiner et al., Nucl.Phys.A489, 525 (1988).

60. A.J.Kreiner, J.Davidson, M.Davidson, P.Thieberger, and E.K.Warburton, Phys.Rev.C42, 878 (1990).
61. D.Santos et al., Phys.Rev.C39, 902 (1989).
62. M.F.Slaughter, R.A.Warner, T.L.Khoo, W.H.Kelly, and Wm.C.McHarris, Phys.Rev.C29, 114 (1984).
63. Ts.Venkova et al., Nucl.Phys.A514, 87 (1990).
64. A.J.Kreiner et al., Phys.Lett.B215, 629 (1988).
65. M.Toki, H.L.Yadav, and A.Faessler, Phys.Lett.B71, 1 (1977).
66. H.L.Yadav, H.Toki, and A.Faessler, Z.Phys.A287, 121 (1978).
67. A.J.Kreiner, M.A.J.Mariscotti, C.Baktash, and P.Thieberger, Phys.Rev.C23, 748 (1981).
68. Table of Isotopes, 7th Ed., edited by C.M.Lederer and V.Shirley (Wiley, New York, 1978).
69. J.P.Schiffer and W.W.True, Rev.Mod.Phys.48, 191 (1976).
70. S.P.Pandya, Phys.Rev.108, 1312 (1957).
71. M.Huyse et al., Phys.Lett.B201, 293 (1988).
72. D.Santos and A.J.Kreiner, to be published.
73. L.W.Nordheim, Phys.Rev.78, 294 (1950).
74. C.Schwartz, Phys.Rev.94, 95 (1954).
75. M.H.Brennan and A.M.Bernstein, Phys.Rev.120, 927 (1960).
76. C.J.Gallagher and S.A.Moszkowski, Phys.Rev.111, 1282 (1958).
77. J.P.Boisson, R.Piepenbring, and W.Ogle, Phys.Rep.26, 99 1976) and references therein.
78. L.Bennour, J.Libert, M.Meyer, and P.Quentin, Nucl.Phys.A465, 35 (1987).
79. R.W.Hoff, A.K.Jain, J.Kvasil, P.C.Sood, and R.K.Sheline, in "Exotic Nuclear Spectroscopy", ed. Wm.C.McHarris (Plenum Press, 1990).
80. A.Ben Braham et al., Nucl.Phys.A482, 553 (1988).
81. A.Ben Braham et al., Nucl.Phys.A (1990), in press.
82. A.J.Kreiner, P.Thieberger, and E.K.Warburton, Phys.Rev.C34 R1150 (1986).
83. J.Davidson et al., Z.Physik A324, 363 (1986).
84. A.J.Kreiner and D.Hojman, Phys.Rev.C36, R2173 (1987).
85. A.Neskakis et al., Nucl.Phys.A261, 189 (1976).
86. B.Roussiere, these, Orsay, 1986, B.Roussiere et al., Nucl.Phys.A438, 93 (1985).
87. H.Sodan et al., Nucl.Phys.A237, 333 (1975).
88. C.Schück et al., Nucl.Phys.A325, 421 (1979).
89. S.André et al., Phys.Rev.Lett.38, 327 (1977).
90. S.André et al., Nucl.Phys.A328, 445 (1979).
91. A.J.Kreiner, M.A.J.Mariscotti, C.Baktash, E.der Mateosian, and P.Thieberger, Phys.Rev.C 148 (1981).

COLLECTIVE AND QUASI-PARTICLE DEGREES OF FREEDOM IN ^{178}Hf (*)

P. Quentin[1,2], S.J. Krieger[1], J. Libert[1,2] and M.S. Weiss[1]

1) Department of Physics, L.L.N.L., Livermore, CA 94550 (U.S.A.)
2) C.S.N.S.M., CNRS-IN2P3, Bât. 104-108, 91405 Orsay Campus (France)

ABSTRACT

A consistent description using the Skyrme SIII force, of both the individual and collective spectroscopic properties at low excitation energy of the ^{178}Hf nucleus, is presented. Along with rotational and vibrational states, the $K^{\pi}=16^{+}$ isomeric state is very well reproduced. A discussion of the crucial importance of pairing correlations is made.

The ^{178}Hf nucleus belongs to the well-known region of permanently deformed rare-earth nuclei. It exhibits a rotational ground band ($E(2^{+})=0.093$ MeV, $E(4^{+})/E(2^{+})=3.29$), a β-vibrational band ($E(0^{+})=1.198$ MeV) and a γ-vibrational band($E(2^{+})=1.175$ MeV) which are rather typical of what is found for neighbouring nuclei. It differs from them however by the very important feature of possessing an isomeric state rather unique by the conjunction of its high spin ($I^{\pi}=16^{+}$), moderate excitation energy (E=2.446 MeV) and intermediate lifetime (τ=31 year).

Its interest has ben considerably enhanced in recent years by the efforts of a Dubna-Orsay collaboration to prepare targets with a sizeable isomeric rate (slightly less than 10% typically)[1]. Coulomb experiments on such a target are under way at GSI. Laser spectroscopy and nuclear orientation experiments will be soon performed in Orsay and Dubna, while the possibility of neutron capture and transfer experiments is currently studied.

The aim of the present calculations is to reproduce within the same framework, both the single particle (s.p.) and collective low excitation energy properties. Taking stock on a possible success in unifying these two aspects of such a nucleus, we also would like to improve our knowledge of the wawefunction of the 16^{+} isomeric state which is understood at first order as a 4-qp state, upon including the residual couplings with collective modes. This constitutes a very difficult task and we will only be concerned here with the first question.

We have computed static properties of the ^{178}Hf nucleus within a Hartree-Fock plus BCS framework using the SIII Skyrme force whose parameters have been fitted about twenty years ago mainly on saturation properties[2]. For the abnormal density and the abnormal part of the total energy however, we have used a seniority force whose intensity has been taken roughly equal to what was used in a previous study of such rare earth nuclei[3].

The dynamical properties have resulted from the diagonalization of a quantized Bohr hamiltonian whose scalar functions of the quadrupole variables have been microscopically deduced from the same interaction within the so-called ATDHFB framework, as presented in ref. 4. One may add that the static calculations underlying such a dynamical treatment have been performed here within an approximate method[5], relying on self-consistent semi-classical calculations in an Extended Thomas-Fermi approach[6].

The static calculations yield a deformed nucleus (Q=7.3 barn for the charge distribution) exhibiting (see Figure 1 and table 1) on both sides of the chemical potentials $9/2^+$n, $7/2^-$n, $9/2^-$p and $7/2^+$p s.p. states whose main components are on the [624]↑, [514]↓, [514]↑ and [404]↓ asymptotic states. This gives rise to a 2-qp neutron (proton resp.) state at E = 1.046 MeV (1.481 MeV resp.), defined both by $K^\pi=8^-$. Out of those configurations one can build a $K^\pi=16^+$ configuration with E=2.527 MeV. At first sight, the lowest 8^- and the 16^+ calculated states are good candidates for the two isomeric states, the higher 8^- state being related to the second $I^\pi=8^-$ state observed at E=1.479 MeV. In fact, transfer reaction data, as well as the mere fact that there is only one 8^- isomeric state, indicate that the two 8^- states are indeed of a mixed 2-qp neutron and 2-qp proton nature. The 16^+ state quite on the contrary has always been considered as being of a pure 4-qp nature, i.e. built on the two 2-qp states yielding the two previously described 8^- states. Phenomenological calculations reproduce this feature with reasonable fits of their parameters, see e.g. ref. 7. It is quite remarkable that with the rather limited phenomenological flexibility allowed by our using the SIII Skyrme force, we get precisely around the chemical potentials at the right nucleon numbers, the correct single particle states.

Table 1. Decomposition of the relevant single particle states on so-called "asymptotic basis" states, labelled as [N n_z Λ] Σ.

$7/2^-$ neutron		$9/2^+$ neutron		$7/2^+$ proton		$9/2^-$ proton	
[303]↑	-0.116	[404]↑	-0.131	[404]↓	0.857	[505]↓	-0.122
[514]↓	0.848	[615]↓	-0.212	[413]↑	0.140	[514]↑	0.813
[523]↑	0.242	[624]↑	0.804	[624]↓	0.191	[734]↑	0.287
[734]↓	0.243	[844]↑	0.310	[604]↓	-0.433	[714]↑	-0.460
[714]↓	-0.347	[604]↑	0.152	[804]↓	0.106	[934]↑	-0.107
[723]↑	-0.107	[815]↓	0.104				
		[824]↑	-0.374				

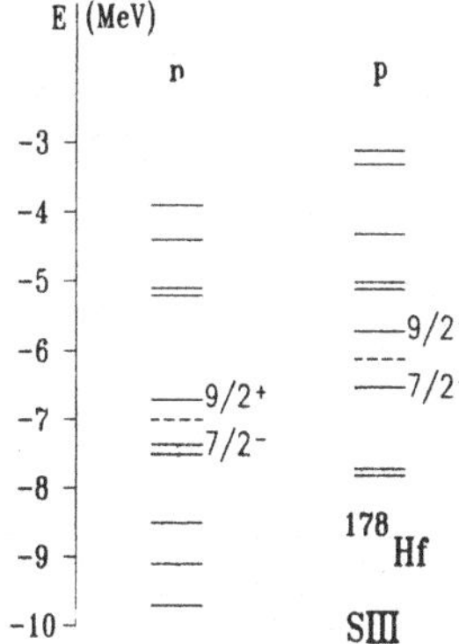

Figure 1. Hartree-Fock neutron and proton spectra near the chemical potentials materialized as dashed lines. The Ω^{π} quantum numbers are also reported. The SIII Skyrme force has been used.

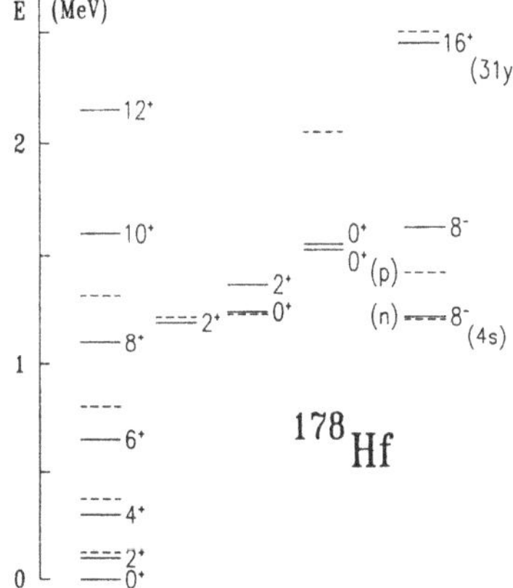

Figure 2. Comparison of experimental (solid lines) and calculated (dashed lines) spectra of ^{178}Hf as calculated with the SIII Skyrme force.

Both the qp-vacuum and the 4-qp states are stable against triaxial and odd-multipole collective deformations. Moreover they exhibit similar energy variations in β and γ around their common equilibrium deformation and are "soft" with respect to odd-multipole modes.

As seen on figure 2, the reproduction of low energy collective properties is amazingly good. Of particular significance is the quality of calculated energies for one β- and γ-phonon vibrational states. Clearly the small underestimation of the moment of inertia for the ground band states should be due to a slightly too high global pairing strength.

The simultaneous description of both individual and collective properties performed here, is to be compared with IBM or more generally shell model calculations where the input of experimental single particle energies is an essential ingredient.

Now, it is interesting to notice that the previous agreement has been obtained with a pairing strength which is significantly too low to reproduce the odd-even binding energy differences generally understood as corresponding to the minimal one quasi particle energies for each charge state (see Table 2). Upon increasing by 10% the pairing strength, one yields such minimal energies which are fairly consistent with the data so interpreted, as shown in Table 2. But in doing so, one completely destroy the agreement previously obtained for the energies of 2-qp , 4-qp and collective states. The latter are then found at a too much high energy.

Table 2. Values of minimal one quasi-particle energies (in MeV) for neutron and proton states for two values of the global pairing strength G1 and G2 related by (G2-G1)/G2 = 10%. The "experimental" figures correspond to odd-even binding energies differences.

	G_1	G_2	"experimental"
E_{qp}^{min} (neutron)	0.520	0.707	0.700
E_{qp}^{min} (proton)	0.746	0.926	0.850

A general discussion on the irrelevant character of odd-even binding energy differences in any reasonable fit of the global pairing strength is currently under completion and will be published elsewhere.

ACKNOWLEDGEMENTS

Interesting discussions with V.G. Soloviev, Ch. Briançon and R Kulessa are gratefully acknowledged. Two of us (P. Q. and J. L.) would like to thank the Department of Phyics of the LLNL for the warm hospitality extended to them during their stay in Livermore and for the access to excellent computing facilities.

(*) Supported in part by the U.S.D.O.E. under Contract # W-7405-ENG-48

REFERENCES

1. Y. Oganessian, S.A. Karamian, Y.P. Gangrski, B. Gorski, B.N. Markov, Z. Szeglowski, Ch. Briançon, D. Ledu, R. Meunier, M. Hussonois, O. Constantinescu and M.I. Subotin, J. Phys. G in press.

2. M. Beiner, H. Flocard, Nguyen Van Giai and P. Quentin, Nuc. Phys, A238:29 (1975).

3. H. Flocard, P. Quentin and D. Vautherin, Phys. Lett. 46B:304 (1973).

4. I. Deloncle, J. Libert, L. Bennour and P. Quentin, Phys. Lett. 233B:16 (1989).

5. K. Bencheikh, I. Deloncle, M. Brack, J. Libert, J. Meyer and P. Quentin, in preparation.

6. M. Brack, C. Guet and H.-B. Håkansson, Phys. Rep. 123:275 (1985).

7. V.G. Soloviev and A.V. Sushkov, J. Phys. G16:L57 (1990).

DIABATIC EFFECTS AND SHAPE COEXISTENCE IN NEUTRON DEFICIENT Pb ISOTOPES

N. Tajima,[1,2] P. Bonche,[3] J. Dobaczewski,[4] H. Flocard,[1] and P.-H. Heenen[5]

[1]Division de Physique Théorique, Institut de Physique Nucléaire, 91406 Orsay, France
[2]Institute of Physics, University of Tokyo, Komaba, Meguroku, Tokyo 153, Japan
[3]Service de Physique Théorique, CEN Saclay, 91191 Gif-sur-Yvette Cedex, France
[4]Institute of Theoretical Physics, Warsaw University, Hoża 69, PL 00681 Warsaw, Poland
[5]Service de Physique Nucléaire Théorique, U.L.B., CP229, 1050 Brussels, Belgium

1. INTRODUCTION

In neutron-deficient lead isotopes, one observes low-lying excited 0^+ levels. Their energy decreases with the neutron number and becomes lower than 1 MeV for $A \leq 194$ isotopes [1]. These states can be regarded as oblate shape isomers while the ground states are spherical [2]. This interpretation, in terms of shape coexistence, is supported by the existence of a rotational band built on top of the isomer [1] and by the observed E0 transitions to the ground states [3] whose values are consistent with the calculated deformations.

The potential energy surfaces (PES) of these nuclei have several minima along the axially symmetric deformation path. In the static mean-field approach, the lowest minimum which is spherical corresponds to the ground state while the oblate minimum corresponds to the isomer. However, this oblate minimum, which one would like to assign to the excited 0^+ band, is so shallow that quantum fluctuations may prevent the localization of a collective wavefunction in its vicinity.

A mechanism involving diabatic effects has sometimes been invoked to resolve this apparent discrepancy [2]. Most PES are calculated within the adiabatic approximation, in which one assumes that, along the collective path, all the degrees of freedom but the collective ones are relaxed. However, the validity of this assumption has never been tested in situations involving single-particle level crossings. At such crossings the collective motion may not stay on the adiabatic path which requires an abrupt change of the occupation probabilities of the crossing orbitals as a function of the collective coordinate. Instead, it may choose to follow the diabatic path for which the occupation probability remains the same before and after the crossing.

The neutron-deficient lead isotopes appear to provide a good test case of the adiabatic hypothesis. The HFBCS solutions near the oblate minimum have a different configuration from those in the spherical potential well. The $(1h_{9/2},|\Omega|=9/2)$ proton orbital is occupied at the oblate minimum while it is empty in the spherical configuration. Because its density is concentrated in the equatorial region of the nuclear surface, this orbital induces strong core-polarization which favors oblate shapes. For this reason, it is often called a deformation-driving orbital.

Using the adiabatic 0qp (quasiparticle) and 2qp PES, Bengtsson and Nazarewicz have proposed a method to define a diabatic potential energy surface in which the occupation probability of the orbital is frozen (0 or 1). They have found that the minimum energy of this obtained diabatic potential at oblate deformation is consistent with the experimental excitation energy of

Nuclear Shapes and Nuclear Structure at Low Excitation Energies
Edited by M. Vergnes *et al.*, Plenum Press, New York, 1992

the isomer. On the other hand, they did not examine the dynamical effects which may still favor the adiabatic path. In this work we investigate such effects in the framework of the Generator Coordinate Method (GCM) including 2qp excitations.

There have been earlier attempts to include single-particle degrees of freedom into the GCM [4,5]. However, they could not analyze the adiabaticity of collective motion in a region of level crossings because the model space was too restricted and they concentrated instead on a study of the pairing vibration in some $A \leq 74$ nuclei.

2. THE GCM BASIS

We use the solutions of Hartree-Fock plus BCS (HFBCS) equation constrained on quadrupole motion as the basis for our GCM calculations. The SkM* Skyrme interaction is employed for the mean-field part of the hamiltonian and a schematic pairing interaction (seniority force) is adopted for the pairing part. Details and parameters of the model are described in ref. [6]. Following ref. [2], we consider only axially symmetric deformations. We choose the average value of the z-component of the mass quadrupole moment $\langle Q_z \rangle$ as the generator coordinate q.

Collective dynamics depends strongly on pairing correlations. This is particularly the case for the problem we are interested in, in which one investigates the possible admixture of 2qp states into the low-lying collective states. Indeed diabatic coupling should decrease when the pairing becomes strong. On the other hand, when the pairing gap is small or vanishes as is the case at shell closure (protons in near spherical shape), one expects dynamical pairing effects to be important. We have treated this effects with the Lipkin-Nogami (LN) prescription[7].

In this work, the GCM basis includes BCS states. In order to keep the average number of protons (Z) and neutrons (N) in the GCM collective states at the correct values, we have replaced the hamiltonian $\hat{H}$ by the routhian $\hat{R}$

$$\hat{R} = \hat{H} - \lambda_{\rm N}(\hat{N} - N) - \lambda_{\rm P}(\hat{Z} - Z), \tag{1}$$

in the Hill-Wheeler equations. In an improved calculation in the future, we will use wavefunctions obtained by projecting BCS states on the correct nucleon numbers.

3. EVALUATION OF GCM KERNELS BETWEEN 0QP AND 2QP STATES

A technical problem appears when the GCM basis includes both ground states and quasi-particle excited states. Since some of the overlaps between 0qp states and 2qp states vanish, one cannot evaluate the GCM kernels with the generalized Wick theorem[8,9]. Indeed this theorem provides only formulae for the ratios of the GCM matrix elements over the overlap.

Given a collective coordinate q, let us consider a set of time-even BCS states $|q\rangle$ which, as is the case for the lowest constrained HFBCS states, are such that all the overlaps $\langle q'|q\rangle$ are non zero. The 2qp excited states associated with an orbital i are $|i,q\rangle = \alpha_i^+(q)\alpha_{\bar{\imath}}^+(q)|q\rangle$, where $\alpha_i^+(q)$ and $\alpha_{\bar{\imath}}^+(q)$ denote the quasiparticle creation operators of the orbital i at the deformation q.

To circumvent the difficulty of computing matrix elements between orthonormal states as for instance $|q\rangle$ and $|i,q\rangle$, we introduce the following family of BCS states,

$$|q,\theta_i\rangle = \cos\theta_i\,|q\rangle + \sin\theta_i\,|i,q\rangle \quad , \tag{2}$$

which depends on the angle θ_i. In terms of the BCS coefficients $u_i(q)$ and $v_i(q)$ of the orbital i in $|q\rangle$, the coefficients $u_i(q,\theta)$ and $v_i(q,\theta)$ for the state $|q,\theta\rangle$ are given by the relations:

$$\begin{aligned} u_i(q,\theta_i) &= \cos\theta_i\, u_i(q) - \sin\theta_i\, v_i(q) \\ v_i(q,\theta_i) &= \sin\theta_i\, u_i(q) + \cos\theta_i\, v_i(q) \quad . \end{aligned} \tag{3}$$

Since in the rest of this letter, we will consider only 2qp states associated with a single pair $\{i,\bar{\imath}\}$, we drop the index i on the rotation angle in order to shorten the notations.

Using eq.(2), states $\{|q\rangle, |i,q\rangle\}$ can be expressed as linear combinations of $\{|q,\theta_1\rangle, |q,\theta_2\rangle\}$ $(\theta_1 \neq \theta_2)$:

$$(|q\rangle, |i,q\rangle) = (|q,\theta_1\rangle, |q,\theta_2\rangle)T\,, \qquad T = \frac{1}{\sin(\theta_2-\theta_1)}\begin{pmatrix} \sin\theta_2 & -\cos\theta_2 \\ -\sin\theta_1 & \cos\theta_1 \end{pmatrix}. \tag{4}$$

Now, if one selects values of θ_1 and θ_2 small enough to ensure that none of the overlaps between the basis states $\{|q,\theta_1\rangle, |q,\theta_2\rangle, |q',\theta_1\rangle, |q',\theta_2\rangle\}$ vanish, one can calculate all GCM kernels by means of the generalized Wick's theorem. Then, the kernels in the basis $\{|q\rangle, |i,q\rangle, |q'\rangle, |i,q'\rangle\}$ are obtained by a linear transformation:

$$\begin{pmatrix} \langle q'|\hat{O}|q\rangle & \langle q'|\hat{O}|i,q\rangle \\ \langle i,q'|\hat{O}|q\rangle & \langle i,q'|\hat{O}|i,q\rangle \end{pmatrix} = T^\dagger \begin{pmatrix} \langle q',\theta_1|\hat{O}|q,\theta_1\rangle & \langle q',\theta_1|\hat{O}|q,\theta_2\rangle \\ \langle q',\theta_2|\hat{O}|q,\theta_1\rangle & \langle q',\theta_2|\hat{O}|q,\theta_2\rangle \end{pmatrix} T. \tag{5}$$

Two requirements help us select an appropriate value for θ's. First, this angle should be small enough so that the overlaps like $\langle q',\theta_1|q,\theta_2\rangle$ do not vanish. Second, it should be sufficiently large to ensure an accurate numerical evaluation of the differences occurring in the transformation of GCM matrix elements (5). Let us note that the transformation (5) also gives the correct phases for the overlap kernels.

In this work, we choose $\theta_1=0$ and for reasons of simplicity and flexibility of our GCM calculations we use a q-independent value for θ_2. The accuracy on the values of the hamiltonian kernels obtained by the transformation (5) is about 10^{-1} keV for $\theta_2=0.2°$. We have found it sufficient for the present study. One can in fact achieve a better accuracy with a symmetric algorithm $\theta_1=-\theta$, $\theta_2=\theta$ and much larger values of θ.

4. MEAN-FIELD AND PAIRING CONTRIBUTIONS TO THE GCM KERNELS

There is another advantage in our prescription. In many HFBCS calculations, such as ours, different interactions are used for calculating the mean-field and the pairing contributions to the energy. The total binding energy is then defined as the sum of the mean-field part of a Skyrme type interaction $\hat{H}_{MF}$ and the pairing-field part of a seniority-force interaction $\hat{H}_{\rm P}$. Let us examine how this separation is done for the case of $\hat{H}_{\rm P}$,

$$\hat{H}_{\rm P} = -G \sum_{jk>0} a_j^\dagger a_{\bar{j}}^\dagger b_{\bar{k}} b_k. \tag{6}$$

In the above definition, $a_j^\dagger$ (b_k) is the creation (annihilation) operator for the lefthand (righthand) state $|{\rm L}\rangle$ ($|{\rm R}\rangle$). We define the matrix element of $\hat{H}_{\rm P}$ as,

$$\langle {\rm L}|\hat{H}_{\rm P}|{\rm R}\rangle = \langle {\rm L}|H_{\rm P}|{\rm R}\rangle_{\rm MF} + \langle {\rm L}|H_{\rm P}|{\rm R}\rangle_{\rm P}, \tag{7}$$

$$\langle {\rm L}|\hat{H}_{\rm P}|{\rm R}\rangle_{\rm MF} = -G\langle {\rm L}|{\rm R}\rangle \sum_{jk>0} \{(a_j^\dagger b_k)(a_{\bar{j}}^\dagger b_{\bar{k}}) - (a_j^\dagger b_{\bar{k}})(a_{\bar{j}}^\dagger b_k)\}, \tag{8}$$

$$\langle {\rm L}|\hat{H}_{\rm P}|{\rm R}\rangle_{\rm P} = -G\langle {\rm L}|{\rm R}\rangle \sum_{jk>0} (a_j^\dagger a_{\bar{j}}^\dagger)(b_{\bar{k}} b_k), \tag{9}$$

where the notation $(\hat{O}) = \langle {\rm L}|\hat{O}|{\rm R}\rangle/\langle {\rm L}|{\rm R}\rangle$ denotes the generalized elementary Wick contractions. Loosely speaking, when $|{\rm L}\rangle = |{\rm R}\rangle$, $\langle {\rm L}|\hat{H}_{\rm P}|{\rm R}\rangle_{\rm MF}$ includes the terms proportional to products v^2, while those proportional to uv's are included in $\langle {\rm L}|\hat{H}_{\rm P}|{\rm R}\rangle_{\rm P}$. A straightforward application of eqs.(8) and (9) is not feasible, however, for the matrix elements between 0qp states and 2qp states. Since the transformation from $|q\rangle$ to $|i,q\rangle$ involves an exchange of the pair (u_i,v_i) into $(-v_i,u_i)$, the distinction between uv-like and v^2-like terms is no longer meaningful for the orbital i. Moreover, when $\langle {\rm L}|{\rm R}\rangle$ goes to zero ($|{\rm L}\rangle = |q,\theta\rangle, |{\rm R}\rangle = |q,0\rangle, \theta \to 90°$), each of the terms, $\langle {\rm L}|\hat{H}_{\rm P}|{\rm R}\rangle_{\rm MF}$ and $\langle {\rm L}|\hat{H}_{\rm P}|{\rm R}\rangle_{\rm P}$, diverges like $\langle {\rm L}|{\rm R}\rangle^{-1}$ while the sum $\langle {\rm L}|\hat{H}_{\rm P}|{\rm R}\rangle$ remains finite. The divergences are due to the terms of $j=k=i$ in the righthand sides of eqs. (8) and (9). The vanishing overlap $\langle {\rm L}|{\rm R}\rangle$ does not cancel the double divergence of the sums which is due to a divergence of each contraction.

This problem is circumvented by applying the transformation (5) for the partial sums of contractions ($\langle {\rm L}|\hat{H}_{\rm P}|{\rm R}\rangle_{\rm MF}$ or $\langle {\rm L}|\hat{H}_{\rm P}|{\rm R}\rangle_{\rm P}$) and then transforming them as if they were the matrix elements of an operator. In principle, the result depends now on the values of θ which are chosen. To suppress this ambiguity, we choose to define the pairing matrix elements as the derivative of the expression (9) at $\theta = 0$ by considering the limit $\theta_1,\theta_2 \to 0$. Generally speaking, we can define the pairing-field part of the matrix elements of an operator $\hat{O}$ between a n quasiparticle-pair state $|{\rm R}\rangle=|i_1,\cdots,i_n,\ q\rangle$ and a n' quasiparticle-pair state $|{\rm L}\rangle=|j_1,\cdots,j_{n'},\ q'\rangle$ as,

$$\langle \mathrm{L}|\hat{O}|\mathrm{R}\rangle_{\mathrm{P}} = \prod_{p'=1}^{n'} \frac{\partial}{\partial\theta^{\mathrm{L}}_{j_{p'}}} \prod_{p=1}^{n} \frac{\partial}{\partial\theta^{\mathrm{R}}_{i_p}} \langle q',\, \theta^{\mathrm{L}}_{j_1},\cdots,\theta^{\mathrm{L}}_{j_{n'}}|\hat{O}|q,\, \theta^{\mathrm{R}}_{i_1},\cdots,\theta^{\mathrm{R}}_{i_n}\rangle_{\mathrm{P}} \Bigg|_{\substack{\theta^{\mathrm{L}}_{j_1} = \cdots = \theta^{\mathrm{L}}_{j_{n'}} = 0 \\ \theta^{\mathrm{R}}_{i_1} = \cdots = \theta^{\mathrm{R}}_{i_n} = 0}} , \quad (10)$$

where $\langle q',\, \theta^{\mathrm{L}}_{j_1},\cdots,\theta^{\mathrm{L}}_{j_{n'}}|\hat{O}|q,\, \theta^{\mathrm{R}}_{i_1},\cdots,\theta^{\mathrm{R}}_{i_n}\rangle_{\mathrm{P}}$ is defined by eq. (9) for small θ's. The mean-field part $\langle \mathrm{L}|\hat{O}|\mathrm{R}\rangle_{\mathrm{MF}}$ is defined similarly.

To calculate the mean-field and pairing GCM kernels $\langle i, q'|\hat{O}|i, q\rangle_{\mathrm{MF\ or\ P}}$ we also use (5) or (10). Since the 2qp states are not orthonormal to each other, a direct evaluation with eq. (8) or (9) is also possible. The results are rather similar. For instance, we find that the values of $\langle i, q=-16\mathrm{b}|\hat{H}|i, q=-12\mathrm{b}\rangle$ in ^{186}Pb calculated with the two methods differ by 300 keV.

5. RESULTS OF GCM CALCULATIONS

In the upper part of fig.1, we show the sections of the PES along the axial quadrupole path for ^{186}Pb and ^{190}Pb. The solid curve with open circles gives the expectation value of $\hat{H}$ in the BCS states obtained with the LN prescription for mass quadrupole moments ranging from -32 barn (b) to 96b [1]. Note that these curves do not include the correction terms obtained by the LN prescription which are the approximations to the energy gain due to the projection on the correct numbers of nucleons. The dashed line expresses the PES including the correction. Compared to the result of Nilsson-Strutinsky method (NS) [2], our PES is stiffer (softer) for oblate (prolate) deformations. While in the NS method the minimum is always at the spherical point, it is located at 20b for ^{186}Pb and at -12b for ^{190}Pb. These conclusions may be modified once the effects of number projection are taken into account.

The solid curve with filled circles gives the energy of 2qp states obtained by breaking the $1h_{9/2}$, $|\Omega|=9/2$ proton pair. They are calculated with eq. (5) using values of $\theta_1=0$ and θ_2 equal to 0.2° for ^{186}Pb and 2.0° for ^{190}Pb. For both nuclei, the excitation energy of the minimum of the 2qp curves is $\sim$3.5MeV. This is consistent with the value of the proton pairing gap around the crossing $\Delta^{\mathrm{P}} = \sim$1.5MeV.

The solid curve with open squares shows the result of a self-consistent calculation of the energy of BCS states in which a pair has been broken in the same orbital. These states are obtained by solving HFBCS (plus LN) equations within a variational space of wavefunctions with the following form,

$$|\Psi\rangle = (-v_i + u_i a_i^\dagger a_{\bar{i}}^\dagger) \prod_{j\neq i} (u_j + v_j a_j^\dagger a_{\bar{j}}^\dagger)|0\rangle. \quad (11)$$

When the values for all the u's and v's are kept positive, the BCS wave-function (11) always describes a 2qp state. With this method, we can take into account the effects of the core polarization and of the reduction of pairing correlations due to the 2qp excitation. We find that the excitation energy is reduced to 2.2 MeV at the crossing. In future work, we plan to use these wavefunctions to determine how core polarization and pairing reduction in the 2qp states influence the adiabaticity of collective motion.

For both nuclei, we have performed GCM calculations with the wavefunctions corresponding to the points marked with open and filled circles in the upper part of fig.1. [2]

In the upper part of the fig.1, the crosses indicate the energy (corrected according to eq. (1)) and the mass quadrupole moment of the lowest GCM states. The collective wavefunction of some GCM states are shown in the lower part of fig.1. These give the amplitudes in the natural basis [10]. The components on the 0qp states are shown with open circles connected with a solid line while those on the 2qp states are given by filled circles connected with dashed lines. The GCM ground states contain very little admixture of 2qp states. The excitation energy of the first excited state is 0.356MeV (0.413MeV) for ^{186}Pb (^{190}Pb). However, the properties of these

[1] For small deformations, a quadrupole deformation parameter can be obtained from the quadrupole moment with the relation: $\beta_2 \simeq \langle Q_z\rangle/65$b.

[2] Since we use the linear transformation (5) for hamiltonian kernels, the result should be the same in principle whether we use the basis $\{|q, 0°\rangle, |q, 90°\rangle\}$ or the basis before the transformation $\{|q, 0°\rangle, |q, 0.2°\rangle\}$, because both of them span the same Hilbert space. In practice, however, the former basis has more merits than the latter in truncating small-norm states and in interpreting the wavefunctions.

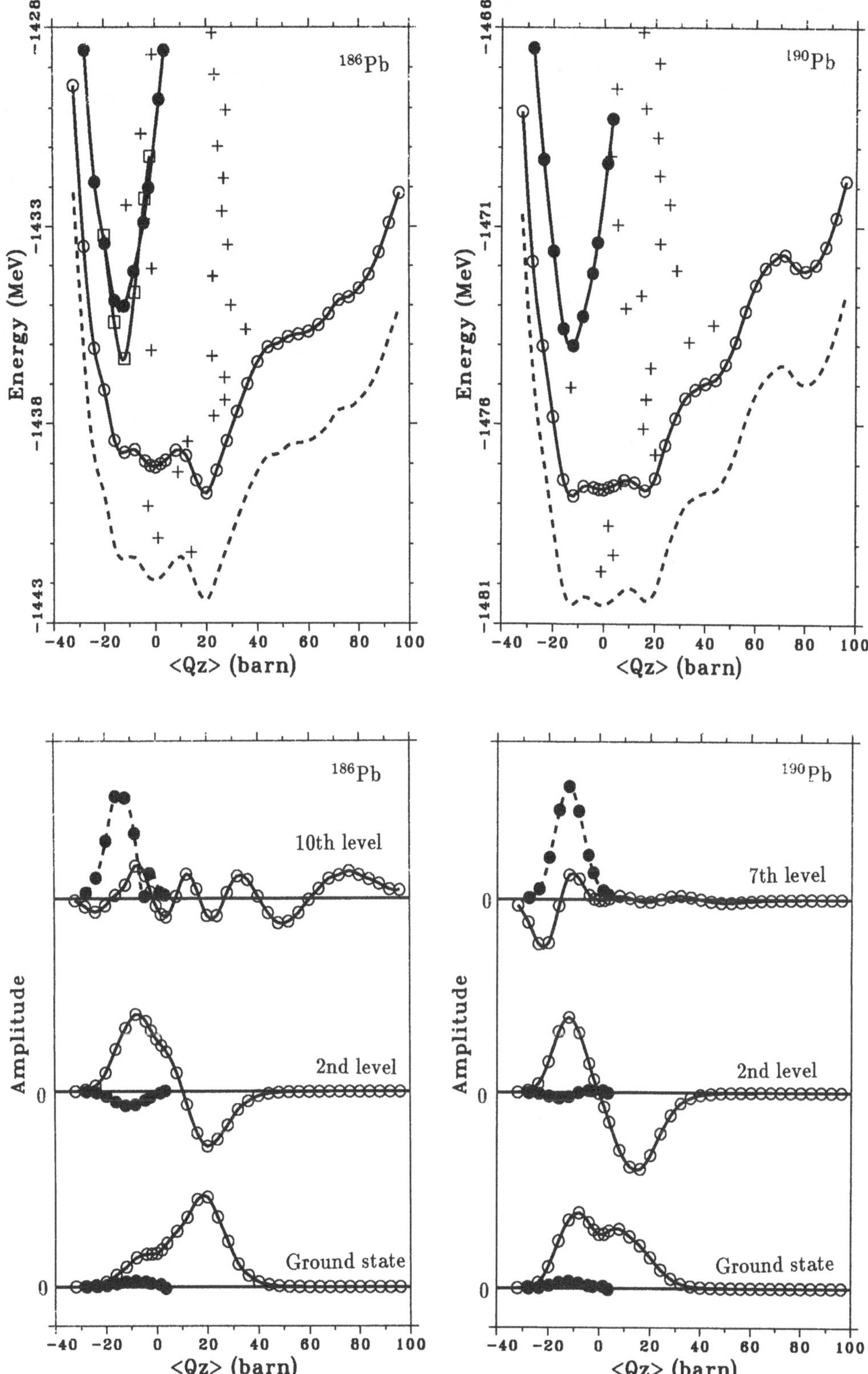

Fig. 1 . Potential energy curve of ^{186}Pb and ^{190}Pb and collective wavefunctions of three GCM eigenstates. See explanations in text.

states are almost not affected by the 2qp coupling. On the other hand, the 10th (7th) GCM level for ^{186}Pb (^{190}Pb) at E^*=5.073MeV (4.601MeV) has a large component in 2qp basis. The wave function is more localized for ^{190}Pb than for ^{186}Pb. This localization can be understood as a trapping of the collective motion within the diabatic potential centered around the oblate minimum which one obtains by keeping the occupation number of the $1h_{9/2}$ $\Omega=9/2$ proton orbital at 1. It does not appear to be trapped on the sole 2qp adiabatic potential considering the large mixing of the 0qp components. In any case, these localized states are too excited to correspond to the observed intruder 0^+ states.

6. CONCLUSIONS

We have developed a prescription which permits the calculation of GCM kernels on a basis including 0qp states and 2qp excited states. We showed that it is always possible to calculate the kernels by means of the generalized Wick theorem although in principle this theorem cannot be used straightforwardly for matrix elements between orthonormal states such as a 0qp and a 2qp state. Our method also provides a convenient way to evaluate separately the mean-field and the pairing part of the matrix element of two-body operators. We can thus generalize to GCM kernels the separation between mean-field and pairing contributions which is used in many HF plus BCS calculations.

We have performed GCM calculations including 2qp excitations for neutron deficient lead isotopes (A=186,190). The shapes of the potential energy curves calculated by HFBCS including the Lipkin-Nogami prescription are different from those obtained with the Nilsson-Strutinsky method. In the present calculation, performed without adjustments of the parameters of the effective interaction, we do not find a low excitation collective wavefunction localized in the diabatic potential.

In the future, this work will be extended in the following directions.

a) Investigation of core-polarization effects with a GCM basis including self-consistent 2qp states. The diabatic coupling may be enhanced since, at least, we have found that such states are less excited than the perturbative 2qp states.

b) Study of the effects of projection of the GCM basis states on the correct nucleon numbers.

c) Investigation of the adequacy of an axial GCM basis by a study with enlarged calculation involving a triaxial quadrupole degree of freedom.

It is also important to study the sensitivity of our results on the the pairing strength and to determine the origin of the difference between our deformation energy curve and those obtained with the Nilsson-Strutinsky approach.

Acknowledgement

The authors would like to thank the CCVR for extended computing facilities. One of the authors (N.T.) was supported in part by the Grant-in-Aid for Scientific Research (No.02740136) from the Ministry of Education, Science and Culture of Japan.

References

[1] P.van Duppen, E.Coenen, K.Deneffe, M.Huyse, K.Heyde and P.van Isacker, Phys.Rev.Lett. **52** (1984) 1974.

[2] R. Bengtsson and W. Nazarewicz, Z.Phys. **A334** (1989) 269.

[3] D. Dendooven, et al., Phys. Lett. **B226** (1989) 27.

[4] M. Didong, H. Müther, K. Goeke, and A. Faessler, Phys. Rev. **C14** (1976) 1189.

[5] H. Müther, K. Goeke, K. Allaart, and A. Faessler, Phys. Rev. **C15** (1977) 1467.

[6] P. Bonche, H. Flocard, P.H. Heenen, S.J. Krieger and M.S. Weiss, Nucl.Phys. **A443** (1985) 39.

[7] Y. Nogami, Phys. Rev. **134** (1964) B313.

[8] N. Onishi and S. Yoshida, Nucl. Phys. **80** (1966) 367.

[9] R. Balian and E. Brezin, Nuo. Cim. **64B** (1969) 37.

[10] P. Ring and P. Schuck, The nuclear many-body problem (Springer, New York, 1980)

[11] P. Bonche, J. Dobaczewski, H. Flocard, P.H. Heenen and J. Meyer, Nucl. Phys. **A510** (1990) 466.

IV - MULTIPHONON STATES

LOW-ENERGY MULTIPHONON STATES IN DEFORMED NUCLEI

Robert Piepenbring

Institut des Sciences Nucléaires
Laboratoire mixte IN2P3-CNRS, UJFG
53 Avenue des Martyrs
38026 Grenoble-Cedex, France

A discussion of the low-energy multiphonon states in deformed nuclei is made based on theoretical results obtained by using the multiphonon method. Comparisons with experimental data show that collective two phonon states exist in deformed nuclei and that the multiphonon method can be used as a dynamical approach to octupole correlations.

1. INTRODUCTION

This paper gives a summary of some major results concerning the low-energy multiphonon states in deformed nuclei, achieved during the last ten years. Obviously, it is not possible to cover all important features in a short paper. Several subjects are only briefly mentioned or not discussed at all. Among these are the detailed comparisons of the results obtained in the framework of the multiphonon method (MPM) with those obtained by other theoretical approaches.

The first part of the paper gives a summary of the MPM for even-even[1] as well as odd-mass nuclei[2]. Since this theory has already been published, some details are omitted. They can be found in the appropriate references. The second part of the paper contains a summary of the results dealing with the problem of the existence of collective two phonon states in even-even nuclei. It is mainly concentrated on the rare-earth mass region, since this is the mass region which is best investigated experimentally and where the application of the MPM is best justified. The third part deals with the results of the application of the MPM to nuclei where octupole correlations are known to be important. Some representative examples are given.

2. SUMMARY OF THE MPM

It is well known that large-basis shell-model calculations in heavy nuclei are practically intractable. All model calculations need therefore to truncate the space used. One way to do this is to take advantage of the collective nature of the non rotational states which appear at low energy in even-even deformed nuclei and to introduce new entities: the phonons $Q_i^\dagger$. These are defined as a superposition of two quasiparticles

$$Q_i^\dagger = \frac{1}{2} \sum_{\mu\nu} (X_i)_{\mu\nu} \, \alpha_\mu^\dagger \, \alpha_\nu^\dagger \tag{1}$$

where

$(X_i)_{\mu\nu} = - (X_i)_{\nu\mu}$ is chosen to be an antisymmetric matrix, and $\alpha_\mu^\dagger$ are creation operators of fermions. Note that these phonons cannot be considered as bosons since their commutation rules are

$$[Q_1, Q_2^\dagger] = - \frac{1}{2} \mathrm{Tr}\,(X_1 X_2) + \sum_{\mu\nu} (X_1 X_2)_{\mu\nu} \, \alpha_\nu^\dagger \, \alpha_\mu \tag{2}$$

In the multiphonon method one selects the collective phonons $Q_i^\dagger$ which are supposed to play an important role at low-energy and builds the MPM states

$$| \mathbf{k}_r > = | k_1, k_2, ... k_r > = \frac{(Q_1^\dagger)^{k1}}{k_1!} \, \frac{(Q_2^\dagger)^{k2}}{k_2\,!} \, \frac{(Q_r^\dagger)^{kr}}{k_r!} \, | 0 > \tag{3}$$

where $| 0 >$ is the fermion ***and*** phonon vacuum, and where the phonons (1) are piled up k_i times, taking the Pauli principle into account through (2). This set (3) of MPM states does not, in general, form an orthogonal basis. In the forthcoming discussion we shall however use the term "basis" for the set (3) in even-even nuclei. (In eq. (3) the k! are introduced for numerical convenience). In odd-mass nuclei the multiphonon states are written

$$| \mathbf{k}_r , \lambda > = \alpha_\lambda^\dagger \; | \mathbf{k}_r > \tag{4}$$

(Note that the operators α and $Q^\dagger$ do not commute). As for the even case, the set (4) does not in general form a "basis" in the mathematical sense.

The main problem of the MPM is the determination of the overlap matrix of the multiphonon states (3) and (4) and the calculation of the matrix elements of the model Hamiltonian in the (collective) subspace spanned by these states. Because of the Pauli principle, this calculation is by no means easy. In ref.[1] two methods have been indicated to carry out this task in the case of an even system of fermions (hereafter referred to as the even case). The first method appears to be a generalization of the Wick's theorem for phonons and defines "new contractions". It is especially suited for studying matrix elements of multiphonon states (3) with a few phonons of many different types. The second method is a recursion formulation of the problem which is more easily handled in the case of numerous phonons of the same type or when only a few (e.g. two) types of phonons are involved. In applications to realistic nuclei the second method has been found to be more useful. For this reason it has been extended[2] to an odd system of fermions (hereafter referred to as the odd case).

Let us note by P_{ij} a product which contains i creation operators $\alpha^\dagger$ and j annihilation operators α. After the Bogoluybov-Valatin transformation, the model Hamiltonian contains terms with P_{11}, P_{22}, P_{31}, P_{13}, P_{40} and P_{04}.

We now introduce the matrix elements $A^{ij}(\mathbf{k'}_r;\mathbf{k}_r)$ of operators P_{ij} between states (3) in the even case.

$$A^{ij}(k'_r;k_r) = < k'_r \mid P_{ij} \mid k_r > \tag{5}$$

For instance :

$$A^{02}_{\mu\nu}(k'_r;k_r) = < k'_r \mid \alpha_\nu \, \alpha_\mu \mid k_r > = < k_r \mid \alpha^\dagger_\mu \, \alpha^\dagger_\nu \mid k'_r > = A^{20}_{\mu\nu}(k_r;k'_r)$$

Similarly, in the odd case, we introduce the notation

$$T^{ij}(k'_r \, \lambda';k_r \, \lambda > = <k'_r \, \lambda' \mid P_{ij} \mid k_r \, \lambda > \tag{6}$$

It has been shown in[1], that A^{11} can be deduced from A^{20}; A^{31} from A^{40}; and A^{22} from A^{20} and A^{40}.

As a consequence, the problem, in the even case, is mainly reduced to the determination of A^{00}, $A^{20}_{\mu\nu}$ and $A^{40}_{\mu\nu\rho\sigma}$, related by some rather sophisticated recursion formula.

In the odd case, the situation is quite similar but slightly more involved as demonstrated in ref.[2]. The overlap matrix T^{00} can be deduced from A^{00} and A^{11}, i.e. from A^{00} and A^{20}; the one body matrix elements T^{20} from A^{20} and A^{31}, i.e. from A^{20} and A^{40}; the one body matrix elements T^{11} from A^{00}, $A^{11}$and A^{22}, i.e. from A^{00}, A^{20} and A^{40}. More generally, the two body matrix elements T^{ij} can be deduced from A^{ij} and $A^{i\pm1,j\pm1}$ which can all be expressed in terms of A^{00}, A^{20}, A^{40} and A^{60}. As a consequence, in the odd case, the problem is mainly to determine $A^{60}_{\lambda\mu\rho\sigma\tau}$ in addition to what is needed in the even case. Here also some sophisticated recursion formulae have been established[2].

Once the model Hamiltonian H has been chosen, the different steps of the MPM go as follows :

- Using some physical arguments, one has to select the MPM "basis", i.e. choose the phonons which are expected to play a major role in the collective subspace. In the odd-case, one has to make the additional choice of the active quasiparticle states. In this step one generally retains the phonons, which lead to the lowest lying collective non rotational states (e.g. β, γ or octupole $K = 0$ phonon), and the lowest lying quasiparticle states in the odd nucleus.
- The second step is to calculate (and to store) the matrix elements A^{00}, A^{20}, A^{40} and, in the odd case, A^{60} in the MPM "basis" (3), using the recursion formulation of the method up to some given total number of phonons

$$n = k_1 + k_2 + k_3 + \dots k_r \tag{7}$$

From these values of A^{i0}, one deduces the matrix elements of the one and two-body operators involved in the model Hamiltonian.

- The next step is the determination of thc matrix of II in the MPM generally non orthogonal basis and its diagonalization.

These steps are repeated for successive increasing values of the total number n of phonons (7). The truncation of the collective space is fixed by the value n_{max} for which one obtains numerical stability of the lowest eigenvalues of H. It is worthwhile to note that the value of n_{max} depends mainly on the collectivity of the building phonons (1), i.e. on the choice of the antisymmetric matrices X_i. In practical situations, where these matrices are taken from the Tamm-Dancoff approximation, the value of n_{max} is of the order of 8 to 12.

To summarize, the multiphonon method is mainly the ***exact*** diagonalization of the ***total*** model Hamiltonian in the space spanned by ***collective*** phonons. It is, to our knowledge, the only method which takes properly and fully (i.e. without any approximation) the Pauli principle into account. Another advantage of the MPM is that the exact diagonalization of the H matrix eliminates all the convergence problems encountered in every perturbative approach (e.g. boson expansion techniques or nuclear field theory). The eigenstates appear as a superposition of the basic multiphonon states. As a consequence there is no restriction to one and two phonon components in the low-lying excited states as in the quasiparticle-phonon nuclear model of Soloviev et al.[3]. It is also interesting to point out that the use of the Tamm Dancoff phonons as building blocks in the MPM avoids the problem related, in the random phase approximation, to the disappearance of the collective root of the secular equation. The fact that the phonon vacuum is chosen to be the same as the quasiparticle vacuum is required if one wants to treat the Pauli principle correctly[1]. It should be added here that this does not at all exclude correlations in the ground state, which, after the diagonalization, appears as a superposition of the multiphonon states with $K^\pi = 0^+$.

Finally, it is necessary to say a few words on the limitations of the MPM encountered in practical applications to realistic situations. They are mainly of numerical order and related to the calculation ***and*** storage of the matrix elements $A^{0i}(\mathbf{k'}_r; \mathbf{k}_r)$. These entities depend :

- on the i indices of the annihilation operators α_μ, where μ labels n_ℓ active quasiparticle states introduced in the BCS calculation (a reasonable choice deals with $n_\ell = 20$, two times degenerate, equally distributed on each side of the Fermi level, for each type of particles)
- on the vectors $\mathbf{k}_r$ and $\mathbf{k'}_r$, which summarize the partitions of the initial and final phonon numbers $n = k_1 + k_2 + k_2 + \dots k_r$

and $$n - \frac{i}{2} = k'_1 + k'_2 + \dots k'_r \qquad (8)$$

where $$\frac{i}{2} \leq n \leq n_{max}$$

To give an idea, if we consider 2 types of phonons (i.e. r = 2), $n_{max} = 8$ and $n_\ell = 20$ active levels, we have 3 x 85 = 255 entities A^{40}. Each of those has $(2 \times 20)^4$ elements. This leads to a total number of $\approx 6.5\ 10^8$ elements. In the same conditions for r, n_{max} and n_ℓ , one gets 4 x 394 = 1576 entities A^{60} with $(2 \times 20)^6$ elements each, i.e.$\approx 6.5\ 10^{12}$ elements.

Even if a lot of these elements vanish according to the conservation of parity and K quantum numbers, their evaluation is very time consuming even on modern large computers.

As a consequence, up to now, the realistic calculations have been done only with r = 2 in the even case, and r = 1 in the odd case. Some representative examples of the results of these applications with these limited values of r are given in the two next sections. We would like to emphasize that the aim of these realistic calculations is therefore limited to the description of the vibrational motions for the degree of freedom corresponding to the case where the first one (or two) collective state(s) in the even nuclei are much lower in energy than any other *intrinsic* excitation. As a consequence of this restricted goal, the model Hamiltonian H can be chosen as simple and realistic as possible. As usual, the residual interaction introduced in H contains a constant pairing force and the first (L = 2 and / or L = 3) terms of the multipole-multipole expansion of a charge independant long range force.

3. ON THE EXISTENCE OF COLLECTIVE TWO PHONON STATES

In this section several applications of the MPM to even-even nuclei are given. They have been selected in order to study the longstanding central problem of the existence of collective two phonon states in deformed nuclei. In these applications r = 2, i.e. the multiphonon "basis" (3) contains two types of phonons $Q_1^\dagger$ and $Q_2^\dagger$.

The first example[4] deals with the γ degree of freedom in some rare earth nuclei, where the γ vibrational states are strongly collective, and well separated from other intrinsic excitations. This application concerns $^{162-164}Dy$ and $^{164-166-168}Er$. The two basic phonons are the γ phonon with K = 2 and its time reversed $\bar{\gamma}$ with K = - 2. The multiphonon states are written

$$| pq > = (p!\, q!)^{-1} Q_1^{\dagger p} Q_2^{\dagger q} \quad | 0 > \tag{9}$$

Note that, in this application, the states (9) are orthogonal, so that the calculation of the overlap matrix < p'q' | p q > is reduced to the determination of the norms of the multiphonon states.

We focus our main attention on the lowest states arising from the γ motion and restrict therefore to states having K = 0, 2 or 4. Accordingly, the basis (9) can be reduced to states where p = q, p = q + 1 and p = q + 2, respectively. It has been found that numerical stability of the energies of the three lowest excited states 2_1, 4_1 and 0_2 is obtained with $n_{max} = 8$. [Note that in the notation K_i, the subindex i labels the ith state with the angular projection K reached within the basis (9) which contains only phonons of the types γ and $\bar{\gamma}$].To reduce as much as possible the number of free parameters in the model Hamiltonian, standard single particle potentials of the Nilsson[5] or Woods-Saxon[6] type have been used with parameters suited for the rare earth region. The pairing strength parameters, which depend on the number n_ℓ of individual levels introduced in the usual BCS calculations, are deduced from the gaps Δ_p and Δ_n evaluated empirically from the experimental masses[7]. To reduce the determination of the A^{40} matrix elements to a tractable numerical problem $n_\ell = 20$ has been chosen. There remains then only ***one free*** dynamical parameter : the strength χ_2 of the multipole-multipole force. It has been adjusted so as to get, ***within the MPM***, the lowest 2^+ energy E (2_1) in the neighbourhood of the intrinsic band-head energy E_{int} of the γ state. [This value of E_{int} can be obtained by subtracting from the observed energy E(γ) the contribution of the rotational motion]. We may also note here that in the evaluation of electromagnetic transitions, effective charges are to be introduced according to the value chosen for n_ℓ.

To discuss the problem of the existence of collective two phonon states, we study the properties of the 4_1^+ and 0_2^+ states. This can be done by looking at the energy ratios

$$R(4) = E\,(4_1^+\,) \,/\, E(2_1^+) \tag{10}$$

and $$R\,(0) = E(0_2^+) \,/\, E(2_1^+) \tag{11}$$

and the E2 intrinsic matrix element ratio

$$Q(K) = \left| \frac{< 2_1^+ \mid E2 \mid K^+ >}{< 0_1^+ \mid E2 \mid 2_1^+ >} \right|^2 \tag{12}$$

which was found practically independant of the effective charges.

In Table 1 the results obtained within the MPM are given for five nuclei of the mass region $162 \leq A \leq 168$.

Table 1

	^{162}Dy	^{164}Dy	^{164}Er	^{166}Er	^{168}Er
R(4)	2.6	2.9	2.7	2.8	2.8
R(0)	3.3	3.9	3.5	3.7	3.5
Q(4)	0.42	0.40	0.37	0.41	0.38
Q(0)	0.036	0.008	0.012	0.014	0.023

Several comments can be made :

The energy ratios R(K) are greater than 2, demonstrating a strong anharmonic character of the γ vibrational motion in these nuclei. The 4_1^+ states present an anharmonicity ratio of the order of 2.6 - 2.9. As a consequence, one can definitely rule out any interpretation of a K = 4^+ with an energy lower than 2 E(γ) as proceeding from the γ degree of freedom. The 0_2^+ states are always predicted higher in energy than their K = 4^+ partner. Note that this finding is at variance with the quasiparticle phonon nuclear model[3].

The results concerning the Q(K) values show that the 4_1^+ states keep some collective character in their decay to the γ band whereas the 0_2^+ do not. On one hand, there is therefore no great chance to identify the 0_2^+ states : in the energy region where they are predicted they cannot be identified in the sea of surrounding non-collective excitations of the nucleus. On the other hand, one should be able to identify the 4_1^+. Successful experimental efforts have been made[8] in that direction for ^{168}Er.

As is well known, the band head of the γ vibration in ^{168}Er has an energy of 821 keV. From the completeness of the low spin energy level spectrum[9] one deduces that the lowest candidate for the 4_1^+ state occurs with a band head at 2055 keV. This implies a major deviation from a harmonic vibration since $E(4_1^+) / E(\gamma) \geq 2.5$, in agreement with the results for R(4) obtained within the MPM. A second much more crucial test is the lifetime measurement of the 4^+ level at 2055 keV. Due to the problem of the effective charge, the only physical reliable quantity, related to E2 transitions, which the MPM can give, is the Q(4) value. If one uses the measured lifetime of the γ state at 821 keV, the value of Q(4) leads to a prediction of a few hundreds of fs for the lifetime τ of the 4_1^+ state. The beautiful measurement[8], also reported by Jan Jolie at this conference, leads to a value $0.24 < \tau < 0.69$ ps. This result leads to definite evidence for the existence of a collective "two phonon state" in ^{168}Er. Note that we use here some quotes, since in the sense of the MPM, the 4_1^+ state is of course not a pure two phonon state. After the diagonalization of the model Hamiltonian, the eigenvector of the 4_1^+ state appears as a superposition of the multiphonon states (9).

Explicitly

$$|4_1^+> = \sum_{i=2}^{n_{max}} c_i\,|\,i,i\text{-}2> = c_2\,|\,20> + c_3\,|\,31> + c_4\,|\,42> + \ldots \tag{13}$$

where c_2 is the predominant component.

Similarly, the correlated ground state eigenfunction is

$$|0_1^+> = \sum_{i=0}^{n_{max}} d_i\,|\,i,i> = d_0\,|\,0\,0> + d_1\,|\,11> + d_2\,|\,22> + \ldots \tag{14}$$

It is worthwhile to note here that, in the MPM description, the 4_1^+ state (13) can decay through the T_{11} part of the Eλ operator (i.e. that which contains $\alpha^\dagger\alpha$) to the correlated ground state. In particular, the isoscalar transition rate B(ISE4) from the ground state to the 4_1^+ level at 2055 keV can be reasonably described[4] in the framework of the MPM. We should also emphasize that, according to (13), the 4_1^+ can decay through the part T_{20} of the Eλ operator (i.e. the part containing $\alpha^\dagger\,\alpha^\dagger$) to any two quasiparticle state. In particular, the observed E1 transition from 4_1^+ to the 4^- state at 1094 keV can be explained in this way.

To be complete, we should, at least briefly, mention here the results of the other theories dealing with the problem of the existence of the two phonon γ states in deformed nuclei. First, the various microscopic theories treating properly the effect of the Pauli principle on nuclear vibrations all agree to say that it is not possible to obtain a compressed energy spectrum (at variance with the sd versions of IBM[10] and the dynamic deformation model of Kumar[11]). Second, the anharmonicity properties of the 4_1^+ state in ^{168}Er have also nicely been predicted in the self-consistent collective-coordinate method by Matsuo and Matsuyanagi[12] and confirmed in the pseudo symplectic model by Castanos et al.[13]. Third, in the Interacting Boson Model, the hexadecapole degree of freedom is needed[14] whereas in the previous mentioned approaches[4,12,13] it is not. Finally, the evidence of the existence of collective two phonon states established through the measurement of the absolute transition rate for the decay of the 4_1^+ vibrational state ^{168}Er definitively rules out the interpretation within the quasiparticle phonon nuclear model[3] which claims that the energy centroid of the collective two phonon states is shifted towards energies larger than 3 MeV, where they loose completely their collective character.

It is of course needless to say that it would be of great interest to measure the lifetime of the 4^+ states which would be good candidates for collective "two phonon states" in the other nuclei mentioned in Table 1 or in other regions[4] (e.g. Ba, Ce, Os) where the γ degree of freedom prevails.

The second example deals with the K = 0 vibrational states in two regions of deformation. In heavy nuclei[15], e.g. in ^{230}Th and $^{232\text{-}234}$U, and in the rare earth[16] nucleus ^{152}Sm. In this case too, r = 2.The multiphonon states are given by eq (9) ; they contain now only phonons with K = 0, p have a positive and q a negative parity. The $K^\pi = 0^-$ phonon is of the octupole type, whereas the $K^\pi = 0^+$ basic phonon is mainly a pairing vibration[17] in the heavy nuclei and a β vibration in ^{152}Sm. The intrinsic K = 0 states are labelled by their parity and by an index indicating the order of their appearance in the energy spectrum. Similarly to the first example, one defines the following anharmonicity ratios

$$R_1 = \frac{E(0_3^+)}{E(0_2^-)} \qquad R_2 = \frac{2E(0_2^-)}{E(0_2^+) + E(0_1^-)} \qquad R_3 = \frac{E(0_4^+)}{E(0_1^-)} \tag{15}$$

where 0_3^+, 0_2^-, and 0_4^+ are the expected "two phonon states", whose main components are respectively $|20\rangle$, $|11\rangle$ and $|02\rangle$. It is also interesting to study the ratios of some specific $E\lambda$ transition matrix elements.

$$Q_1 = \left| \frac{\langle 0_2^+ | E2 | 0_3^+ \rangle}{\langle 0_1^+ | E2 | 0_2^+ \rangle} \right|^2$$

$$Q_2 = \left| \frac{\langle 0_1^- | E1 | 0_4^+ \rangle}{\langle 0_1^+ | E1 | 0_1^- \rangle} \right|^2$$

$$Q_3 = \left| \frac{\langle 0_1^- | E2 | 0_2^- \rangle}{\langle 0_1^+ | E2 | 0_2^+ \rangle} \right|^2 \tag{16}$$

$$Q_4 = \left| \frac{\langle 0_2^+ | E1 | 0_2^- \rangle}{\langle 0_1^+ | E1 | 0_1^- \rangle} \right|^2$$

We remind the reader that in a harmonic situation the energy ratios R_i, Q_1 and Q_2 equal 2, whereas Q_3 and Q_4 equal 1. In Table 2 the results obtained within the MPM are given for the four nuclei mentioned.

Several remarks can be made. From the values of R_3 and Q_2, it clearly appears that the octupole degree of freedom leads to strong anharmonicities. The "two phonon states" keep some collective character in their electric transitions to the "one phonon states". This property should be of some help in finding some experimental evidence for their existence. From the values of R_1 and Q_1, it is seen that the $K^\pi = 0^+$ phonons are nearer to the harmonic situation specially in the case of ^{152}Sm where the β degree of freedom plays a role. Furthermore, a comparison with the results of the MPM with $r = 1$ (i.e. where only the octupole $K^\pi = 0^-$ phonon is introduced), shows that the $K^\pi = 0^+$ mode is only weakly coupled to the octupole mode, i.e. the octupole vibrational mode is only very weakly

Table 2

	^{230}Th	^{232}U	^{234}U	^{152}Sm
R_1	2.3	2.3	2.3	2.1
R_2	2.3	2.3	2.3	2.6
R_3	3.4	2.9	2.5	2.6
Q_1	2.37	2.44	2.16	1.95
Q_2	0.98	1.21	1.35	1.17
Q_3	0.88	0.93	0.92	0.73
Q_4	0.89	0.73	0.77	0.75

influenced by the $K^\pi = 0^+$ mode, whatever the nature of the latter. The 0^-_2 states resulting mainly from "one $K^\pi = 0^+$ phonon and one $K^\pi = 0^-$ phonon" are also of special interest since they decay to both "one phonon states". The systematic experimental search for these states remains also to be undertaken.

The main conclusion of this section is that the anharmonicities found within the MPM for the octupole vibrations $K^\pi = 0^-$ and the γ motion are quite similar. The anharmonicity ratios increase with increasing collectivity of the one phonon state. The situation is intermediate between a harmonic vibrator and a static pear-shaped or triaxially deformed rotator. It corresponds to a large amplitude collective motion which appears near or beyond the critical strength parameter value of the RPA.

The existence of low-lying collective "two phonon states" in such nuclei seems now quite well established from the theoretical as well as from the experimental point of view.

4. ON A DYNAMICAL APPROACH OF OCTUPOLE CORRELATIONS IN SOFT DEFORMED NUCLEI

This section is devoted to several applications of the MPM to nuclei where the octupole correlations are known to be very strong. (e.g. neutron rich Ba, and actinides around $Z \approx 90$ and $N \approx 134$). It has been universally recognized that a deformation of the Y_{30} type is needed to understand the low lying level structure of these nuclei. A longstanding problem is to know whether the strong octupole correlations lead to a stable deformation or to a dynamical deformation (i.e. vibrations).

In the case of a reflection asymmetric shape, with a very high barrier between the shape and its mirror image, one expects a single band with spin sequence 0^+, 1^-, 2^+, 3^-... in even nuclei and rotational bands with doubly degenerate states of same spin and opposite parity in odd-mass nuclei. Such an extreme situation has never been observed experimentally.

What is usually seen at low energy in even-even nuclei, is : two bands of opposite parity one with even spin and parity levels 0^+, 2^+, 4^+... and one with odd spin and parity states 1^-, 3^-, 5^-,.... Similarly, in odd mass nuclei one encounters some couples of rotational bands with the same spin sequence but with opposite parities.

At *low* energy, these observations can be explained[18] either in the framework of a stable octupole deformation $\varepsilon_3 \neq 0$ with a rather low potential barrier, or in the case of a $\varepsilon_3 = 0$ symmetry where the octupole degree of freedom leads to K = 0 vibrations.

Since no experimental evidence for two octupole phonon states with $K = 0^+$ in the energy area around ***twice*** the energy of the one phonon state has been found[19] in light actinides, the explanation in the framework of $\varepsilon_3 \neq 0$ deformation has been favoured by many authors[20].

A first aim of the present section is to show that, assuming a $\varepsilon_3 = 0$ symmetry, the MPM is capable of treating such genuine non linear vibrations which lead to strong anharmonicities. On the one hand, since the K = 0 octupole one phonon state is very low in energy and well separated from all other *intrinsic* excitations, the restriction to r = 1 (i.e. only one type of basic phonon) can be assumed in the basis (3). It has been shown[21] that in this case the recursion formulation of the MPM is rather simple. On the other hand, to obtain the stability of the numerical results one needs to enlarge the basis (3) to $n_{max} = 10$ (for the Ba) and $n_{max} = 12$ (for the light actinides).

This version of the MPM has been applied in two regions of the periodic table : for light even Ra and Th isotopes[21] and for ^{146}Ba in ref.[22]. The most interesting information consists in the location of the first excited 0^+ states, whose energy $E(0_2^+)$ is compared to the observed one whenever these data have been confirmed. Note that the experimental intrinsic energy $E(0_1^-)$ for the first excited state is deduced from the band head energy $E(1^-)$ of the $I = 1^-$ state, which contains some rotational contribution roughly estimated to 10 keV in light Ra or Th nuclei.

In Table 3, the results for three Ra isotopes, ^{226}Th and ^{146}Ba are given. They include also the anharmonicity ratio $R = E(0_2^+) / E(0_1^-)$.

Since the gap parameters cannot be deduced here from the odd-even mass differences we use some standard values for G_p and G_n recommended by Nilsson and Prior[23]. It should be noted that the χ_3 value, fitted to $E(0_1^-)$ in the MPM, is such that the secular equation of the RPA has no real collective solution. The calculated $E(0_2^+)$ energies of the first excited 0^+ state due to the octupole K = 0 degree of freedom, are systematically larger than $2E(0_1^-)$ showing an extremely strong anharmonicity.

We also note that the $E_{MPM}(0_2^+)$ are systematically slightly larger than the measured $E_{exp}(0_2^+)$. According to the philosophy behind the MPM approach, which does not try to force the parameters to provide a best fit to the experimental data, these deviations are not very relevant. Therefore, the MPM calculations suggest that the first excited 0^+ state, at least in ^{222}Ra and ^{224}Ra can practically be assumed to be of pure octupole nature, and still referred to as "two phonon states", despite the large observed anharmonicity. (Note the quotes, which are here even more relevant than in section 2).

One may add a few more comments on ^{146}Ba. The set of pairing parameters used in Table 3 results from an extrapolation of the values suggested by Nilsson and Prior. Their values may therefore not correspond exactly to the physical situation. Therefore, the calculated $E(0_2^+)$ should not be used as a precise location of the "two phonon state" in ^{146}Ba.

Table 3. Calculated and experimental energies, in keV, of the first two-excited levels of octupole nature

	^{222}Ra	^{224}Ra	^{226}Ra	^{226}Th	^{146}Ba
$E_{MPM}(0_1^-)$	232	203	246	221	716
$E_{exp}(1^-)$	242	216	254	230	739
$E_{MPM}(0_2^+)$	926	964	997	939	1730
$E_{exp}(0_2^+)$	914	916	824	805	
R_{MPM}	4.0	4.7	4.0	4.2	2.4

Other sets[22] may lead to $E(0_2^+)$ varying from 1.6 to 2 MeV. It is however clear that the octupole correlations appear weaker in ^{146}Ba than in the light Ra-Th region. This fact may be related to the different opposite parity orbitals responsible for the strong octupole interactions involved in each region.

To summarize the results presented in this part of section 3, we could say that the MPM is able to describe the spectroscopic properties of nuclei in which the octupole correlations are very strong, starting with the $\varepsilon_3 = 0$ reflection symmetry. This result should however not be used, *per se*, as an argument against stable octupole deformation. Models[20] starting with $\varepsilon_3 \neq 0$ lead to similar quality for the description of these kind of nuclei. The theoretical potential energy surfaces obtained in these models present, in these nuclei, only shallow minima for octupole deformed shapes.

One can conclude that the whole difference between the two approaches is not so fundamental. It may even simply be reduced to using different representations.

It seems therefore worthwhile to extend[24] this simple version of the MPM (where r = 1), to the odd-mass neighbouring nuclei and to see if it is also able to furnish an alternative approach to the models starting with $\varepsilon_3 \neq 0$. In this context, special attention will be paid to various arguments usually invoked to assert that the odd-mass nuclei of the region of light actinides have stable octupole deformation. First, it may be useful to remind the reader some of the changes expected, *a priori*, when one switches from even-even to odd mass nuclei. Due to the Pauli principle, the odd particle implies some blocking effects. It can also polarize the core, change the quadrupole ε_2 deformation and affect the pairing calculations. Second, a better knowledge of the single particle potential is required in the description of odd mass nuclei.

It is generally admitted that the single proton states are better determined than the neutron states in the transitional Ra and Th region. We shall therefore restrict the application[25] of the MPM (with r = 1) to odd-Z nuclei $^{223\text{-}225\text{-}227}Ac$, for which some detailed experimental data exist. In the multiphonon states (4) we restrict ourselves to only one doubly degenerate quasiparticle state for each given Ω and parity. More precisely we introduce the orbitals 1/2 + 660, 3/2 + 651 and 5/2 + 642 from the unique parity subshell

$i_{13/2}$ and the negative parity states 1/2 - 530, 3/2 - 532 and 5/2 - 523. It can be shown[24-25] that with this choice, the states (4) are orthogonal so that standard techniques can be used for the diagonalization procedure. The number of single particle states introduced in the BCS calculation is $n_\ell = 30$ and the numerical stability of the lowest eigenstates is obtained with $n_{max} = 12$ as for the even-even neighbouring nuclei. Special attention has been paid to three observables : the existence of parity doublets (PD) bands, the decoupling factors of the K = 1/2 bands and the E1 transitions with $\Delta K = 0$, whose properties have often been used[26] to assert stable octupole deformation.

Small energy splitting of observed PD has often been considered[26] as a signature of stable $\varepsilon_3 \neq 0$ deformation, since in this case the PD arise from a single parity mixed Nilsson orbital. It has been shown[25] that the small observed energy splitting $\Delta E(K) = E(K^+) - E(K^-)$ of the PD can also be obtained within the framework of the MPM based on reflection symmetry. As a consequence, it becomes clear that the existence of these narrow energy separations is *not* typical of a stable octupole deformation, but simply characteristic of strong octupole correlations.

A similar conclusion can be drawn for the experimental values of the decoupling parameters *a*. In the models starting with $\varepsilon_3 \neq 0$, the K = 1/2 bands of opposite parities should exhibit opposite decoupling factors. This property has also been given[26] as a criterium for stable octupole deformation, since it is quite different from what is obtained within a *pure* Nilsson calculation where *no* octupole correlations are taken into account. Here too the MPM calculation[25], starting with $\varepsilon_3 = 0$, but treating the strong octupole correlations, leads to several K = 1/2 states of opposite parities and nearly opposite decoupling factors *a*.

In ^{227}Ac the values of *a* extracted from experiment are :
$4.4 < a < 5.6$ for the $1/2^+$ band
and $-2.5 < a < -1.8$ for the $1/2^-$ band
With a stable $\varepsilon_3 \neq 0$ deformation Sheline et al.[26] obtain $a = \pm 3.1$ for these two bands, and the MPM respectively 6.5 and - 3.3. From these values it is certainly hasardous to conclude which approach is the best ! It is however expected that an enlargement of the basis of diagonalization (4), including for example 1/2 - 541 (which in the Nilsson model has a large positive decoupling factor), may have some non negligible effects on the MPM results, which would probably go in the right direction.

Finally, let us discuss briefly the criterium of the enhancement of the observed E1 transition linking members of the PD bands used in ref.[26] to assert $\varepsilon_3 \neq 0$. Here too, the enhancement is defined relative to a single particle description of the Nilsson type. But very early results in the rare earth region have shown[27] that a small coupling of the odd quasiparticle to the K = 0 octupole vibration of the even core can explain such observed enhancements. The MPM confirms these findings for the nuclei under consideration here.

In Table 4 absolute values of E1 matrix elements with $\Delta K = 0$ between intrinsic states are given for ^{223}Ac. The first column gives their values in a pure single particle Nilsson model, the second column indicates their values between pure quasiparticle states (i.e. when the pairing correlations are introduced). Finally the last column gives the results of the MPM calculation. Compared to a Nilsson calculation, an enhancement of two order of magnitude for the matrix element, i.e. a factor of 10^4 for the transition probability is obtained. It is very important to emphasize the following point : at variance with the pure individual Nilsson model, *with or without* $\varepsilon_3 \neq 0$, *with or without* the introduction of the pairing effect, the MPM allows for a contribution of the neutrons to the E1 transition. This has an important consequence. On the one hand, for the three $^{223-225-227}$Ac, the proton mean field is practically the same and the E1 transition probability calculated within a model with $\varepsilon_3 \neq 0$

Table 4. Absolute values of three E1 matrix elements with $\Delta K = 0$ between single particle, quasiparticle and MPM states in ^{223}Ac

	Nilsson	Nilsson + BCS	MPM
$\vert\langle 1/2^+ \vert E1 \vert 1/2^- \rangle\vert$	0.0115	0.000878	0.136
$\vert\langle 3/2^+ \vert E1 \vert 3/2^- \rangle\vert$	0.00585	0.00134	0.153
$\vert\langle 5/2^+ \vert E1 \vert 5/2^- \rangle\vert$	0.00468	0.00221	0.296

will not differ from one isotope to the next. On the other hand, the difference of the neutron situation from one odd Ac to the other, allows *within the MPM* a modification of the core which is large enough to modify the E1 matrix element by a factor 10 from one isotope to another. This gives a possible explanation of the lowering by two orders of magnitude of the B(E1) between the K = 3/2 parity doublets when one goes from ^{225}Ac to ^{227}Ac.

In conclusion, the analysis of the properties of the three studied observables, clearly shows that the related criteria often used as an evidence for stable octupole deformation are necessary, but not at all sufficient to assert that $\varepsilon_3 \neq 0$. They are simple signatures of strong octupole correlations, and do not furnish any definite test to choose between the two assumptions $\varepsilon_3 = 0$ or $\varepsilon_3 \neq 0$. Here too the difference between the two approaches does not seem to be so essential.

We should like to mention briefly a striking feature obtained in the MPM calculation which may lead to a possible test able to distinguish between the two approaches.

In the MPM energy spectrum one notes the systematic appearance at rather low energy (say below 1 MeV) of two parity doublets for each value of K, i.e. of a parity quadruplet PQ. The second parity doublet of a given K has a different order of the parities compared to the first PD. In a PQ the levels appear in parity order (+, -, -, +) or (-, +, +, -). (Exceptions to this rule were only observed in the case of very small PD energy splitting). Such a feature seems not to be obtained (e.g. ref.[28]) in microscopic calculations based on stable octupole deformation. Further experimental work or analysis and other calculations in the frame of $\varepsilon_3 \neq 0$, in this direction, would be therefore of great interest to give evidence for this proposed test.

5. SUMMARY AND CONCLUSION

The application of the multiphonon method to even-even deformed nuclei has clearly shown that this approach is able to describe not only weakly anharmonic vibrations but also strongly anharmonic motions which prominently appear near or beyond the critical strength of the random phase approximation. It also demonstrates that *collective* two phonon excitations exist and that experimental evidence can be given for those of these states which keep some collective character in their decay to the one phonon state.

The multiphonon method also offers a dynamical approach to octupole correlations. It leads to a description of the spectroscopic properties of nuclei where strong octupole correlations are expected, comparable in quality to that obtained in the framework of models based on a stable octupole deformation. As a consequence, it has been demonstrated that the existence of parity doublets, enhancement of the intra parity doublet E1 transitions and the empirical values of the decoupling factor of the K = 1/2 bands can only be used as probes of strong octupole correlations.

Finally, we would like, once more, to emphasize that the MPM results presented here are not to be used for *fine* predictions of the multiphonon states at low energy. Some effects not yet introduced in the method (e.g. coupling with the non-collective modes, non conservation of the number of particles), may slightly influence the results.

The present applications of the MPM are limited to nuclei where one or two types of phonons are predominant. Applications to other cases require the introduction of other phonons in the MPM basis. This leads to more complex recursion formulae and exponentially increases the storage capacity and computer time needed. An application of the MPM with two basic phonons to odd-mass nuclei seems still tractable and is in progress.

ACKNOWLEDGEMENTS

The author is very grateful to Drs M.K. Jammari, J. Leandri and B. Silvestre-Brac for their precious collaboration in the different steps of development and applications of the multiphonon method.

REFERENCES

1. B. Silvestre-Brac and R. Piepenbring, Phys. Rev. C26 : 2640 (1984)
2. R. Piepenbring, Phys. Rev. C42 : 1453 (1990)
3. V.G. Soloviev and N. Yu Shirikova, Z. Phys. A301 : 263 (1981), Yad Fiz 36 : 1976 (1982), Sov. J. Nucl. Phys. 36 : 799 (1982)
4. R.Piepenbring and M.K. Jammari, Nucl. Phys. A481 : 81 (1988), A487 : 77 (1988) A510 : 339 (1990)
5. I.L. Lamm, Nucl. Phys. A125 : 504 (1969)
6. W. Ogle, S. Wahlborn, R. Piepenbring and S. Frederiksson, Rev. Mod. Phys. 43 : 424 (1971)
7. A.H. Wapstra and G. Audi, Nucl. Phys. A432 : 1 (1985)
8. H.G. Börner, J. Jolie, S.J. Robinson, B. Krusche, R. Piepenbring, R.F. Casten, A. Aprahamian and J.P. Draayer, Phys. Rev. Lett. 66 : 691 (1991)
9. W.F. Davidson, D.D. Warner, R.F. Casten, K. Schreckenbach, H.G. Börner, J. Simic, M. Slojanovic, M. Bogdanovic, S. Koicki, W. Gelletly, G.B. Orr and M.L. Stelts, J. Phys. G7 : 455,843 (1981)
10. F. Iachello and A. Arima, *The interacting boson model*, (Cambridge Univ. Press, Cambridge, 1987) and references quoted therein
11. K. Kumar, Prog. Part. Nucl. Phys. 9 : 233 (1983) *Nuclear Models and the Search for Unity in Nuclear Physics* (Universiteitforlaget, Bergen, Norway, 1984)
12. M. Matsuo and K. Matsuyanagi, Progr. Theor. Phys. 74 : 1227 (1985) 76 : 93 (1986), 78 : 591 (1987)
13. O. Castanos, P. Hess, J.P. Draayer and P. Rochford, to be published
14. N. Yoshinaga, Y. Akiyama and A. Arima, Phys. Rev. Lett. 56 : 1116 (1986)
15. J. Leandri and R. Piepenbring, Phys. Rev. C36 : 1235 (1987), C 37 : 2779 (1988)
16. J. Leandri and R. Piepenbring, Phys. Lett. B232 : 437 (1989)
17. R.R. Chasman, Phys. Rev. Lett. 42 : 630 (1979)
18. R. Piepenbring, XVIII Mikolajki Summer School on Nuclear Physics, *Coherent effects in highly excited nuclei*, Harwood Academic Publishers in Nuclear Science Research Conference Series, 12 : 36 (1986)
19. W. Kurcewicz et al., Nucl. Phys. A270 :175 (1976), A289 : 1 (1977)
20. G. Leander et al., Nucl. Phys. A388 : 452 (1982), A413 : 375 (1984) W. Nazarewicz et al., Phys. Rev. Lett. 52 : 1272(1984) Nucl. Phys. A429 : 269 (1984)
21. R. Piepenbring, Phys. Rev. C27 : 2968 (1983) *Capture gamma-ray spectroscopy and related topics*, 1984 International Symposium Knoxville, Tennessee, S. Raman, American Institute of Physics, New-York, 1985, p.38
22. R. Piepenbring, Z. Phys. A322 : 495 (1985)
23. S.G. Nilsson and O. Prior, Mat. Fys. Medd. Dan. Vid. Selsk 32 : n° 16 (1961)

24. B. Silvestre-Brac and R. Piepenbring, Z. Phys. A323 : 331 (1986)
25. R. Piepenbring, Z. Phys. A323 : 341 (1986)
26. R. Sheline et al., Phys. Rev. Lett. 51 : 359 (1983),
Phys. Lett. 133B : 13 (1983), B197 : 500 (1987), Nucl. Phys. A486 : 306 (1988)
I. Ahmad et al., Phys. Rev. Lett. 49 : 1758 (1982), 52 : 503 (1984) and references quoted therein
27. G. Monsonego and R. Piepenbring, Nucl. Phys. 78 : 265 (1966)
R. Piepenbring, Thèse d'Etat, Université de Strasbourg (1966)
28. R.R. Chasman, Phys. Rev. Lett. 42 : 630 (1979),
Phys. Lett. B96 : 7 (1980), 175 : 254 (1986), J. Phys. C6 : 167 (1984)

LOW-LYING VIBRATIONAL STATES IN DEFORMED NUCLEI

V.G. Soloviev, A.V. Sushkov and N.Yu. Shirikova

Joint Institute for Nuclear Research
Dubna, USSR

INTRODUCTION

Vibrational states in doubly even nuclei have been calculated in the framework of the Quasiparticle–Phonon Nuclear Model (QPNM) (Soloviev, 1987; 1989) with inclusion of ph and pp isoscalar and isovector multipole and ph spin–multipole magnetic interactions with wave functions consisting of one– two– and in some cases of three– RPA phonon components. A new phonon operator has been introduced by Soloviev (1991) consisting of the electric and magnetic parts. Among numerous results obtained in our calculations we have chosen for this report the following two: The contribution of doubly gamma vibrational components to the wave function of the first $K^\pi = 4^+$ state in ^{168}Er and the M2 and M3 strength distribution in ^{164}Dy and ^{168}Er.

ON CONTRIBUTION OF TWO–PHONON CONFIGURATIONS TO WAVE FUNCTIONS OF LOW–LYING STATES

Based on QPNM calculations of the energy centroids of two–phonon collective states (Soloviev and Shirikova, 1981, 1982) it was concluded that collective two–phonon states cannot exist in deformed nuclei. If the contribution of the two–phonon component to the wave function normalization exceeds 50%, this state is determined as a two–phonon state. According to the Multi–Phonon Method (MPM) (Piepenbring and Jammari, 1988) and sdg IBM (Yoshinaga et al., 1988) first 4^+ state in ^{168}Er should be a collective two–phonon state. New experimental investigation by Börner et al. (1991) has established a large double gamma vibrational component in the first $K^\pi = 4^+$ state in ^{168}Er.

New experimental data (Börner et al., 1991) revived our interest in studying the contribution of two–phonon components to the wave functions of low–lying states. Three–phonon terms in the wave functions of excited states were taken into account by Soloviev and Shirikova (1982). In the diagonal approximation the inclusion of three–phonon terms leads to an additional shift of the two–phonon pole towards lower energies. This shift differs from zero even if the Pauli principle is not taken into account in three–phonon terms. The shift due to the three–phonon terms decreases by

Nuclear Shapes and Nuclear Structure at Low Excitation Energies
Edited by M. Vergnes *et al.*, Plenum Press, New York, 1992

Table 1. Contribution of double gamma vibrational components to the wave function of the first $K^\pi = 4^+$ state in ^{168}Er.

κ_0^{22}	κ_0^{44}	$\frac{G^{\lambda\mu}}{\kappa_0^{\lambda\mu}}$	$K^\pi = 2_1^+$ E_1	$K^\pi = 2_1^+$ B(E2)↑	$K^\pi = 4_1^+$ E_1	$K^\pi = 4_1^+$ B(E4)↑	$\vert\gamma,\gamma\vert$	$\frac{B(E2;4_1^+ \to 2_1^+)}{B(E2;2_1^+ \to 0_{g.s.}^+)}$
fm^{2}MeV^{-1}	fm^{2}MeV^{-1}		MeV	s.p.u.	MeV	s.p.u.	%	
0.0225	0.0310	0.9	0.89	5.1	2.3	1.2	4.3	0.1
0.0223	0.0310	1.0	0.82	4.3	2.1	1.1	16.4	0.45
0.0222	0.0305	1.0	0.84	4.2	2.2	1.0	10.8	0.3
0.0222	0.0305	1.03	0.79	3.8	2.2	1.0	18.8	0.5
0.0223	0.0310	1.05	0.73	3.7	2.1	1.1	22.5	0.6
Experiment (Govil et al., 1986; Börner et al., 1991)			0.82	4.7	2.06	0.6	10–26	0.5–1.6

(10–20)% the shift due to the Pauli principle taken into account in two–phonon terms. The inclusion of many–phonon terms does not change the shift of two–phonon poles but changes the state energy. If the root of the secular equation is strongly lowered with respect to the pole, a two–phonon state turns out to be strongly fragmented. In the framework of MPM Piepenbring and Jammari (1988) have taken into account interactions corresponding to the creation of two phonon from vacuum and absorption of two phonon by it. N.N. Bogolubov called the relevant diagrams "dangerous" and by excluding them he derived equations of the superconductivity theory. For the RPA solutions "dangerous" diagrams are automatically excluded. To take them into account in the QPNM means to renormalize a function describing the quasiparticle–phonon interaction; this renormalization is small and we disregard it.

We have made new calculation of the first $K^\pi = 2^+, 3^+$ and 4^+ states in ^{168}Er. The equilibrium deformation of one–quasiparticle excited states may differ from the one of the ground state. According to Arseniev et al. (1972) the proton one–particle state $1/2^-[541]$ has a large quadrupole deformation $\Delta\beta_2 = 0.024$ as compared to the ground state equilibrium deformation. Therefore, the energy of the $1/2^-[541]$ state is shifted from -4.66 to -5.4 MeV. The monopole and quadrupole pairing was taken into account. The energies of two–quasiparticle poles were calculated by taking account of the blocking effect and the Gallagher–Moszkowski corrections. Calculations were made for several values of the ph isoscalar $\kappa_0^{\lambda\mu}$ and pp $G^{\lambda\mu}$ constants with the isovector ph constant $\kappa_1^{\lambda\mu} = -1.5\kappa_0^{\lambda\mu}$. Shift of the double gamma vibrational pole of the $K^\pi = 4^+$ state equals (1.15 – 1.25)MeV. With three–phonon terms taken into account this shift decreases up to 1 MeV. The experimental data (Börner et al., 1991 and Govil et al., 1986) and the calculations within the QPNM are shown in table 1, together with the contribution (in per cent) of the double vibrational gamma component $\{\gamma, \gamma\}$ to the normalization of the wave function of the 4^+ state. The result of calculation within the QPNM do not contradict the available experimental data for $K^\pi = 2^+, 3^+$ and 4^+ states. The best description was obtained for $\kappa_0^{22} = 0.0223$ fm^{2}MeV, $\kappa_0^{44} = 0.031$ fm^{2}MeV, $G^{\lambda\mu}/\kappa_0^{\lambda\mu} = 1.0$.

According to MPM calculations, the first 4^+ state is a double gamma vibrational state but B(E2; $2_1^+ \to 0_{g.s.}^+$) is three times (Piepenbring and Jammari, 1988) and B(E4; $0^+ \to 4^+$) is twenty times (Jammari and Piepenbring, 1990) smaller than the

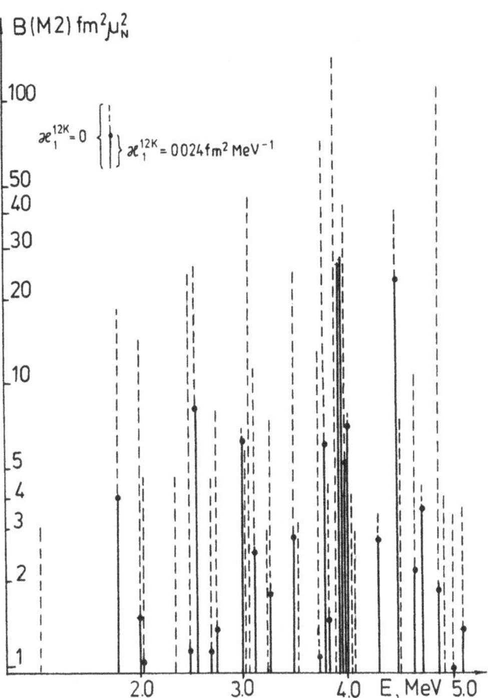

Fig. 1. B(M2) values in ^{164}Dy for transitions from the ground to excited states with $K^{\pi} = 1^{-}$ and 2^{-}, calculated in the RPA with octupole electric isoscalar and isovector constants $\kappa_0^{3K} = -0.015\,\text{fm}^2\text{MeV}^{-1}$, $\kappa_1^{3K} = -1.5\kappa_0^{3K}$ of ph and $G^{3K} = 0.9\kappa_0^{3K}$ pp interactions with the inclusion of the spin-quadrupole magnetic ph interaction with $\kappa_1^{12K} = 0.024$ fm^{2}MeV^{-1} (solid vertical lines ending by filled circles) and without it at $\kappa_1^{12K} = 0$ (solid and dashed vertical lines). The states with B(M2) $> \mu_N^2$ and in some cases with B(M2) $< \mu_N^2$ at $\kappa_1^{12K} - 0.024$ fm^{2}MeV^{-1} are shown.

experimental values (Govil et al., 1986). The QPNM calculations show that a small decrease in B(E2; $0^+_{g.s.} \rightarrow 2^+_1$) value considerably increases the contribution of the double gamma vibrational component to the first 4^+ state. According to calculations by Yoshinaga et al. (1988) within the sdg IBM the first $K^\pi = 4^+$ state is a two–phonon one; however, for excitation of the $I^\pi K^\pi = 4^+3$ state B(E4; $0^+_{g.s.} \rightarrow 4^+3_1$) = 50 s.p.u. whereas D. Burke estimates experimental data (Govil et al., 1986) as 1 – 2 s.p.u. We think that results cannot be taken as satisfactory when some experimental data are described correctly and others are not.

We can state that the available experimental data do not contradict the conclusion about the absence of collective two–phonon states in deformed nuclei. Further experiments are needed to find the contribution of two–phonon configurations to wave functions of low–lying states in deformed nuclei.

MAGNETIC M2 AND M3 STRENGTH DISTRIBUTION IN DEFORMED NUCLEI

The discovery of orbital M1 transitions in deformed nuclei create interest in searching for collective M3 excitations (Bohle et al., 1987) and theoretical investigations Lo Iudice, 1988). We have calculated in the RPA the M2 and M3 strength distribution in ^{164}Dy and ^{168}Er up to the excitation energy of 6 MeV. We took into account ph and pp quadrupole and ph spin–octupole magnetic or ph and pp octupole and ph spin–quadrupole magnetic interactions. The state structure is mainly determined by electric quadrupole or octupole interactions. Inclusion of spin–multipole magnetic interactions almost does not change the energies, B(Eλ) values and the largest components of the wave functions of states with energies smaller than 3 MeV. Spin–multipole magnetic interactions decrease B(M2) and B(M3) values 3–10 times. They shift part of the M2 and M3 strength from low–lying states towards giant isovector M2 and M3 resonances. There are no correlations between B(Eλ) and B(Mλ) values. Calculation with g_s(eff)=0.8g_s(free) increase B(Mλ) values by several times of magnitude as compared to g_s(eff)=0. In the energy interval of 3–6 MeV there are states with B(M3)=(1–5) $10^3 \mu_N^2$fm^4 and B(M2)=(10–30) μ_N^2fm^2. The M2 strength distribution in ^{164}Dy is exemplified in fig. 1. It shows the decrease in B(M2)–values with inclusion of a spin–quadrupole magnetic interaction. It is to be noted that quasiparticle–phonon interactions will lead to further fragmentation of the M2 and M3 strength over excited states. One can conclude that experimental study of the Mλ strength distribution in doubly even deformed nuclei is of great interest.

REFERENCES

Arseniev, D.A., Fedotov, S.I., Pashkevich, V.V., and Soloviev, V.G., 1972, The effect of changes in equilibrium deformations of the excited states of odd–A deformed nuclei on their energy and structure, Phys.Lett 40B:305.

Bohle, D. Richter, A., de Jager, C.W., and de Vries, H., 1987 , Search for collective M3 excitations, to low lying states in ^{164}Dy, Z.Phys. A – Atomic Nuclei 328:463.

Börner, H.G., Jolie, J., Robinson, S.I., Krusche, B., Pieperbring, R., Casten, R.F., Aprahamian, A., and Draayer, J., 1991, Evidence for the existance of two–phonon collective excitations in deformed nuclei, Phys.Rev.Lett, 66:694.

Govil, I.M., Fulbright, H.V., Cline, D., Wesolovski, E., Kotlinski, B., Backlin, A., and Gridnev, K., 1986, Multipole collectivits in ^{168}Er from inelastic scattering,Phys.Rev. C33:793.

Jammary, M.K., and Piepenbring, R., 1990, The influence of hexadecapole forces on γ–vibrations in ^{168}Er, Nucl. Phys. A510:339.

Lo Iudice, N., 1988, M3 transition to the scissors mode in the two–rotor model and random–phase approximation, Phys.Rev. C38:2895.

Piepenbring, R., and Jammari, M.K., 1988, Anharmonicities of γ–vibrations in ^{168}Er, Nucl.Phys. A481:81.

Soloviev, V.G., and Shirikova, N.Yu., 1981, On collective two–phonon states in deformed nuclei, Z.Phys.A–Atoms and Nuclei, 301:203

Soloviev, V.G., and Shirikova, N.Yu., 1982, Collective two–phonon states in deformed nuclei, Yad.Fiz. 36:1376.

Soloviev, V.G., 1987, Fragmentation of single–particle and collective motions in quasiparticle–phonon nuclear model, Prog.Part.Nucl.Phys., 19:107.

Soloviev, V.G., 1989, Theory of atomic nuclei. Quasiparticle and phonons. Energoatomizdat, Moscow.

Soloviev, V.G., 1991, Unified description of electric and magnetic excitations in deformed nuclei, Z.Phys.A – Hadrons and Nuclei, 338:271.

Yoshinaga, N., Akiyama, Y., and Arima, A., 1988, sdg interacting– boson model in the SU(3) scheme and its application to ^{168}Er, Phys.Rev. C38:419

V - SHAPES AND COEXISTENCE

ALGEBRAIC APPROACHES TO NUCLEAR STRUCTURE

Richard F. Casten

Physics Department
Brookhaven National Laboratory
Upton, New York 11973

Institute for Kernphysik
University of Köln
Köln, Germany

ABSTRACT

The concept and application of algebraic models to nuclear structure is surveyed with emphasis on interpretations in terms of traditional geometric models. Recent studies relating to the relationship between β and γ vibrations in deformed nuclei, to the rigidity or softness of axial asymmetry, to signature splitting in octupole excitations, to O(6) quantum numbers and multi-phonon excitations, and to pseudo-SU(3) are discussed.

INTRODUCTION

The last fifteen years have seen a remarkable renaissance in nuclear structure, and a notable part of this has been motivated and mediated through the development and application of a new generation of algebraic or group theoretic models. Such approaches are founded in the geometrical symmetry properties of the nucleus and are appealing because of their power, simplicity, and success. The best known of these today is, of course, the Interacting Boson Approximation (IBA) model, but other approaches, based on pseudo SU(3) and symplectic group theory, are achieving new breakthroughs and, though often less well known, bear careful attention and scrutiny as well. The present discussion is, in no sense, a review of the field (recent overviews already exist[1,2]) but rather a short synopsis of some of the basic ideas and results of these models along with a brief summary of some recent work. Most of the emphasis will be on the IBA which is the only sub-area of this growing field where the author has a chance of adequately conveying a sense of the field.

Nuclear Shapes and Nuclear Structure at Low Excitation Energies
Edited by M. Vergnes *et al.*, Plenum Press, New York, 1992

ALGEBRAIC MODELS

Algebraic models share a common approach which is simply described. One starts with a "parent" group embodying, in group theoretic terms, the most general symmetry structure of the system of interest. This group characterizes all states of this system. Successive subgroups of this group define systems with successively lower symmetry, that is, they can describe particular states or, better, define quantum numbers that act as labels for individual states. Each group, or subgroup, is defined in terms of operators called generators and has associated with it certain other operators, called Casimir operators, that commute with all the generators of the group. If a Hamiltonian, H, is written as a linear combination of Casimirs of a group and subgroups then the eigenvalues can be written, analytically, in terms of the eigenvalues of the Casimirs. In practice, this simply means that one can immediately write down the energies, without diagonalization, and, moreover, H then describes a system with particular symmetry and, hence, particular transition selection rules (e.g., for E2 transitions) and simple analytic relationships between allowed transitions (e.g., E2 branching ratios). Often, the effects of symmetry breaking (e.g., adding a new term in H that is not a Casimir operator of the group chain) can be estimated in perturbation theory or by a rather simple diagonalization. In this way, seemingly complex phase transitions can often be treated in terms of the variation of a single parameter.

THE IBA

Probably the best known current example of this process in nuclear physics is the IBA (refs. 3 and 4). Here, the system is defined in terms of correlated *valence* nucleon pairs, coupled to total angular momentum 0 or 2 and denoted s and d bosons, respectively. The degrees of freedom embodied in the model, through the five magnetic substates of the d boson and the one for the s boson, define a 6-dimensional space whose symmetry is denoted by the group U(6). There are three subgroup chains of U(6) that preserve rotational invariance, namely

$$U(6) \supset \begin{array}{ll} SU(3) \supset O(3) & \quad SU(3) \\ O(6) \supset O(5) \supset O(3) & \quad O(6) \\ U(5) \supset O(5) \supset O(3) & \quad U(5) \end{array} \qquad (1)$$

where the labels at the right are the conventional names for these "dynamical symmetries" (dynamical because each describes a system complete with ground and excited states of varying structure).

Geometrically, SU(3) describes a particular kind of symmetric rotor, O(6) a completely γ-soft axially asymmetric rotor, and U(5) a spherical vibrator which can be either harmonic or anharmonic (within certain restrictions). Noting that the γ-instability of O(6) stems from the O(5) subgroup, we recognize that U(5) is also γ-unstable. This should not be surprising: a spherical vibrator is trivially γ-independent.

It is of the utmost importance to recognize that the IBA explicitly takes account of the finite number, N_B, of valence nucleons and that, for finite N_B, these geometrical analogies are only approximate. We shall see implications of this below.

Empirical Manifestations of the Dynamical Symmetries

There is no a priori reason that any of these symmetries need exist in real nuclei. It is already sufficient that the IBA structure embodying them provides a powerful means of calculating nuclei (as we shall see) that deviate from the strictures of the pure symmetries. Nevertheless, in fact, nuclei close to each of these symmetries are known. The nucleus ^{196}Pt and those in the Xe-Ba region near A=130 are excellent examples[5] of O(6). SU(3) is most closely approximated[4,6] in ^{156}Gd and the N~102 nuclei of Yb and Hf, while a near harmonic realization of U(5) occurs in the Cd isotopes, especially in ^{118}Cd where mixing with intruder degrees of freedom is minimal[7]. The evidence for each of these has been detailed many times and, hence, rather than repeat it in the limited space here, it is of more interest to focus on a few specific issues of recent importance.

β and γ Vibrations

In traditional geometrical models, the β and γ vibrations are completely independent quadrupole excitations of the ground state, generated by $Y_{2,0}$ and $Y_{2,\pm 2}$ operators respectively. The former represents a vibrational extension in deformation (preserving axial symmetry), while the latter corresponds to an oscillation in γ (with a γ_{ave} of zero but with finite γ_{rms}). E2 transitions between β and ground bands, or γ and ground bands, are allowed and collective, while $\beta \leftrightarrow \gamma$ transitions are forbidden, since they correspond to the creation of one phonon and the destruction of another.

The SU(3) limit of the IBA is radically different. Here, β and γ bands occur in the *same* excited representation [(λ,μ) values] which is different from that of the ground state. The E2 selection rule, $\Delta(\lambda,\mu) = (0,0)$ then forbids *both* $\beta \rightarrow g$ and $\gamma \rightarrow g$ transitions, while $\beta \leftrightarrow \gamma$ transitions are allowed and collective. SU(3) is also a special case of the rotor in that the β and γ vibrations, again because they occur in the same representation, must be degenerate, $E(J_\beta) = E(J_\gamma)$, for $J_\beta = J_\gamma$.

Clearly, then, real deformed nuclei, with collective $\beta \rightarrow g$ and, especially, $\gamma \rightarrow g$ transitions *in addition to* the more recently discovered $\beta \leftrightarrow \gamma$ transitions of collective strength, and with $E_\beta \neq E_\gamma$, must represent deviations from SU(3). Indeed, the deviations (i.e., the mixing of SU(3) intrinsic excitations in calculations for realistic nuclei) are large, and it is only in the few isolated nuclei listed above where all the SU(3) signatures are approached empirically. What happens in these deviations is of interest. First $\beta \rightarrow g$ and $\gamma \rightarrow g$ E2 transitions become collective ($\beta \leftrightarrow \gamma$ ones remain so) and, interestingly, the IBA *automatically* predicts that $B(E2:\gamma \rightarrow g) >> B(E2:\beta \rightarrow g)$, as observed. Secondly, *intra*band, $\gamma \rightarrow g$ or $\beta \rightarrow g$, and $\beta \leftrightarrow \gamma$ transitions depend functionally on N_B as a constant, as $1/N_B$ and as $1/N_B^2$, respectively, so that, as $N_B \rightarrow \infty$, the traditional results of highly collective intraband transitions, collective $\beta \rightarrow g$ and $\gamma \rightarrow g$ decays and forbidden $\beta \leftrightarrow \gamma$ matrix elements is indeed recovered. For finite and realistic N_B, however, this is far from realized and experiment clearly supports the IBA scenario.

Linkage between β- and γ-vibrations
(Collaboration with W.-T. Chou and P. von. Brentano)

There is an even more striking consequence of the IBA for the β and γ band structures. In SU(3), as noted, these modes are intimately linked. When SU(3) is broken, as in calculations for realistic deformed nuclei, this linkage is maintained. Specifically, SU(3) is broken by strong $\Delta K = 0$, and much weaker $\Delta K = 2$, matrix elements. Thus, the SU(3) β and g bands mix (and repel). Then, the $\Delta K = 2$ mixing of β and γ bands leads indirectly to γ-g mixing, that is, to the bandmixing that is so well known in actual deformed nuclei. But this implies that there must be a relationship between β and γ band properties in deformed nuclei and, specifically, between the energy of the β band, relative to the γ band, and the decay properties of the γ band. Such a relationship is completely foreign to traditional geometric models and unexpected empirically. It has also never been systematically sought in the data. However, it marks a telling test of the IBA and its recognition in the data would revise our understanding of the E2 collective modes of deformed nuclei.

To address this issue, we use the simple IBA Hamiltonian

$$H = a_2 Q \cdot Q \qquad (2)$$

where $$Q = (s^\dagger d + d^\dagger s) + \chi (d^\dagger d)^{(2)} \qquad (3)$$

and $$T(E2) = e_B Q \qquad (4)$$

with the same Q operator (same χ) as in the Hamiltonian. This is known as the Consistent Q Formalism[8] or CQF. a_2 is only a scale factor on energies and is of no consequence here. e_B is a boson effective charge and is also of no interest for branching ratios. In SU(3) $\chi = -\sqrt{7}/2 = -1.32$. SU(3) symmetry breaking proceeds by changing χ, towards 0 (which gives the O(6) limit). Note that this approach to transition regions is a characteristic one in the IBA, whose "structure" can be viewed in terms of the symmetry triangle of fig. 1. Here, symmetries occupy the vertices, and transition regions the legs. Transit through such a region is then a 1-dimensional (i.e., 1-parameter) problem specified by the position [e.g., χ value for SU(3)→O(6)] along the leg. For deformed nuclei, which are intermediate between SU(3) and O(6), χ can be fixed from γ band E2 decay properties. Then, the β/γ band energy ratio can be inspected to see if linkage does occur. Since the only relevant parameter is χ, we can see the predicted relation between these observables easily in universal contour plots against N_B and χ such as shown in fig. 2 for several observables involving the β and γ bands.

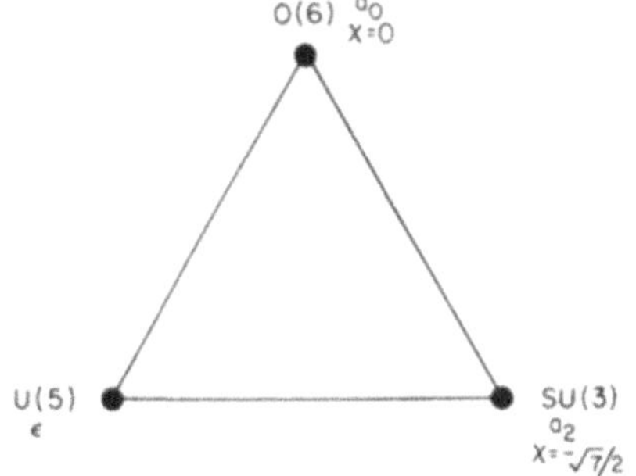

Fig. 1. Symmetry triangle of the IBA. The labels on the vertices refer to the coefficients of the standard IBA multipole Hamiltonian (see ref. 1) that produce each symmetry.

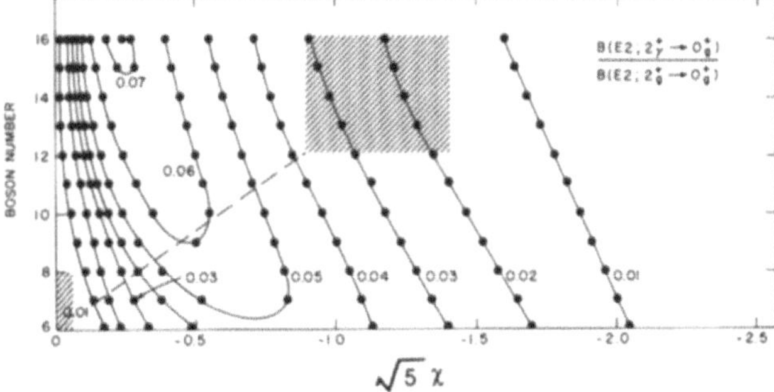

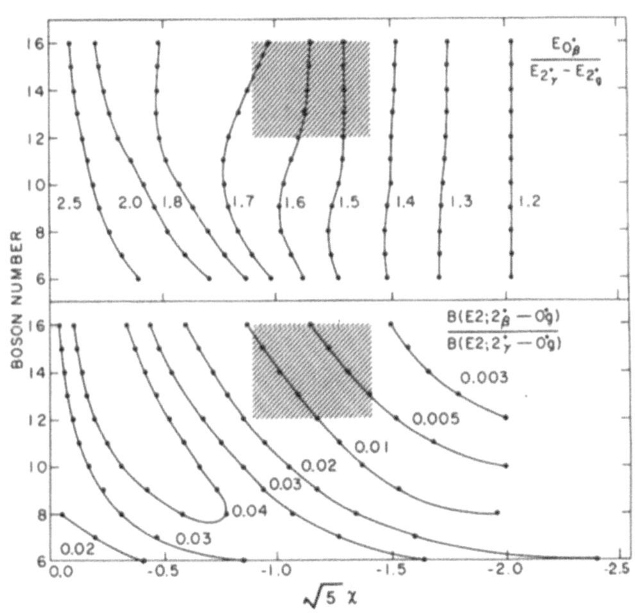

Fig. 2. Contour plots of the energy ratio of β and γ bands (top) and of two B(E2) ratios as functions of N_B and χ. (From ref. 8.)

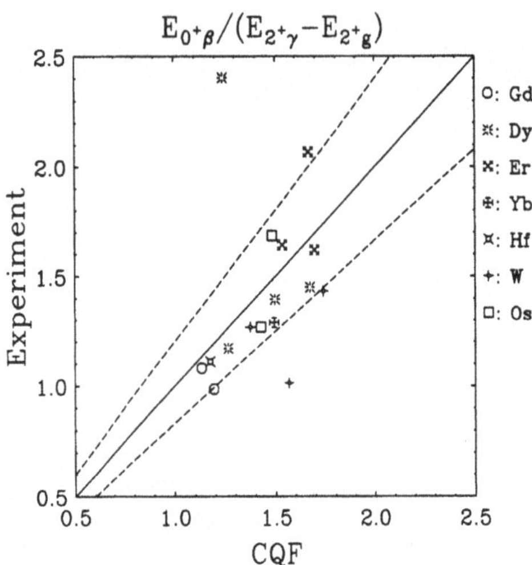

Fig. 3. Plot of experimental vs. theoretical values of $R_{\beta\gamma}$ for CQF IBA calculations. The dashed lines correspond to deviations of $\pm$20% from perfect agreement.

We are now in a position to inspect the actual data. To do so, we define

$$R_{\beta\gamma} = E_{0^+_\beta}/(E_{2^+_\gamma} - E_{2^+_g}) \tag{5}$$

(This is identical to the ratio in the middle of fig. 2.) We fix χ from the E2 decay properties of the γ-band, specifically using empirical values of the bandmixing parameter, Z_γ, and calculate $R_{\beta\gamma}$ and compare with experiment. This is shown[9] in fig. 3 where $R_{\beta\gamma}$ (IBA) is plotted against $R_{\beta\gamma}$ (Exp): agreement corresponds to points along the 45° diagonal. While, clearly, perfect agreement is elusive, there is, just as clearly, an evident correlation which goes counter to the prevailing understanding of these excitations and which supports the linkage embodied in the IBA by virtue of the algebraic structure of SU(3) and the way the structure is perturbed en route from SU(3) to O(6). This empirical linkage needs to be understood microscopically.

Axial Asymmetry in the IBA

There is another interesting aspect of an SU(3)$\leftrightarrow$O(6) region that illustrates nicely the geometrical content of the algebraic IBA. SU(3) is an axially symmetric rotor while O(6) is γ-unstable with the shape oscillating uniformly from $\gamma = 0°$ to 60°, giving $\gamma_{rms} = 30°$. Hence, somehow, the $\chi = -1.32 \rightarrow \chi = 0$ transition leg must correspond to the introduction of axial asymmetry. This has been studied[10] by equating observables calculated as a function of χ in the IBA to those obtained in the Davydov model[11] to obtain a $\gamma\leftrightarrow\chi$ correlation. [Even though there is no triaxiality in the IBA, that is, $V(\gamma)$ never has a

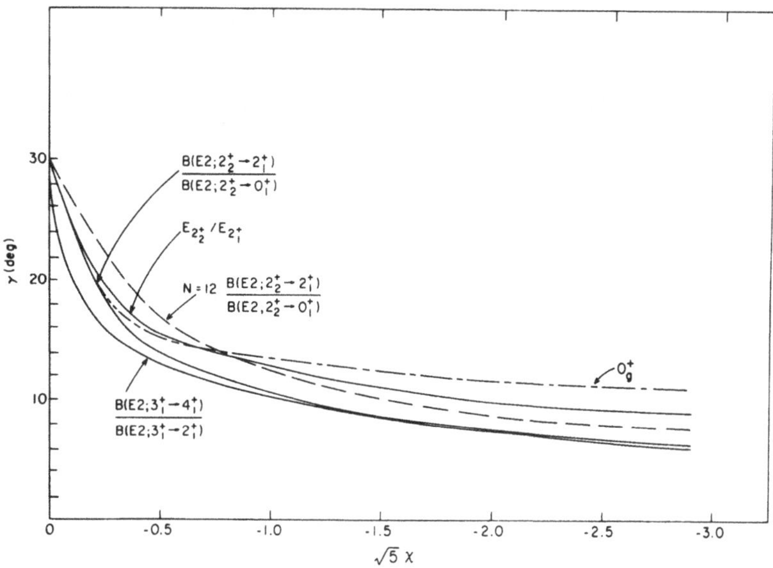

Fig. 4. γ-χ relation in the CQF (for N=16 except where indicated). The curves correspond to different observables. That labelled 0^+_g refers to an analytic shape analysis of the ground state in ref. 12. (Based on ref. 10.)

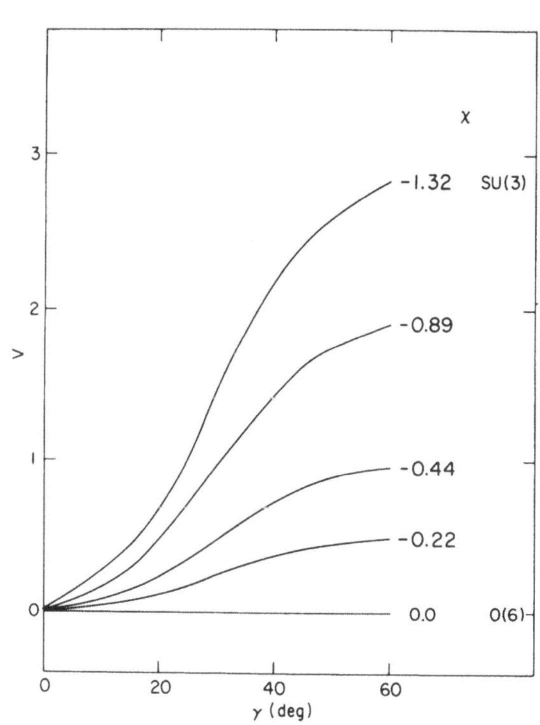

Fig. 5. The IBA potential $V(\gamma)$ as a function of γ for several values of χ. (From ref. 10.)

minimum for $\gamma \neq 0°$ and all axial asymmetry is dynamic, the use of the Davydov model in this way has been shown[12] to be valid.] The results are given in fig. 4. As expected, $\gamma_{rms} = 30°$ in O(6) where $V(\gamma)$ is flat from 0° to 60°, and decreases as SU(3) is approached. Interestingly, SU(3) corresponds, not to $\gamma = 0°$ but to $\gamma \sim 10°$. The reason is finite N_B. γ [SU(3)] $\rightarrow 0$ as $N_B \rightarrow \infty$ but, for finite N_B, though $V(\gamma)_{SU(3)}$ has a minimum at $\gamma = 0°$, it is parabolic and finite in γ so that $\gamma_{rms} \neq 0$ (see fig. 5).

Empirical Axial Asymmetry: Soft or Hard? (Collaboration with N. Zamfir)

An on-going issue for decades has been whether empirical axial asymmetry is rigid triaxial or soft. The low lying energy levels of the γ-band provide a simple and readily available signature that can be inspected, and the IBA, in the CQF, can be exploited[13] to assess the degree of softness. Rigid triaxial and γ-unstable models have opposite γ-band energy staggering as illustrated in fig. 6. The staggering in the Davydov[11] picture, $(2^+_\gamma,3^+_\gamma)$, $(4^+_\gamma,5^+_\gamma)$, ... arises because, for finite γ, the *even* spin γ-band members mix with the ground band and are repelled upward in energy towards the next *higher* odd spin level (e.g., 4^+_γ close to 5^+_γ). In contrast, for the Wilets-Jean (γ-unstable)[14], or O(6) picture[3], the levels of ground and γ-bands are arranged in degenerate multiplets whose origin is related to those of the harmonic vibrator. In the latter, the 1-, 2-, and 3-phonon states multiplets are: (2^+), $(0^+,2^+,4^+)$, $(0^+,2^+,3^+,4^+,6^+)$, respectively. The difference for the γ-unstable rotor is that the finite β (i.e., the rigidity in β) forces the "β" band (that is, the 0^+_2, 2^+_3 .....) levels up in energy, leaving behind the Wilets-Jean multiplets (2^+_g), $(2^+_\gamma,4^+_g)$, $(0^+,3^+_\gamma$, $4^+_\gamma,6^+_g)$, where the γ or g nature has been labelled where appropriate. Hence, in the γ-band, each even spin level is associated with the next *lower* odd spin state (e.g., 4^+_γ with 3^+_γ).

To assess the data, we use the signature[15]

$$S(J,J\text{-}1,J\text{-}2) = \frac{(E_J - E_{J-1}) - (E_{J-1} - E_{J-2})}{E_{2^+_1}} \qquad (6)$$

where the energy levels in the numerator are those of the γ-band. A rigid triaxial rotor has positive S (for example, $S(4,3,2) = +1.67$ for $\gamma = 30°$) while the γ-unstable rotor has negative S ($S(4,3,2) = -2$). The data for S(4,3,2) and S(6,5,4) are displayed in fig. 7 and compared[13] with several models, in particular the Davydov model as a function of γ, and O(6). Clearly, all the data fall either very close to the symmetric rotor value (hardly surprising for well deformed regions) or below it in the direction of γ-softness, especially in nuclei such as Xe, Ba, Pt which have already been described[5] as O(6)-like. Now, to understand these results, we need to know how S changes with the introduction of γ-dependence in the potential. To do this, we exploit a simple extension to the IBA in which a cubic term [of the form $H_{cubic} = \{(d^\dagger d^\dagger d^\dagger)^{(3)}(ddd)^{(3)}\}^{(0)}$] is introduced[16]. This adds a term to the potential $V(\gamma)$ with a minimum at $\gamma = 30°$. We take

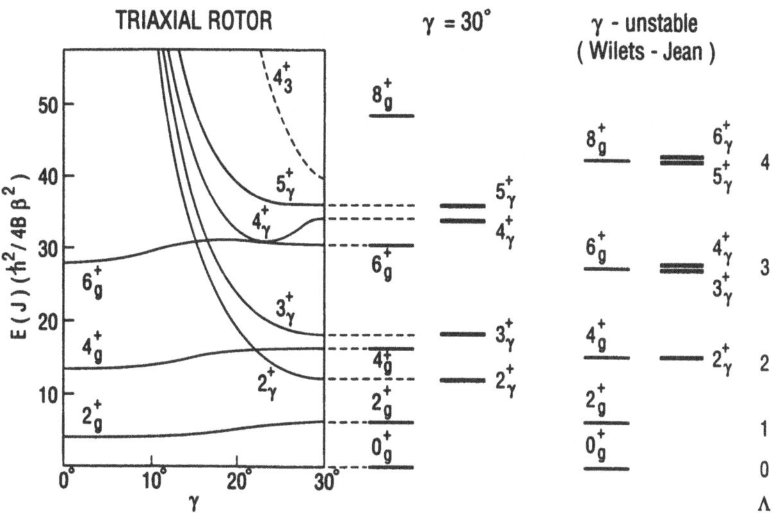

Fig. 6. Comparison of the Davydov rigid triaxial rotor and Wilets-Jean or γ-unstable rotor pictures. The Davydov results for γ=30° are shown explicitly near the middle for easier comparison. The γ-band levels are given as thicker lines. (From ref. 13.)

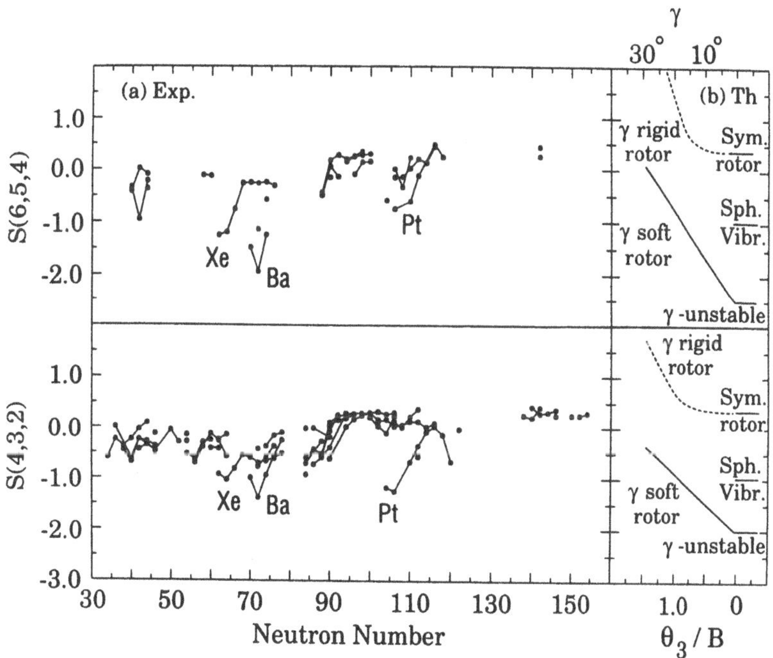

Fig. 7. γ-band energy staggering signatures S(4,3,2) and S(6,5,4) for N>30. On the right are the predictions of several models. (From ref. 13.)

$$H = H[0(6)] + \theta_3 H_{cubic} \quad (7)$$

The energy spacings in the γ-band, in O(6), are proportional to a parameter B contained in H[O(6)]. Hence the degree of γ-dependence is given simply by θ_3/B. Large values of θ_3/B reduce the typical O(6) or Wilets-Jean γ-band energy staggering, in the direction of a triaxial rotor. The question is, how much γ-dependence is needed to change S by a given amount? The answer is startling, as shown by the line on the lower right in fig. 7 giving calculated values of S (for $N_B = 6$) for several θ_3/B. It is seen that values of θ_3/B as small as unity can account for the data. However, $\theta_3/B = 1$ corresponds to a minimum in V(γ) at γ = 30° which, for $N_B = 6$, amounts to only 69 keV (it is somewhat larger for larger N_B). This γ-dependence can be expressed by a $\delta V/V$ ratio of 2.2% where δV is the maximum change in V(γ) induced by the finite θ_3 for $\theta_3/B = 1$. Thus, we reach two striking conclusions[13]. First, *real nuclei with large asymmetry are extremely γ-soft*, and, secondly, S values between O(6) and Davydov, do *not at all* correspond *to potentials intermediate* in γ-dependence between γ-flat and γ-rigid: S is an extremely sensitive indicator of even very weak γ-dependence.

Octupole Excitations and the IBA
(Collaboration with N. Zamfir, P. von Brentano, F. Seiffert, and W. Lieberz)

In the U(5) symmetry, the levels of the quasi-γ-band can be described in terms of the weak coupling of a quadrupole excitation to the yrast states. This is consistent with the γ-band staggering: to make either a 3^+_γ or 4^+_γ excitation requires the coupling $2^+_g \otimes 2^+_\gamma$, while both 5^+_γ and 6^+_γ states are of the form $4^+_g \otimes 2^+_\gamma$. Although it is not as intuitively obvious, the "γ-band" in O(6) is also a quadrupole excitation weakly coupled to the ground state band. (This *is* as clear from the wave functions as in U(5) when $N_B \to \infty$.)

This leads to an interesting prediction. Suppose one asks what energy staggering would be expected of an octupole or f-boson excitation in an axially asymmetric nucleus. If the asymmetry is rigid (Davydov rotor), the γ-band, like the ground band, is a complex mixture of K = 0 or 2 components and provides no particular clue: detailed calculations of the negative parity states would be required for an answer. The weak coupling character of the γ-band in a γ-soft nucleus, though, provides an immediate intuitive expectation: a weakly coupled f-boson, with $l = 3$, should have energy staggering exactly analogous to that of the γ-band, except shifted upwards one unit in spin (i.e., the staggering would be out of phase with that of the γ-band): each even spin level would be degenerate with the next higher, rather than lower, odd spin state.

This simple prediction can be tested since there is much recent medium and high spin data in the A = 130, Ba and Xe nuclei[17]. To do this, we utilize another staggering index S(J), which is slightly different, but analogous, to S(J,J-1,J-2) defined earlier. S(J) is defined[18] by

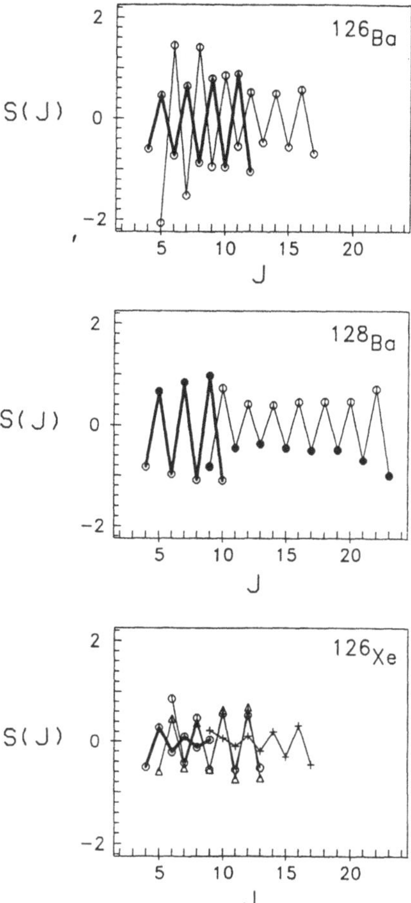

Fig. 8. S(J) plots for several axially asymmetric nuclei near A=130. The thick lines are for the γ-bands, the thin ones for negative parity bands. (From ref. 19.) In the middle panel the data for J odd are marked by solid dots to highlight the fact that $S(J_{odd})$ is a maximum for the γ band and a minimum for the negative parity bands.

$$S(J) = \frac{R(E_J)}{R(E_J)_{rotor}} - 1 \quad (8)$$

where $$R(E_J) = \frac{2[E_J - E_{J-1}]}{E_J - E_{J-2}}, \quad R(E_J)_{rotor} = \frac{J}{J-\frac{1}{2}} \sim 1 \quad (9)$$

For the γ band in O(6), and J_{odd}, we have $E_{J-1} \sim E_{J-2}$ so that $R(E_{J_{odd}}) \sim 2$. Hence, $S(J_{odd})_\gamma \sim +1$. Analogously, $S(J_{even})_\gamma = -1$. Our expectation is that S(J) will have the opposite phase for an octupole or single f-boson excitation, that is $S(J_{even})_{oct} \sim +1$, $S(J_{odd})_{oct} = -1$. [This prediction is verified by detailed calculations with the Hamiltonian $H = H[O(6)] + \varepsilon f^\dagger f + \kappa Q \cdot Q_f$.] The data for several A~130 nuclei are summarized[19] in fig. 8 and show precisely this feature. These data represent the first clear evidence of the effects of an O(6)-f boson coupling scheme and highlight another illustration of a situation where traditional models do not provide an intuitive solution, and detailed calculations are difficult and not yet practical, but where an algebraic approach gives an immediate prediction, verified by simple calculations, and in agreement with experiment.

Quantum Numbers of O(6)

There are two quantum numbers (besides the total spin J) that characterize O(6), namely σ, which labels different representations of the subgroup O(6), and τ, which labels levels within an O(6) representation according to O(5) and which resembles a phonon-like quantum number. The E2 selection rules are $\Delta\tau = \pm 1$ and $\Delta\sigma = 0$. The first has been thoroughly tested as have the predictions of particular branching ratios and absolute B(E2) values within a given σ representation. However, although several levels have been assigned[5] to excited ($\sigma < N$) representations in ^{196}Pt, and branching ratios are reasonable, there has not been, until recently, any quantitative test of the σ quantum number. What is needed is an absolute B(E2) value that *violates* the $\Delta\sigma = 0$ rule. If O(6) is a reasonable approximation, this B(E2) value should be much less than allowed B(E2) values.

Unfortunately, such B(E2) values, for high lying states weakly connected to the lower levels, are extremely difficult to measure with techniques such as Coulomb excitation or in decay experiments.

However, this issue has been recently addressed and solved[20] using a new technique, called GRID, in which the miniscule Doppler broadening, from decay in flight, of a nucleus *recoiling from prior emission of a γ-ray*, is observed. This remarkable technique[21], developed and installed at the ILL in Grenoble, France, is described in another contribution[22] to these proceedings by Jan Jolie. [That paper[22] also discusses other recent GRID experiments, including one[23] that established proof of the collectivity of 2-phonon excitations in deformed nuclei, and another measurement, in progress, of the collectivity of possible 3-phonon states in spherical nuclei.] GRID accesses level lifetimes shorter than a few ps. In an O(6) nucleus like ^{196}Pt the expected lifetime for the decay of the "bandhead" of the excited 0^+ state with σ = N-2 = 4 at 1402 keV, would be ~ 0.2 ps *if* the $0^+ \rightarrow 2_1^+$ E2 transition were collective. The measured[24] value, $\tau_{1/2}$ (1402) > 1.86 ps is an order of magnitude larger. Since this is a conservative limit, it establishes that the B(E2:0^+_{1402} (σ=4) $\rightarrow 2^+_{355}$ (σ=6)) is at least an order of magnitude smaller than for allowed transitions in O(6) and establishes the goodness of the σ quantum number.

PSEUDO-SU(3) AND SYMPLECTIC MODELS

The IBA is a valence space boson model. Algebraic techniques can also be applied directly to *fermion* excitations (e.g., Elliott[25] SU(3)) and, in symplectic models, need *not* be limited to a single shell. Such approaches are at once more microscopic but also generally more complicated. Despite the added complexity, important recent breakthroughs have been achieved. Two are worth citing.

The SU(3) symmetry that has been successful[25] in light nuclei runs into difficulties in heavier species because one orbit from each shell (the j_{max} = N + 1/2 orbit) is pushed down into the next lower shell by the l^2 and $\vec{l} \cdot \vec{s}$ forces. This badly breaks the oscillator symmetry and precludes the use of SU(3) in such mass regions. This difficulty can be circumvented by the so-called pseudo-SU(3) scheme[26]. Here, briefly, a set of orbits, such as the positive parity states of the 50-82 shell $s_{1/2}$, $d_{3/2}$, $d_{5/2}$, $g_{7/2}$ (with l = 0,2,4 but lacking the $g_{9/2}$ orbit) is treated *as if* they comprised a pseudo-(N=3) shell with j = 1/2, 3/2, 5/2, 7/2, constructed from pseudo l = 1,3 and a nuclear spin of 1/2. The unique parity $1h_{11/2}$ orbit in the 50-82 shell is treated passively, generally with pair occupation in seniority zero configurations. The advantage of pseudo-SU(3) stems from the following point. In the normal shell model Hamiltonian, the $\vec{l} \cdot \vec{s}$ (and $\vec{l}^2$) terms break SU(3). However, when the pseudo-spin transformation is made, most of the effects of the $\tilde{l} \cdot \tilde{s}$ term are incorporated into the new Hamiltonian. The quadrupole-quadrupole interaction that is added in the Nilsson model nearly preserves pseudo-SU(3) and therefore, now, the only major pseudo-SU(3)-breaking term is l^2 which is overwhelmed by the Q.Q interaction for large deformation. When this happens the pseudo-SU(3) quantum numbers, and associated selection rules, decoupling parameters, and so on, become excellent approximations. It thus turns out that the pseudo-SU(3) scheme gives an excellent treatment of the normal parity levels of the Nilsson model in medium and heavy nuclei and, in fact, gives asymptotic predictions that are closer to those of Nilsson wave functions for realistic deformation than are the asymptotic Nilsson wave functions themselves. The reader is referred to ref. 27 for elegant summaries of the basic ideas of pseudo-SU(3).

One of the predictions of pseudo-SU(3) is that K = 1/2 bands have decoupling parameters of $a = \pm 1$. Whether this prediction is taken alone[28], or embodied in a supersymmetry (SUSY) scheme[29] via the IBA and IBFA, it can

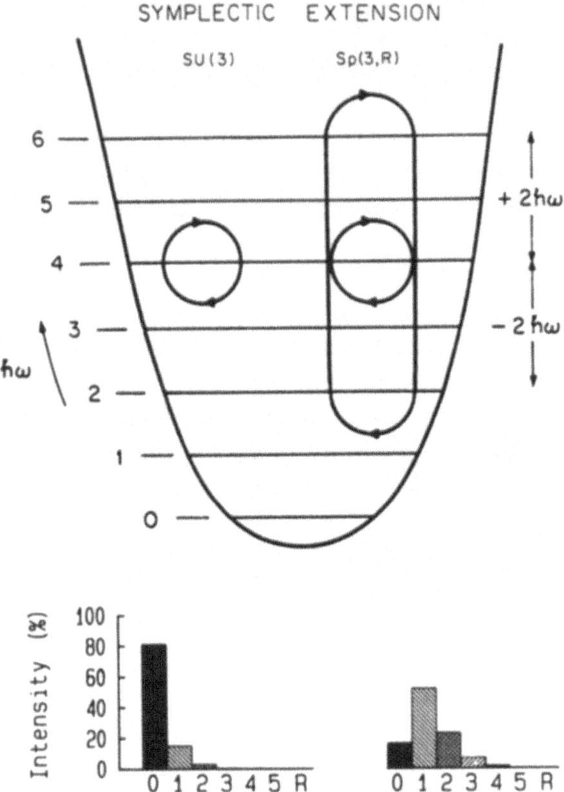

Fig. 9. (Top) Schematic illustration of the differences between the normal SU(3) approach and the Sp(3,R) symplectic theory. (Based on the second of ref. 27.) (Bottom) Example of the % distribution of wave function amplitudes over different shells for a symplectic model calculation of the ground state (G.S.) and E2 giant resonance state (R.S.) of ^{238}U. The abscissa labels the number of $2\hbar\omega$ boson excitations. (Based on the first of ref. 31.)

directly give predictions in accord with recent observations[30] of the so-called "identical superdeformed bands" in adjacent even and odd mass nuclei.

The second illustration embodies pseudo-SU(3) into a multi-shell (multi-$\hbar\omega$), or "vertical", space, treated by exploiting the properties of symplectic groups such as Sp(3,R). This approach, which is the subject of the excellent recent reviews in refs. 2, 27 and 31, has the appealing feature that it simultaneously incorporates intra- and inter-shell B(E2) calculations, that is, it is capable of simultaneously treating the collectivity of E2 giant resonances and low lying quadrupole collectivity on the same unified footing.

The multi-$\hbar\omega$ approach is illustrated schematically at the top of fig. 9. The lower part of the figure shows a typical recent example of an application of the pseudo-symplectic model to a heavy deformed nuclei, ^{238}U. While the ground state is comprised predominantly of $0\hbar\omega$ components, $2\hbar\omega$ excitations are not at all negligible. Even more dramatic is the structure of the E2 giant resonant state (R.S.) which contains major amplitudes beyond the traditional description in terms of $2\hbar\omega$ excitations, including amplitudes for $0\hbar\omega$ and extending to $12\hbar\omega$. Work in fermionic algebraic models such as this is now breaking new ground and achieving important results which auger well for the future of this approach and which offer the hope of a group theoretical avenue to understanding the microscopic structure of rotational motion and quadrupole collectivity.

CONCLUSIONS

In these few pages we have tried to outline some of the features of algebraic approaches to nuclear structure, both in their pure algebraic or symmetry limits and when used as a starting point for symmetry breaking calculations. Emphasis has been on the geometric content of the different symmetries and on their relationships to other collective models. Results focussed on the relation of β and γ vibrations in deformed nuclei, on the softness of axially asymmetry in nuclei, on signature splitting in octupole excitations of axially asymmetric nuclei, on recent tests of the σ quantum number of O(6), on pseudo-SU(3) and identical superdeformed bands in adjacent odd and even nuclei, and on symplectic approaches to the quantum rotor.

ACKNOWLEDGEMENTS

I am grateful to my collaborators in much of this work, in particular to W.-T. Chou, N. Zamfir, A. Aprahamian, D. D. Warner, P. von Brentano, J. A. Cizewski, A. Gelberg, F. Seiffert, W. Lieberz, H. Börner, and J. Jolie. For work which is not yet published, these collaborators are explicitly noted in the text. In addition, many useful discussions with K. Heyde, D. S. Brenner, I. Talmi, F. Iachello, J. P. Draayer, J.-Y. Zhang, and W. Nazarewicz are thankfully acknowledged. Work has been supported by the United States Department of Energy under contract No. DE-AC02-76CH00016, and by the BMFT.

REFERENCES

1. R. F. Casten and D. D. Warner, Rev. Mod. Phys. 60:389 (1988).
2. J. P. Draayer, Nucl. Phys. A520:259c (1990).
3. A. Arima and F. Iachello, Phys. Rev. Lett. 35:1069 (1975).
4. F. Iachello and A. Arima, "The Interacting Boson Model," Cambridge University Press, Cambridge (1987).
5. J. A. Cizewski et al., Phys. Rev. Lett. 40:167 (1978); R. F. Casten and P. von Brentano, Phys. Lett. 152B:22 (1985).
6. R. F. Casten, P. von Brentano, and A.M.I. Haque, Phys. Rev. C31:1991 (1985).
7. A. Aprahamian et al., Phys. Rev. Lett. 59:535 (1987).

8. D. D. Warner and R. F. Casten, Phys. Rev. Lett. 48:1385 (1982); Phys. Rev. C28:1798 (1983).
9. W.-T. Chou, R. F. Casten, and P. von Brentano, to be published.
10. R. F. Casten, A. Aprahamian, and D. D. Warner, Phys. Rev. C29:356 (1984).
11. A. S. Davydov and G. F. Filippov, Nucl. Phys. 8:237 (1958).
12. O. Castanos, A. Frank, and P. Van Isacker, Phys. Rev. Lett. 54:263 (1984).
13. N. V. Zamfir and R. F. Casten, Phys. Lett., in press.
14. L. Wilets and M. Jean, Phys. Rev. 102:788 (1956).
15. R. F. Casten, "Nuclear Structure from a Simple Perspective," Oxford University Press, Oxford (1990), p. 195.
16. K. Heyde et al., Phys. Rev. C29:1420 (1984).
17. U. Neuneyer et al., Z. Phys. A336:2461 (1990); W. Lieberz et al., Z. Phys. A330:221 (1988); K. Schiffer et al., Z. Phys. A327:265 (1987); W. Lieberz et al., Phys. Lett. 240B:38 (1990); W. Zipper et al., Nucl. Phys. A504:36 (1989).
18. P. von Brentano et al., in: Proc. Int. Conf. on High Spin Physics and Gamma-soft Nuclei, J. X. Saladin, R. A. Sorensen, and C. M. Vincent, eds., World Scientific, Singapore (1991), p. 344; P. von Brentano et al. in Proc. Predeal Intern. Summer School on Recent Advances in Nuclear Structure, World Scientific, Singapore, in press.
19. R. F. Casten, N. V. Zamfir, P. von Brentano, F. Seiffert, and W. Lieberz, to be published.
20. H. G. Börner, J. Jolie, S. Robinson, R. F. Casten, and J. A. Cizewski, Phys. Rev. C42:R2271 (1990).
21. H. G. Börner et al., Phys. Lett. 215B:45 (1988); H. G. Börner, J. Jolie, F. Hoyler, and S. Robinson in Contemporary Topics in Nuclear Structure Physics, R. F. Casten, A. Frank, S. Pittel, and M. Moshinsky, eds., World Scientific, Singapore (1988), p. 27: J. Jolie, S. J. Robinson, H. G. Börner, and P. Schillebeeckx in Spectroscopy of Heavy Nuclei, IOP Conf. Ser. 105:179 (1990).
22. J. Jolie, these proceedings.
23. H. G. Börner et al., Phys. Rev. Lett. 66:691 (1991).
24. H. G. Börner et al., Phys. Rev. C42:R2271 (1990).
25. J. P. Elliott, Proc. R. Soc. London, Ser. A245:128 (1958) and Ser. A245:562 (1958).
26. R. D. Ratna-Raju, J. P. Draayer, and K. T. Hecht, Nucl. Phys. A202:433 (1973).
27. J. P. Draayer and H. A. Naqui in High Spin Physics and Gamma-Soft Nuclei, J. X. Saladin, R. A. Sorensen, and C. M. Vincent, eds., World Scientific, Singapore (1991), p. 94; J. P. Draayer in Understanding the Variety of Nuclear Excitations, A. Covello, ed., World Scientific, Singapore (1991), p. 439.
28. P. J. Nolan and P. J. Twin, Ann. Rev. Nucl. Part. Sci. 38:533 (1988), and references therein; W. Nazarewicz, P. J. Twin, P. Fallon, and J. D. Garrett, Phys. Rev. Lett. 64:1654 (1990); K. Zuber et al., Phys. Lett., to be published.
29. A. Gelberg, P. von Brentano, and R. F. Casten, J. Phys. G16:L143 (1990).
30. T. Byrski et al., Phys. Rev. Lett. 64:1650 (1990).
31. O. Castanos, P. O. Hess, J. P. Draayer, and P. Rochford, Nucl. Phys. A524:469 (1991); J. P. Draayer, Inst. Phys. Conf. Ser. 105:445 (1989).

SHAPE COEXISTENCE IN THE PB REGION

Mark Huyse, Patrick Decrock, Peter Dendooven, Geert Reusen, Piet Van Duppen and Jan Wauters

LISOL, Instituut voor Kern– en Stralingsfysica, K. U. Leuven Celestijnenlaan 200D, B–3001 Leuven, Belgium

INTRODUCTION

By studying the β^+/EC and α decay of mass–separated neutron–deficient Hg to Fr nuclei, the spectroscopy group at the Leuven Isotope Separator On Line (LISOL) project collected in recent years extensive systematics on shell–model intruder states in the Pb region [1 - 10]. These shell–model intruder states, leading to shape coexistence, have been identified in odd–proton At, Bi and Tl nuclei [3 - 4], odd–neutron Pb nuclei [8 9], even – even Pb nuclei [1 2 5 7] and odd–odd Tl nuclei [6 10]. A remarkable systematic behavior of the excitation energy of the intruder–based states as a function of neutron number is now evident and several theoretical descriptions have been developed to reproduce this specific trend (see the references [11] to [15] and the contributions to this conference by K. Heyde, R. Bengtsson and N. Tajima).

Besides the energy systematics, it is of interest to deduce transition probabilities between the intruder and normal states. These transition probabilities will not only teach us about the structures of the two connected states but can also give information on the mixing between them. Extensive information on the intruder–state half life in the odd Bi and Tl isotopes is already available (for a compilation see reference [16]). More recently, we have reported on half–life measurements of excited 0^+ states in the $^{190\text{-}192\text{-}194}$Pb nuclei populated via the α decay of $^{194\text{-}196\text{-}198}$Po [7]. The deduced mixing of intruder and normal states obtained from these results and from B(E2) ratios [5] is discussed in [17].

In this contribution, we will concentrate on new results obtained in α–decay studies. First of all, a reference frame will be set by studying the α decay from groundstate to groundstate in even–even nuclei. Then, the influence of shape changes

Nuclear Shapes and Nuclear Structure at Low Excitation Energies
Edited by M. Vergnes *et al.*, Plenum Press, New York, 1992

on the α–decay rate will be discussed with the $^{191}Bi \rightarrow {}^{187}Tl \rightarrow {}^{183}Au$ decay chain as an example. Finally, a search for fine structure in the α decay of even–even nuclei will lead to the identification of new intruder states.

THE α DECAY OF EVEN–EVEN NUCLEI IN THE Z=82 REGION

The alpha decay between the ground states of even–even nuclei around Z=82 (lead) has been studied extensively: a large systematics on half lives, α–decay energies and α–branching ratios (α_{br}) exists nowadays [18]. With these data one can calculate the reduced α widths (δ^2) using e.g. the method of Rasmussen [19]. Reduced α widths contain most of the nuclear structure information and are, in case of s–wave groundstate to groundstate transitions between even–even nuclei, generally taken as a reference for unhindered α decay. The behavior of the reduced α widths for the translead nuclei around the double magic shell closure N=126 and Z=82, as a function of neutron number, is in general terms well understood. The δ^2 values show a pronounced minimum at N=126 and are largest a few nucleons above the closed shell. However, the reduced widths for the Pb isotopes behave totally differently.

The first detailed α–branching ratio measurements of even $^{188-190-192}Pb$ by Hornshoj et al. [20] – based on a comparison between α intensities and X–ray intensities – together with α–branching ratios deduced from cross–section estimates, revealed a very peculiar behavior for the deduced δ^2 values of the Pb isotopes. A strong increase in the δ^2 values going from ^{186}Pb to ^{192}Pb was observed, completely in contradiction with an expected decrease if one approaches the N=126 neutron–shell closure. Later on, Toth et al. measured (partly remeasured) the α–branching ratios of the even $^{184-186-188-190-194}Pb$ isotopes (see reference [21] and references therein) based on a comparison between the α intensity and the β^+/EC–delayed gamma–ray intensity. They constructed level schemes for the even $^{188-194}Tl$ nuclei populated via the beta decay of the corresponding Pb isotopes. From these decay schemes, they deduced the absolute gamma–ray intensity and were able to extract – in their opinion – a more precise value for the α–branching ratio since they did not depend on corrections that Hornshoj et al. [20] had to make in order to relate the measured X–ray intensity to the EC+β^+ intensity. The two α_{br} measurements agreed for the ^{192}Pb nucleus within the experimental uncertainties, but were in strong disagreement in the case of $^{188-190}Pb$. The results from Toth et al. showed a completely different behavior for the δ^2 values of the Pb isotopes: the lead anomaly was no longer present. However, another peculiar behavior of the δ^2 values of the Pb isotopes became clear: the Pb α decay was less hindered than the Hg α decay. According to the authors it appeared that midway between N=82 and N=126 the proton number Z=82 is not magic. This assumed disappearance of the Z=82 shell gap would lead to totally unexpected decay– and groundstate properties of the neutron deficient isotopes around Z=82.

Since all these conclusions are based on one set of measurements – only the α_{br}

values of Toth et al. are used – we remeasured the α_{br} ratios for $^{188-190-192}Pb$. Therefore we used three methods: i) comparison of the α intensity and the β^+/EC–delayed gamma–ray intensity using the absolute gamma–ray intensities reported in literature [$\alpha\leftrightarrow\gamma$], ii) comparison of the α intensity and the X–ray intensity using the correction factors deduced from the Pb–decay schemes [$\alpha\leftrightarrow$X] and iii) the comparison of the α decay of the feeding parent nucleus and the daughter nucleus [$\alpha\leftrightarrow\alpha$]. The third method is in principle the most interesting one since it requires only one α detector for which the absolute efficiency need not be known; neither is the knowledge of the other decay channels necessary. Only a correction for the recoil losses after parent α decay has to be made. In a recent experiment we have studied this correction procedure in detail using an α–decay chain for which the α_{br} is precisely known [22]. Unfortunately, the parent–daughter α–decay method is not generally applicable: it requires relatively strong sources of the parent nucleus which have to be relatively pure with respect to α–emitting contaminants. For methods i) and ii) where a two–detector set–up is used, information on the efficiency ratio of both detectors is needed and the β–decay schemes have to be known in detail. We have performed [$\alpha\leftrightarrow\gamma$] and [$\alpha\leftrightarrow\alpha$] in the decay of ^{192}Pb, [$\alpha\leftrightarrow\gamma$,X] in the decay of ^{190}Pb and [$\alpha\leftrightarrow$X] in the decay of ^{188}Pb. More details on the experiments carried out at the LISOL separator and at the GSI mass separator can be found in [23]. With the α_{br} determined from this work and the α–decay energies and half lives from [21], we can calculate the reduced α widths using [19]. Results are shown in table 1. For comparison the δ^2 values of Hg are included. It is clear that as far as the $^{188-190-192}Pb$ isotopes are concerned, the Pb α decay is not faster than the Hg α decay: there is no need to assume a disappearance of the Z=82 shell gap halfway between N=82 and N=126, in order to explain the behavior of the δ^2 values of $^{188-190-192}Pb$.

THE INFLUENCE OF SHAPE CHANGES ON THE α DECAY IN THE $^{191}Bi^{m\,g}$ ⟶ ^{187}Tl ⟶ ^{183}Au CHAIN

As can be seen in reference [21] and table 1, the reduced α widths of most even–even nuclei in the region of $Z \leq 82$ are lying within a factor of two around 60 keV. Could differences in shape between the groundstate of the mother nucleus and the groundstate of the daughter nucleus influence the α–decay rate? Hornshoj et al. already discussed this question in 1974 [20] and their conclusion, on the basis of the reduced decay rates for thorium and radium isotopes, was that *"shape forbiddenness does not manifest itself strongly in a decay"*. Shape changes are a common feature in the neutron–deficient Pb region: a survey on shape coexistence in odd–mass nuclei is given in [16]. For instance, the groundstate of the odd Tl nuclei is a $\pi s_{1/2}$ hole configuration and is nearly spherical, while the isomeric state in the odd Tl nuclei is a $\pi h_{9/2}$ one–particle, two–hole intruder configuration and oblately deformed ($\beta_2 \approx$ 0.18) [16 24]. In the odd Au isotopes the $\pi h_{9/2}$ –based states become

Table 1. The adopted α–branching ratios and the calculated reduced α widths. For comparison the reduced α widths of the corresponding Hg isotopes are shown. The α energies and half lives are from [20 - 21].

A	E_α (MeV)	$T_{1/2}$ (s)	α_{br}	δ^2 (^{A}Pb) (keV)	δ^2 ($^{A-2}$Hg) (keV)
192	5.112	210(6)	$6.2(6)10^{-5}$	67(7)	
190	5.580	72(6)	$4.0(4)10^{-3}$	48(6)	41(9)
188	5.982	24(1)	0.03 – 0.10	16 – 53	59(19)

lowest–lying but laser–spectroscopic studies indicate that they now have a strong prolate deformation ($\beta_2 \approx 0.25$) [25-26]. The situation is less clear for the odd Bi isotopes as there are no laser data available. The groundstate is known as a $\pi h_{9/2}$ one–particle state and is probably spherical; the isomeric state is a $\pi s_{1/2}$ two–particle, one–hole intruder state and most likely deformed [16].

The ^{191}Bi$^{m\ g}$ → ^{187}Tl → ^{183}Au α–decay chain has been studied at the LISOL on–line mass separator with the same experimental techniques as described in [3]. The α–branching ratios of ^{191}Bi$^{m\ g}$ have been taken from [3]. The value for ^{191}Bi^g (40 – 77 %)

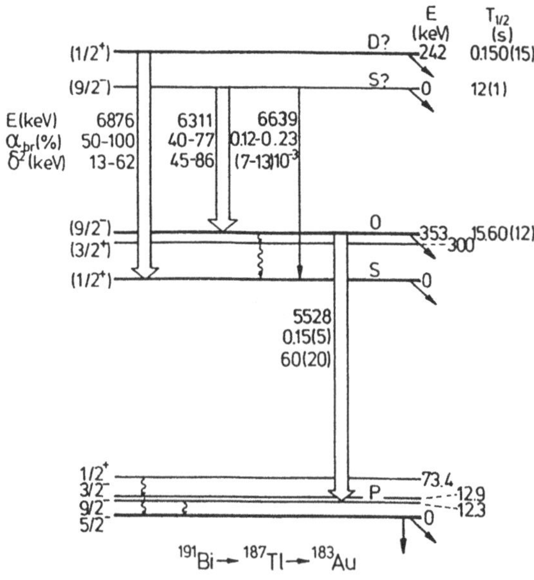

Figure 1. The ^{191}Bi$^{m\ g}$ ⟶ ^{187}Tl ⟶ ^{183}Au α–decay chain. "S" stands for spherical, "D" for deformed, "O" for oblate deformation and "P" for prolate deformation.

is in agreement with an unpublished value of Leino (85 $^{+15}$,$_{-35}$ %) [27]. By comparing the intensity of the α line from the decay of the $^{187}Tl^m$ isomeric state to the intensity of the feeding α line from the $^{191}Bi^g$ decay, it is possible to obtain the α–branching ratio of the isomeric state in ^{187}Tl, provided the necessary corrections for recoil losses have been made [22]. A value of 0.15 (5) % is obtained. Only one α line from the $^{187}Tl^m$ decay is known [28] but the isomeric state in ^{187}Tl is 9/2$^-$, while the groundstate in ^{183}Au is 5/2$^-$ [29]. We therefore have searched, at the mass separator on line to the heavy–ion accelerator UNILAC at GSI [30], for fine structure in the decay of $^{187}Tl^m$. More details on these experiments can be found in reference [31]. The whole α–decay chain, together with the measured α–branching ratios and deduced δ^2 values, is given in figure 1. The three unhindered α lines have all reduced α widths around 60 keV, very similar to the reduced widths of the neighboring even–even nuclei and they do not show any influence of the difference in size or in sign of the shape of the connected states. As we are dealing with $\Delta I = 0$ transitions ($I \neq 0$) the speed of the transition will be mainly determined by the $L = 0$ component, which is isotropic and thus most probably not sensitive to non–sphericity. The L=2 admixture could be an indicator of deformation effects and a technique that is highly sensitive to these admixtures is nuclear orientation on line (see reference [32] and the contribution of J. Wouters and H. Postma to this conference). Experiments on the anisotropic α emission in this and similar decay chains are planned.

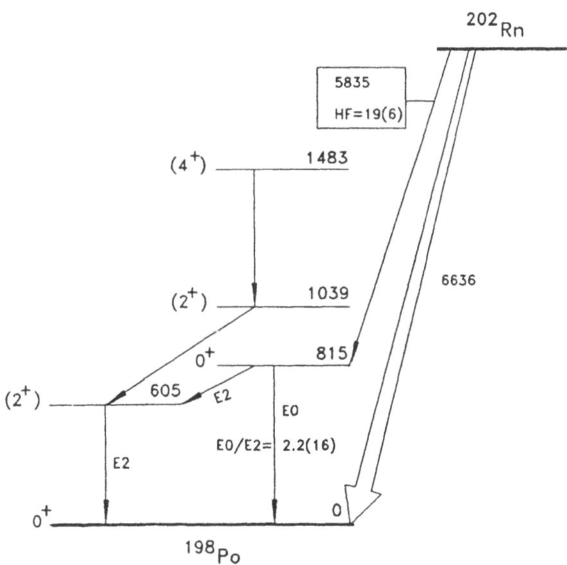

Figure 2. The α–decay scheme of ^{202}Rn, together with first evidence for a rotational band structure in ^{198}Po based on this work and on reference [33]. The energies are given in keV and HF stands for hindrance factor relative to the groundstate to groundstate transition.

The (α–e–t) coincidence technique, applied on even–even decays, is extremely powerful to detect weak branches to excited 0^+ states, due to the high sensitivity and strong background reduction of the particle–particle coincidence measurements. We have used this technique to identify the 0^+ intruder state in ^{190}Pb [2] and to determine the half lives of the 0^+ intruder states in $^{190-192-194}$Pb [5]. Recently the same technique has been used at ISOLDE, CERN and GSI, Darmstadt to search for 0^+ states in ^{198}Po, ^{184}Hg and $^{176-178}$Pt. Although the analysis is not yet fully completed, we can already report the first identification of a low–lying 0^+ state in the even–even Po isotopes. Figure 2 summarizes this preliminary result: a 0^+ state was found in ^{198}Po at 815 keV which is probably the bandhead of a rotational band in ^{198}Po from which the 2^+ and 4^+ members have recently been found in Berlin [33].

ACKNOWLEDGEMENTS

We wish to thank R. Kirchner, E. Roeckl and O. Klepper from the GSI on–line mass separator, Darmstadt and P. Lievens and the ISOLDE collaboration for their help and hospitality.

REFERENCES

1. P. Van Duppen et al., Phys. Rev. Lett. **52** (1984) 1974
2. P. Van Duppen et al., Phys. Lett. **B154** (1985) 354
3. E. Coenen et al., Phys. Rev. Lett. **54** (1985) 1783
4. E. Coenen et al., Z. Phys. **A324** (1986) 485
5. P. Van Duppen et al., Phys. Rev. **C35** (1987) 1861
6. M. Huyse et al., Phys. Lett. **B201** (1988) 293
7. P. Dendooven et al., Phys. Lett. **B226** (1989) 27
8. J. Vanhoorenbeeck et al., accepted for publication in Nucl. Phys.
9. J. C. Griffin et al., submitted to Nucl. Phys.
10. J. Wauters et al., accepted for publication in Nucl. Phys.
11. K. Heyde et al., Nucl. Phys. **A466** (1987) 189
12. K. Heyde et al., Nucl. Phys. **A484** (1988) 275
13. G. E. Arenas Peris and P. Federman, Phys. Rev. **C38** (1988) 493
14. R. Bengtsson and W. Nazarewicz, Z. Phys. **A334** (1989) 269
15. M Yamamura et al., Prog. of Theor. Phys. **82** (1989) 744
16. K. Heyde et al., Phys. Rep. **102** (1983)291
17. P. Van Duppen et al., J. Phys. **G16** (1990) 441
18. B. Al–Bataina and J. Jänecke, Phys. Rev. **C37** (1988) 1667
19. J. O. Rasmussen, Phys. Rev. **113** (1959) 1593
20. P. Hornshoj and al., Nucl. Phys. **A230**, (1974) 365
21. K. Toth et al., Phys. Rev. Lett. **53**, (1984) 1623
22. J. Wauters et al., accepted for publication in Nucl. Instr. and Meth.
23. J. Wauters et al., to be published
24. J. A. Bounds et al., Phys. Rev. Lett. **55** (1985) 2269 and Phys. Rev. **C36** (1987) 2560
25. U. Krönert et al., Z. Phys. **A331** (1988) 521
26. K. Wallmeroth et al., Nucl. Phys. **A493** (1989) 224
27. M. Leino, Ph. D thesis, University of Helsinki (1983) HU–P–D37, unpublished
28. U. J. Schrewe et al., Phys. Lett. **91B** (1980) 46
29. M. I. Macias–Marques et al., Nucl. Phys. **A427** (1984) 205
30. R. Kirchner et al., Nucl. Instr. Meth. **B26** (1987) 235

31. J. Wauters et al., accepted for publication in Z. Phys.
32. J. Wouters et al., Phys. Rev. Lett. **56** (1986) 1901
33. D. Alber et al., accepted for publication in Z. Phys.

POPULATION OF THE OBLATE AND PROLATE SYSTEMS IN ^{187}Au PRODUCED AT VARIOUS BEAM ENERGIES IN TWO DIFFERENT REACTIONS

A. Korichi,[1] N. Perrin,[1] C. Bourgeois,[1] F. Hannachi,[2] D. Hojman,[1] D.G Popescu,[1] M.G. Porquet,[2] and H. Sergolle[1]

[1]Institut de Physique Nucléaire
91406 Orsay Cedex, France

[2]Centre de Spectrométrie Nucléaire et de Spectrométrie de Masse
91405 Orsay Cedex, France

Abstract: the population of the oblate and the prolate systems in ^{187}Au has been studied through fusion- evaporation reactions induced by two different heavy ions at various beam energies. Data from both reactions established that the prolate system is less populated than the oblate one with increasing energy .

INTRODUCTION

The coexistence of oblate and prolate nuclear shapes was established in ^{187}Au through the properties of the lower excited states [1]). Later, theoretical calculations [2]) showed that this shape coexistence survives up to spin 50 $\hbar$. A recent experiment [3]) clearly established that both prolate and oblate shapes survive well separated up to spin 57/2 $\hbar$. The high spin states of ^{187}Au were populated in the ^{172}Yb(^{19}F, 4n) reaction at 95 MeV and studied with in-beam spectroscopic techniques using the " Château de Cristal " 4 π multidetector array.

In order to study the population of the prolate and the oblate systems, we carried out several experiments using two different reactions at various beam energies. All these experiments were achieved using the "Château de Cristal" array and in-beam spectroscopic techniques.

EXPERIMENTAL METHODS

Levels of ^{187}Au were populated in two reactions at the MP tandem accelerator at Orsay :

^{172}Yb(^{19}F,4n) at 90, 95 and 100 MeV
^{160}Gd(^{31}P, 4n) at 140, 145 and 150 MeV

Nuclear Shapes and Nuclear Structure at Low Excitation Energies
Edited by M. Vergnes *et al.*, Plenum Press, New York, 1992

Targets were 1mg/cm^2, 93% enriched ^{172}Yb, and deposited on 0.1 mm lead backing and 1.2 mg/cm^2, 98% enriched ^{160}Gd deposited on 0.1 mm lead backing. Gamma rays were detected by the "Château de Cristal" set-up consisting of 12 Compton-suppressed high-purity germanium detectors (8 Ge 70% efficiency and 4 Ge 20%) for discrete γ-rays spectroscopy together with 26 BaF_2 counters acting as total-energy and multiplicity filter. With the requirement that at least 2 BaF_2 and 2 Ge had fired, the events were stored on exabyte tapes. Details of the experimental set-up and data analysis can be found in refs. [4]) and [5]).

EXPERIMENTAL RESULTS

The analysis of γ-γ coincidence data showed a difference between the relative populations of the oblate and the prolate systems. The prolate system is less populated than the oblate one with increasing energy and therefore angular momentum.
γ rays intensities were measured on gated spectra because these measurements are very difficult to perform on singles spectra or on total projections of γ-γ coincidence spectra ; this is mainly due to the high density of the γ lines in the different opened channels in fusion-evaporation reactions. Examples of gated spectra are presented in figure 1.

The γ ray intensities were measured on the same gates corresponding to the different beam energies in the two reactions.

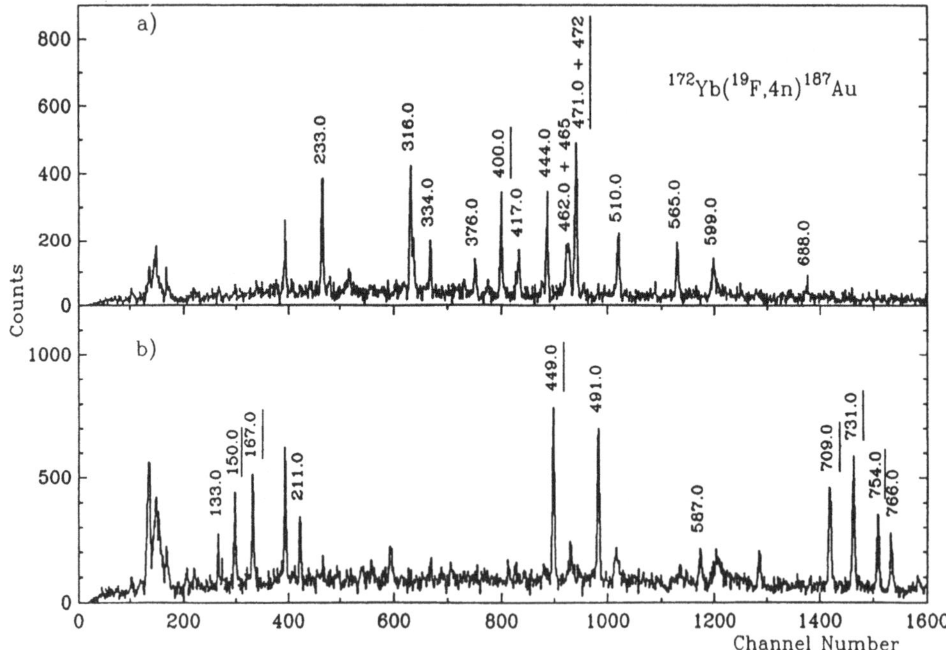

Figure 1. Examples of background corrected γ-γ coincidence spectra showing the γ cascades of the prolate (a) and the oblate (b) bands of ^{187}Au.The energies (in keV) of the gates used are underlined.

The results of these experiments are shown on figs. 2, 3 and 4 for both ^{19}F and ^{31}P projectiles at various beam energies. In all the cases, for an easier comparison, we normalized the data at the lowest energy to 1 for each reaction.

In figure 2, we can observe that the intensities of the γ-rays belonging to the oblate and the prolate systems both increase in the interval of 90 - 95 MeV though the former increases much faster than the later. In the interval of 95 - 100 MeV the prolate system γ-

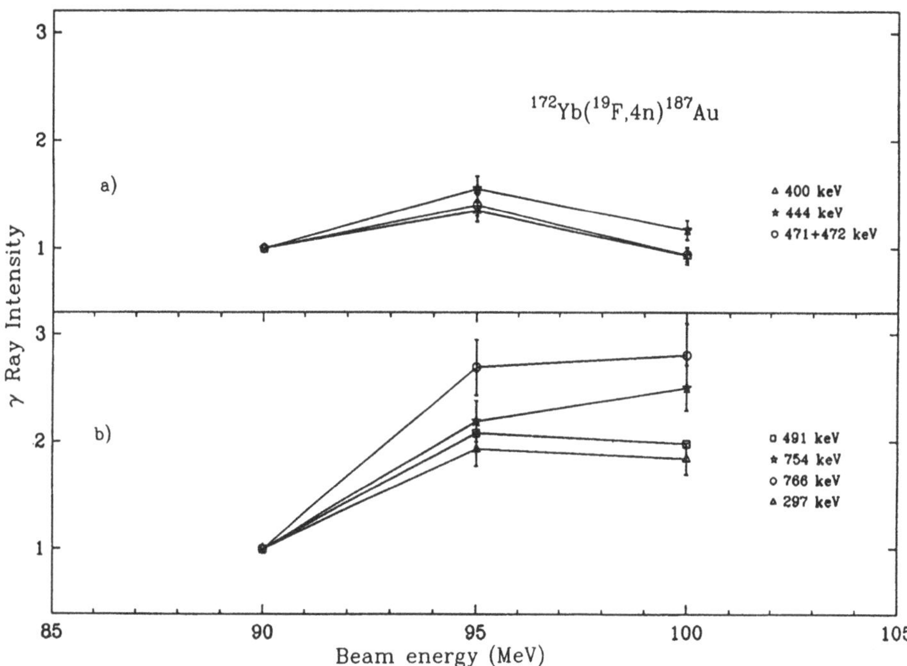

Figure 2. γ-ray intensities in the $^{172}Yb(^{19}F,4n)$ reaction as a function of the beam energy, for the prolate system (a) and the oblate one (b).

ray intensities decrease while the ones of the oblate system remain approximately constant. The same tendancy is obtained for the ^{31}P reaction (figure 3). We can observe that while the intensity of the prolate system decreases, the oblate one increases between 140 and 145 MeV. Between 145 and 150 MeV both the prolate and the oblate system intensities decrease as the rate of ^{187}Au production also comes down .

In figure 4 we present the intensities of several γ rays belonging to the prolate and the oblate systems of ^{187}Au produced in ^{19}F + ^{172}Yb reaction. All the intensities are taken equal to 1 at 90 MeV and normalized to the 449 keV(15/2-11/2) one. In order to compare the results at the different beam energies, we normalized the intensities at the coulomb excitation wich was normalized to the Rutherford scattering [5]).

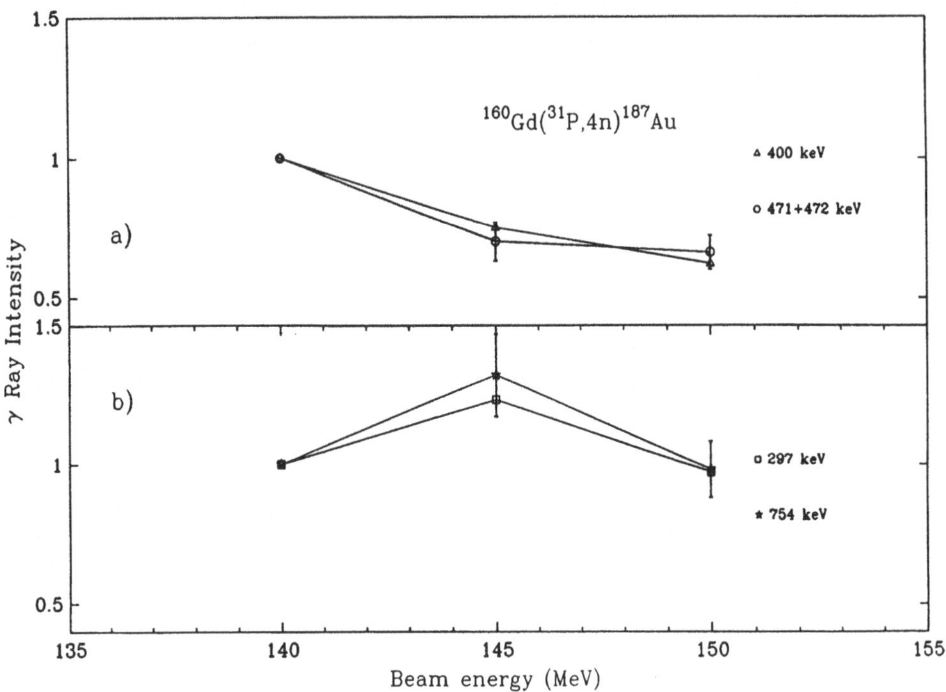

Figure 3. γ-ray intensities in the ^{160}Gd(^{31}P,4n) reaction as a function of the beam energy, for the prolate system (a) and the oblate one (b).

DISCUSSION

Such a population difference has to our knowledge never been observed and two explanations can be invoqued :

1- This population difference may be due to the change of the entry point in the (E^*,I) plane as the energy of the projectile increases. It is known that below spin 57/2 $\hbar$, both oblate and prolate bands are near the yrast line. The fact that the prolate system is less populated than the oblate one could be explained if the oblate band becomes yrast at higher spins. Further experiments should be carried out to verify this assumption. In particular the high performance of the EUROGAM array will be very usefull for solving this problem.

2- One can wonder whether or not fission plays a role in the population of the prolate and the oblate systems : the fission channel increases with bombarding energy and correlatively with increasing angular momentum and then could compete selectively with the evaporation channel leading to the prolate system in ^{187}Au . We have measured [5] the fission cross section for the two reactions at various beam energies (between 135 and 150 MeV for the ^{31}P projectile and between 95 and 105 MeV for the ^{19}F one)). We only present the results for the energies of interest in the context of this work in figure 5. Of particular interest are the values of the fission cross sections at 145 MeV in the ^{31}P reaction

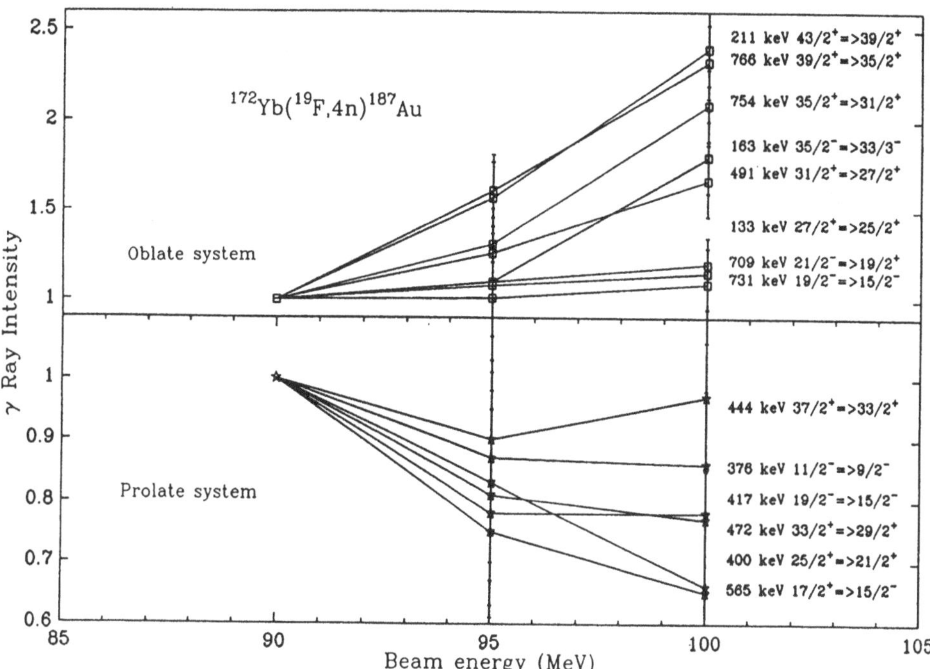

Figure 4. γ ray intensities normalized to the 449 keV intensity in the ^{172}Yb(^{19}F,4n) as a function of the beam energy. This figure shows that the intensities of the γ rays belonging to the oblate system increase with energy and spin as expected but those of the prolate system are clearly below 1.

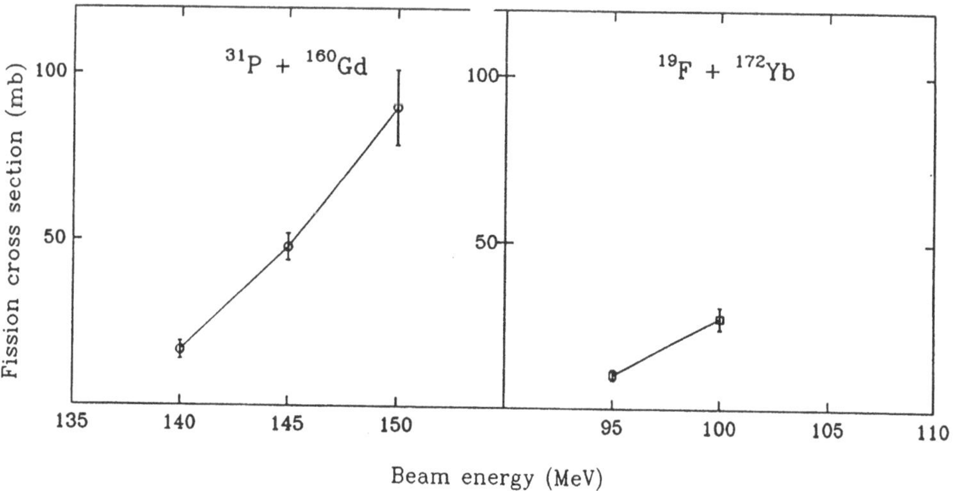

Figure 5. Fission cross section measurements in the two reactions as a function of the beam energy.

and the one at 95 MeV in the ^{19}F reaction ; these energies approximately correspond to the maximum rate of ^{187}Au production. It appears that , the fission cross section is four times higher for the ^{31}P reaction compared to the ^{19}F one , and this could be related to the weaker population of the prolate bands. We observe that while the intensity of the prolate system decreases and the oblate one increases, fission becomes very important in the two reactions, in particular between 95 and 100 MeV for ^{172}Yb(^{19}F,4n) and between 145 and 150 MeV for ^{160}Gd(^{31}P,4n). Unfortunately, there is no experimental test allowing us to firmly conclude that the increase of the fission is responsible of the weaker population of the prolate evaporation channel.

REFERENCES

1 M.A. Deleplanque et al., J. de Phys.36, L205(1975)
2 T. Bengtsson,I. Ragnarsson, Nucl. Phys.A436, 14(1985)
3 C. Bourgeois et al., Z.Phys. A 333, 5(1989)
4 F. Hannachi, Thèse de l'Université Paris Sud, Orsay(1987)
5 A. Korichi, Thèse de l'Université Paris Sud, Orsay(1991)

THE SHELL MODEL DISTRIBUTION OF NUCLEAR SHAPES

J. Carvalho# and D. J. Rowe†

Department of Mathematics, Physics, and Computer Science
Ryerson Polytechnical Institute
Toronto, Ont. M5B 2K3, Canada

† Department of Physics
University of Toronto
Toronto, Ont. M5S 1A7, Canada

INTRODUCTION

The recent observation of superdeformed states has revived interest in the co-existence of different deformation shapes in the low energy spectra of nuclei. The question arises as to what deformation shapes are available to a given nucleus and where in the energy spectrum they should make their appearance.

We give a precise definition of intrinsic quadrupole shapes and show that it is meaningful to consider the set of quadrupole shapes available to a particular nucleus. Although infinite, the set becomes countable upon factoring out the renormalizing effects of the giant resonance degrees of freedom. Moreover, it can be partially ordered. We discuss suitable criteria for ordering the available deformation shapes according to their importance for the description of low energy nuclear states.

DEFINITION OF INTRINSIC SHAPES

The mass quadrupole operators for a nucleus, $\{Q_\nu; \nu = 0, \pm 1, \pm 2\}$, span a five-dimensional Abelian Lie algebra, labelled $\mathbb{R}^5$. An intrinsic state of a rigid rotor band is defined to be an eigenstate of these operators. It follows that the states of a rigid rotor band have shapes characterized by well-defined quadrupole moments.

The (total) angular momentum operators, $\{J_k; k = 0, \pm 1\}$, obey su(2) commutation relations, expressed in spherical tensor notation,

$$[J, J] = -\sqrt{2}\, J, \tag{1}$$

where the commutator on the left is the $j = 1$ spherical tensor with components $[J, J]_{1m} = \sum_{m_1 m_2} (1m_1, 1m_2|1m)[J_{m_1}, J_{m_2}]$. The quadrupole operators transform as rank two spherical tensors under rotations. They satisfy the commutation relations

$$[J, Q] = -\sqrt{6}\, Q. \tag{2}$$

Together, the angular momentum and quadrupole operators span a dynamical algebra

Nuclear Shapes and Nuclear Structure at Low Excitation Energies
Edited by M. Vergnes *et al.*, Plenum Press, New York, 1992

for the rigid rotor model.[1,2] This Lie algebra is called $[\mathbb{R}^5]\mathrm{su}(2)$ or simply the rotor algebra. A rigid rotor band spans an irreducible unitary representation of the rotor algebra. Such a representation can be induced, in accordance with the rotational model, from a single intrinsic state; an eigenstate of the quadrupole operators. Conversely, all states of a rigid rotor band share a common intrinsic (quadrupole) shape.

An important property of the rotor algebra is that its elements, the angular momentum and quadrupole moment operators, have simple shell model expressions as sums of single-nucleon moments. Thus, in principle, it is possible to expand arbitrary shell model wave functions in terms of states of well-defined intrinsic shape, i.e., in terms of wave functions belonging to irreducible rigid rotor representations. Conversely, one can question which rigid rotor representations and, hence, which shapes feature in the expansion of specific shell model wave functions. This question is well-defined without reference to a model and without assumptions about the ways nuclei rotate.

It is useful to observe that the (J-rotor) algebra $[\mathbb{R}^5]\mathrm{su}(2)$ is isomorphic to the (L-rotor) algebra $[\mathbb{R}^5]\mathrm{so}(3)$, in which the total angular momenta, $\{J_k; k = 0, \pm 1\}$, are replaced by orbital angular momenta $\{L_k; k = 0, \pm 1\}$. Representations of the two algebras are simply related. An intrinsic state for a representation of the L-rotor algebra, can be converted to an intrinsic state for the J-rotor algebra by combining it with appropriate intrinsic nucleon spin wave functions. In other words, one strongly couples the intrinsic nucleon spins to the orbital rotor. Since the intrinsic spin degrees of freedom carry no quadrupole moments, they do not change the quadrupole moments of the intrinsic state. This means that the set of shapes available to the J-rotor is the same as the set available to the L-rotor. Thus, to investigate the distribution of shapes available to the nucleus, we are free to use either the J-rotor algebra or the L-rotor algebra. The latter is usually simpler.

Another simplification results from factoring out the giant monopole and quadrupole degrees of freedom; they can be restored subsequently by renormalizing the results. The mass quadrupole operators of the nucleus have three components

$$Q_\nu = \mathcal{Q}_\nu + Q_\nu^+ + Q_\nu^-, \tag{3}$$

where $\mathcal{Q}_\nu$ is the component of Q_ν that conserves the number of quanta of any (harmonic oscillator shell model) state it operates on and $Q_\nu^\pm$, respectively, change the number of quanta by ± 2. We find that $\mathcal{Q}_\nu$ is an su(3) quadrupole operator and $Q_\nu^\pm$ are, respectively, raising and lowering operators for $2\hbar\omega$ giant quadrupole resonance excitations. A similar expansion can be made for the mass monopole operator $M = \sum_{i=1}^{A} r_i^2$. Let P denote the operator that projects the full shell model Hilbert space to the effective shell model subspace of states annihilated by the $2\hbar\omega$ giant monopole and quadrupole lowering operators. Under this projection, the rotor operators project to the operators

$$Q_\nu \to PQ_\nu P = \mathcal{Q}_\nu, \quad L_k \to PL_kP = L_k. \tag{4}$$

The remarkable fact is that the projected operators close on Elliott's su(3) algebra.[3] In other words, the projection P is a map from the L-rotor algebra to the su(3) algebra;

$$P: \ [\mathbb{R}^5]\mathrm{so}(3) \to \mathrm{su}(3). \tag{5}$$

A rigorous treatment of this projection procedure,[4,5] based on the representation theory of the sp(3,$\mathbb{R}$) Lie algebra,[6,7] reveals the following:

(i) The L-rotor algebra is a subalgebra of sp(3,$\mathbb{R}$), the symplectic algebra generated by the su(3) operators and the giant monopole/quadrupole moment operators.
(ii) All states of an sp(3,$\mathbb{R}$) irrep project to a single irrep of su(3) $\subset$ sp(3,$\mathbb{R}$).
(iii) It follows from (i) and (ii) that all irreps of the L-rotor algebra, contained in an sp(3,$\mathbb{R}$) irrep, project to a single su(3) irrep.
(iv) It also follows that a rigid rotor band is an su(3) band renormalized by coupling to giant resonance degrees of freedom.

(v) If a realistic Hamiltonian is diagonalized within an sp(3,$\mathbb{R}$) irrep, one obtains a low energy band of renormalized su(3) states and higher bands of states involving giant resonance excitations.

(vi) Physical states are combinations of renormalized su(3) states projected from different sp(3,$\mathbb{R}$) irreps.

The intrinsic state of a (λ, μ) su(3) irrep has quadrupole moments given[8,9], in harmonic oscillator units, by

$$Q_0 = 2\lambda + \mu + 3, \quad Q_2 = \sqrt{\tfrac{3}{2}}\, \mu. \tag{6}$$

A self-consistency argument shows that the renormalizing effects of the giant resonance degrees of freedom double these quadrupole moments. Therefore, the charge quadrupole moments to be associated with the corresponding rotor bands are given by

$$Q_0^e = e_{\text{eff}}(2\lambda + \mu + 3), \quad Q_2^e = e_{\text{eff}}\sqrt{\tfrac{3}{2}}\, \mu, \tag{7}$$

where, on the basis of the renormalization argument, $e_{\text{eff}} \approx 2Ze/A$.

Since experimentally observed states are mixtures of states from different su(3) irreps, the values of λ and μ inferred from measured quadrupole moments and $B(E2)$ transitions are weighted averages of the contributing representations. Some of the average values, deduced from bands observed in erbium and tungsten isotopes by Jarrio *et al.*,[10] are shown in Table 1.

Table 1. Average values of $2\lambda + \mu$ and μ inferred from experiment.

	$\langle 2\lambda + \mu \rangle$		$\langle \mu \rangle$
Nucleus	$K^\pi = 0^+$	$K^\pi = 2^+$	$K^\pi = 2^+ \to K^\pi = 0^+$
^{160}Er	92–136		
^{162}Er	107	135 ± 44	
^{164}Er	108–155	181 ± 20	10
^{166}Er	136–173	165 ± 14	14–21
^{168}Er	151–183	110-233	7–33
^{170}Er	112–165	151 ± 26	8–17
^{182}W	48–173		15–18
^{184}W	117–159		13
^{186}W	108–158	93 ± 16	15

THE SHELL MODEL DISTRIBUTION OF SHAPES

After factoring out the shape fluctuations associated with giant resonance degrees of freedom, the nuclear shapes of the shell model Hilbert space are in one-to-one correspondence, by eq. (6), with *core* su(3) representations; core irreps are defined to ones whose states are annihilated by the giant resonance lowering operators.

The wave functions for a single nucleon in the N'th harmonic oscillator shell belong to an su(3) irrep with $\lambda = N, \mu = 0$. The su(3) irreps associated with a single-shell N^n configuration are therefore contained in the su(3) Kronecker product

$$(N,0) \times (N,0) \times \ldots \times (N,0), \quad (n \text{ copies}). \tag{8}$$

The desired irreps are those that respect the antisymmetry constraints of nucleons. Programs exist for constructing these.[11]

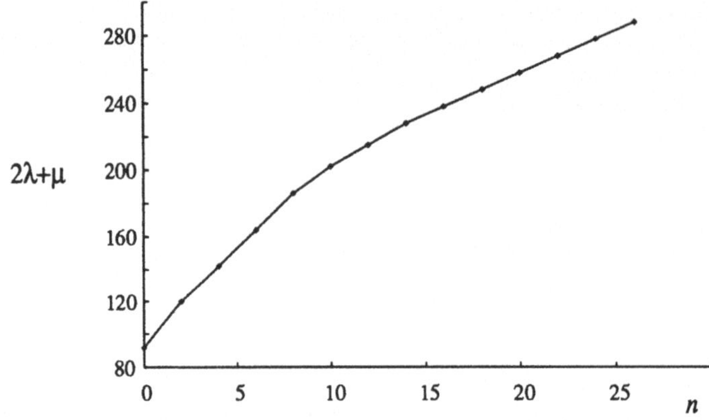

Fig. 1. A plot of $2\lambda + \mu$ for the leading irrep of each $n\hbar\omega$ shell model space for ^{152}Dy as a function of n.

The su(3) irreps associated with a multi-shell configuration are given by products of single shell results. Among these, we retain only the core irreps. However, the number of core su(3) irreps associated with any given shell-model configuration remains, in general, a very large number. Fortunately, we are only interested in those irreps that are likely to contribute to the states observed in the low energy spectrum of the corresponding nucleus. What we need is a physically relevant criterion for selecting these.

First observe that the states of a given shell model configuration, in a harmonic oscillator basis, are eigenstates of the spherical harmonic oscillator Hamiltonian. The lowest-energy shell model configuration is conveniently referred to as the $0\hbar\omega$ configuration. A configuration of harmonic oscillator energy $n\hbar\omega$ above the lowest energy, is correspondingly described as an $n\hbar\omega$ configuration.

Now, the states of an su(3) irrep are also eigenstates of the su(3) Hamiltonian

$$H = H_{HO} - \tfrac{1}{2}\chi Q \cdot Q = H_{HO} - \tfrac{1}{2}\chi C^{(2)} + \tfrac{3}{2}\chi L^2, \tag{9}$$

where H_{HO} is the spherical harmonic oscillator Hamiltonian and $C^{(2)}$ is the su(3) quadratic Casimir operator. The Casimir operator has eigenvalues equal to $4(\lambda^2 + \mu^2 + \lambda\mu + 3\lambda + 3\mu)$. This suggests that we order irreps by increasing values of

$$E(n, \lambda, \mu) = N_0\hbar\omega - 2\chi(\lambda^2 + \mu^2 + \lambda\mu + 3\lambda + 3\mu), \tag{10}$$

where $N_0\hbar\omega$ is the harmonic oscillator energy of the su(3) states. The value of $2\lambda + \mu$ for the leading irrep of each $n\hbar\omega$ shell model space (i.e., the irrep with largest value of $\langle C^{(2)} \rangle$) is plotted for ^{152}Dy as a function of n in Fig. 1.

One sees that, even with a factor of two allowance for the renormalizing effects of higher shells (and this is probably an overestimate), the values of $2\lambda + \mu$ (and hence Q_0) occurring in the low spherical shells are much too small to explain observed low lying bands. The values indicated by experiment for ground state bands, $2\lambda + \mu \approx 150$, are not to be found below the sixth spherical harmonic oscillator shell. This implies that it is meaningless to attempt a spherical shell model description of rotational states in rare-earth nuclei. No matter how much configuration mixing takes place or how much renormalization of the effective quadrupole moment operators result from coupling to the giant resonance degrees of freedom, the low lying shell model states remain orthogonal to those required; they belong to different sp(3,$\mathbb{R}$) irreps.

A plot of $E(n, \lambda, \mu)$, with $\chi = 0.36\hbar\omega/N_0$, is shown in Fig. 2. An su(3) irrep from the $18\hbar\omega$ spherical shell now lies lowest in energy and, with an ordering of irreps by

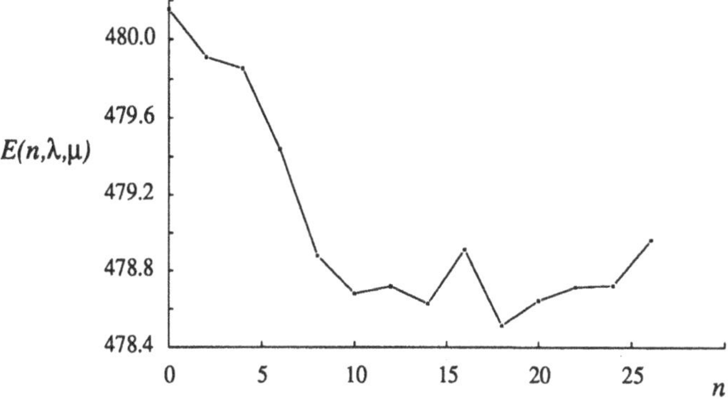

Fig. 2. A plot of $E(n, \lambda, \mu)$ for ^{152}Dy, given by eq. (10), as a function of n.

increasing values of $E(n, \lambda, \mu)$, there is no longer a problem in explaining experimental deformations.

NILSSON MODEL CONSIDERATIONS

Investigations of nuclear shapes have traditionally been explored in the Nilsson model (with Strutinski corrections).[12] This model considers only intrinsic states that are Slater determinants. Nevertheless, it has a strong intuitive appeal and makes sense for the leading representations, which usually are Slater determinants. We therefore consider ordering core su(3) irreps according to their Nilsson energies.

First observe that a deformed Nilsson model state is easily projected to a spherical shell model state. This is because a spherical shell model wave function can be transformed continuously into a deformed shell model wave function by a scale transformation of the (x, y, z) Cartesian coordinates; i.e.,

$$\psi_{\text{sph}}(x, y, z) \rightarrow \psi_{\text{def}}(x, y, z) = \psi_{\text{sph}}(ax, by, cz). \tag{11}$$

Conversely, the spherical shell model wave function, to which a Nilsson model wave function projects, is given by the inverse scaling.[10]

A (many-nucleon) Nilsson model state generally has overlaps with several (renormalized) su(3) irreps. Nevertheless, as shown by Jarrio *et al.*,[9] an asymptotic Nilsson model wave function gives a well-defined value of $2\lambda + \mu$. Thus, an asymptotic Nilsson model state has a well-defined value of Q_0 (apart from giant resonance vibrational fluctuations). Hence, it is rigid with respect to (low energy) beta vibrations. (If it also had a well-defined value of μ, which is generally not the case, it would be rigid with respect to gamma vibrations and correspond to a renormalized su(3) irrep.) Thus, for each spherical shell model configuration, we find the lowest energy asymptotic Nilsson model state and the corresponding value of $2\lambda + \mu$. Fig. 3 shows a plot of the lowest Nilsson model energies and $2\lambda + \mu$ values obtained in this way for ^{168}Er. Observe that the lowest energy deformations now occur at $n \approx 16$.

CONCLUDING REMARKS

We have shown that to obtain the large deformations needed to explain observed rotational bands, it is necessary to consider highly excited spherical shell model configurations. We have also found that the su(3)-mixing terms of the Nilsson model reorder su(3) irreps such that the most deformed, within an $n\hbar\omega$ shell, are not necessarily always the lowest. Pairing and other residual interactions are expected to cause further

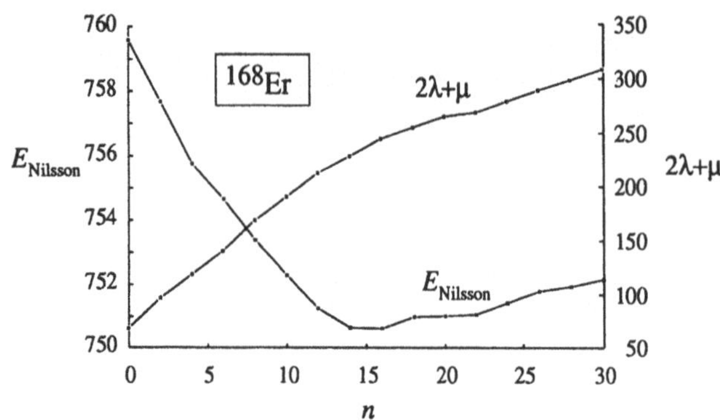

Fig. 3. Energies and $2\lambda + \mu$ values of asymptotic Nilsson model states as functions of n.

representation mixing and, hence, average $2\lambda + \mu$ values that are lower for each $n\hbar\omega$ shell than those predicted without mixing. It is also clear that superdeformed bands must have their origins in extraordinarily high-lying spherical shell model configurations which we plan to identify.

The results presented here are preliminary. A more extensive analysis will be published in the regular journals. We also intend to investigate the validity of the pseudo-su(3) and pseudo-symplectic models[11] from the present perspective.

REFERENCES

1. H. Ui, Quantum mechanical rigid rotator with an arbitrary deformation, Prog. Theor. Phys. **44**:153 (1970).
2. L. Weaver, L. C. Biedenharn and R. Y. Cusson, Rotational bands in nuclei as induced group representations, Ann. Phys. (N.Y.) **77**:250 (1973).
3. J. P. Elliott, Collective motion in the nuclear shell model, Proc. Roy. Soc. **A245**:128, 562 (1958).
4. J. Carvalho, R. Le Blanc, M. Vassanji, D. J. Rowe and J. B. McGrory, The symplectic shell-model theory of collective states, Nucl. Phys. **A452**:240 (1986).
5. R. Le Blanc, J. Carvalho, M. Vassanji and D. J. Rowe, An effective shell-model theory of collective states, Nucl. Phys. **A452**:263 (1986).
6. G. Rosensteel and D. J. Rowe, Nuclear Sp(3,ℝ) model, Phys. Rev. Lett. **38**:10 (1977).
7. D. J. Rowe, Microscopic theory of the nuclear collective model, Rep. Prog. Phys. **48**:1419 (1985).
8. D. J. Rowe, R. Le Blanc and J. Repka, A rotor expansion of the su(3) Lie algebra, J. Phys. A: Math. Gen. **22**:L309 (1989).
9. D. J. Rowe, M. G. Vassanji and J. Carvalho, The coupled-rotor-vibrator model, Nucl. Phys. **A504**:76 (1989).
10. M. Jarrio, J. L. Wood, The su(3) structure of rotational states in heavy deformed nuclei, Nucl. Phys. (in press).
11. J. P. Draayer, Y. Leschber, S. C. Park and R. Lopez, Representations of U(3) in U(N), Comput. Phys. Commun. **56**:279 (1989).
12. S. G. NIlsson *et al.*, On the structure and stability of heavy and superheavy elements, Nucl. Phys. **A131**:1 (1969).
13. Castaños. J. Draayer and Y. Leschber, Towards a shell-model description of the low-energy structure of deformed nuclei, Ann. Phys. (N.Y.) **180**:290 (1987).

MAGNETIC DIPOLE EXCITATIONS OF HEAVY NONSPHERICAL NUCLEI

D. Zawischa[1] and J. Speth[2]

[1] Institut für Theoretische Physik
Universität Hannover
W-3000 Hannover, Germany

[2] IKP (Theorie), Forschungszentrum Jülich
W-5170 Jülich, Germany

In the past few years, considerable experimental effort has been dedicated to the detection and investigation of magnetic dipole excitations of deformed nuclei. Orbital magnetic dipole excitations have been found at energies around 3 MeV in the rare earths[1,2]; and recently also spin-flip magnetic dipole strength has been located between 5 and 10 MeV in this region of nuclei[3].

ORBITAL EXCITATIONS

Writing the magnetic dipole operator as

$$\vec{\mathcal{M}}(\mathrm{M1}) = \sqrt{\frac{3}{4\pi}}\,\mu_\mathrm{N}\,(g_l\,\mathbf{J} + (g_s - g_l)\,\mathbf{S}) \tag{1}$$

we see that strong transitions due to the orbital part are only possible between states of the same j-shell, therefore they do not occur in spherical nuclei. It is easy to estimate orbital transition strengths in the nonspherical shell model of free particles: For small deformations, the wave functions do not change much, so the matrix elements may be approximated by the spherical ones. For the excitation of a proton in the $\mathrm{h}\frac{11}{2}$-multiplet from the $|k| = \frac{5}{2}$ substate to the $|k| = \frac{7}{2}$ one, one estimates a $B(\mathrm{M1})\uparrow$-strength of $\approx 5\mu_\mathrm{N}^2$ by evaluating the matrix element of $\mathbf{J}$ and ignoring the spin part in eq. (1). Other transitions possible in Gd and Dy are in the $\mathrm{g}\frac{7}{2}$ subshell or in the $\mathrm{d}\frac{5}{2}$ subshell and have strengths between 1 and 2 μ_N^2.

The magnetic transition strength is strongly quenched by pairing, and including the residual interaction in the RPA, the total orbital transition strength is reduced still more. Using a realistic effective interaction[4,5] or a reasonably chosen separable schematic force[6–8] it is possible to reproduce the data in microscopic calculations to a fair degree of accuracy. The remaining discrepancies are thought to be primarily due to deficiencies of the single particle energies which always have to be adjusted to the experimental data.

The interpretation of the orbital excitations (experimental as well as theoretical results) has been strongly biased by predictions of a special type of collective modes: "scissors modes" or "isovector rotational vibrations", see ref.[2] for references. This picture originates from the model of rigid rotors where the bulk of protons on the one hand and neutrons on the other hand are assumed to move like interpenetrating rigid bodies.

Nuclear Shapes and Nuclear Structure at Low Excitation Energies
Edited by M. Vergnes *et al.*, Plenum Press, New York, 1992

In real rotations, the nuclear fluid does not move like a rigid body. From microscopic calculations[5] one obtains a flow pattern as shown schematically in fig. 1 (b): in the central part of the nucleus the flow is smaller than in the case of rigid motion.

The transition current densities of the low-energy orbital M(1) excitation obtained in RPA calculations[5] resemble neither of the two types of rotation (rigid or 'quantum mechanical'). Instead, the flow is large in the inside of the nucleus, with vortices, and near the surface it is either small or tangential to the surface. In contrast to the quantum mechanical rotation, the flow is large near the 'equator' and small near the poles in the examples investigated so far. Thus, to our opinion, it is misleading to call this type of motion a 'rotational vibration' or a 'scissors mode'.

To understand how collective states arise in microscopic calculations, the schematic model introduced by Brown and Bolsterli[9] is very useful. (In this discussion we ignore the influence of pairing.) Approximating the realistic particle-hole interaction by the quadrupole-quadrupole force $v_{mjin} = \lambda Q_{mi} Q^*_{nj}$, with $Q_{mi} = \langle m|r^2 Y_{21}|i\rangle$, Y being the spherical harmonic, and admitting different force constants $\lambda^{(l)}$, $\lambda^{(u)}$ for the interaction between like and unlike particles, respectively, one gets two classes of states (of isoscalar and isovector type). The coefficients of the state vector expanded in particle-hole states $a^+_m a_i|0\rangle$ (Tamm-Dancoff-Approximation) are

$$C^{\nu}_{m,i} = \mathcal{N} \frac{s_{mi} Q_{mi}}{E_\nu - \epsilon_m + \epsilon_i} \tag{2}$$

where $\mathcal{N}$ is a normalization factor and s_{mi} a phase which for isovector-type solutions $s_{mi} = 1$ if m and i refer to proton states, $s_{mi} = -1$ for neutron particle-hole amplitudes, while it is unity for isoscalar-type solutions. The excitation energies are obtained from the secular equation, the graphical solution of which can be found in the original paper and in textbooks[10]. (Distinguishing protons and neutrons and not coupling to a given isospin, the solution of this secular equation gets slightly more involved.)

From (2) the transition matrix element of the quadrupole operator and of the magnetic dipole operator are obtained (omitting the tensor indices of the operators):

$$\langle 0|Q|\nu\rangle = \mathcal{N} \sum_{mi} \frac{Q^*_{mi} Q_{mi}}{E_\nu - \epsilon_m + \epsilon_i}, \qquad \langle 0|\mathcal{M}^{\mathrm{orb}}|\nu\rangle = \mathcal{N} \sum_{mi} \frac{\mathcal{M}^{\mathrm{orb}*}_{mi} Q_{mi}}{E_\nu - \epsilon_m + \epsilon_i}. \tag{3a, b}$$

The sums in eqs. (3 a, b) are assumed to run over proton excitations only. We see that the excitations with unperturbed energies below E_ν contribute all with the same sign to $\langle 0|Q|\nu\rangle$, those with higher energies all with the other sign, and as strongly collective states exist, we know that by this mechanism the isovector quadrupole strength in the energy range of the unperturbed excitations is quenched to almost zero. Since the matrix elements of Q and $\mathcal{M}^{\mathrm{orb}} \propto g_l L$ are not proportional, the magnetic dipole strength can not be as well uniformly quenched. Apparently, depending on the details of the level scheme, in some cases there occurs partially constructive interference with regard to the orbital magnetic transition operator.

The microscopic results do not appeal to classical imagination. In view of the common tendency to interpret the low energy M1 states in terms of classical collective motion (scissors modes), we now deal with the model of a liquid drop having a nonspherical (prolate) equilibrium shape. We assume that $N = Z$ only to keep the expressions simple; it in not a real restriction. For the same reason we only consider isovector modes. Instead of considering pure rotations we include as a second degree of freedom the quadrupole mode of fig. 1 (a). The displacement of a fluid element from its position at rest $\mathbf{r}$ is

$$\mathbf{u}(\mathbf{r}) = \alpha\, \mathbf{e}_x \times \mathbf{r} + \chi\, \nabla yz = \alpha\,(0, -z, y) + \chi(0, z, y). \tag{4}$$

for proton matter, and that of neutron matter is $-\mathbf{u}$. One obtains for the kinetic energy of the motion

$$T = \frac{1}{2}\left(\dot{\alpha}^2 I_1 + 2\dot{\alpha}\dot{\chi}(I_3 - I_1) + \dot{\chi}^2 I_1\right) \tag{5}$$

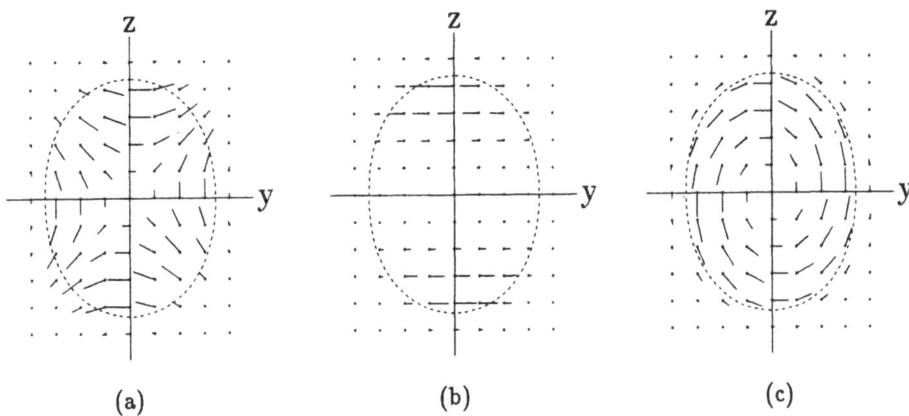

Fig. 1. Current patterns. (a) The component of the quadrupole vibration which from its symmetry can mix with rigid rotation around the x-axis. (b) Schematic drawing of a quantum mechanical rotation. (c) Flow pattern of the low energy mode of eq. (14)

where I_1 is the rigid body moment of inertia around the x-axis, I_3 that around the z-axis. These quantities are used to denote the well known integrals which appear also in the case of the quadrupole motion.

The restoring force is derived from the asymmetry energy[11,12] which in the Bethe-Weizsäcker mass formula is given by

$$E_{as} = \frac{K_{as}(N-Z)^2}{A} = K_{as} \int \frac{(\rho^n - \rho^p)^2}{(\rho^n + \rho^p)} d^3r \tag{6}$$

We have $\delta\rho = \delta\rho^p - \delta\rho^n = -\mathbf{u} \cdot \nabla(\rho^p + \rho^n)$. The nuclear density is parametrized by a Fermi function $\rho(\mathbf{r}) = \rho_0(1 + \exp((r - R(\theta,\phi))/a))^{-1}$ and the nuclear shape by

$$R(\theta,\phi) = \bar{R}\,(1 + \frac{\delta}{2}\cos(2\theta) + \frac{\delta}{6}) \tag{7}$$

which defines the nuclear deformation parameter δ. After some approximations we get with $\delta' := \delta\,(42 + 4\delta)/(42 - 5\delta)$ and $C := 16 K_{as} \pi \rho_0 \bar{R}^4\,(1 - \frac{5}{42}\delta)^2/(15a)$

$$V(\alpha,\chi) = \frac{1}{2} C(\chi - \delta'\,\alpha)^2. \tag{8}$$

This result is remarkable: the two modes which we consider are coupled with each other and are therefore not the normal modes of the system, and, moreover, one of the normal modes can be obtained by inspection: If rotation and vibration occur in the special mixture given by

$$\chi = \delta'\,\alpha, \tag{9}$$

from eq. (8) we see that there is no restoring force. The reason is that the flow is tangential to the nuclear surface, the density distribution and the surface are not changed, and no separation of protons and neutrons occurs. Thus one of the normal modes is a zero energy mode which can not be excited. (The high energy mode deviates only very little from a pure quadrupole vibration.)

There is a way to include quantum corrections in the classical treatment: The method has been worked out more than ten years ago[13] in the frame of 'nuclear fluid dynamics'. The idea is the following: In a vibrating nucleus, the change of shape occurs with a certain speed. In order to always remain in the lowest energy state at any deformation, particles occupying single particle states which move upwards in the Nilsson level scheme would have to make transitions to energetically more favourable states. The assumption of nuclear fluid dynamics has been that the collective motion is so fast that these transitions do not occur. This is a limiting case, the other limit being the adiabatic case when the system sticks to the lowest energy level, avoiding level crossing. Without giving any quantitative argument, we may assume that the truth is somewhere in between, but in the following we assume the maximum effect.

The difference of nonequilibrium and equilibrium kinetic energies acts as an additional restoring potential. The nonequilibrium distribution can be obtained by simple reasoning in the Fermi gas model in the limit of a large system, assuming a volume element which is a cube and periodic boundary conditions on its surface. If now the cube undergoes a volume conserving distortion (the principal axes being parallel to the cube's edges), keeping the boundary conditions fixed and not allowing transitions, the non-equilibrium form of the Fermi surface is easily calculated.

For the nuclear droplet this yields

$$\Delta T_{\mathrm{F}} = \frac{3(3\pi^2)^{\frac{1}{3}}\hbar^2 A}{20\, m\, r_0^2}\chi^2 = \frac{1}{2} D\,\chi^2 \tag{10}$$

A being the nuclear mass number, m the nucleon mass, and $r_0 \approx 1.1$ fm the nuclear radius constant. The second equality defines the additional restoring "potential" with force constant D. — From T, eq. (5) and V, eq. (8) and (10) follow the equations of motion:

$$\begin{aligned} &I_1\ddot{\alpha} - (I_1 - I_3)\ddot{\chi} + C\delta'^2\alpha - C\delta'\chi = 0 \\ &-(I_1 - I_3)\ddot{\alpha} + I_1\ddot{\chi} - C\,\delta'\alpha + (C + D)\,\chi = 0 \end{aligned} \tag{11}$$

Proceeding in standard fashion we set

$$\begin{pmatrix}\alpha \\ \chi\end{pmatrix} = \begin{pmatrix}\alpha_0 \\ \chi_0\end{pmatrix} e^{i\omega t} \tag{12}$$

to get an eigenvalue equation. — In order to keep things simple we restrict ourselves to very small deformation and keep only the lowest powers of δ', though later for the numerical estimates $\delta \approx 0.25$ will be used. — The eigenfrequencies are

$$\omega_1^2 \approx \frac{D}{I_1}\cdot\frac{\delta'^2}{1 + D/C} \qquad \text{and} \qquad \omega_2^2 \approx \frac{C + D}{I_1} \tag{13}$$

Thus for $D \neq 0$ the low frequency mode now has a finite frequency. The corresponding normal modes are

$$(\alpha_0, \chi_0) = (\alpha_0,\ \delta'(1 + D/C)^{-1}\cdot\alpha_0) \qquad \text{and} \qquad (\alpha_0, \chi_0) = (0,\ \chi_0) \tag{14}$$

for the low energy and for the high energy mode, respectively.

The normal vibrations can now be quantized in the usual way. Then one obtains the matrix elements of the transition operator and finally the $B(\mathrm{M1})\!\uparrow$-values

$$B(\mathrm{M1})\!\uparrow = \frac{3}{16\pi\hbar^2}\mu_{\mathrm{N}}^2\hbar\omega\,\Theta \qquad \text{with} \qquad \Theta = \frac{(\alpha_0 I_1 - \chi_0(I_1 - I_3))^2}{\alpha_0^2 I_1 - 2\alpha_0\chi_0(I_1 - I_3) + \chi_0^2 I_1} \tag{15}$$

This formula encompasses the rigid rotor model (setting $\alpha_0 = 1$, $\chi_0 = 0$). In this case, our expressions agree exactly with eqs. (6), (8), and (32) of Lo Iudice et al.[12] if the integral I_{symm} of eq. (33)[12] is approximated by $\int_0^\infty dr\, r^2 (d\rho/dr)^2/\rho \approx R^2 \int_0^\infty dr\,(d\rho/dr)^2/\rho \approx \frac{\rho_0 R^2}{2a}$.

For numerical estimates, we use $A = 164$, $\delta = 0.256$, $K_{as} = 28.1$ MeV, $r_0 = 1.12$ fm, $a_0 = 0.54$ fm, $\rho_0 = 0.17\,\mathrm{fm}^{-3}$, and in the evaluation of the moments of inertia assume a sharp nuclear surface. We get $D/C = 0.128$. For comparison we also give the results for the two–rigid-rotors model as obtained with the same set of constants, and for the high energy vibrational mode:

Low energy mode: $\hbar\omega = 2.23$ MeV, $B(\mathrm{M1})\uparrow = 8.23\,\mu_N^2$,
Rigid rotors (scissors mode): $\hbar\omega_{\mathrm{rig}} \approx 7.5$ MeV, $B(\mathrm{M1})\uparrow_{\mathrm{rig}} \approx 30\,\mu_N^2$,
High energy vibrational mode: $\hbar\omega_{\mathrm{vib}} = 26$ MeV, $B(\mathrm{M1})\uparrow_{\mathrm{vib}} = 5.9\,\mu_N^2$.

The classical model which correctly includes the coupling to the vibrational degree of freedom does not yield a low energy rotational vibration, but instead a zero energy mode with zero excitation probability. The flow pattern differs from that of rigid rotation only by a small admixture of quadrupole vibration — the overlap of quantum mechanical state vectors would be close to unity — yet this small deviation is quite important and makes a qualitative difference.

Only after the introduction of quantum corrections a finite energy results, but the mechanism supplying the restoring force is completely different from that assumed previously. We get essentially the same proportionality of the $B(\mathrm{M1})$-value and excitation energy as in the two rotor model, and so while the excitation energy is slightly underestimated, too much transition strength is obtained by a factor of ≈ 2 (if we compare with the total strength at low energy). In the microscopic calculation, pairing strongly quenches the strength without much change of the excitation energies. Pairing effects are missing in our semiclassical treatment.

The flow pattern of the low energy mode is sketched in fig. 1 (c), it differs only very slightly from that of the classical zero mode, cf. eq. (9) and (14). It is remarkable that it exhibits some of the features which have been found in the transition densities calculated microscopically[5], viz. that at the surface the flow is almost tangential, and in contrast to the 'qantum rotation' (see fig. 1 (b)), it is not small in the inside of the nucleus. Of course we can not get details like the vorticity which depends on the single particle states involved; but apart of this the agreement with microscopic results seems to be more than an accident. The M1 form factors derived from this current density should deviate only very little from the corresponding rigid rotor result.

SPIN-FLIP EXCITATIONS

The spin-flip strength distribution exhibits a two-hump structure[3] which is easily reproduced in microscopic calculations[14]. In contrast to the electric dipole GR, these two humps are definitely not caused by a separation of $K = 0$ and $|K| = 1$ components due to deformation. An interplay of two tendencies can be observed: by the deformation, the distribution in energy of the noninteracting spin-flip particle hole excitations is widened as a consequence of the "Nilsson-model" level splitting. The interaction tends to single out one collective isovector state at an energy higher than that of the unperturbed spin-flip excitations, but is not strong enough to redistribute all the strength. Of course, it is the spin- and the spin-isospin dependent part of the force which is important here; but with the proper replacements the schematic model as discussed above (eqs. (2) and (3 a)) here too accounts for the qualitative features. Approximately half of the transition strength remains in the energy region of the unperturbed excitations, if the interaction is switched on. This is illustrated by fig. 2. The double-hump structure thus reflects the intermediate situation where only part of the strength is collected into a 'giant resonance in statu nascendi' (upper hump) while the remaining strength is distributed among 'noncollective states' (lower hump).

It has been proposed that the structure might be explained by the separation of proton and neutron spin flip excitations [15]. To check this, we have plotted our results also in another form (fig. 3): In the lower part, the proton- and neutron- $B_\sigma(\mathrm{M1})\uparrow$ distributions are plotted, in the upper part the sum of these (dotted line) and the total $B_\sigma(\mathrm{M1})\uparrow$ strength distribution. Gaussians with energy dependent widths have been folded into the discrete distributions[14] to improve the optical impression. The difference between the two lines in the upper part of the figure is due to interference between the proton- and neutron-components of the single states. For a pure isovector state the enhancement would be by a factor of 1.92; for pure proton- and

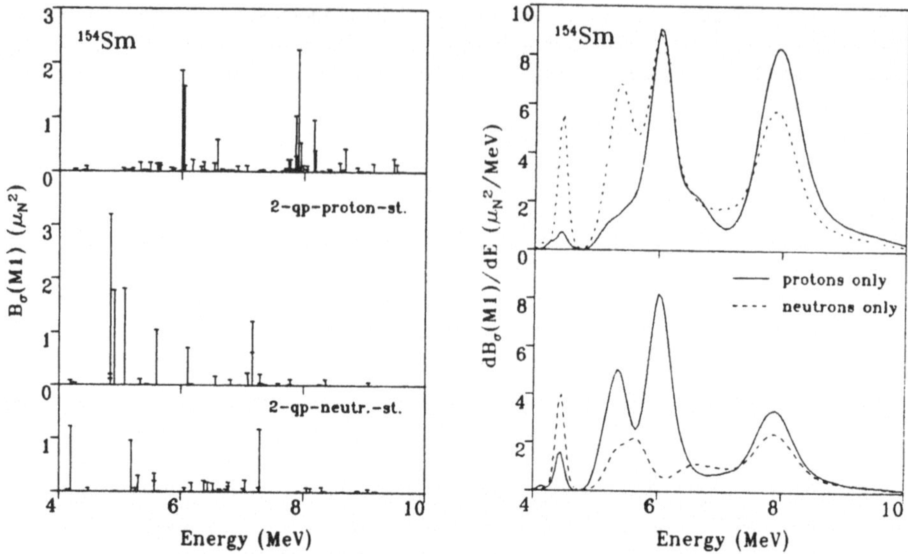

Fig. 2 (left). Top: The strength distribution of spin-flip M1 transitions for ^{154}Sm obtained in quasiparticle-RPA. Middle and bottom part: M1-transition strength for the unperturbed two-quasiparticle states.

Fig. 3 (right). The data of fig. 2 have been folded with gaussians[14] to obtain the continuous distribution shown in the upper part, solid line. The lines of the lower part have been obtained by alternatively putting equal to zero the g_s-factors of neutrons and protons. The sum of these contributions is given by the dotted line in the upper part.

neutron states (regardless if separated or interlacing in energy) there would be no interference, whereas for an isoscalar state the enhancement (or rather depletion) would be by a factor of 0.08. The real situation is again the intermediate one, with an enhancement factor of almost 1.5 for the high energy bump.

REFERENCES

1. D. Bohle et al., Phys. Lett. B **137** : 27 (1984)
2. A. Richter, pp. 127 – 164 in: Contemporary Topics in Nuclear Structure Physics, edited by F. Casten et al., World Scientific, Singapore (1988)
3. D. Frekers et al., Phys. Lett. B **244** : 178 (1990)
4. J. Speth and D. Zawischa, Phys. Lett. B **211** : 247 (1988) and B **219** : 529 (1989)
5. D. Zawischa and J. Speth, Z. Physik A, Spring 1991 (in press)
6. A. Faessler and R. Nojarov, Phys. Lett. B **215** : 439 (1988)
7. F.G. Scholtz, R. Nojarov, A. Faessler, Phys. Rev. Lett. **63** : 1356 (1989)
8. R. Nojarov and A. Faessler, Z. Phys. A **336** : 151 (1990)
9. G.E. Brown and M. Bolsterli, Phys. Rev. Lett. **3** : 472 (1959)
10. P. Ring and P. Schuck, The Nuclear Many-Body Problem. Springer, New York - Heidelberg - Berlin (1980)
11. R. Nojarov, Z. Bochnacki, and A. Faessler, Z. Phys. A **324** : 289 (1986)
12. N. Lo Iudice, F. Palumbo, A. Richter, and H.J. Wörtche, Phys. Rev. C **42** : 241 (1990)
13. G. Holzwarth and G. Eckart, Nucl. Phys. **A325** : 1 (1979)
14. D. Zawischa and J. Speth, Phys. Lett. B **252** : 4 (1990)
15. C. De Coster and K. Heyde, Phys. Rev. Lett. **63** : 2797 (1989)

VI - OCTUPOLES

REFLECTION-ASYMMETRIC SHAPES IN ATOMIC NUCLEI

W. Nazarewicz*

Joint Institute for Heavy-Ion Research
Holifield Heavy Ion Research Facility
P.O. Box 2008, Oak Ridge, Tennessee 37831, U.S.A.

ABSTRACT

Can atomic nuclei be unstable with respect to deformations that break intrinsic parity? On the theoretical side calculations indicate the existence of stable octupole deformations. There is also vast supporting experimental evidence for the presence of very collective low-energy dipole and octupole modes. In this contribution, recent advances in the physics of nuclear reflection-asymmetric shapes are discussed in terms of underlying shell effects. Particular attention is given to the recently predicted octupole excitations at superdeformed shapes.

INTRODUCTION

Why can certain nuclei be described in terms of intrinsic shapes with parity-breaking static moments? At first glance such a violation of a very fundamental symmetry is astonishing since strong interactions do actually conserve parity. One has to bear in mind, however, that the description in terms of the *intrinsic system* automatically involves the symmetry breaking mechanism. Of course, the fundamental quantum numbers, such as angular momentum, parity, baryon number, etc., associated with the basic space-time symmetries, are conserved in the laboratory system. On the other hand, in the body-fixed system of a nucleus the many-body wave function is, usually, not an eigenstate of fundamental symmetry operators. From this viewpoint the concept of nuclear stable octupole deformation is analogous to the idea of permanently quadrupole-deformed nuclear shapes. Therefore, if one accepts the possibility of stable quadrupole deformation in the ground state of the even-even nucleus, one cannot reject the picture of nuclear shapes that do not conserve *intrinsic parity*.

Microscopically, the symmetry breaking mechanism is always associated with pairs of quantum-mechanical states that are (almost) degenerate in energy. In quantal systems such a hybridization leads to reduced stability and even an infinitely small perturbation of a degenerate system produces a final response in the system due to the rearrangement of many close states. For instance, quadrupole deformation can be immediately associated with the residual interaction acting between spherical single-

*On leave from the Institute of Theoretical Physics, Warsaw University

Nuclear Shapes and Nuclear Structure at Low Excitation Energies
Edited by M. Vergnes *et al.*, Plenum Press, New York, 1992

particle states of the isotropic harmonic oscillator. Indeed, as soon as the Fermi level enters the valence shell the (ℓ, j) degeneracy is removed by the quadrupole–quadrupole force and the nucleus becomes deformed.

An analogous coupling between intrinsic states of opposite parity is produced by the long-range octupole-octupole residual interaction. In some cases the mixing is so strong that the nucleus acquires stable octupole deformation in the body-fixed frame. For normally-deformed systems the condition for strong octupole coupling occurs for particle numbers associated with the maximum $\Delta N=1$ interaction between the intruder subshell (ℓ, j) and the normal parity subshell $(\ell - 3, j - 3)$. The regions of nuclei with strong octupole correlations correspond to particle numbers near 34 ($g_{9/2} \leftrightarrow p_{3/2}$ coupling), 56 ($h_{11/2} \leftrightarrow d_{5/2}$ coupling), 88 ($i_{13/2} \leftrightarrow f_{7/2}$ coupling), and 134 ($j_{15/2} \leftrightarrow g_{9/2}$ coupling), i.e. the tendency towards octupole deformation occurs just above closed shells.

The main goal of this presentation is not to cover the whole field of nuclear octupole and dipole modes. For this we rather refer the reader to recent reviews.[1-7] Instead, I would like to emphasize selected aspects of octupole collectivity. In particular, I will discuss "hidden" symmetries of the anisotropic harmonic oscillator and their apparent consequences for the presence of superdeformed reflection–asymmetric shapes.

REFLECTION–ASYMMETRIC SHAPES AT LOW EXCITATION ENERGY

For the Ra-Th (Z~88, N~134) and Ba-Sm (Z~56, N~88) nuclei, the features of stable octupole deformation, namely low-lying negative-parity states, parity doublets, alternating parity bands with enhanced E1 transitions have been established (for a recent review see, e.g., ref. [7]).

Many mean-field calculations predict reflection instability for nuclei around ^{222}Th and ^{146}Ba. There are calculations based on the Nilsson-Strutinsky approach with folded-Yukawa and Woods-Saxon deformed potentials[8-14] as well as Hartree–Fock calculations based on the Skyrme or Gogny forces.[15-20] All these models yield similar results for stable octupole deformations, but give slightly different predictions for the height of the octupole barrier.

The calculated octupole minima are usually very shallow with octupole barriers varying between 0.5 and 2 MeV, depending on the model. Consequently, dynamic fluctuations are expected to play a significant role. Some attempts towards including dynamic corrections by means of parity projection or gaussian overlap approximation have been made in refs.[17-20]

Deformation selfconsistent cranking calculations confirm empirical results, i.e. the stability of octupole deformation increases at medium spins.[21-23] The enhancement of octupole strength with rotation is caused by (i) weaker pairing correlations for the octupole shape, which increase the moments of inertia, and (ii) by the octupole mixing between single-particle states of opposite parity approaching each other with frequency.

In the presence of low-lying octupole excitations a large E1 moment may arise in the intrinsic frame due to a shift between the centre of charge and the centre of mass. Such a dipole moment manifests itself by very enhanced electric dipole transitions between opposite parity members of quasi-molecular rotational bands. Many properties of low-lying E1 modes in the transitional nuclei near ^{146}Ba and ^{224}Th can be explained by the reflection-asymmetric deformed shell model theory based on the Strutinsky renormalisation procedure.[22,24-26] The calculations reproduce the systematic trends seen in the experimental data and the vanishing values of the intrinsic dipole moments, D_o, in certain nuclei. In particular, it has been demonstrated[26] that the macroscopic contribution to the intrinsic dipole moment based on the droplet model is very small for lanthanides and becomes important in actinide nuclei, thus providing a consistent description of experimental data in both mass regions. This effect is partly related to the presence of the neutron-skin contribution to D_0. Recently, first attempts to calculate intrinsic dipole moments within the full selfconsistent theory have been made. Egido and Robledo used the parity-projected HF+BCS model with Gogny interaction to reproduce B(E1) rates in the Ra-Th [19] and Ba [20] nuclei. They obtained excellent agreement with experimental data. For more discussion related to the question of "collective" intrinsic dipole moments I refer the reader to contribution by P. Butler.[27]

The complex interplay between rotation and reflection-asymmetric and triaxial degrees of freedom is most pronounced in transitional nuclei, like ^{218}Ra or ^{148}Sm, which do not have well developed ground-state deformation but still exhibit "quasirotational" bands of alternating parity and large B(E1) rates. According to the calculations[22, 23] octupole instability in these nuclei persists up to very high rotational frequencies. Such a pattern with low quadrupole- and strong octupole- and dipole collectivity has recently been observed in ^{218}Ra (ref. [28]) and ^{148}Sm (ref. [29]).

Other candidates for the strong octupole–quadrupole coupling can be find in medium-mass nuclei. In the $A\sim70$ mass region the shell correction calculations[10] predict octupole softness only in the transitional isotopes of Zn-Se with $N\leq36$. A recent study of the nucleus ^{64}Ge [30] suggests that a coupling between reflection-asymmetric and triaxial degrees of freedom takes place in this exotic $N=Z$ system†.

The nonaxial, $K\neq0$, octupole components should certainly play a significant role at nearly spherical shapes. Very good testing-grounds for investigation the coupling between octupole excitations and triaxiality are the nuclei around ^{96}Zr, see discussion in refs.[7, 32]

NUCLEAR SHELL STRUCTURE AND OCTUPOLE DEFORMATION

Since the normal-parity single–particle states form a pseudo–oscillator pattern reflecting the presence of the pseudo-SU(3) symmetry[33-36] one can argue that many properties of deformed and superdeformed states should reflect basic features of the harmonic oscillator. In this section I shall discuss some properties of the single-particle shell structure of a three-dimensional harmonic oscillator with frequencies in rational ratios (RHO). The RHO classification scheme leads to the geometrical "multicluster" picture of nuclear deformed shapes. This simplistic picture is strongly supported by the behaviour of the octupole shell–forces of the RHO and the presented results of microscopic calculations.

Quantum Numbers of the Rational Harmonic Oscillator

The RHO exhibits the strongest level degeneracy that results in an appearance of spherical and deformed magic gaps. The problem of unusual degeneracies of the single–particle levels in the RHO has been extensively discussed in the literature, see reviews [37, 38] and refs. therein.

The single-particle Hamiltonian of the three–dimensional oscillator potential is given by

$$\hat{H} = \frac{1}{2}\sum_{i=1}^{3}(p_i^2 + \omega_i^2 x_i^2) = \frac{1}{2}\sum_{i=1}^{3}\omega_i\{a_i, a_i^+\}, \tag{1}$$

where we assume $M=1$, $\hbar=1$. The single-particle energies of (1) are

$$e_{n_1,n_2,n_3} = \sum_{i=1}^{3}\omega_i\left(n_i + \frac{1}{2}\right), \tag{2}$$

†In the recent work[31] Skalski employed parameterisation of nuclear shapes involving both β_3 and γ and confirmed octupole instability and γ-softness in ^{64}Ge.

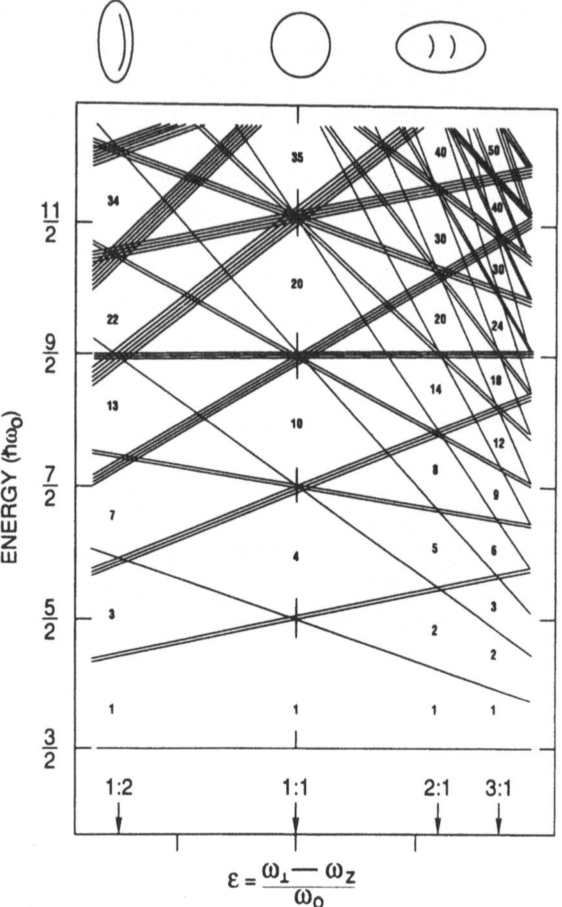

Fig. 1. Single-particle level spectrum of the axially symmetric harmonic oscillator shown as a function of quadrupole deformation ε. Here $\omega_o = \frac{1}{3}(2\omega_\perp + \omega_3)$. The orbital degeneracy is $n_\perp+1$, which is illustrated by artificially splitting the lines. The arrows indicate the characteristic deformations corresponding to the ratio of $\omega_\perp : \omega_3 =$ 1:2, 1:1, 2:1 and 3:1.

where n_i are numbers of oscillator quanta in three spatial directions. The single-particle diagram of the axially-deformed harmonic oscillator is displayed in fig. 1. Here $\omega_1 = \omega_2 = \omega_\perp$ and the single–particle energies depend only on two quantum numbers: $n_\perp = n_1 + n_2$ and n_3. Due to the axial symmetry each level is $(n_\perp + 1)$ times degenerate.

For the RHO the ratios of frequencies are rational numbers. This condition can be expressed in terms of three integers k_i, i=1,2,3:

$$\omega_i k_i = \tilde{\omega}, \tag{3}$$

where $\tilde{\omega}$ can be calculated from the volume conservation condition, $\omega_1\omega_2\omega_3 = \tilde{\omega}_o^3$, and is equal to

$$\tilde{\omega} = \sqrt[3]{k_1 k_2 k_3} \cdot \tilde{\omega}_o, \quad \tilde{\omega}_o = \frac{41}{A^{1/3}}. \tag{4}$$

For example, the spherical shape corresponds to $k_1=k_2=k_3=1$, while the axial shapes have $k_1=k_2=k_\perp$. In the axial case it is convenient to write the single-particle energies in terms of the shell frequency, ω_{shell}, and the shell principal quantum, N_{shell}, defined by [39]:

$$\omega_\perp n_\perp + \omega_3 n_z = \omega_{shell} N_{shell}, \quad N_{shell} = n_\perp k_3 + n_3 k_\perp, \quad \omega_{shell} = \frac{\tilde{\omega}}{k_\perp k_3}. \tag{5}$$

Let us now introduce new quantum numbers ν_i and λ_i:

$$n_i = k_i \nu_i + \lambda_i, \qquad \nu_i = \left[\frac{n_i}{k_i}\right], \qquad \lambda_i = n_i \,(\mathrm{mod}\, k_i), \tag{6}$$

where the symbol $[x]$ stands for the integer part of x. Single-particle energies of the RHO can be thus written as

$$e_{M;\lambda_1\lambda_2\lambda_3} = \tilde{\omega} M + \tilde{\omega} \sum_{i=1}^{3} \frac{\lambda_i + \frac{1}{2}}{k_i}, \tag{7}$$

where $M = \nu_1 + \nu_2 + \nu_3$ is the new principal quantum number. At fixed values of λ_i the level degeneracy is equal to $\frac{1}{2}(M+1)(M+2)$, i.e. it corresponds exactly to the degeneracy of a spherical oscillator with principal quantum number M. In the next step one can introduce new ladder operators[40-42]

$$A_i^{\{\lambda\}} = \frac{1}{\sqrt{k_i}} (a_i)^{k_i} \left(\frac{\hat{n}_i - \lambda_i}{\hat{n}_i(\hat{n}_i - 1)...(\hat{n}_i - k_i + 1)} \right)^{\frac{1}{2}} = (a_i)^{k_i} \left(\left[\frac{\hat{n}_i}{k_i}\right] \frac{(\hat{n}_i - k_i)!}{\hat{n}_i!} \right)^{\frac{1}{2}}, \tag{8}$$

where $\hat{n}_i = a_i^+ a_i$ is the boson number operator and $\{\lambda\} \equiv \{\lambda_1\lambda_2\lambda_3\}$. It is easy to verify that operators (8) indeed fulfil the standard boson commutation rules, i.e.

$$\left[A_i^{\{\lambda\}}, A_j^{\{\lambda\}+}\right] = \delta_{i,j}. \tag{9}$$

The new boson operator $A_i^{\{\lambda\}}$ acts only on the quantum number ν_i *leaving* λ_i *unchanged*,

$$A_j^{\{\lambda\}+} | \cdots, \nu_i k_i + \lambda_i, \cdots\rangle = \sqrt{\nu_i + 1} | \cdots, (\nu_i + 1) k_i + \lambda_i, \cdots\rangle. \tag{10}$$

Now, analogously to the well-known spherical case, one can construct eight generators linear in $\left\{A_i^{\{\lambda\}}, A_i^{\{\lambda\}+}\right\}$ that fulfil the commutation rules of SU(3) and commute with the RHO Hamiltonian. This proves that the dynamical symmetry in question is SU(3). The RHO eigenstates belonging to the same $\{\lambda\}$ family form for a given M the basis of an irreducible symmetric representation (irrep) of SU(3), $(M, \{\lambda\})$. Each family has a corresponding ground state belonging to the one-dimensional representation of SU(3) for $M=0$, which is the vacuum for the new bosons, eq. (8). Since $0 \le \lambda_i < k_i$ the number of $\{\lambda\}$-families is equal to $k_1 k_2 k_3$.

At the spherical shape, $k_1=k_2=k_3=1$, there is only one family present, labelled by $\{\lambda\}=(000)$. The degeneracy of each level is $\frac{1}{2}(M+1)(M+2)$ and the magic gaps occur at particle numbers

$$N_M = \frac{1}{6}(M+1)(M+2)(M+3) = 1,\ 4,\ 10,\ 20,\ 35,\ 56,\ 84,\ \cdots \tag{11}$$

Other cases of significant physical interest are the superdeformed prolate ($k_1=1$, $k_2=1$, $k_3=2$) and the hyperdeformed prolate ($k_1=1$, $k_2=1$, $k_3=3$) shapes. Here, since

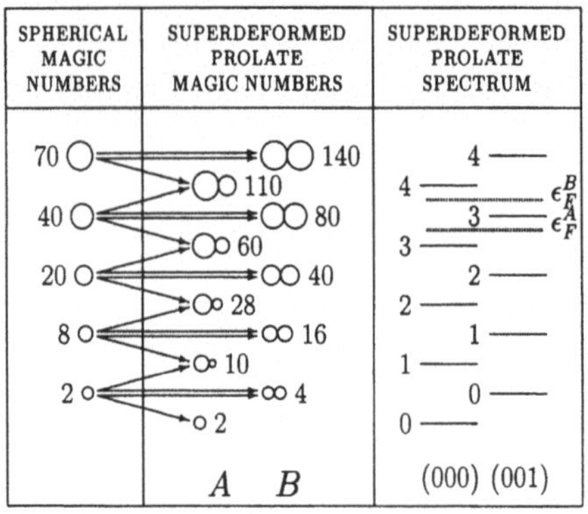

SPHERICAL MAGIC NUMBERS	SUPERDEFORMED PROLATE MAGIC NUMBERS	SUPERDEFORMED PROLATE SPECTRUM
70	140	4
	110	4 ϵ_F^B
40	80	3 ϵ_F^A
	60	3
20	40	2
	28	2
8	16	1
	10	1
2	4	0
	2	0
	A B	(000) (001)

Fig. 2. Magic numbers and spectrum of the prolate superdeformed (k_1=1, k_2=1, k_3=2) harmonic oscillator. The arrows indicate from which spherical oscillator representations a given superdeformed representation is built (see text). The sizes of circles schematically illustrate the dimensions of spherical representations. The harmonic oscillator degeneracies have been doubled to take into account the spin. Two different positions of the Fermi level, ϵ_F^A and ϵ_F^B, are indicated in the spectrum. The quantum numbers $(\lambda_1, \lambda_2, \lambda_3)$ of two λ-families are shown below the spectrum. (Taken from ref. [38].)

$\lambda_1=\lambda_2=0$, the number of independent SU(3) irreps for a given M is equal simply to k_3 and they can be easily distinguished by means of λ_3.

As seen in fig. 2 there are two kinds of closed-shell systems that are expected at superdeformed shapes. In the "asymmetric" case, indicated as A, the number of filled shells within the family (λ)=(000) is larger by one than that within the family (λ)=(001). In fig. 2 this is illustrated by the position of the Fermi level ϵ_F^A. Consequently the magic numbers are then equal to sums of two consecutive spherical magic numbers and read $N_{M+1}^{(000)}+N_M^{(001)}$=1, 5, 14, 30, 55, etc. The numbers shown in the figure have been doubled to include the spin degeneracy. In the "symmetric" variant B the missing (001) shell is filled and the magic numbers are equal to doubled spherical oscillator magic numbers, $N_M^{(000)}+N_M^{(001)}$=2, 8, 20, 40, 70, etc.

The situation becomes slightly more complex at hyperdeformed shapes, see fig. 3. In the "strongly asymmetric" variant A the number of filled shells within the family (000) is larger by one than those of the families (001) and (002), which leads to magic numbers $N_{M+1}^{000}+N_M^{001}+N_M^{002}$=1, 6, 18, 40, 75, etc. In the variant B the occupation of the family (002) is lower than those of the families (000) and (001) and the resulting magic numbers are $N_{M+1}^{000}+N_{M+1}^{001}+N_M^{002}$=2, 9, 24, 50, 90, etc. Finally, in the "symmetric" case C the occupations of families (000), (001) and (002) are identical and the magic numbers are equal to tripled spherical oscillator magic numbers, $N_M^{000}+N_M^{001}+N_M^{002}$=3, 12, 30, 60 and 105, etc.

The examples discussed so far represent the simplest possible situation in which the numbers k_i are relatively prime. The degeneracy pattern becomes, however, rather

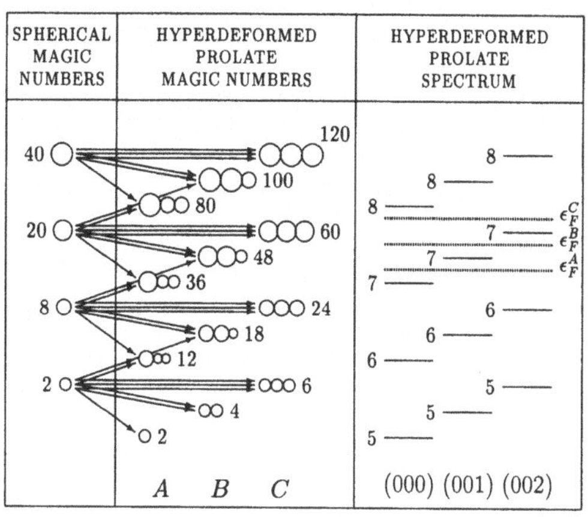

Fig. 3. Similar to fig. 2 but for the prolate hyperdeformed ($k_1=1$, $k_2=1$, $k_3=3$) harmonic oscillator. Three different positions of the Fermi level, ϵ_F^A, ϵ_F^B, and ϵ_F^C, are indicated in the spectrum. The quantum numbers $(\lambda_1\lambda_2\lambda_3)$ of three λ-families are shown below the spectrum. (Taken from ref. [38].)

complicated in the case when two k_i's have a common multiplier. Such a situation is present where different irreps of SU(3) are degenerate, e.g., on the oblate side, $k_1=k_2=k_\perp>1$, $k_3=1$.

In order to label the oblate superdeformed shells of the RHO, one can introduce quantum numbers $\lambda_\perp$ and $\tilde{N}$ defined in a similar way as λ_i (see e.g. ref.[43]):

$$\lambda_\perp = \nu_\perp (\mathrm{mod}\, k_\perp), \quad \tilde{N} = \nu_\perp + \nu_3, \tag{12}$$

where $\nu_\perp=[n_\perp/k_\perp]$. Contrary to the case of prolate shapes the shell degeneracy depends explicitly on $\lambda_\perp$:

$$n(\tilde{N},\lambda_\perp) = \frac{1}{2}(\tilde{N}+1)[k_\perp\tilde{N}+2(\lambda_\perp+1)]. \tag{13}$$

It can be shown (see detailed discussion in ref.[38]) that (i) the degeneracy pattern of the RHO on the oblate side represented by labelling (12) corresponds in fact to reducible representations of SU(3) [44], and (ii) in the particular case of superdeformed oblate nuclei, $k_\perp=2$, $k_3=1$, the dynamical symmetry is O(4) [45].

Multicluster Model

According to eq. (5) the energy difference between neighbouring oscillator shells, $\tilde{\omega}_{shell}$, decreases smoothly with deformation. This indicates that the overall magnitude of the shell effects is expected to be strongest at the spherical shape. Below are discussed some examples that indicate that even at very strong elongations the appearing shell structure leads to an enhanced stability similar to that observed for spherical shell gaps. Moreover, we will describe the deformed shell-stabilized systems in terms of "multiclusters" of spherical subsystems (clusters), as dictated by the decomposition of the RHO representations into the isotropic ones, as described in the previous section. Of course, the term "cluster" should not be understood in the most direct sense of a spatial spherical cluster, since in medium mass and heavy nuclei the probability of clustering into large fragments is strongly inhibited by the Pauli principle. However, it turns out that the group-theory symmetries of these clusters induce some properties of superdeformed states as if the clustering occurred in the real space.

> The main assumption of the "cluster" model is that every $\{\lambda\}$–family (an SU(3) oscillator) should correspond to an independent fragment. The number of fragments is then equal to the number of one-dimensional irreps of SU(3), i.e. it is equal to $k_1k_2k_3$.

Octupole Shell Force of the RHO

The degenerate shell of the RHO consists, in general, of states having different parities. Indeed, the generalized Bose operators (8) are parity–even for even values of k_i and parity–odd for odd values of k_i. Consequently, the total parity of single–particle state $|\hat{n}\rangle=|n_1n_2n_3\rangle$ can be written as

$$\pi_{\hat{n}} = (-1)^N = (-1)^{\lambda_1+\lambda_2+\lambda_3}(-1)^{k_1\nu_1+k_2\nu_2+k_3\nu_3} = \pi_\lambda\pi_\nu. \tag{14}$$

The above expression can be given a simple interpretation. The parity π_λ is the intrinsic parity of the corresponding bosonic vacuum whilst π_ν represents the parity of an excited mode. In the case of superdeformed prolate shapes with even values of k_3, eq. (14) reduces to

$$\pi_{\hat{n}} = (-1)^{\lambda_3}(-1)^{n_\perp} = (-1)^{\lambda_3+\Lambda}, \tag{15}$$

where Λ is the projection of the orbital angular momentum on the symmetry axis (z-axis).

The fact that degenerate single–particle orbitals have different parities has interesting consequences for the octupole mode, $Q_{3K}=r^3Y_{3K}$, since the optimum condition for the level hybridization is met. Table 1 shows the energies of particle–hole excitations associated with various components of octupole tensor. Let us first consider the superdeformed shape with $k_\perp=1$ and $k_3=2$. The $K=1$ and $K=3$ octupole components conserve intrinsic parity π_λ. Interestingly, since $\Delta E=2\omega_3-\omega_\perp=0$, for $K=1$ there exist non–vanishing matrix elements between states belonging to the same supershell. This suggests that superdeformed magic prolate nuclei are potential candidates for stable "banana shapes" [46, 47], $K=1$ octupole deformations[‡]. The $K=0$ and $K=2$ interactions act only between states with opposite values of π_λ. At the superdeformed oblate shape with $k_\perp=2$, $k_3=1$, this scenario is reversed: the $K=0$ and $K=2$ modes conserve π_λ. By

Table 1. Energies of the particle–hole excitation, ΔE, associated with the octupole interaction Q_{3K}.

K	$\Delta E/\hbar$	Optimal conditions for instability
0	ω_3, $2\omega_\perp$-ω_3, $2\omega_\perp+\omega_3$, $3\omega_3$	superdeformed oblate shapes
1	$\omega_\perp$, $2\omega_3$-$\omega_\perp$, $2\omega_3+\omega_\perp$, $3\omega_\perp$	superdeformed prolate shapes
2	ω_3, $2\omega_\perp$-ω_3, $2\omega_\perp+\omega_3$	superdeformed oblate shapes
3	$\omega_\perp$, $3\omega_\perp$	no instability

inspecting Table 1 one can immediately conclude that the superdeformed oblate nuclei should be unstable with respect to K=0 and K=2 octupole fields (ΔE=$2\omega_\perp$-ω_3=0!).

In order to analyze the octupole couplings in the RHO model, the doubly-stretched octupole interaction,

$$Q''_{3K} = r''^3 Y_{3K}(\Omega''), \quad x''_i \equiv \frac{\omega_i}{\omega_o} x_i, \ (i = 1,2,3) \tag{16}$$

of Sakamoto and Kishimoto [48, 49] has been used. This interaction can be viewed as an improved conventional multipole–multipole force. Firstly, it satisfies the nuclear-selfconsistency§ rigorously even if the system is deformed. Secondly, it yields the zero-energy RPA spurious modes, i.e. it automatically separates the translational and reorientation modes. Last but not least, for the doubly-stretched interaction the coupling between octupole and dipole modes disappears.

Let us first discuss properties of low-lying octupole modes within the RPA formalism. The RPA equation for the excitation energy, ω, is given by the dispersion relation

$$\frac{1}{\kappa_{3K}^{self}} - R_{3K}(\omega) = 0, \tag{17}$$

where κ_{3K}^{self} is the self-consistent octupole coupling strength[48],

$$\begin{aligned} \frac{4\pi M\omega_o^2}{7\kappa_{3K}^{self}} &= \langle (r^4)''\rangle_0 + \frac{2}{7}(4-K^2)\langle (r^4 P_2)''\rangle_0 \\ &+ \frac{1}{84}\left[K^2(7K^2-67)+72\right]\langle (r^4 P_4)''\rangle_0, \end{aligned} \tag{18}$$

and

$$R_{3K}(\omega) = \sum_i \frac{E_i - E_0}{(E_i - E_0)^2 - \omega^2}\left[|\langle i|Q''_{3K}|0\rangle|^2 + |\langle i|Q''^{*}_{3K}|0\rangle|^2\right] \tag{19}$$

is the RPA response function. The value of $R_{3K}(\omega=0)/2$ is the inverse energy-weighted sum rule S_{-1}, which can be related to the *microscopic* interaction strength, κ_{3K}^{mic}, by [50]

$$\kappa_{3K}^{mic} = (2S_{-1})^{-1}. \tag{20}$$

‡Strictly speaking the instability is not expected for doubly magic systems since the corresponding coupling disappears. The optimal situation is expected to occur in nuclei that are singly magic. In this case the superdeformed shape is stabilized by the magic neutron/proton gap whilst the instability is caused by valence protons/neutrons.

§The shape of the potential and that of the density must be the same [39].

Consequently, if the quantity

$$I_{3K} \equiv (\kappa_{3K}^{self})^{-1} - (\kappa_{3K}^{mic})^{-1} \tag{21}$$

is negative (positive), then the lowest energy octupole mode is unstable (stable) with respect to permanent deformation. In order to check the susceptibility of the RHO to octupole distortions one can thus calculate I_{3K} as a function of the shell filling. For the RHO this can be done analytically. Below are presented selected result of the calculations for the $K=0$ mode (and closed-shell systems).

(i) **Spherical case,** $k_1=k_2=k_3=1$. Here

$$I_{30}^{sph} = \frac{7}{4\pi}\frac{1}{\omega_o^4}A\frac{5}{2} > 0, \tag{22}$$

where $A=N_M$ is the particle number, cf. eq. (11), and M is the principal quantum number of the last occupied shell. It is seen that the spherical magic harmonic oscillator is *stable* with respect to octupole distortion and I_{30} does not exhibit any shell fluctuations¶.

(ii) **Superdeformed case,** $k_1=k_2=1$, $k_3=2$. Here, the result depends on the value of N_{shell}, i.e.

$$I_{30}^{SD} = \frac{7}{40\pi}\frac{\omega_3^2}{\omega_o^6}A \times \begin{cases} 37 & \text{asymmetric system, case } A \text{ of fig. 2} \\ 73 & \text{symmetric system, case } B \text{ of fig. 2.} \end{cases} \tag{23}$$

(iii) **Hyperdeformed case,** $k_1=k_2=1$, $k_3=3$. Here we consider three positions of the Fermi level, see fig. 3. Also in this case the result depends on N_{shell}:

$$I_{30}^{HD} = \frac{7}{1120\pi}\frac{\omega_3^2}{\omega_o^6}A \times \begin{cases} 2252\left(1 - \frac{0.58}{M+2}\right) & \text{asymmetric system, } \epsilon_F^A \\ 2252\left(1 + \frac{0.58}{M+2}\right) & \text{asymmetric system, } \epsilon_F^B \\ 4316 & \text{symmetric system, } \epsilon_F^C. \end{cases} \tag{24}$$

As seen from eqs. (23-24) there exists a correlation between the predictions of the geometrical multi-cluster model and the underlying single-particle picture. Namely, for the systems expected to be asymmetric the value of I_{30} is small, and it increases for more symmetric multicluster configurations. This result is already quite encouraging. However, no octupole instability has been predicted by the RPA since in all cases $I_{30}>0$. On the other hand, it is well known that the deformed shell model alone (here: the RHO) is not able to predict correctly the nuclear binding and deformation energies since it partly neglects the interaction energy arising from the two-body effective forces [50]. According to the Strutinsky energy theorem the fluctuating part of the total energy, the shell-correction, is, however, reproduced fairly well by the independent single-particle model.

Since we know that the smooth energy of the harmonic oscillator is a very poor approximation to the liquid drop energy we should not expect the RPA result discussed above to be very accurate‖. In the next step, therefore, we calculate the shell driving force associated with the doubly-stretched octupole interactions.

¶It is worth noting that the exact result (22) comes from from the delicate cancelation between the M-dependent terms. In ref. [48], where the the terms of $O(1/M^2)$ were neglected, the authors obtained $I_{30}=0$ also for spherical closed-shell systems.

‖The RPA is the harmonic expansion around the equilibrium point and its predictions depend crucially on the curvature of the total potential energy.

In the presence of the small perturbing potential, V, the total shell correction can be written as

$$\delta E_{shell} = \delta E^{(0)}_{shell} + \delta E^{(1)}_{shell} + \delta E^{(2)}_{shell} + ... \tag{25}$$

where $\delta E^{(0)}_{shell}$ is the unperturbed shell correction,

$$\delta E^{(1)}_{shell} = \sum_{\alpha=1}^{A} V_{\alpha\alpha} - \sum_{\alpha=1}^{\infty} V_{\alpha\alpha}\tilde{n}_{\alpha} \tag{26}$$

is the first order correction to δE_{shell} [51], and

$$\delta E^{(2)}_{shell} = \sum_{\alpha=1}^{A}\sum_{\beta=A+1}^{\infty} \frac{|V_{\alpha\beta}|^2}{E_0 - E_{\alpha\beta}} - \sum_{\alpha=1}^{\infty}\sum_{\beta=1}^{\infty} \frac{|V_{\alpha\beta}|^2}{E_0 - E_{\alpha\beta}}(\tilde{n}_{\alpha} - \tilde{n}_{\beta}) \tag{27}$$

is the second order contribution to the shell energy. In eqs. (26-27) $\tilde{n}_{\alpha}$ is the smoothed occupation number of the single–particle state $|\alpha\rangle$ and $E_{\alpha\beta}$–E_0 is the particle–hole excitation energy.

For the octupole field, $V{=}\beta_{3K}Q''_{3K}$, the first order term (26) vanishes and the shell driving force is solely determined by the second order correction $\delta E^{(2)}_{shell}$, which is proportional to the square of the corresponding deformation β_{3K},

$$\delta E^{(2)}_{shell} = C_{3K}\beta^2_{3K}. \tag{28}$$

The shell–energy octupole–stiffness coefficient, C_{3K}, given by eq. (27) ($V \rightarrow Q''_{3K}$), determines the octupole susceptibility of shell energy. If C_{3K} is negative then there exists a shell force favouring stable deformations**. On the other hand, if C_{3K} is positive, the shell correction tends to restore reflection symmetry.

The results of calculations for C_{3K} are displayed in figs. 4-6. For the spherical shape, fig. 4, the octupole–driving shell force is positive, i.e. there is no tendency to develop stable octupole deformations. Of course, in this case all octupole modes are degenerate.

The situation at superdeformed prolate shape is shown in fig. 5. For particle numbers representing the asymmetric case A of fig. 2 (N_{shell}–even) the C_{3K} is negative for the K=0,1 and 3 modes. For the symmetric systems (case B) there is no shell octupole driving force towards reflection–asymmetric shapes.

Finally, the hyperdeformed case is illustrated in fig. 6. As expected, for the systems representing the asymmetric case A of fig. 3 the shell correction decreases with octupole deformations for the K=0,1 and 3 modes, whilst no octupole–driving tendency is predicted for the symmetric case C.

In summary, the role of spherical clusters in defining properties of superdeformed states becomes more clear when one considers the shell energy of the RHO. For magic numbers given by two unequal spherical clusters (case A in fig. 2, i.e., N or Z equal to 28, 60, 110 etc.), the shell energy decreases with increasing reflection asymmetry. On the other hand, for the particle numbers 40, 80, 140 (case B), the nuclear shape is expected to be fairly rigid with respect to reflection asymmetric distortions††. For hyperdeformed shapes, fig. 3, the harmonic oscillator model suggests that the strongest tendency for reflection asymmetry should be expected in case A, i.e. for the particle

**The liquid drop model energy never favours reflection–asymmetric shapes. This means that stable octupole shapes can only arise from shell effects, i.e. from the shell driving force.

††The relation between spherical and superdeformed magic numbers was discussed by Bengtsson et al. in 1981 [43]. In order to understand the alternating behaviour of microscopic octupole shell correction they introduced the model of two touching harmonic oscillators, i.e. two spherical clusters.

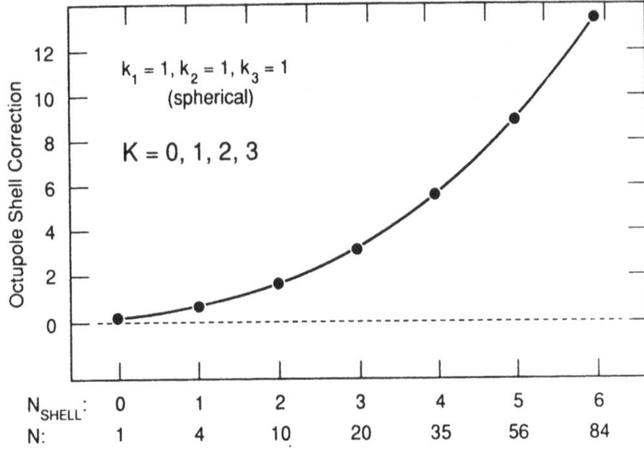

Fig. 4. Shell-correction octupole-stiffness coefficient C_{3K} (in units of $\frac{7}{4\pi}\tilde{\omega}_o^{-4}$ for the closed-shell configurations of the spherical harmonic oscillator.

numbers 12, 36, 80, 150. Particle numbers that stabilize reflection-symmetric shapes (case C) are equal to 24, 60, 120.

Reflection-Asymmetric Shapes in Light Nuclei

In light nuclei the spin-orbit interaction is relatively weak and, in addition, the diffuseness of the nuclear surface is comparable with the nuclear radius. Consequently, the harmonic oscillator model gives a fairly good approximation to the nuclear average potential. Among many well-deformed configurations in light nuclei there are several good examples that nicely illustrate the simple oscillator (multicluster) scheme.

A classic example is the ground-state of ^{20}Ne, which can be well described as arising from an ^{16}O–^{4}He di-nucleus configuration[52-54]. According to the RHO scheme this *reflection asymmetric superdeformed* system can be viewed as a combination of two spherical "clusters" with particle numbers 2 (alpha particle) and 8 (spherical ^{16}O).

There are several candidates for reflection-asymmetric structures among the so called quasi-molecular resonances. For instance, the alternating parity band built on the 0_2^+ state in ^{18}O can be well described in terms of α+^{14}C dipole molecular band [55]. Observed resonances in the asymmetric fission ^{24}Mg→^{16}O + ^{8}Be (or ^{20}Ne+α) can be attributed to the calculated low-lying reflection-asymmetric hyperdeformed minimum in ^{24}Mg [56]. According to the RHO scheme this configuration can be associated with the symmetric α+^{16}O+α (see [57]) or asymmetric ^{16}O+α+α or ^{16}O+^{8}Be structures.

Octupole Correlations in Superdeformed Nuclei

The microscopic mechanism behind reflection asymmetry at certain superdeformed shapes is twofold. The octupole interaction Y_{30} couples the orbitals with asymptotic quantum numbers $[N\ n_3\ \Lambda]\Omega$ and $[N+1\ n_3 \pm 1\ \Lambda]\Omega$. The largest number of such matrix elements corresponds to states with the highest possible value of $n_\perp$, i.e.

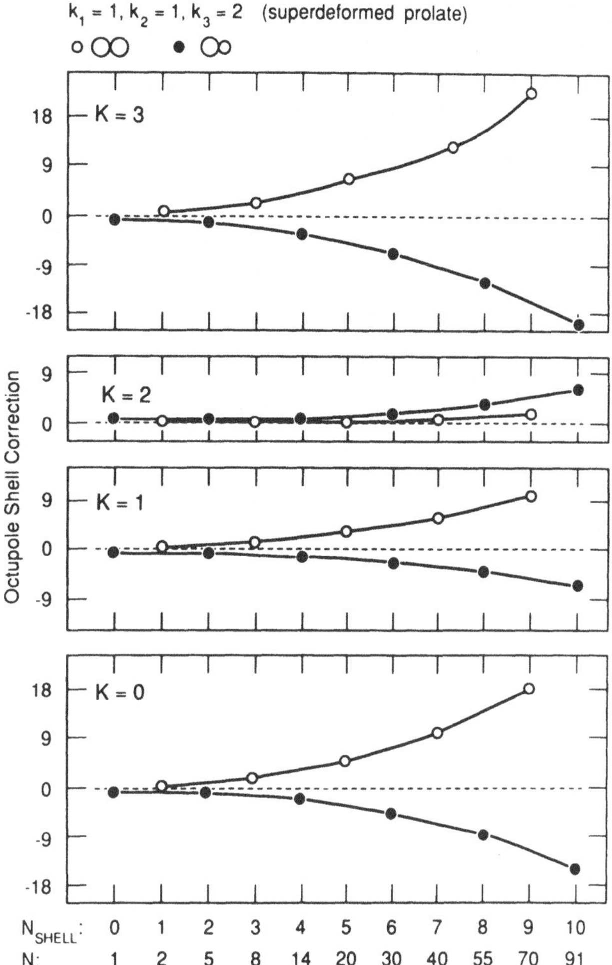

Fig. 5. Similar to fig. 4 but for the superdeformed closed–shell configurations of the RHO.

for n_3=0. This tendency has been discussed long ago in the context of the fission barrier asymmetry [58, 59]. The second mechanism behind the octupole instability in superdeformed states is the octupole interaction between the high-N intruder orbitals and specific pseudo-oscillator levels. For example, the same pairs of orbitals, such as ([660]1/2–[530]1/2) or ([770]1/2–[640]1/2), which are responsible for octupole deformations in the light actinides, appear close to the Fermi level in superdeformed SD configurations around ^{148}Gd and ^{192}Hg.

Calculations based on the realistic mean-field potentials confirm the prediction of the RHO, i.e. regions of particle numbers, which favour reflection-symmetric or reflection-asymmetric shapes alternate [43,60-63] and the tendency towards mass-asymmetry is strongly favoured at particle numbers around 28, 64 and 114 whilst for particle numbers around 38, 84 and 144 the minimum shell correction energy is found at reflection-symmetric shapes.

For superdeformed bands around ^{152}Dy the low-energy octupole collectivity can be attributed to the "octupole-driving" proton number Z=64. On the other hand

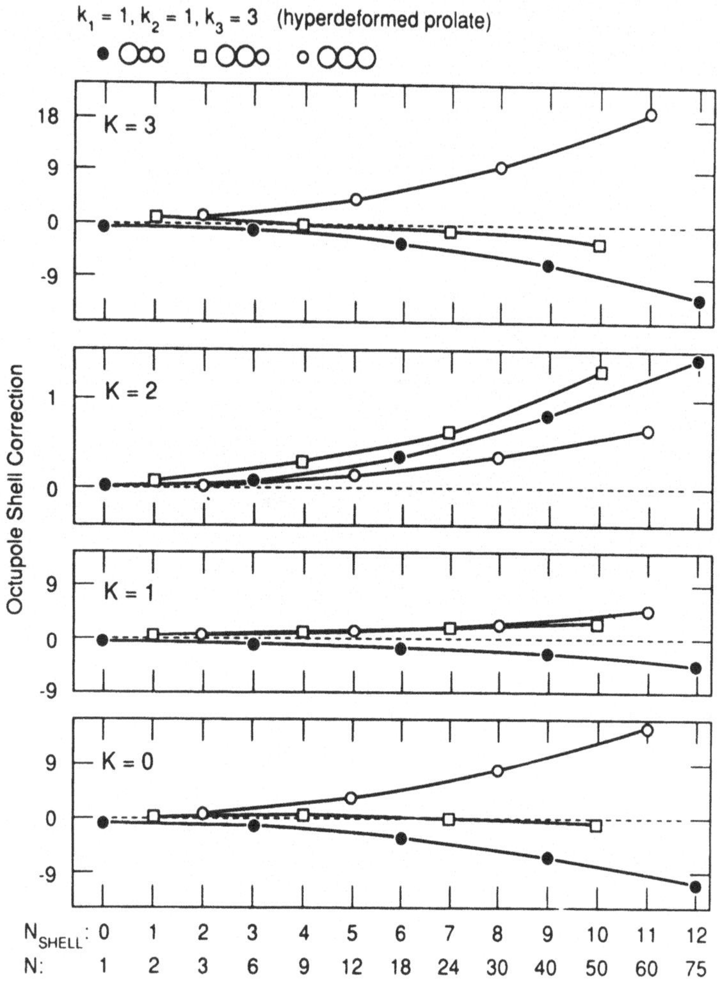

Fig. 6. Similar to fig. 4 but for the hyperdeformed closed-shell configurations of the RHO.

there is no such a tendency for the neutron system. But as predicted, the particle number N=84 has been found to strongly favour reflection-symmetric shapes. The opposite is true for superdeformed configurations in the Hg-Pb region, i.e. octupole correlations have neutron origin (because of the "optimal" neutron number N=114). As an illustrative example the results of Woods-Saxon-Strutinsky calculations for ^{192}Hg are shown in fig. 7. Whilst the neutron shell correction strongly favours octupole distortions the proton shell correction drives the system towards β_3=0. The resulting shell correction is almost insensitive to β_3 and, thanks to the very shallow macroscopic energy, the total potential energy reveals a pronounced octupole softness.

Recently, the octupole susceptibility in superdeformed configurations has been investigated within the parity-projected Skyrme-Hartree-Fock model [64]. The selfconsistent calculations do confirm the predictions of models based on the shell-correction approach, i.e. they indicate quite a sizeable lowering of the octupole excitations built

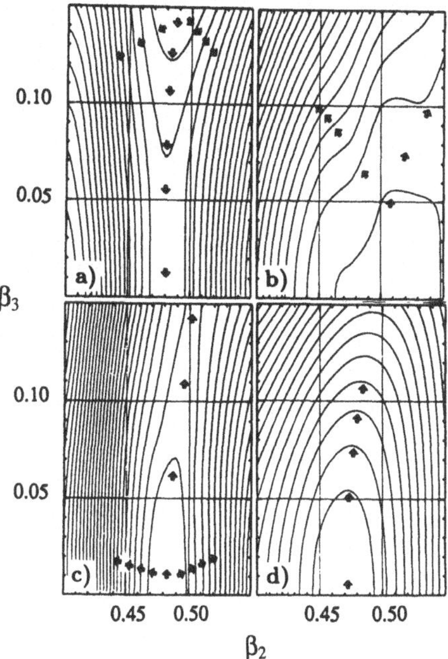

Fig. 7. The neutron shell correction energy (a), the proton shell correction energy (b), the total shell energy (c) and the total potential energy (d) for ^{192}Hg as a function of β_2 and β_3. Arrows indicate the direction of growing energy. The distance between the contour lines is 200 keV. (From [62].)

on the superdeformed intrinsic state. Recent calculations [65] in terms of the Generator Coordinate Method (GCM) confirm this tendency.

Properties of low-frequency octupole vibrations built upon superdeformed shapes have been analysed recently by Mizutori et al. [47,66,67] in terms of the RPA method with the cranked Nilsson potential and the doubly-stretched octupole residual interaction. They found rather strong K-dependence of superdeformed octupole modes. In particular, the $K=1$ "banana" mode turned out to be very collective. At normal deformations, the spatial difference between doubly-stretched and normal multipole-multipole forces is rather small. At large deformations, however, these interactions give markedly different predictions. For instance, the $K=3$ octupole vibrations are very collective in terms of doubly-stretched coordinates, but they are relatively weak when expressed through normal octupole interactions [67]. This means that the experimental observation (or: non-observation) of very collective $K=3$ octupole vibrations will tell us which of these two residual interactions is more realistic.

The deformation-driving shell-forces of $K=0,1,2,3$ superdeformed octupole modes have been recently discussed in ref. [68] within the deformed Woods-Saxon model. The calculations indicate a correlation between the shell forces of different K-modes. This result is in a qualitative agreement with the predictions of the RHO model.

Octupole softness (but not *octupole instability*) in superdeformed nuclei in the

A$\sim$190 region is expected to persist at high angular momenta.[47,60,61,63,66,67]. Consequently, collective octupole vibrational excitations can be mixed with low-lying one- and two-particle states thus modifying the excitation pattern near the yrast line. According to the calculations the first excited state in the doubly-magic superdeformed configurations in ^{152}Dy and ^{192}Hg should have a collective octupole character, in a nice analogy to the well-known collective 3^- state in the doubly-magic spherical nucleus ^{208}Pb. Moreover, the B(E1) rates for depopulating the superdeformed octupole band should be markedly enhanced because of (i) the reduced excitation energy of the giant dipole resonance built on the superdeformed state [25] and (ii) the large macroscopic contribution to the intrinsic dipole moment, which is proportional to the product of $\beta_2\beta_3$. The presence of large dipole moments (or enhanced B(E1) rates) is a direct consequence of the doubly–stretched octupole force. The K=0 and K=1 $(r^3Y_{3K})''$ operators are linear combinations of the ordinary octupole fields, r^3Y_{3K} and the compressional dipole fields, r^3Y_{1K} [67]. Again, if the doubly–stretched residual interaction is realised in nature, strong dipole transitions de-exciting superdeformed octupole states should be present.

Recent experimental data on ^{193}Hg [69] show a low-frequency pseudo crossing in one of the observed superdeformed bands as well as dipole transitions (most likely in one direction only) between one superdeformed band and another. An admixture of an octupole phonon built on the [624]9/2 ground-state into the [512]5/2 band provides a possible explanation of these effects [69]. The experimental data for ^{193}Hg, together with the observed reduction in alignments and the unusual similarity of superdeformed bands in the A$\sim$190 region, are the first pieces of experimental evidence supporting the presence of strong octupole correlations in superdeformed configurations.

As far as hyperdeformed shapes are concerned, a third hyperdeformed minimum around the fission barrier has been calculated [70-72] for neutron-rich nuclei around Z=86, N=148, i.e. exactly around the octupole–driving particle numbers 80 and 150). Experimentally, the third minimum shows up as an alternating-parity microstructure of resonances [73, 74].

SUMMARY

The field of octupole deformation is among the most quickly expanding areas of nuclear structure. Experimental discovery of nuclear quasimolecular bands, parity doublets and collective intrinsic dipole moments certainly gave this subject a strong push. Theoretically, many properties of low–lying octupole and dipole modes have been successfully described using reflection-asymmetric mean field approach and collective models like the GCM or RPA.

Recent discovery of discrete superdeformed rotational bands in heavy nuclei opened up a whole new field, namely the near yrast superdeformed spectroscopy. Especially exciting is the question of low–energy collective excitations at very deformed configurations. Among them, the octupole modes have been predicted to be particularly favoured. In this contribution, overall features of octupole correlations have been described in terms of the RHO model. In particular, we have demonstrated that there exists an apparent relation between the multiple irreps of SU(3) and the tendency (susceptibility) to cluster into spherical fragments. The simple relation between the cluster size, the number of clusters (equal to the number of irreps of SU(3)), and the overall equilibrium shape is supported by the microscopic calculations.

Acknowledgements This work has been partly done in collaboration with J. Dobaczewski and P. Van Isacker. Useful discussions with K. Matsuyanagi and R. Robinson are gratefully acknowledged. The Joint Institute for Heavy Ion Research has as member institutions the University of Tennessee, Vanderbilt University, and the Oak Ridge National Laboratory; it is supported by the members and by the Department of Energy through Contract Number DE-FG05-87ER40361 with the University of Tennessee. This project was supported in part by the Polish Ministry of National Education under Contract G-MEN-147/90.

REFERENCES

1. G.A. Leander, *in*: "Nuclear Structure 1985", R.A. Broglia, G.B. Hagemann and B. Herskind, eds., North-Holland, Amsterdam (1985) p. 249.
2. W. Nazarewicz, *in*: "Nuclear Structure 1985", R.A. Broglia, G.B. Hagemann and B. Herskind, eds., North-Holland, Amsterdam (1985) p. 263.
3. J. Żylicz, *in*: "Nuclear Structure, Reaction and Symmetries", R.A. Meyer and V. Paar, eds., World Scientific, Singapore (1986) p. 79.
4. R.R. Chasman, *in*: "Nuclear Structure, Reaction and Symmetries", R.A. Meyer and V. Paar, eds., World Scientific, Singapore (1986) p. 5.
5. P. Butler, *in*: "Heavy Ions in Nuclear and Atomic Physics", Z. Wilhelmi and G. Szeflińska, eds., Adam Hilger, Bristol and Philadelphia (1989) p. 295.
6. S.G. Rohoziński, *Rep. Prog. Phys.* 51:541 (1988).
7. W. Nazarewicz, *Nucl. Phys.* A520:333c (1990).
8. P. Möller and J.R. Nix, *Nucl. Phys.* A361:117 (1981).
9. G.A. Leander, R.K. Sheline, P. Möller, P. Olanders, I. Ragnarsson and A.J. Sierk, *Nucl. Phys.* A388:452 (1982).
10. W. Nazarewicz, P. Olanders, I. Ragnarsson, J. Dudek, G.A. Leander, P. Möller and E. Ruchowska, *Nucl. Phys.* A429:269 (1984).
11. R.R. Chasman, *J. Physique* 45:C5 (1984).
12. A. Sobiczewski, Z. Patyk, S. Ćwiok and P. Rozmej, *Nucl. Phys.* A485:16 (1988).
13. S. Ćwiok and W. Nazarewicz, *Phys. Lett.* 224B:5 (1989).
14. S. Ćwiok and W. Nazarewicz, *Nucl. Phys.* A496:367 (1989); *Nucl. Phys.* A (1991) in press.
15. P. Bonche, P.H. Heenen, H. Flocard and D. Vautherin, *Phys. Lett.* 175B: 387 (1986).
16. L.M. Robledo, J.L. Egido, J.F. Berger and M. Girod, *Phys. Lett.* 187B:233 (1987).

17. P. Bonche, *in*: "The Variety of Nuclear Shapes", J.D. Garrett et al., eds., World Scientific, Singapore (1988) p.302.

18. L.M. Robledo, J.L. Egido, B. Nerlo-Pomorska and K. Pomorski, *Phys. Lett.* 201B:409 (1988).

19. J.L. Egido and L.M. Robledo, *Nucl. Phys.* A494:85 (1989).

20. J.L. Egido and L.M. Robledo, *Nucl. Phys.* A518:475 (1990).

21. W. Nazarewicz, P. Olanders, I. Ragnarsson, J. Dudek and G.A. Leander, *Phys. Rev. Lett.* 52:1272 (1984); 53:2060E (1984).

22. W. Nazarewicz, *in*: "Proc. Int. Conf. on Nuclear Structure Through Static and Dynamic Moments", H.H. Bolotin, ed., Conference Proceedings Press, Melbourne, (1987) p. 180.

23. W. Nazarewicz, G.A. Leander and J. Dudek, *Nucl. Phys.* A467:437 (1987).

24. G.A. Leander, *in*: "AIP Conf. Proc. 125", American Institute of Physics, New York (1985) p. 125.

25. G.A. Leander, W. Nazarewicz, G.F. Bertsch and J. Dudek, *Nucl. Phys.* A453:58 (1986).

26. P. Butler and W. Nazarewicz, submitted to *Nucl. Phys.* A.

27. P. Butler, these proceedings.

28. N. Schulz, V.R. Vanin, M. Aiche, A. Chevallier, J. Chevallier, J.C. Sens, Ch. Briançon, S. Ćwiok, E. Ruchowska, J. Fernandez-Niello, Ch. Mittag and J. Dudek, *Phys. Rev. Lett.* 63:2645 (1989).

29. W. Urban, R.M. Lieder, J.C. Bacelar, P.P. Singh, D. Alber, D. Balabanski, W. Gast, H. Grawe, G. Hebbinghaus, J.R. Jongman, T. Morek, R.F. Noorman, T. Rząca-Urban, H. Schnare, M. Thoms, O. Zell and W. Nazarewicz, *Phys. Lett.* B (1991), in press.

30. P.J. Ennis, C.J. Lister, W. Gelletly, H.G. Price, B.J. Varley, P.A. Butler, T. Hoare, S. Ćwiok and W. Nazarewicz, to be published.

31. J. Skalski, *Phys. Rev.* C43:140 (1991).

32. H. Mach, S. Ćwiok, W. Nazarewicz, B. Fogelberg, M. Moszyński, J. Winger and R.L. Gill, *Phys. Rev.* C42:R811 (1990).

33. R.D. Ratna-Raju, J.P. Draayer and K.T. Hecht, *Nucl. Phys.* A202:433 (1973).

34. A. Bohr, I. Hamamoto and B.R. Mottelson, *Phys. Scr.* 26:267 (1982).

35. J.P. Draayer, *Nucl. Phys.* A520:259c (1990).

36. J. Dudek, W. Nazarewicz, Z. Szymański and G.A. Leander, *Phys. Rev. Lett.* 59:1405 (1987).

37. H.V. McIntosh, *in*: "Group Theory and Its Applications", E.M. Loebl, ed., Academic Press, New York and London (1971), vol. II, p.75.

38. W. Nazarewicz, J. Dobaczewski, P. Van Isacker, *in*: APS Proc. Press (1991), in press; Preprint JIHIR 91-02 (1991).

39. A. Bohr and B.R. Mottelson, "Nuclear Structure", vol. 2, W.A. Benjamin, New York (1975).

40. F. Duimio and G. Zambotti, *Nuovo Cimento* 48A: (1966) 1203.

41. I. Vendramin, *Nuovo Cimento*, 54A:190 (1968).

42. R.A. Brandt and O.W. Greenberg, *J. Math. Phys.* 10:1168 (1969).

43. T. Bengtsson, M.E. Faber, G. Leander, P. Möller, M. Płoszajczak, I. Ragnarsson and S. Åberg, *Phys. Scr.* 24:200 (1981).

44. G. Maiella and G. Vilasi, *Lett. Nuovo Cimento*, 1:57 (1969).

45. A. Cisneros and H.V. McIntosh, *J. Math. Phys.* 11:870 (1970).

46. B.R. Mottelson, Symp. on Nucl. Struc., Argonne 1988; Copenhagen Workshop 1988.

47. S. Mizutori, K. Matsuyanagi and Y.R. Shimizu, *in:* APS Proc. Press (1991), in press; preprint KUNS 1069 (1991).

48. H. Sakamoto and T. Kishimoto, *Nucl. Phys.* A501:205 (1989).

49. H. Sakamoto and T. Kishimoto, *Nucl. Phys.* A501:242 (1989).

50. P. Ring and P. Schuck, "The Nuclear Many-Body Problem", Springer-Verlag, New York (1980).

51. M. Brack, J. Damgård, A.S. Jensen, H.C. Pauli, V.M. Strutinsky and C. Y. Wong, *Rev. Mod. Phys.* 44:320 (1972).

52. F. Nemoto and H. Bando, *Prog. Theor. Phys.* 47:1210 (1972).

53. S. Marcos, H. Flocard and P.-H. Heenen, *Nucl. Phys.* A410:125 (1083).

54. D. Provoost, F. Grümmer, K. Goeke and P.-G. Reinhardt, *Nucl. Phys.* A431:139 (1984).

55. M. Gai, M. Ruscev, A.C. Hayes, J.F. Ennis, R. Keddy, E.C. Schloemer, S.M. Sterbenz and A.A. Bromley, *Phys. Rev. Lett.* 50:239 (1983).

56. G.A. Leander and S.E. Larsson, *Nucl. Phys.* A239:93 (1975).

57. H. Flocard, P.H. Heenen, S.J. Krieger and M. Weiss, *Prog. Theor. Phys.* 72:1000 (1984).

58. S.A.E. Johansson, *Nucl. Phys.* A22:529 (1961).

59. C. Gustafsson, P. Möller and S.G. Nilsson, *Phys. Lett.* 34B:349 (1971).

60. S. Åberg, *Nucl. Phys.* A520:35c (1990).

61. J. Dudek, T. Werner and Z. Szymański, *Phys. Lett.* 248:235 (1990).

62. W. Satuła, S. Ćwiok, W. Nazarewicz, R. Wyss and A. Johnson, *Nucl. Phys.* A (1991), in press.

63. J. Höller and S. Åberg, *Z. Phys.* A336:363 (1990).

64. P. Bonche, S.J. Krieger, M.S. Weiss, J. Dobaczewski, H. Flocard and P.-H. Heenen, *Phys. Rev. Lett.* 66:876 (1991).

65. P. Bonche, S.J. Krieger, M.S. Weiss, J. Dobaczewski, H. Flocard and P.-H. Heenen, to be published.

66. S. Mizutori, Y.R. Shimizu, and K. Matsuyanagi, *Prog. Theor. Phys.* 83:666 (1990).

67. S. Mizutori, Y.R. Shimizu, and K. Matsuyanagi, *Prog. Theor. Phys.* 85:559 (1991); *Prog. Theor. Phys.* 86 (July, 1991).

68. J. Dudek, *in*: "High Spin Physics and Gamma-Soft Nuclei", J.X. Saladin, R.A. Sorensen and C.M. Vincent, eds., World Scientific, Singapore (1991) p. 146.

69. D.M. Cullen, M.A. Riley, A. Alderson, I. Ali, C.W. Beausang, T. Bengtsson, M.A. Bentley, P. Fallon, P.D. Forsyth, F. Hanna, S.M. Mullins, W. Nazarewicz, R.J. Poynter, P.H. Regan, J.W. Roberts, W. Satuła, J.F. Sharpey-Schafer, J. Simpson, G. Sletten, P.J. Twin, R. Wadsworth and R. Wyss, *Phys. Rev. Lett.* 65:1547 (1990).

70. V.V. Pashkevich, *Nucl. Phys.* A169:275 (1971).

71. P. Möller, *Nucl. Phys.* A192:529 (1972).

72. R. Bengtsson, I. Ragnarsson, S. Åberg, A. Gyurkovich, A. Sobiczewski and K. Pomorski, *Nucl. Phys.* A473:77 (1987).

73. J. Blons, C. Mazur, D. Paya, M. Ribrag and H. Weigmann, *Phys. Rev. Lett.* 41:1282 (1978).

74. B. Fabbro, J. Blons, A. Greiner, J.M. Hisleour, C. Mazur, Y. Patin, D. Paya and M. Ribrag, *J. Physique Lett.* 45:L-843 (1984).

ELECTRIC TRANSITION STRENGTHS IN OCTUPOLE NUCLEI MEASURED BY COULOMB EXCITATION

P. A. Butler[a], N. Clarkson[a], T. Czosnyka[b], R. Ibbotson[c], C. A. White[a], B. Kotlinski[c], D. Cline[c], R. A. Cunningham[d], M. Devlin[c], K. G. Helmer[c], T. H. Hoare[a], J. R. Hughes[e], G. D. Jones[a], A. E. Kavka[c], I-Y. Lee[f], R. J. Poynter[e], J. Srebrny[g], W. Urban[h], E. G. Vogt[c], R. Wadsworth[e], D. L. Watson[e], C.-Y. Wu[c]

[a]Physics-University of Liverpool, [b]HIL University of Warsaw, [c]NSRL University of Rochester, [d]SERC Daresbury Laboratory, [e]Physics-University of York, [f]HHIRF-ORNL, [g]Physics-University of Warsaw, [h]KFA Jülich

INTRODUCTION

There is now a wealth of experimental and theoretical evidence, e.g.[1,2], that nuclei with $Z \approx 60, N \approx 88$ and $Z \approx 88, N \approx 134$ are reflection-asymmetric in the intrinsic frame. The usual experimental signature for such pear-shaped nuclei is the observation of low-lying negative parity states which become interleaved with the yrast positive parity states at medium values of angular momentum. These data have been interpreted using geometric models within the mean field approach as arising from the presence of static octupole deformation, i.e. $\overline{\beta}_3 \neq 0$. As with quadrupole degrees of freedom, the most reliable measurements of collective octupole strength are the $E\lambda$ matrix elements connecting the rotational states in the octupole and ground state bands. While many $E1$ transition strengths have been determined in 'octupole' nuclei, they are strongly affected by fluctuating single particle contributions and are not a reliable indicator of octupole strength. While observation of real photon $E3$ emission is not yet technically possible, the determination of $E3$ matrix elements from Coulomb excitation (Coulex) is feasible. Results of such measurements of $E1$, $E2$ and $E3$ matrix elements in ^{148}Nd are presented here, and are used to extract deformation parameters which can be compared with theoretical predictions. Recent measurements in other nuclei are also briefly discussed.

COULOMB EXCITATION MEASUREMENTS IN ^{148}Nd

In ^{148}Nd many transitions have already been identified, e.g. the octupole band observed by Urban et al.[3], and extended to high spin using fission spectroscopy[4] and the many low lying states populated by β-decay[5]. The decay scheme for states populated by Coulex with ^{92}Mo ions is shown in fig.1. In order to determine the many electric matrix elements connecting the states in ^{148}Nd populated following Coulex it is necessary to perform many independent measurements of observables which are sensitive to these matrix elements. The following experiments were carried out: i) Particle-γ and particle-γ-γ coincidences were measured following bombardment with 330 MeV ^{92}Mo ions, provided by the NSF at the Daresbury Laboratory.

Nuclear Shapes and Nuclear Structure at Low Excitation Energies
Edited by M. Vergnes *et al.*, Plenum Press, New York, 1992

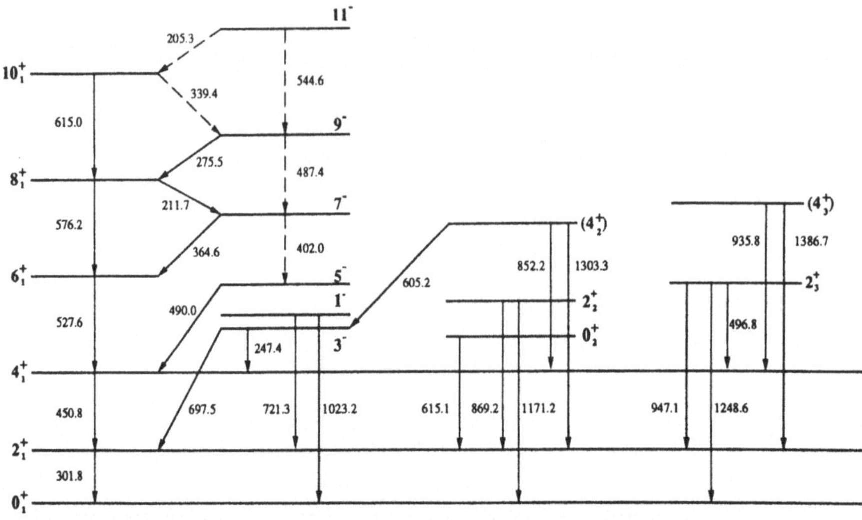

Figure 1. Level scheme of ^{148}Nd as populated by Coulex with 330 MeV ^{92}Mo ions.

The γ-rays were detected in POLYTESSA consisting of 20 Compton suppressed Ge spectrometers (CSS) and the scattered ions were detected in position sensitive avalanche detectors (PSAD). ii) Gamma-ray intensities were measured in coincidence with the detection of ^{58}Ni ions, provided by the NSRL, University of Rochester, using 5 CSS's and a PSAD system. The primary beam energy was 200 MeV. iii) Lifetimes of excited states in ^{148}Nd were measured using the recoil distance method, following inelastic scattering of ^{58}Ni ions at a bombarding energy of 210 MeV. In this experiment, which has been described in more detail elsewhere[6], the backscattered ions were detected in a PSAD, and the γ-rays were detected in a CSS placed at 0°.

Measurements have also been carried using the inverse reaction 700 MeV ^{148}Nd + ^{208}Pb, using the 16 ESS array and PSAD within the Spin Spectrometer at the Holifield Facility, Oak Ridge. The analysis of these data is not yet complete.

DETERMINATION OF MATRIX ELEMENTS

The γ-ray transition intensities and mean lifetimes were fitted using the semi-classical Coulomb-excitation,-dexcitation least-squares search code, GOSIA[7], to extract the $E1$, $E2$ and $E3$ matrix elements for ^{148}Nd. In the fitting procedure the predictions of the rigid rotor model were used to constrain the possible values of matrix elements with $\lambda = 1, 2, 3$ between states with $I > 9$. The signs of the $E2$ matrix elements were chosen to be consistent with the assumption of a prolate rotor. The results are rather insensitive to the relative phase of the $E1$ and $E3$ matrix elements, and the error bars have been increased to incorporate an estimated uncertainty due to this effect. The extracted values for the $E1$, off-diagonal $E2$, and $E3$ matrix elements are given in fig.2. There is good agreement between our measured values and those reported elsewhere for the $< 0^+ \| E1 \| 1^- >$[8], $< 0^+ \| E2 \| 2^+ >$[9] and $< 0^+ \| E3 \| 3^- >$[9] matrix elements.

COMPARISON WITH THEORETICAL CALCULATIONS

The values of the $E1$, $E2$ and $E3$ matrix elements are given, in the rotational model, by

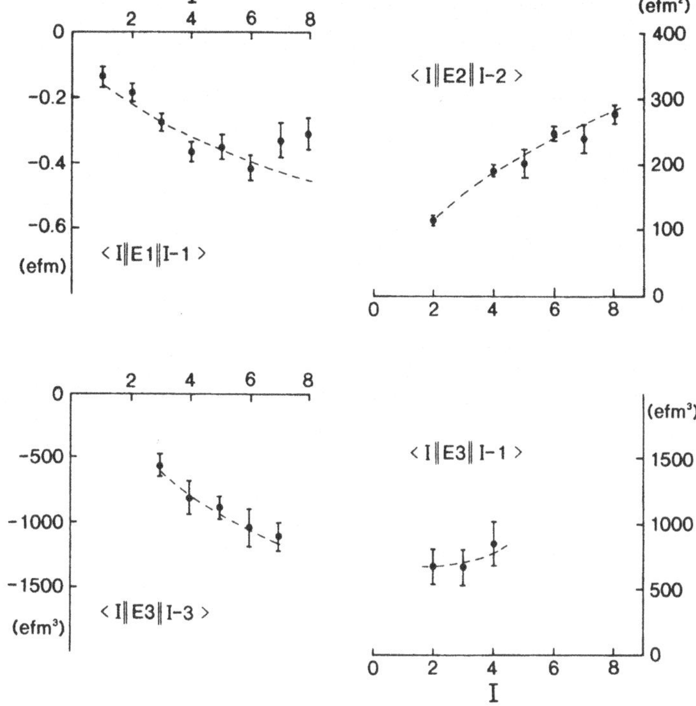

Figure 2. Experimental matrix elements for transitions in ^{148}Nd. The dashed line is the fit to the data assuming a constant value of $\mathcal{Q}_{\lambda 0}$ as described in the text.

the relationship:

$$< I||E\lambda||I' >= (2I+1)^{\frac{1}{2}}(I0\lambda 0|I'0)\mathcal{Q}_{\lambda 0}a_{\lambda}$$

where $\mathcal{Q}_{\lambda 0}$ is the intrinsic $E\lambda$ moment and $a_1 = \sqrt{3/4\pi}$, $a_{\lambda \geq 2} = \sqrt{(2\lambda+1)/16\pi}$. It can be seen from fig. 2 that for states up to $I = 8$ the values of $\mathcal{Q}_{\lambda 0}$ remain roughly constant as a function of spin. This behaviour of the charge distribution seems to be quite different to that of the energy levels, especially for the negative parity band at low spin (see fig. 1). The measured values of the matrix elements were fitted to the rotational model formula assuming constant intrinsic electric moments, and the *preliminary* fitted values of $\mathcal{Q}_{\lambda 0}$ were found to be: $\mathcal{Q}_{10} = 0.33 \pm 0.05$ efm, $\mathcal{Q}_{20} = 380 \pm 20$ efm^2, and $\mathcal{Q}_{30} = 1650 \pm 300$ efm^3. The values of the matrix elements calculated using the fitted values of $\mathcal{Q}_{\lambda 0}$ are shown as the dashed lines in fig. 2. The values of the deformation parameters β_λ are related to $\mathcal{Q}_{\lambda 0}$ by the following expression:

$$\mathcal{Q}_{\lambda 0} = b_\lambda Ze R_0^\lambda(\beta_\lambda + \text{higher order terms})$$

where $b_\lambda = 3/\sqrt{(2\lambda+1)\pi}$. Assuming a constant value of $\mathcal{Q}_{40}$ of 8500 efm^4 taken from the ground state $E4$ transition matrix element previously measured[9], and assuming no higher terms than $\lambda = 4$, we estimate, using $R_0 = 1.2A^{\frac{1}{3}}$, values of $\beta_2 = 0.17$, $\beta_3 = 0.14$ and $\beta_4 = 0.10$. The expressions for the higher order terms in the expression for $\mathcal{Q}_{\lambda 0}$ were taken from reference 10. Mean field calculations which allow octupole deformation have been made[3] for ^{148}Nd which give an octupole deformed configuration at spin 7-8 with mean values of the deformation parameters of $\overline{\beta}_2 = 0.20$, $\overline{\beta}_3 = 0.075$. It should be emphasised that the experimental value of β_3 extracted using the above expressions depends not only on the value of $\overline{\beta}_3$ but also on the curvature of the nuclear potential with respect to octupole deformation.

As yet, no attempts have been made to calculate this effect, although calculations of the $B(E3; 0^+ \rightarrow 3^-)$ in Ba isotopes have been recently reported[11].

The experimental value of the intrinsic electric dipole ($D_0 = Q_{10}$ in the above nomenclature) is in agreement with the value calculated, 0.34, using the shell correction method based on the reflection-asymmetric Woods-Saxon model[12].

MEASUREMENTS IN OTHER NUCLEI

We have also carried out Coulex yield measurements on ^{150}Nd, using ^{58}Ni and ^{92}Mo induced reactions, and on ^{150}Sm, using ^{58}Ni and ^{208}Pb induced reactions. The analysis of these data are in progress. Coulex data are also available for ^{226}Ra, following bombardment by ^{16}O, ^{32}S, and ^{208}Pb projectiles[13,14]. Preliminary analysis of these data using the GOSIA code indicates that the quadrupole and octupole deformation parameters in ^{226}Ra have similar values to those in ^{148}Nd.

In summary, a comprehensive set of $E1$, $E2$ and $E3$ matrix elements are presented for ^{148}Nd. The electric quadrupole and octupole matrix elements are consistent with predictions of a model which assumes a rotating shape having quadrupole and octupole deformation with $\beta_2 \approx 0.2$ and $\beta_3 \approx 0.1$.

Acknowledgement

We are very grateful to W. Nazarewicz for stimulating discussions. This work was supported by the U.K. Science and Engineering Research Council, by the U.S. National Science Foundation, and the Polish Ministry of Science.

References

1. W. Nazarewicz et al., Nucl. Phys. A429:269 (1984)
2. P.A. Butler, in: Heavy Ions in Nuclear and Atomic Physics, ed., Z. Wilhelmi and G. Szeflińska, Adam Hilger, Bristol and Philadelphia (1989)
3. W. Urban et al., Phys. Lett. B200:424 (1988)
4. J.L. Durell et al. Daresbury Annual Report p.28 (1987/88)
5. T. Karlewski et al., Z.Phys.A - Atom.Nucl. 330:55 (1988)
6. R. Ibbotson et al., to be published in Nucl. Phys. A
7. T. Czosnyka et al., Bull. Am. Phys. Soc. 28:745 (1983)
8. H.H. Pitz et al., Nucl. Phys A509:587 (1990)
9. A. Ahmad et al., Phys. Rev. C37:1836 (1988)
10. G.A. Leander and Y.S. Chen, Phys. Rev. C37:2744 (1988)
11. J.L. Egido and L.M. Robledo, Nucl. Phys. A524:65 (1991)
12. P.A. Butler and W. Nazarewicz, to be published in Nucl. Phys.
13. C. Fleischmann et al., Munich Annual Report p.42 (1984)
14. H.-J. Wollersheim, 3rd Int. Spring Seminar on Nuclear Physics, Ischia (1990)

OPEN QUESTIONS ON THE LOW-ENERGY NEGATIVE-PARITY EXCITATIONS

I.N.Mikhailov[1)] and Ch.Briançon[2)]

1) JINR, Dubna, USSR
2) Centre de Spectrométrie Nucléaire et de Spectrométrie de Masse, CNRS-IN2P3, Orsay, France

ABSTRACT

A reexamination of nuclear collective phase space is made, and some new types of collective motion are identified.

The nature of the negative-parity collective degrees of freedom and especially the role of dipole isoscalar modes in the structure of heavy nuclei are discussed. The possibility to interprete the unusual spectrum of ^{218}Ra, admitting the active role of dipole phonons with low excitation energy is shown.

The open problems arising from the recent experimental findings and from their theoretical interpretation are discussed.

1. The nature of low-lying negative-parity collective excitations

The negative-parity correlations leading to the collective phenomena are usually associated with the the multiple deformations of the nuclear surface which is produced by the coherent motion of nucleons. This picture is admitted in the unified nuclear model. In spite of the great success of this model, one may wonder if it represents the true image of what is happening in reality. The active role of degrees of freedom different from the multipole deformation of the nuclear surface shows itself in many ways and can be learned of from the presence of several concentration regions of the strength of E3 transitions which manifests the coupling of the octupole vibrations with some other types of motion.

One may question oneself to what extend the difficulties of the phenomenological collective models show the limitations of the fluid-dynamical picture. An answer to the question may be obtained starting from the classification of collective motions by the pattern of the collective velocity field. The hypotheses on it may be used to find the changes in the shape and in other collective nuclear properties on the basis of the dynamic equations. Such approaches have a long history, especially due to the Riemann- Lebovitz-Chandrasekhar developement of methods used in astrophysics for the study of the shapes of gravitating bodies[1]. In nuclear physics this way is taken in the theory of collective motion developed in Dubna[2].

The classification of the collective excitations can be started by considering the displacements of "liquid elements". The displacement $\zeta(\mathbf{r},t)$ is related to the collective velocity field: $\zeta(\mathbf{r},t) = \mathbf{u}(\mathbf{r},t)\cdot\delta t$. The velocity field $\mathbf{u}(\mathbf{r},t)$ can be parametrised as follows :

Nuclear Shapes and Nuclear Structure at Low Excitation Energies
Edited by M. Vergnes *et al.*, Plenum Press, New York, 1992

$$u_i(\mathbf{r},t)=\dot{L}_i+\dot{L}_{i,j}x_j+\dot{L}_{i,jk}x_jx_k+\dot{L}_{i,jkl}x_jx_kx_l+\cdots \tag{1.1}$$

where $L_i, L_{i,j}, \ldots$ are functions of the time only and the dots over L's indicate the time derivatives.

Due to the continuity equation, the displacement changes the density of the liquid matter $\rho(\mathbf{r},t)$ by $\delta\rho = -div(\rho\cdot\zeta)$. Thus, the nuclear integral characteristics

$$Q(t) = \int d\mathbf{r} \;\; q(\mathbf{r})\cdot\rho(\mathbf{r},t) \tag{1.2}$$

changes due to the collective flow by

$$\delta\mathbf{q} = -\int d\mathbf{r}\; q(\mathbf{r})\cdot div(\rho\cdot\zeta) = \int d\mathbf{r}\; \rho(\mathbf{r},t)\;(\zeta(\mathbf{r},t)\cdot\nabla)q(\mathbf{r}) \tag{1.3}$$

To find the restoring force driving the system to the equilibrium the virial theorems method by Chandrasekhar[1] may be used. It consists in integrating the dynamic equation for the density matrix with appropriate weights. In the Fermi-liquids the restoring forces are not associated entirely with the potential energy. They arise also because of the perturbation of the particle distribution in the momentum space. In the case of a stepwise distribution of fermions in the momentum space this produces the deformation of the Fermi-surface. In nuclei having a finite size, deformations of the Fermi-surface are dynamically coupled with deformations in the configuration space. This produces the coupling of collective variables having a geometrical nature with some new collective variables.

2. Dipole modes in hydrodynamics

The changes of the nuclear dipole integral characteristics (*i.e.* of the integrals (1.2) with vectorial functions $\mathbf{q}(\mathbf{r})$) may be associated with the velocity fields parametrised as follows:

$$\mathbf{u}(\mathbf{r},t) = \dot{\mathbf{a}}(t) + \dot{\mathbf{b}}(t)\cdot r^2 + (\dot{\mathbf{c}}(t)\cdot\mathbf{r})\cdot\mathbf{r} + \ldots \tag{2.1}$$

Such a velocity field containing only three vectorial time dependent parameters $\mathbf{a}(t)$, $\mathbf{b}(t)$ and $\mathbf{c}(t)$ describes three different types of collective flow.

The first of the modes corresponds to the case when $\mathbf{b}=0$ and $\mathbf{c}=0$: this is a translation of the system as a whole in the space. The others may be defined as "solenoidal" ($div\mathbf{u}=0$) and irrotational ($rot\mathbf{u}=0$) modes.

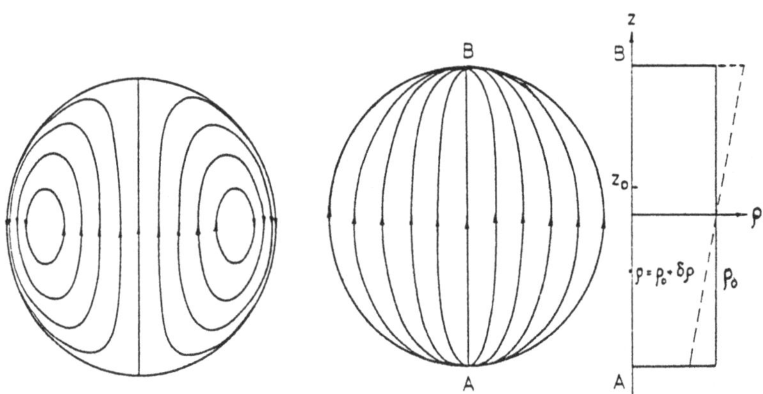

Fig 1. *Lines of the current corresponding to the solenoidal (left side) and irrotational (center) modes and the density variation created by the irrotational flow (right side).*

The velocity field corresponding to the solenoidal mode is given by the formula :

$$\mathbf{u}^{sol} = \dot{\mathbf{L}}_{sol} \cdot (r^2 - \tfrac{1}{2}R^2) - \tfrac{1}{2} (\dot{\mathbf{L}}_{sol} \cdot \mathbf{r}) \cdot \mathbf{r} \tag{2.2}$$

The third (irrotational) mode contained in the formula (2.1) has the velocity field :

$$\mathbf{u}^{irr} = \dot{\mathbf{L}}_{irr} \cdot (r^2 - 3R^2) + 2(\dot{\mathbf{L}}_{irr} \cdot \mathbf{r}) \cdot \mathbf{r} \tag{2.3}$$

Each of the three modes describes the transformation of a sphere of radius R into another sphere with the same radius. The irrotational flow shifts the center of the sphere by $\delta\mathbf{R} = 4 \cdot \mathbf{L}_{irr} \cdot R^2$ from the position of the center of mass. The lines of the current associated with these two modes and the density variation induced by the irrotational flow are shown in Fig. 1.

In the system composed of particles of the same nature the translational mode must be excluded from the modes of intrinsic excitations, whereas in nuclei it constitutes the most important component of the collective motion which manifests itself as the giant dipole resonance.

In the homogeneous liquid the solenoidal mode describes the motion which does not affect the density distribution. However, it generates the toroidal moment of the system:

$$\mathbf{T}(t) = \frac{1}{10c} \int d\mathbf{r}\ \rho_{ch}(\mathbf{r},t) \cdot \Big((\mathbf{r} \cdot \mathbf{u}(\mathbf{r},t)) \cdot \mathbf{r} - 2 \cdot r^2 \cdot \mathbf{u}(\mathbf{r},t) \Big) \tag{2.4}$$

ρ_{ch} being the electric charge density and c bering the light velocity. The physical meaning of this integral will be breafly discussed later on (see also ref.3). In the case of a spherical homogeneous distribution of the matter inside the sphere of radius R one has:

$$\mathbf{T}(t) = -(1/70) \cdot (Ze/Ac) \cdot R^4 \cdot q \cdot \dot{\mathbf{L}}_{sol} \tag{2.5}$$

The irrotational flow changes the density by $\delta\rho = -10\ (\mathbf{L}_{irr}\mathbf{r})\rho_0$ and is associated with the variations of the quantity:

$$\mathbf{r}_d^{(2)}(t) = \int d\mathbf{r}\ \rho_{ch}(\mathbf{r},t) \cdot r^2 \cdot \mathbf{r}. \tag{2.6}$$

which is called in Ref. 3 the "quadratic dipole radius". At the same conditions as before, the variation of this quantity is equal to :

$$\delta\mathbf{r}_d^{(2)}(t) = (4/7) Ze/A) \cdot q \cdot R^4 \cdot \dot{\mathbf{L}}_{irr} \cdot \delta t \tag{2.7}$$

3. Dipole modes in the Fermi-Liquid Droplet (FLD)

The dynamic equations for the usual and toroidal moments and for the quadratic dipole radius include in addition to the parameters of the velocity field also the integrals[2] :

$$\sum_{k=1}^{3} \Pi^k_{ik}, \quad \sum_{k=1}^{3} \Pi^i_{kk}, \quad \text{where } \Pi^i_{j,k} = \frac{1}{m} \cdot \int d\mathbf{r} d\mathbf{p} f(\mathbf{x},\mathbf{p},t) \cdot x_i p_j p_k,$$

$$\text{and } \sum_{k=1}^{3} \Pi_{ikk} = \frac{1}{m^2} \cdot \int d\mathbf{r} d\mathbf{p} f(\mathbf{x},\mathbf{p},t) \cdot p_i \cdot p^2$$

Altogether, the set of equations for all these variables describes three isoscalar and four isovector modes.

The dipole isoscalar modes were analysed[2] assuming the nuclear matter to be homogeneous and incompressible and the surface to be spherical with a sharp edge. Only one physical toroidal mode could be described in this way: the oscillations of the quadratic dipole radius were excluded by the assumption of the incompressibility, and the other isoscalar modes degenerated in a spurious motion. The theory gives the prediction for the excitation energy of the toroidal mode (the giant dipole toroidal resonance) :

$$\hbar\,\omega = \hbar\, V_F\,/(A^{1/3} r_0)\sqrt{11/5} = 65.6/\,A^{1/3}\ \text{Mev} \qquad (3.1)$$

Here r_0=1.2 fm is the nuclear size parameter and V_F is the Fermi-velocity. The toroidal transition strength is equal to :

$$S = \hbar\omega\cdot|T_z|^2_{1^-,0} = 0.0557\cdot\left(\frac{e\hbar}{2mc}\right)^2\cdot\frac{\hbar^2}{m}\cdot\frac{Z^2}{A^{1/3}} \qquad (3.2)$$

The position of the dipole irrotational mode is presumably comparable with, or higher than, that of the breathing mode.

Realistic assumptions on the diffuseness of the nuclear surface lead to the coupling of the modes considered before. The solenoidal mode which is spurious in the case of a sharp surface becomes physical because such flow changes the density distribution near the diffused surface. The role of the surface diminishes with increasing nuclear size, and one may expect the presence of a collective mode with the excitation energy decreasing fast with increasing nuclear mass number.

Notice, that the nuclear softness in respect to the irrotational motion is determined by essentially the same conditions which determine the nuclear softness in respect to the octupole deformation. This provides an argument to expect the appearence of low-lying dipole modes in nuclei near the regions of stable "octupole" deformation.

4. Experimental evidence for dipole correlations

Experiments show a large increase of the probability of dipole electric transitions between the yrast states of opposite parities in precisely those even-even nuclei which are soft or unstable in respect to the parity-breaking deformations as compared to the probabilities in the neighbouring "rigid" nuclei. In the case of a pointlike (dimensionless) object this observation would lead to the conclusion that a mode of excitation generating the charge-dipole moment is activated with the mode generating the octupole moment. The origin of the nuclear dipole moment in the octupolly soft nuclei is widely discussed in the litterature. It may appear as a result of combined quadrupole (hexadecapole) and octupole (λ = 5,7) deformation. A striking example of an important role of unusual negative-parity correlations was found in the combined experimental and theoretical studies of the structure of the nucleus ^{218}Ra. The most recent experimental results concerning this nucleus were obtained using the nuclear fusion reactions to produce this nucleus excited to very high angular momentum and the γ-ray multidetector "Chateau de Cristal"[4]. A new type of spectrum and pattern of the γ-decay were found in this experiment.

The experimental results can be summarised as follows :

- The level scheme was extended to much higher spin (I_{max} = 36) and is composed of hybridised bands that have a vibrational type with $I^{\pm}$ parity states almost equidistant in energy;

- Peculiar yrare bands running closely to the ground yrast band have been discovered. At variance with the yrast band consisting of + parity even spin and - parity odd spin, the yrare band consists of + parity odd spin and - parity even spin states and seems to exist only above spin 12;
- The $B(E1,I\Rightarrow I-1)/B(E2,I\Rightarrow I-2)$ branching ratios are large relative to normal values. At high spins they show a tendency to decrease and a second tendency to be higher for even I than for odd I in the yrast band.

5. Dipole collective degrees of freedom within the interacting boson approach

The interacting multiboson model[5], which is the IBM model generalised for the case of the bosons $\lambda^{\pi} = 0^{+}, 1^{-}, 2^{+}$, and 3^{-} (**s**, **p**, **d**, **f**-bosons), has been applied in its vibrational and intermediate limits to the results of this experiment [6].

A few words concerning the model :

- The total boson number rule $N = n_0+n_1+n_2+n_3$ is modified. In the case of ^{218}Ra, N may change from 5 corresponding to the usual rule to 12 by including the protons below the 82 gap from the 2d3/2, 3s1/2 sub-shells and neutrons below the 126 gap from the 2f5/2, 3p1/2 sub-shells.
- The vibrational limit can be described by two parameters or, in more detail, by 3 parameters using the hamiltonian :

$$\hat{H} = \frac{\varepsilon}{3}\cdot\left(\hat{n}_1+2\hat{n}_2+3\hat{n}_3\right) + \beta\cdot\hat{\omega}_3^3 + 2\alpha\cdot\left(\hat{I}_{1,2}\cdot\hat{I}_3-(\hat{n}_1+2\hat{n}_2)\cdot 3\hat{n}_3\right) \qquad (5.1)$$

Here, $\hat{I}_{i,(j)}$ are contributions to the angular momentum operator coming from the phonon(s) defined by the index (i or i,j) and ω_j^n is the second rank Casimir operator of the group U(n) built from the j-type phonon operator.

The first parameter ε acts alone at spins I<N and gives the linear part of the energy of yrast states. The second parameter β gives small nonlinear corrections in the I>N yrast line and the third parameter α acts in the yrare line only giving a small deviation of its (I-1) state from the yrast I state.

One could reproduce with an excellent agreement the $\simeq$ 50 levels with the three parameters (ε=585 keV, β=4,2 keV, α=3 keV) with χ=49.03 keV. This agreement could be achieved with an essential participation of dipole and octupole bosons. More precisely, the numbers of **p**-bosons (n_1) and of **f**-bosons (n_3) prevail over the number of **d**-bosons (n_2) which remains at a low level $n_2 \simeq 0,1$.

To summarise the analyses made using the multiboson model, we state that according to it, the nucleus ^{218}Ra resembles a multi-dimensional slightly unharmonic vibrator in which the yrast line is composed of fully aligned states with the spin $I^{\pi}=(n_1+2n_2+3n_3)^{\pi}$ and parity $\pi=(-)^{n_1+n_3}$. In the states of yrare lines the alignment is not complete, and the less it is the further the levels are shifted from the yrast line. The states closest to the yrast line are only one Planck unit short of a complete alignment. They exist where at least two types of bosons participate. According to the model, the **f**-bosons are missing in the states with I < N which explains the absense of states of the first yrare sequence with I<12. Whithin the model one is able to fit also the branching ratios. One may add that the spectrum below spin 4 seems to indicate a transition from a spherical to a γ-soft unaxially deformed shape.

6. Electromagnetic properties of dipole modes

In the studies of the role of nuclear dipole degrees of freedom the knowledge of their electromagnetic properties may be decisive. For this reason we return to the topics discussed in Sections 2 and 3 and present some additional information on the dipole modes of a liquid droplet.

The expression for the the hamiltonian describing the interaction of the system with the dipole-γ-radiation field contains the usual and the toroidal dipole moments. The probability of electromagnetic dipole transitions should be found calculating the matrix elements of the operator[3]:

$$(\mathbf{d}^{eff})_{fi} = \mathbf{d}_{fi} - i \cdot E_\gamma / \hbar c \cdot T_{fi} \tag{6.1}$$

Notice, that the toroidal moment has, in addition to the orbital part considered in Section 3, also a spinor part. Both parts contribute to the electron penetration factor accounting for anomalies in internal conversion processes[7]. In the interaction of the nucleus with charged particles, the dipole square radius may be also important.

A tentative estimation of the role of the dipole toroidal moment in γ-transitions may be obtained on the basis of the expression for the toroidal strength given in eq. (3.2). Whithin the model leading to this expression[2] GDTR is the only state coupled with the ground state by the toroidal operator, and we consider expression (3.2) as an estimation of the "toroidal dipole sum rule". The contribution of the toroidal moment to the matrix element of the γ-transition is determined by the ratio

$$X = (E_\gamma/\hbar c)^2 \cdot |T/D|^2 \tag{6.2}$$

where T and D are the matrix elements of the toroidal and usual dipole moments respectively.

Using eq. (3.2) and the notation $f^{tor} = T^2 \cdot E_\gamma / S_{tor}$ for the fraction of the toroidal strength of a particular γ-transition one obtains:

$$X = 1.26 \cdot 10^{-6} \cdot \frac{Z^2}{A} \cdot f^{tor} \cdot E_\gamma \; / \; B(E1)_{WU} \tag{6.3}$$

In an extreme situation, when about $A^{-1/3}$ of the toroidal strength is concentrated on a single nuclear level with the excitation energy close to 1 Mev, the retardation of the dipole transition by a factor of 10^{-5} is sufficient to make the toroidal contribution to the radiation probability be competitive with the part associated to the dipole moment. One example of an important (destructive) participation of the toroidal moment in the γ-transition is given in Ref.7: according to the data on the anomaly in the internal conversion electron in the transition $1^- \Rightarrow 0^+$ with, $E(\gamma)$=687.7 KeV, in ^{236}U nucleus the toroidal contribution to d^{eff} reduces the probability of γ-transition by more than 40 %.

7. Conclusion

The aim of this report is to show the existence of rather fundamental open problems in the nuclear structure physics. Under examination is the nature of nuclear collective degrees of freedom. Both theoretical studies and the experimental findings show limitations of the description of nuclear collectivity in terms of shape parameters.

These problems are pertinent to the studies of negative-parity excitations in heavy nuclei. Much attention is paid to the nuclear dipole modes. Some hypotheses concerning their participation in the low-energy negative-parity excitations are formulated. To know the real status of low-energy dipole modes new experiments and deeper theoretical analyses are needed.

References

1. S.Chandrasekhar, "Ellipsoidal Figures of Equilibrium", 1969, Yale University Press, New Haven and London
2. E.B. Balbutsev and I.N. Mikhailov, "Description of collective motion in nuclei within the method of moments", in "Collective nuclear dynamics", 1990, "Nauka", Leningrad
3. V.M. Dubovik and A.A. Cheshkov, Sov.Part.and Nuclei **5** (1974) 791
4. N.Schulz, V.Vanin, M.Aïche et al, Phys.Rev.Lett. **63** (1989) 2645
5. E.G.Nadjakov and I.N.Mikhailov, J.Phys.G: Nucl.Phys. **13** (1987) 1221; Izv.A.N. SSSR (ser. fiz.) **52** (1988) 111
6. I.N.Mikhailov, E.G.Nadjakov, M.Aïche, Ch.Briançon, N.Schulz and V.Vanin, J.Phys.G: Nucl.Phys. **15** (1989) L19
7. M.A. Listengarten, In "Modern methods of nuclear spectroscopy", 1985, Leningrad, USSR, p.142

VII - SUPERDEFORMATION

MICROSCOPIC DESCRIPTION OF SUPERDEFORMATION AT LOW SPIN

P. Bonche[(a)], H. Flocard[(b)], P.-H. Heenen[(1,c)], J. Meyer[(d)], J. Dobaczewski[(e)], S.J. Krieger[(f)] and M.S. Weiss[(f)]

(a) *DSM-SPhT, CEN Saclay, 91191 Gif sur Yvette Cedex, France*
(b) *Division de Physique Théorique[(2)], IPN Orsay, 91406 Orsay Cedex, France*
(c) *Physique Nucléaire Théorique, ULB, CP229, 1050 Brussels, Belgium*
(d) *IPN, Université Lyon 1, 69622 Villeurbanne Cedex, France*
(e) *Institute of Theoretical Physics, Hoża 69, PL00681 Warsaw, Poland*
(f) *Lawrence Livermore National Laboratory, Livermore, CA 94550, USA*

INTRODUCTION

Nuclear Hartree-Fock (HF)+BCS calculations have led to predictions of shape isomerism in isotopes of Pt, Hg and Os nuclei[1]. These have been confirmed through the observation of superdeformed rotational bands in $^{190,\cdots,194}$Hg. Encouraged by these measurements and similar observations in ^{194}Pb, we have extended these calculations to a wide range of contiguous nuclei. These HF results have been employed in a Generator Coordinate Method (GCM) calculation utilizing the quadrupole deformation as the generating variable[2]. The resulting spectra confirm the conclusions drawn from the HF results and agree with experiments.

With a phenomenological assumption for the moments of inertia of the GCM states, we construct the radiative transitions in and out of the superdeformed band. The results account for the rapid depopulation of the superdeformed band which occurs in two to three transitions, and reproduce quantitatively the values of the lowest angular momentum at which these bands are seen.

We also study the influence of pairing vibrations on the stability of the superdeformed state in ^{104}Hg. We find an increase of the density of collective states at small deformation above 5MeV and a large mixing of the superdeformed state with states of the first well. Compared with calculations without pair vibrations, our results predict an earlier depopulation of the superdeformed band by about six units of angular momentum, improving the agreement with experiment.

Finally, using parity projection and the generator coordinate method we investigate the left-right asymmetric softness of the superdeformed minimum of ^{194}Pb. The collective wave-functions exhibit large octupole dispersions. Our calculation suggests that

[1]Directeur de Recherches FNRS.
[2]Unité de recherches des Universités Paris XI et Paris VI associée au C.N.R.S.

Nuclear Shapes and Nuclear Structure at Low Excitation Energies
Edited by M. Vergnes *et al.*, Plenum Press, New York, 1992

above the ground-state superdeformed band there should exist a negative parity band of similar quadrupole deformation with an excitation energy close to 2 MeV.

MEAN FIELD CALCULATIONS

Our description is based upon the Hartree Fock (HF) mean field theory which can be derived from a variational principle where the energy

$$E = \langle\Psi|\hat{H}|\Psi\rangle / \langle\Psi|\Psi\rangle \tag{1}$$

is minimized for a given set of trial many-body wave-functions $|\Psi\rangle$. The HF approximation corresponds to variations restricted to the Hilbert space built on Slater determinants only.

From an effective two-body force V_{ij} which includes among other terms a spin-orbit force, a pairing interaction and the coulomb repulsion between the protons, the HF approximation constructs the mean one-body potential which averages the interactions of a given nucleon with all the others. The full many-body problem is reduced to the solution of the one-body HF equations,

$$h\,\varphi^{\lambda} = e^{\lambda}\,\varphi^{\lambda} \quad . \tag{2}$$

The HF hamiltonian, $h = t + U$, is the sum of a kinetic energy term and a mean field,

$$U(\mathbf{r}) = \int d^3r'\, v(\mathbf{r} - \mathbf{r}')\, \rho(\mathbf{r}') \quad , \tag{3}$$

which depends upon the solution of the HF equations through the density ρ. Due to this dependence the HF equations are non-linear. In addition, the mean field may be non local, depending upon the choice of the effective interaction. The solution of these equations yields both the single particle spectrum $\{e^{\lambda}, \varphi^{\lambda}\}$, and the total energy of the nucleus, $E = \sum_{\lambda} e^{\lambda} - \frac{1}{2}\sum_{\lambda\mu} V_{\lambda\mu}$, where the sum runs over occupied states only. The density matrix, $\rho_{\lambda\mu}$, built from the single particle spectrum, is used to calculate expectation values of one-body observables: $\langle\hat{A}\rangle = A_{\lambda\mu}\,\rho_{\mu\lambda}$. The many-body wave function representing the nucleus is a Slater determinant constructed on the occupied single particle orbitals φ_{λ}.

The HF energy being variational is guaranteed to be an extremum. However it may be only a local minimum. As the HF equations are non linear, they are solved iteratively from a suitable first guess of the converged solution. Depending on the starting point, it is a matter of experience that besides the absolute minimum, it is often possible to generate secondary local minima corresponding for instance to collective excitations of the nucleus.

To explore a given collective degre of freedom, such as the quadrupole deformation, it is convenient to generate a sequence of shapes evolving continuously from the ground state to an eventual secondary local minimum. For that purpose the two-body Hamiltonian $\hat{H}$ is replaced in the variational principal by the operator sometimes called the routhian $\hat{H} - \lambda\hat{Q}$. The collective operator, $\hat{Q}$, generates the different shapes and λ is the corresponding Lagrange mutiplier. In general we choose for $\hat{Q}$ the quadrupole operator $\hat{Q} = \sum_i (3z_i^2 - \mathbf{r}_i^2)$. The resulting new HF equations read as:

$$(h - \lambda\hat{Q})\,\varphi^{\lambda} = e^{\lambda}\,\varphi^{\lambda} \quad . \tag{4}$$

The value of λ is adjusted so that the expectation value of the collective operator $\hat{Q}$ is equal to a given value $q = \langle \hat{Q} \rangle$. For each choice of q, one obtains a Slater determinant which minimizes the energy. In particular, the energy is minimal with respect to any other unconstrained degrees of freedom, at least locally. This implies that no parametrization of higher multipoles of the nuclear shape has to be done.

In the calculations presented below, we have used Skyrme-type two-body effective interactions[3]. This particular choice of interaction leads to a local mean field. In addition, we include in the Hamiltonian a term which describes the pairing correlations since they are known to be crucial for a description of nuclear collective motion. We choose the standard constant G pairing matrix element ansatz.

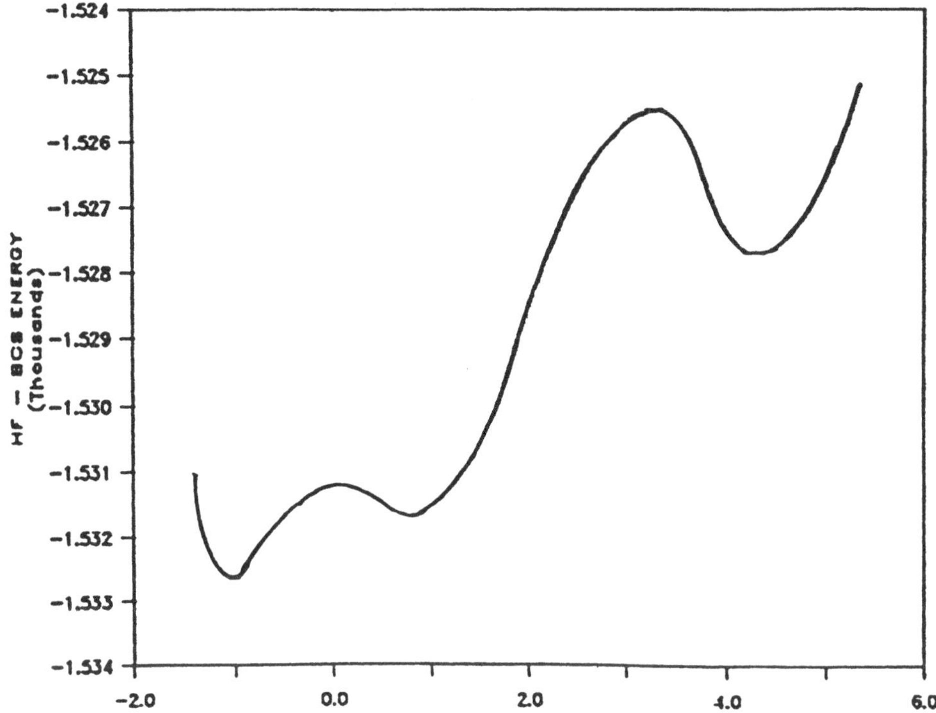

Fig. 1. HF+BCS deformation energy curve of ^{194}Hg as a function of the quadrupole moment (in thousands of Fm^2) with the SkM* interaction.

We solve the HF equations in coordinate space according to the discretization procedure described in Ref. 4. The collective space is mapped through doubly contrained HF calculations defined by the quadrupole deformation q and the asymmetry angle γ.

Figure 1 shows the quadrupole deformation energy curve for ^{194}Hg with the Skyrme interaction SkM*[5]. Only the axial degree of freedom is investigated so that $\gamma = 0^o$ for the prolate deformations ($q > 0$) and $\gamma = -180^o$ for oblate ones ($q < 0$). The curve exhibits three local minima. The absolute minimum is oblate and slightly deformed ($q \approx -10$ b). On the prolate side ($q \approx 8$ b), there is a second minimum whose excitation energy is about 1.0 MeV. At a much larger deformation ($q \approx 44$ b), there is a third minimum 5.0 MeV above the absolute minimum and separated from the main well by a barrier of 2.2 MeV (see Ref. 6).

These predictions have been verified through the observation of superdeformed (SD) bands extending to low angular momentum in the Hg nuclei[7, 8]. They are qualitatively

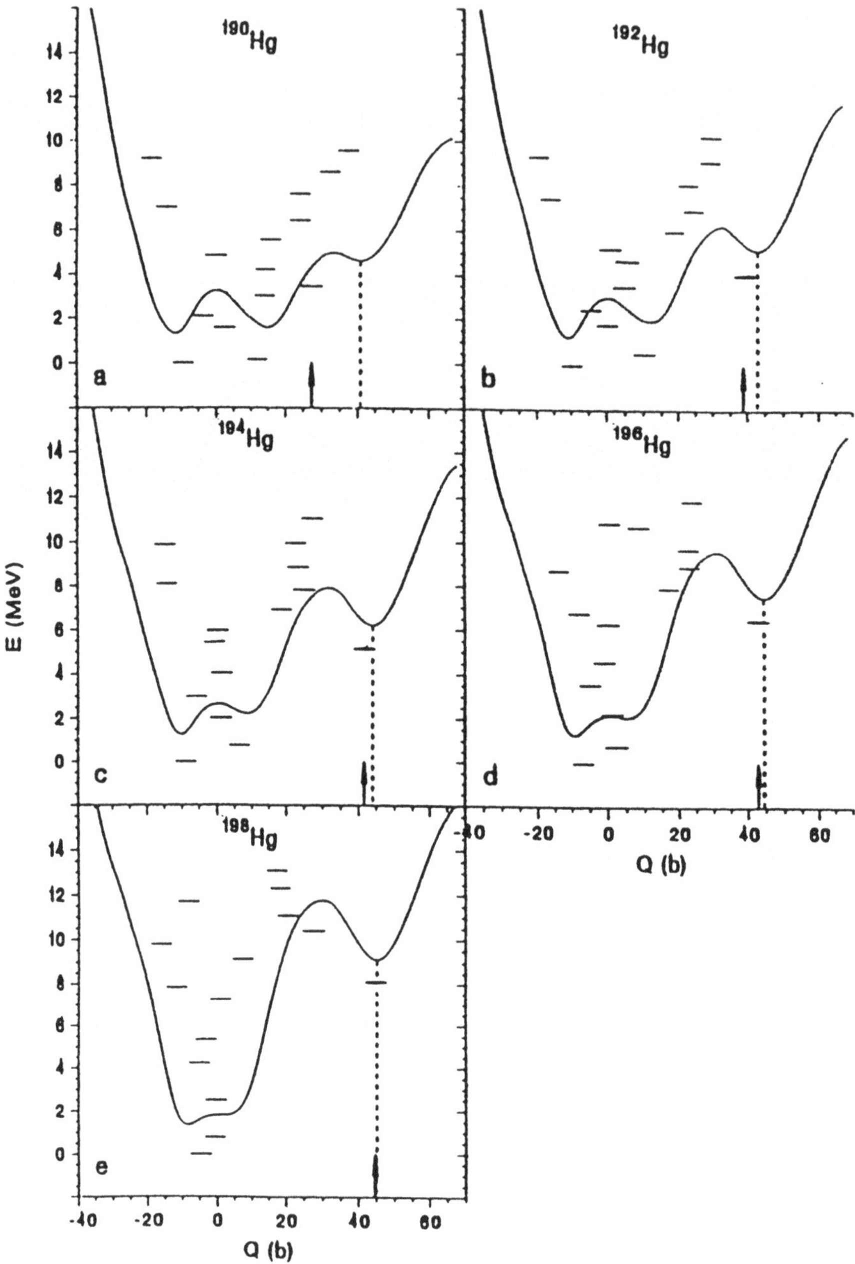

Fig. 2. HF+BCS deformation energy curves as a function of the quadrupole moment. The ordinate of the horizontal bars are the energies of the GCM states. The mid-segment abscissae give the GCM mean values of the quadrupole monent. The dashed vertical lines point to the HF+BCS SD minimum and the arrows to the GCM SD state. The interaction is SkM*.

consistent with the model calculation of Pashkevitch in 1969 (Ref 9), with more recents calculations[10, 11] and with the mean field calculations of the Gogny group[12].

Figure 2a,e displays the deformation energy curves (solid lines) as a function of the axial quadrupole constraint for five Hg isotopes. The ground state configuration is associated with the absolute minimum. For each five isotopes, a secondary minimum occurs at a rather large deformation corresponding to a mass quadrupole moment of approximately 44 b. This deformation corresponds to a typical value of the β_2 parameter of 0.5, or equivalently to a 1.6 to 1 axis ratio of the ellipsoidal shape of the nuclear density. On the basis of these static calculations, one can identify these nuclei whose second minimum corresponds to an isomeric state which will ultimately decay electromagnetically.

The indirect confirmation of this assumption[7, 8] through the experimental observation of the SD decay led us to make an extensive study of the mass region ranging from Gadolinium nuclei up to Radium[13]. The choice of this mass region is twofold. First these nuclei are heavy enough so that shape isomerism will occur at low spin: no additional centrifugal force is required to stabilize them as in the case of those Dysprosium isotopes which exhibit SD rotational bands which terminate at angular momentum above $20\hbar$. On the other hand, they must not be too heavy otherwise spontaneous fission will take over and we would enter the known region of fission isomers. The occurence of a second well deformed minimum for many nuclei in this mass region is a theoretical common pattern: Fig. 3a and 3b presents our results for the region which we call the isthmus of shape isomerim. It may be used as an indication of candidate nuclei for the experimental observation of shape isomerism. Our calculations give predictions similar to those based upon the Strutinsky prescription.[11, 14, 15, 16]

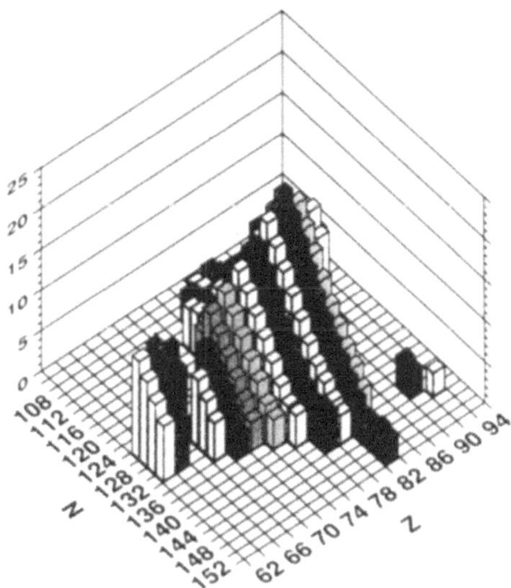

Fig. 3a. Isthmus of shape ismerism: excitation energies of secondary minima for various isotopes of even nuclei ranging from Gadolinium to Radium. Some isotopes of Thorium and Uranium are also shown for comparison.

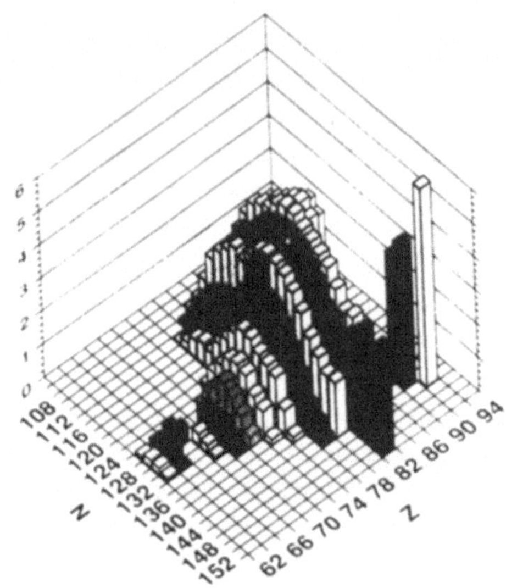

Fig. 3b. Isthmus of shape ismerism: well depths of secondary minima for the same isotopes.

BEYOND THE MEAN FIELD

The generator coordinate method (GCM) is the next term of the Hartree-Fock theory which incorporates collective and single-particle dynamics into a coherent quantum mechanical formulation. Given a family of N-body wave functions $|\Phi(q)\rangle$, depending on a collective continuous variable q, the GCM determines approximate eigenstates of the Hamiltonian $\hat{H}$ having the form:

$$|\Psi_k\rangle = \int dq\, f_k(q)\, |\Phi(q)\rangle \quad . \tag{5}$$

The weight functions f_k are found by requiring that the mean energy (1) calculated with $|\Psi_k\rangle$ is stationary with respect to arbitrary variations δf_k. This prescription leads to the Hill-Wheeler (HW) integral equation[17]

$$\int (\mathcal{H}(q,q') - E_k\, \mathcal{J}(q,q'))\, f_k(q')\, dq' = 0 \quad , \tag{6}$$

in which the Hamiltonian ($\mathcal{H}$) and the overlap kernel ($\mathcal{J}$) are:

$$\mathcal{H}(q,q') = \langle\Phi(q)|\hat{H}|\Phi(q')\rangle \ , \qquad \mathcal{J}(q,q') = \langle\Phi(q)|\Phi(q')\rangle \ . \tag{7}$$

The HW equation is solved by using the generating states $|\Phi(q)\rangle$ defined for a finite set of deformations q_i[6]. This yields a discrete approximation to the GCM, where kernels become matrices, and the integral equation (6) a matrix equation.

All the numerical problems related to the solution of the HW equations have been addressed to in Ref. 6. The same two-body Hamiltonian is used in the HF+BCS calculations and in the construction of the kernel (7). In this way, the residual interaction which is not included in the mean field, is taken into account in a consistent manner. This would not be the case if the single particle basis which generates the HF Slater determinant were replaced by any phenomenological single particle basis. Since HF+BCS wave functions are not eigenstates of the particle number operator, the GCM states may not have the correct particle number. To restore it, one introduces additional constraints on the neutron and proton particle number operators in the GCM Hamiltonian kernel, with appropriate Lagrange multipliers. Let us note that the GCM incorporates automatically the effects of ingredients of the Bohr collective Hamiltonian[18, 4] method such as the zero point motion energy and the effective collective masses. Moreover, it does not rely on assumptions such as the gaussian overlap approximation nor on the validity of quadratic expansions of the kernels[19] which are often used to derive microscopically these ingredients[12].

The short horizontal bars in Fig. 2 show the results of the diagonalization of the HW equations for the five Hg isotopes studied in the previous section. The ordinate is the energy of the states (the GCM ground state is taken as zero) and the abscissa of the center of the bar, the average of the quadrupole moment for that particular state. The energy shifts of the two HF minima relative to their corresponding GCM states are nearly equal. The excitation energy of the shape isomer, as predicted from static HF calculations, is not significantly modified by the collective correlations induced by configuration mixing. This validates the predictions of the sole mean field calculation.

The modifications of the deformation energy curves as a function of N strongly affect the shape isomeric state. In Fig. 2 the vertical dotted lines point to the mass quadrupole moment of the second well minima. The mean value $\langle\hat{Q}\rangle$ of the best SD candidate amongst the GCM eigenstates (vertical arrow on the Q axis) is close to that of the HF+BCS minima at large N and slowly moves toward lower Q as neutrons are removed. This can be related to the changes as a function of N in the thickness and height of the barrier separating the two wells. The smaller of either of these quantities, the more likely the GCM will mix HF states of comparable energy in the both wells, shifting the shape isomer toward smaller deformations.

Assuming that the moment of inertia of the shape isomer is given by the rigid body value, our calculations reproduce the band head moment of inertia inferred from the recent measurements on $^{190-194}$Hg. Moreover, our results are compatible with the observation of ^{192}Hg and ^{194}Hg having the same moment of inertia while that measured in ^{190}Hg is slightly smaller[20]. Concomitantly, we predict that the moment of inertia for the SD band in ^{196}Hg and ^{198}Hg should be the same as observed in ^{192}Hg and ^{194}Hg.

DEPOPULATION OF THE SUPERDEFORMED BANDS

From this study, it is possible to make predictions on the de-excitation rates in and out of the SD band. A necessary ingredient for that purpose is the set of matrix elements of the charge quadrupole operator. It can be calculated within the GCM, assuming that the rotational states can be described as a product of a collective rotor times an intrinsic state.

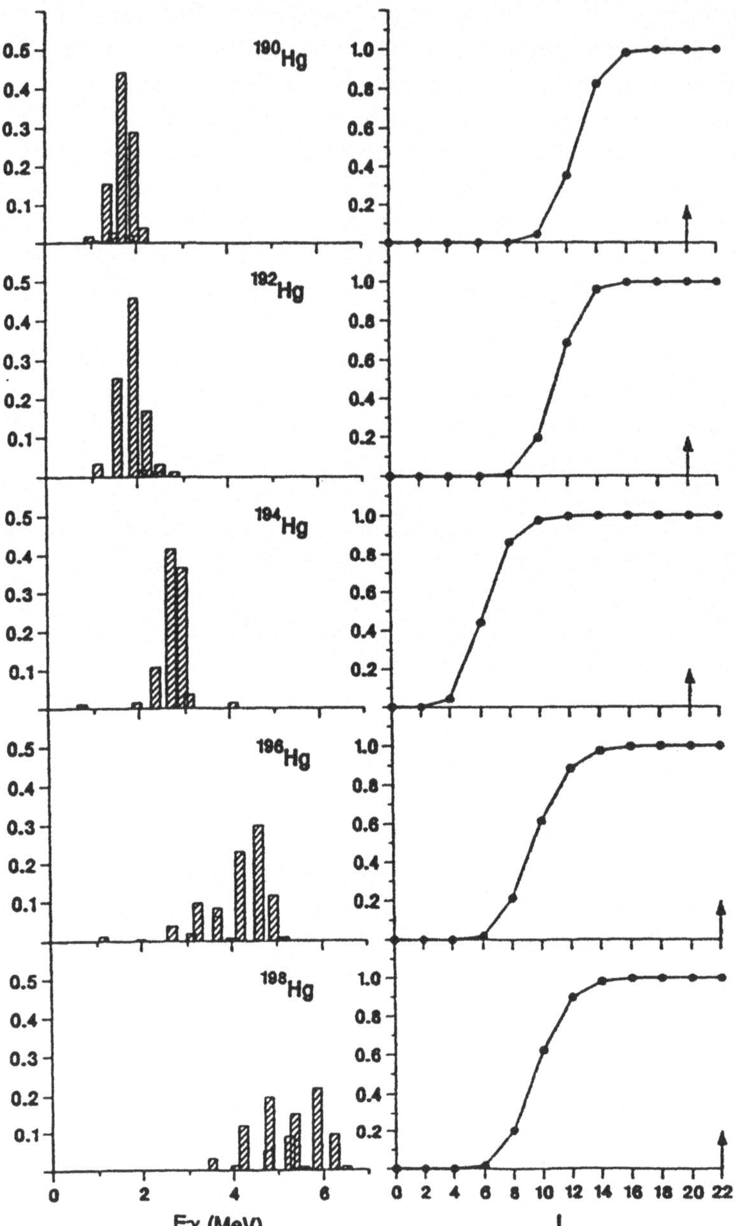

Fig. 4. Decay of the SD band for five Hg isotopes. Left side: relative intensities of the depopulating E2 gamma rays as a function of their energies. Right side: evolution of the population of the SD band as a function of angular momentum. The vertical arrow indicates the spin below which the SD band is excited with respect to at least one positive parity band.

In addition, we introduce phenomenological estimates of the moments of inertia of the bands in order to allow an extrapolation of the GCM results to non-zero spin states. For the SD band, we choose the rigid body moment of inertia[21, 22]. We then assume that the collective GCM states at smaller values of the quadrupole moment are band heads for rotational bands which function as doorways at the appropriate values of the angular momentum. From the choice of the moment of inertia for each band, one constructs a rotational spectrum based on GCM states in the first well. The moment of inertia $\mathcal{J}_k$ of a given band k is chosen phenomenologically[2] according to:

$$\mathcal{J}_k = \sqrt{a^2 + b^2(Q_k/Q_{SD})^2} \quad , \tag{8}$$

where $Q_k = \langle \Psi_k | \hat{Q} | \Psi_k \rangle$ is the quadrupole moment of the corresponding band head. The coefficients $a = 12.5\hbar^2/\text{MeV}$ and $b = 100\hbar^2/\text{MeV}$ have been selected to reproduce the rigid body moment of inertia at large Q and to have the order of magnitude suggested by experiment[23] at small Q. The last ingredient of the model is the energy $E_k^{0^+}$ of the first state of each band. It has been determined from the corresponding GCM intrinsic state energy E_k by subtracting the rotational energy:

$$E_k^{0^+} = E_k - \frac{\hbar^2}{2\mathcal{J}_k} \langle \hat{J}^2 \rangle \quad . \tag{9}$$

In this formula $\langle \hat{J}^2 \rangle$ denotes the HF+BCS expectation value of the total angular momentum for the band head quadrupole moment.

From the calculated spectrum and using standard formulae for transition rates[19], we can follow the population of the SD band as a function of angular momentum, assuming it to be unity at $30\hbar$. Our results are shown on the right side of Fig. 4 for the five Hg isotopes. A nearly complete depopulation of the SD bands occurs in two to three transitions and begins at about 8 to $14\hbar$. Both results appear qualitatively consistent with experiments in the Hg isotopes. The left side of Fig. 4 displays the spectrum of depopulating γ-rays expected from our calculated transitions. The pattern changes systematically as a function of neutron number, with the predicted spectrum hardening with increasing N. If this is the correct mechanism for the emptying of the SD rotational band, then perhaps these γ-rays could be observed. However, the actual γ-ray spectrum may be more complicated since we expect the states in the first well to dilute their strength within states of the non-collective continuum. The resulting spreading will render more difficult the identification of the depopulating γ-rays.

The mechanism responsible for the sudden depopulation of the SD bands has its origin in the structure of the transition rates which are proportional to the square of the quadrupole transition matrix element and to the fifth power of the energy of the transition. The former dependence favors in band transitions whereas the second is responsible for the depopulation as soon as the energy of an out of band transition is large enough to overcome the smallness of the off-diagonal matrix elements of $\hat{Q}$.

It is relevant to be concerned that other than quadrupole transitions will alter our conclusions. However we believe that direct out of band $E1$ transitions will not be important as most of the observed dipole strength lies at too high an excitation energy to affect this process in the Hg isotopes. The situation might be quite different for the SD bands observed in the Dysprosium region since they have been observed to disappear at a much higher angular momentum and absolute energy.

STABILITY OF THE SD BAND AGAINST PAIRING VIBRATIONS

In the previous section, we have studied the direct depopulation of SD states via $E2$ transitions to collective states in the first well. These states are generated from the superposition of Slater determinants having different quadrupole deformation. However this degree of freedom is not the only one and pairing correlations are known to be important as the single-particle configuration rearranges approximately ten times[24, 25] for a variation of the quadrupole moment as experienced by a Hg nucleus in the transition from the ground state to the SD configuration.

Coherent pairing fluctuations can be described in closed analogy to the quadrupole fluctuations by the pair-transfer operator $\hat{P}^+ + \hat{P}$ ($\hat{P}^+ = \sum_{\mu>0} a^+_\mu a^+_{\bar{\mu}}$) which breaks the particle number invariance, playing the role of the quadrupole moment operator which destroys the rotational symmetry. One could then introduce a constraint on the average value $\langle \hat{P}^+ + \hat{P} \rangle = 2\sum_{\mu>0} v_\mu u_\mu$, which appears as the natural generator coordinate for this type of collective motion. However, for a BCS state, the pairing gap Δ is a monotonous function of this average value, so that we decided to use it as an equivalent generator coordinate. Hence, the quadrupole collective variable q in the HW equ. 6 is replaced by the triplet of coordinates $\{Q, \Delta_N, \Delta_Z\}$ composed of the axial mass quadrupole moment together with the neutron and proton pairing gaps.

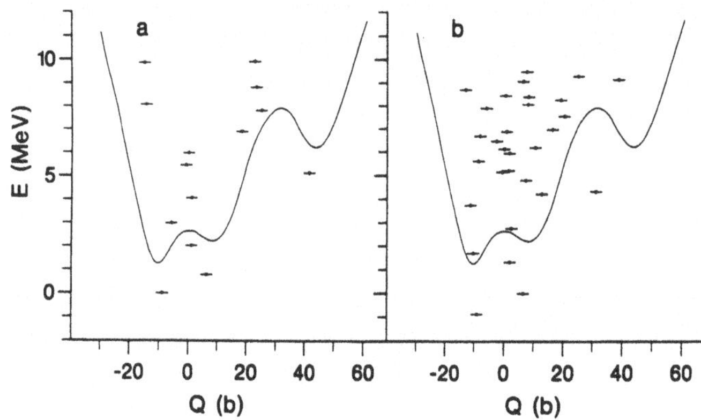

Fig. 5. The solid curves show the HFBCS deformation energy. The dot at the center of the horizontal bars gives the energy and mass quadrupole moment of the GCM states. Left: no pairing correlations are included. Right: full GCM calculation with pairing vibrations. The origin of the energy is chosen at the lowest GCM state without pairing effects included (left).

At a given deformation Q, pairing vibrations take place around the HFBCS solution which corresponds to a minimum of the Δ-constrained energy. Let us denote $\Delta_N^{(0)}(Q)$ and $\Delta_Z^{(0)}(Q)$ the values of the HFBCS neutron and proton gaps solutions of the gap equation. For each selected value of Q, we have dicretized the gaps around these values, as they are our additional generating variables. A complete study of the convergence as well as the stability of the resulting GCM equations has been done[26]. It turns out that accurate results can be obtained with a fixed $d\Delta$ of 0.6 MeV and a set of 28 values of Q evenly spaced between -40b and 70b, i.e. a domain which spans the physically relevant range of deformations.

The energies and quadrupole moments of the lowest GCM states calculated without and with the coupling to pairing vibrations are shown in Fig. 5. From this comparison, the following observations can be made:

- The spectrum below 5MeV is almost unaffected by pairing vibrations. This remark concerns both the relative energies and the deformations. The absolute energies are shifted downward by approximately 0.9MeV. This quantity provides a measure of the magnitude of ground-state dynamical pairing correlations.

- The excitation energy of the SD state ($\approx$5MeV) is not modified by pairing vibrations. By contrast, its mass quadrupole moment is significantly reduced, from 42 to 31b. However, the quadrupole moment provides only an averaged information on the deformation properties of the SD state. The observed decrease does not imply a global shift of the collective SD wave-function to smaller deformations. It is a signature of the fact that the SD GCM state has acquired a component in the first well.

- There is an important increase of the collective level density above 5MeV. The number of GCM states between 5 and 10MeV, grows from 7 in the pure quadrupole calculations, to 21 once the Δ degrees of freedom are taken into account.

- This tripling of the level density concerns only wave-functions in the first well. None of the states above 5MeV has a quadrupole moment large enough to qualify as a SD state. Therefore, from this calculation, which we recall is performed at zero spin, we infer that a SD β-band in ^{194}Hg should not survive at low angular momentum.

- In the ^{194}Hg nucleus studied here, the energy of the SD state happens to coincide with the beginning of the region of increased level density. This circumstance enhances the possibility of a mixing with levels in the first well which is at the origin of the decrease of the SD quadrupole moment.

The implications of these results on the depopulation rate via direct $E2$ decay are as follows. According to the simple phenomenological model described in the previous section[2], to each GCM state is asssociated a rotational band whose moment of inertia is determined by the quadrupole moment of that state following the phenomenological prescription given by eq. 8. The transition rates within and out of the SD band are calculated assuming a factorization of the total wave function as the product of the GCM intrinsic state by a rotational-model wave function describing the global rotation. The $E2$ decay rate is thus obtained as the GCM quadrupole transition matrix element squared times the fifth power of the transition energy and the usual Clebsh-Gordan prefactors[19].

We find that the larger probability at small Q of the SD collective wave-function carries over into an increase of the quadrupole transition matrix elements towards lower GCM states. This leads to a more rapid depopulation of the SD band. Compared with results obtained without coupling to pairing vibrations, the critical angular momentum around which the SD band disappears is raised by six units. On the other hand, the overall pattern of the depopulation with the remarkable vanishing of the band over two transitions, is not significantly modified. This can be understood from the results of ref.[2], where it is shown that the rapidity of depopulation is governed by the ratio between quadrupole matrix elements within and out of the SD band. This study also indicates that as long as the $2/5^{th}$-power of this ratio remains larger then 10, the shape of the depopulation curve is almost unchanged. This is indeed the case here, since although pairing fluctuations decrease the ratio, its value stays above 11.

The present work suggests that the major modification of the collective spectrum density due to pair vibrations, concerns only states in the first well above 5MeV. If we assume that this will be also true for neighbouring isotopes, we find that the nucleus ^{194}Hg marks a sort of boundary: for heavier Hg, the energy of the SD state will be within the region of increased collective level density while it will remain below it, for the lighter ones. Thus, for these latter isotopes, the SD state should not be modified compared with previous GCM calculations. By contrast, for the SD states of ^{196}Hg and ^{198}Hg, one can expect an increased mixing with states of the first well, resulting in a larger component at small Q. As a consequence, coupling with pairing modes should not change the depopulation curve calculated in Ref. 2 for ^{190}Hg and ^{192}Hg and shift to higher values the spin at which the SD band of heavier isotopes disappears.

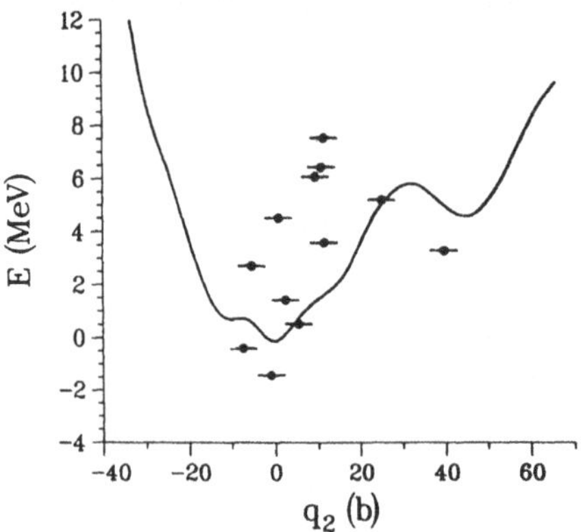

Fig. 6. HFBCS deformation energy curve of ^{194}Pb as function of the quadrupole deformation q_2 shown as solid line. The energies and quadrupole moments of GCM states are shown as short horizontal bars.

OCTUPOLE DEGREE OF FREEDOM

The octupole degree of freedom is known to be important for the description of fission of actinide elements. In many nuclei the fission path is unstable against octupole deformation when the quadrupole deformation becomes very large[27-29]. Several theoretical analyses[30-36] have also suggested an octupole softness of the SD minima found in the energy surfaces of light Pb and Hg isotopes which have been associated with recently discovered SD bands. Moreover, several excited SD bands observed[35] in ^{193}Hg have been considered as providing evidence for the presence of octupole correlations in the SD states. As of today, the position of these bands relative to the Yrast configuration has not been measured.

In this section we present a calculation of the octupole collective wave-functions of the intrinsic states of both the positive and negative parity SD states[37]. We concentrate our analysis on the nucleus ^{194}Pb where SD bands have been found[20, 38]. It belongs to the region of superdeformation centered around the "magic" SD nucleus ^{192}Hg. Our choice of the nucleus ^{194}Pb is motivated by the fact that triaxial quadrupole deformation ($\gamma \neq 0$) plays a smaller role in this nucleus than in the mercury isotopes; as seen in Fig. 6, the HFBCS ground state of ^{194}Pb is indeed spherical. This allows us to restrict the analysis to axial shapes of ^{194}Pb.

The deformation energy curve is constructed with a constraint on the axial quadrupole moment for values ranging from -34b to 66b. In addition to the spherical ground state, one finds a SD minimum at $q_2 \approx 46$b with an excitation energy of about 4.7 MeV. The barrier between the SD and the main well is located at 34b, and is about 1.3 MeV above the SD minimum. The properties of this minimum are therefore similar to those found[2] in ^{194}Hg, except for the excitation energy which is higher in the latter nucleus.

Following the same procedure as in Ref. 6, we have solved the GCM equations using the average value of quadrupole moment q_2 as the collective variable. The GCM spectrum is reported in Fig. 6 where each state is indicated by a short horizontal bar drawn at the corresponding energy and a dot at the average quadrupole moment. The lowest state is spherical and 1.5 MeV below the HFBCS minimum. As in the Hg isotopes, we find that compared to the mean-field predictions, the energy of the SD GCM state is lowered by about the same amount (1.15 MeV), and that its quadrupole moment is slightly reduced ($q_2 \approx 40$b).

Octupole deformation energy curves for fixed q_2 are constructed with an additional constraint on the mean square of the scalar constructed with the average values of components of the octupole operator:[39] $q_3=\sqrt{\langle\hat{Q}_3\rangle^2}$, with $\langle\hat{Q}_3\rangle^2=\sum_{\mu=-3}^{3}(-1)^{\mu}\langle\hat{Q}_{3\mu}\rangle\langle\hat{Q}_{3-\mu}\rangle$ and $\hat{Q}_{3\mu}=r^3Y_{3\mu}(\theta,\phi)$. This choice has the advantage of not precluding nonaxial octupole deformations. Since we enforce a symmetry of the HFBCS states with respect to the x and y planes, the contributions of odd-μ components are suppressed. In addition, the axial constraint along the z-axis generated by the operator $\hat{Q}_2$ almost eliminates all of the $\mu=\pm 2$ components. Therefore, our octupole constraint turns out to be equivalent to a single one on $\langle\hat{Q}_{30}\rangle$ for all practical purposes.

The solid line in Fig. 7a shows the HFBCS energy as a function of q_3 for the zero value of the quadrupole moment (q_2=0). This curve illustrates the octupole stiffness at the spherical minimum. For each value of q_3 we have projected states with good parity out of the HFBCS solution, according to the method exposed in Refs. 40-41. The energies of projected states are shown in Fig. 7a as long-dashed and dashed lines for positive and negative parity, respectively. Projecting the HFBCS state with an octupole moment q_3=750 Fm^3 leads to a 1.05 MeV energy lowering with respect to the HFBCS minimum. The lowest point on the negative parity curve lies 3.8 MeV above the positive minimum, and corresponds to q_3=1400 Fm^3.

It is important to keep in mind that the q_3 values of the minima of parity projected curves should not be interpreted as static octupole deformations, and that the value 1.05 MeV mentioned above does not represent a barrier between symmetric and asymmetric configurations. After parity projection, the resulting multi-determinant wave-functions include correlations beyond the mean-field, and both positive and negative parity projected states have a zero expectation value of $\hat{Q}_{30}$. At these minima, q_3 is to be viewed as a measure of octupole fluctuations present in the optimal intrinsic state pertaining either to the positive or to the negative parity bands.

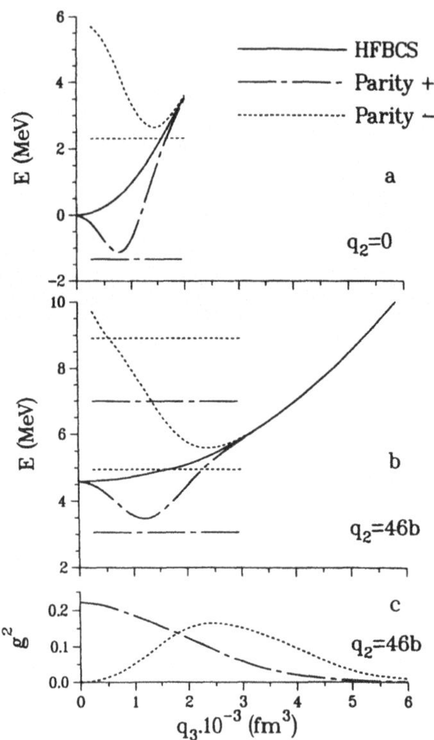

Fig. 7. Octupole deformation properties of ^{194}Pb at the ground state (q_2=0): (a), and at the SD minimum (q_2=46b): (b). The solid lines show the HFBCS energies as a function of the octupole moment q_3. The long-dashed (dashed) lines give the energies of the positive (negative) parity projected at a given q_3. The long-dashed and dashed horizontal lines indicate the energies of the GCM positive and negative parity states. The lower part of the figure (c) shows the probability distributions of the lowest GCM collective wave-functions of both parities at the quadrupole deformation of the SD minimum.

Fig. 7b shows the energy curve as a function of q_3 at $q_2 = 46$b, which is the quadrupole deformation of the SD minimum. The HFBCS curve (solid line) shows that even though the SD state remains stable against octupole excitation, it is much softer than that at q_2=0. An energy increase of 1 MeV is now obtained for q_3=2750 Fm^3, while the same excitation requires only q_3=1000 Fm^3 at the spherical minimum. The positive parity minimum is about 1.15 MeV below the HFBCS SD minimum. This value is similar to that obtained at q_2=0, but the minimum is obtained from an intrinsic state with a much larger value of q_3 (1250 Fm^3). Similarly the minimum of the negative parity curve occurs at q_3=2500 Fm^3 instead of q_3=1400 Fm^3 at q_2=0. In addition it lies only 2.1 MeV above the positive minimum. This value, which is almost a half of that at q_2=0, indicates the increased octupole softness near the SD minimum and provides us with a first estimate of the excitation energy of a possible octupole collective SD band of negative parity.

Since the HFBCS curve is rather flat as a function of q_3, one should expect that many configurations with different q_3 values will contribute to the octupole collective function. This was tested in GCM calculations using q_3 as the collective variable while keeping q_2 constant. As for the case of quadrupole case, we have included a constraint on neutron and proton numbers to ensure that final states have the correct N and Z values. As expected, the lowest GCM state has a positive parity. Its energy is 1.53 MeV below that of the HFBCS SD minimum, it is presented as the lowest long-dashed horizontal line in Fig. 7b. This value is larger than the 1.15 MeV lowering due to the quadrupole correlations (see SD GCM state at $q_2 \approx 40$b, Fig. 6). Therefore, the effect of the octupole deformation on the energy of the shape isomer turns out to be more important than the quadrupole one. The energy of the first negative parity excited state (2nd GCM state) is indicated by the lowest dashed line and is only 1.9 MeV above that of the positive parity GCM state. The energies of a two more GCM octupole states are also indicated in Fig. 7b by long-dashed and dashed lines according to their parities.

The GCM probability densities for the lowest states of both parities are presented in Fig. 7c. One sees that both distributions are rather wide, indicating that many different configurations enter the GCM collective states. On the other hand, the fact that the GCM octupole excitation energy (1.9 MeV) is almost the same as that obtained by a simple parity projection (2.1 MeV) indicates that positive and negative parity states are affected similarly by the configuration mixing. One notes also that the maximum of the distribution for the negative parity state occurs at 2500 Fm^3, namely the q_3 value at the minimum of the negative parity projected energy curve.

So far we have only performed octupole GCM calculations for fixed values of q_2. Since the effect of octupole and quadrupole correlations appear to be of similar magnitude, a two dimensional (q_2–q_3) GCM calculation, which would allow an investigation of the coupling of both modes, appears very desirable.

CONCLUSION

Microscopic mean field theory calculations have been performed for nuclei ranging from Gadolinium to Radon and a large number of these isotopes are predicted to exhibit shape isomerism. Where experiments have been performed within this isthmus of isomerism through observation of SD bands built upon these predicted shape isomers, they are completely consistent with the measurements. We have compared the energy and position in collective space of the GCM eigenstates with those obtained in static

HF+BCS calculation. We conclude from this study that the excitation energy of the SD band head or shape isomer predicted from HF+BCS methods is almost unaltered by the introduction of quadrupole correlations.

The next step in this theory, GCM, yields subtle variation in the moment of inertia of the observed SD rotational bands in the Hg isotopes which are consistent with experiment and correctly reproduces the observed rapid depopulation and limiting angular momentum. This explanation of the sudden depopulation of the SD band makes it unlikely that the SD band head can be reached through the direct feeding of high lying rotational states at high angular momentum. Alternative experiments involving a direct feeding of SD states at low spin will have to be implemented.

The further analysis of the role of the pairing vibration in ^{194}Hg has improved the qualitative agreement with experiment, predicting a larger angular momentum for the rapid transition out of the SD band. As in the previous calculation, only one SD state is found and its energy is not modified by coupling to coherent pairing modes.

Finally, our analysis of the octupole mode in ^{194}Pb has shown the softness of this nucleus with respect to this degree of freedom and given further insight to odd parity excited SD band in this mass region. Two dimensional GCM calculations in the (q_2-q_3) plane are under progress for this isotope.

ACKNOWLEDGEMENT

This work was supported in part by the NATO grant RG 85/0195, in part by the U.S. Department of Energy under Engineering Contract NW 7405-ENG-48 and in part by the Polish Ministry of National Education under Contract CPBP 01.09.

REFERENCES

1. P. Bonche, S.J. Krieger, P. Quentin, M.S. Weiss, J. Meyer, M. Meyer, N. Redon, H. Flocard and P.-H. Heenen, Nucl. Phys. **A500** 308 (1989).

2. P. Bonche, J. Dobaczewski, H. Flocard, P.-H. Heenen, S.J. Krieger, J. Meyer and M.S. Weiss, Nucl. Phys. **A519** 509 (1990).

3. D. Vautherin and D.M. Brink, Phys. Rev. **C5** 626 (1972).

4. P. Bonche, H. Flocard, P.-H. Heenen, S.J. Krieger and M.S.Weiss, Nucl. Phys. **A443** 39 (1985).

5. J. Bartel, P. Quentin, M. Brack, C. Guet and H.-B. Håkanson, Nucl. Phys. **A385** 269 (1982).

6. P. Bonche, J. Dobaczewski, H. Flocard, P.-H. Heenen and J. Meyer, Nucl. Phys. **A510** 466 (1990).

7. E.F. Moore et al, Phys. Rev. Lett. **63** 360 (1989).

8. E.A. Henry, Seventh International Symposium on Capture Gamma-Ray Spectroscopy and related topics, Asilomar, 14-19 Oct. 1990.

9. R.R. Pashkevitch, preprint JINR (Dubna) P4-4383 (1969) unpublished.

10. See for instance S. Åberg, Phys. Scr. **25** 23 (1982).

11. R.R. Chasman, Phys. Lett. **B219** 227 (1989).

12. M. Girod, J.P. Delaroche, D. Gogny and J.F. Berger, Phys. Rev. Lett. **62** 2452 (1989); J.P. Delaroche, M. Girod, J. Libert and I. Deloncle, Phys. Lett. **B232** 145 (1989).

13. S.J. Krieger, P. Bonche, M.S. Weiss, J. Meyer, H. Flocard and P.-H. Heenen, LLNL preprint, unpublished.

14. M.A. Riley et al, Nucl. Phys. **A512** 178 (1990).

15. C.F. Tsang and S.G. Nilsson, Nucl. Phys. **A140** (1970) 275.

16. W. Satuła, S. Ćwiok, W. Nazarewicz, R. Wyss and A. Johnson, Nucl. Phys. in press.

17. D.L. Hill and J.A. Wheeler, Phys. Rev. **89** 1102 (1953).

18. Å. Bohr and B. Mottelson, "Nuclear Structure", volume 2, W. Benjamin, New York (1975).

19. P. Ring and P. Schuck, "The Nuclear Many-Body Problem", Springer Verlag, Berlin (1980).

20. M.J. Brinkman et al, Z. Phys. **A336** 115 (1990); J.A. Becker et al, Proc. Conf. on nuclear structure in the nineties, Oak Ridge, Tennessee, ed. N. Johnson, Nucl. Phys. to be published.

21. H.J. Specht, J. Weber, E. Konecny and D.Heunemann, Phys. Lett. **B41** 43 (1972).

22. V. Metag, D. Habs and H.J. Specht, Phys. Reports **65** 1 (1980).

23. B. Singh, Nucl. Data Sheets **56** 75 (1989).

24. G. Bertsch, Proc. of tht Int. School of Physics Enrico Fermi CIV, (Soc. Ital. di Fisica, Bologna, 1988) 41.

25. G. Bertsch and H. Flocard, Phys. Rev. Lett. in press.

26. J. Meyer, P. Bonche, J. Dobaczewski, H. Flocard and P.-H. Heenen, IPN-Orsay preprint, unpublished.

27. J.F. Berger, M. Girod and D. Gogny, Nucl. Phys. **A502** 85c (1990).

28. J. Blons, Nucl. Phys. **A502** 121c (1990).

29. J.R. Nix, Nucl. Phys. **A502** 609c (1990).

30. T. Bengtsson *et al.*, Phys. Scr. **24** 200 (1981).

31. J. Dudek, in *The Variety of Nuclear Shapes*, eds. J.D. Garrett et al. (World-Scientific, Singapore, 1987) p.195.

32. J. Höller and S. Åberg, Z. Phys. **A336** 363 (1990).

33. S. Mizutori, Y.R. Shimizu and K. Matsuyanagi, in *Proceedings of Int. Conf. on Nuclear Structure in the Nineties*, Oak Ridge, April 1990, p.28.

34. S. Åberg, Nucl. Phys. A, in press.

35. J. Dudek, T. Werner and Z. Szymański, Phys. Lett. **B248** 235 (1990).

36. D.M. Cullen *et al.*, Phys. Rev. Lett. **65** 1547 (1990).

37. P. Bonche, S.J. Krieger, M.S. Weiss, J. Dobaczewski, H. Flocard and P.-H. Heenen, Phys. Rev. Lett. **66** 876 (1991).

38. K. Theine *et al.*, Z. Phys. **A336** 113 (1990).

39. P. Bonche, P.-H. Heenen, H. Flocard and D. Vautherin, Phys. Lett. **B175** 387 (1986).

40. P. Bonche, in *The Variety of Nuclear Shapes*, eds. J.D. Garrett et al. (World-Scientific, Singapore, 1987) p.302.

41. J.L. Egido and L.M. Robledo, Nucl. Phys. **A518** 475 (1990).

SUPERDEFORMATION: AN EXPERIMENTAL REVIEW

R. V. F. Janssens

Argonne National Laboratory
9700 S. Cass Ave., Physics Division
Argonne, IL 60439, USA

INTRODUCTION

The discovery of a superdeformed (SD) band of 19 discrete transitions in the nucleus ^{152}Dy[1] is at the origin of the formidable experimental and theoretical effort which has allowed for much progress in the unraveling of the main properties of superdeformation. Over the last five years, many unexpected features have been discovered. In fact, just as soon as a new feature has been explained invariably the next observation has highlighted another surprising phenomenon. The net result of all the activity has been a dramatic increase of our understanding of rapidly rotating nuclei.

The organizers of this conference have asked me to review the present status of the field: this is clearly an impossible task within the allowed space and time. Fortunately, the subject has recently been reviewed on several occasions[2-5] and, therefore, I will only attempt to present here some of the general features that have emerged thus far, and discuss some of the important open questions.

The discussion will be restricted to SD nuclei in the A=150 and A=190 regions, i.e. to the nuclei with the largest quadrupole deformations ($\beta_2 \sim 0.6$ and 0.5, respectively) and will address four main issues. First, the evidence for the stiffness of the SD minimum will be reviewed. This section will be followed by discussions on the role of pairing and of specific quasiparticle configurations in the second well. Evidence for other effects at large deformation will be addressed in this section as well. The discussion will then address the issue of the so-called "identical bands" and will conclude with a brief presentation on the feeding and decay of SD nuclei.

STABILITY OF THE SUPERDEFORMED MINIMUM

Figures 1 and 2 present the dynamic moments of inertia $J^{(2)}$ of all SD bands observed so far in the A=150 and A=190 regions. SD bands have now been identified in all Gd isotopes with A=146-150,[6-11] in $^{150,151}Tb$ [9,10,12] as well as in $^{151-153}Dy$[1,13,14]. Reports of similar band structures in ^{145}Gd and ^{142}Eu have also become available[15,16], but these are not shown in fig. 1. For the A=190 region, bands have been identified

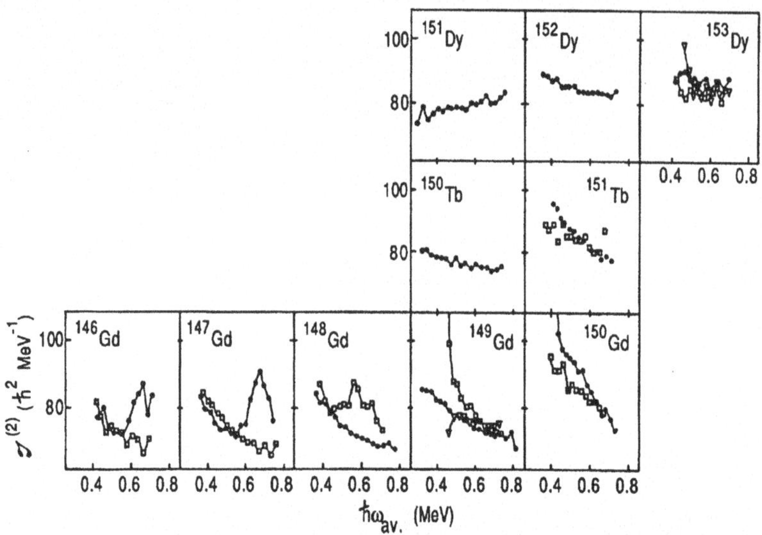

Fig. 1. Dynamic moments of inertia $J^{(2)}$ for all SD bands in the A=150 region.

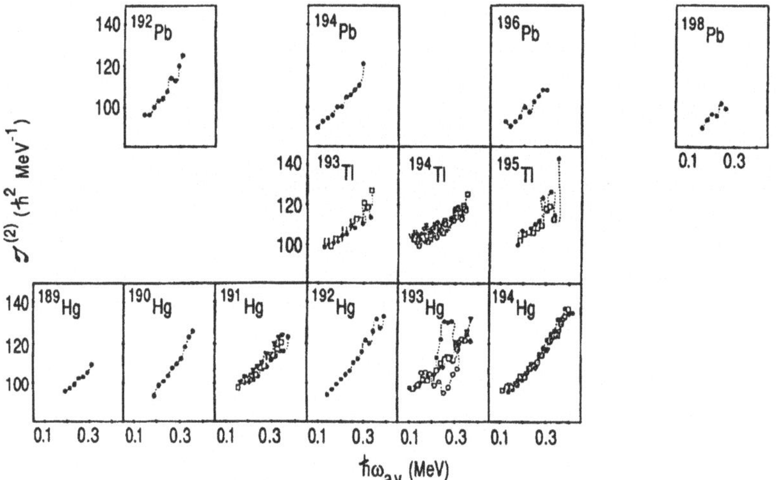

Fig. 2. Dynamic moments of inertia $J^{(2)}$ as a function of rotational frequency for all SD bands in the A=190 region.

in all Hg isotopes with A=189-194[17-22], in $^{193-195}$Tl[23-25] and in the even Pb isotopes with A=192-198[26-28]. Furthermore, in many nuclei several SD bands have been reported. Thus, in both regions it has been possible to perform detailed spectroscopy in the SD well.

From the two figures, it is apparent that the $J^{(2)}$ values exhibit variations with $\hbar\omega$. Mean-field calculations which attempt to reproduce variations in $J^{(2)}$ have suggested that such variations may be attributed to three major factors which could contribute separately or cooperatively:

(i) shape changes as a function of $\hbar\omega$ (e.g. centrifugal stretching), (ii) changes in pairing at large deformations[29], and (iii) occupation of specific high-N orbitals[29,30] (i.e. high-N orbitals from two major shells higher which plunge down as a function of deformation and come towards the Fermi surface at very large deformations, in this case $i_{13/2}$ protons and $j_{15/2}$ neutrons). The first of these possibilities, i.e. the changes in deformation, can be ruled out on the basis of measurements of transition quadrupole moments. As an example for the A=150 region, the case of the lowest SD band in ^{150}Gd is discussed here: in this nucleus $J^{(2)}$ is not only seen to decrease with $\hbar\omega$, but also exhibits a dramatic fall off at the lowest frequencies (fig. 1). This rapid change is also accompanied by an almost instantaneous decay out of the band. The DSAM lifetime measurements of ref. 10 have shown convincingly that there is no evidence for a change in the inband transition strengths for all transitions down to the lowest data point: a constant transition quadrupole moment Q_t=17 $\pm$ 3 eb fits the data extremely well and a sudden change in the nuclear shape can be ruled out. As an example for the A=190 region, the SD band in ^{192}Hg can be considered. In this case, as in most SD bands of this region (fig. 2), a smooth rise in $J^{(2)}$ of about 40% over the frequency range is observed. In contrast with measurements for SD states near A=150, where only fractional Doppler shifts $F(\tau)$ have been measured so far, one has been able to analyze detailed lineshapes for individual transitions between SD states in this case[31]. Such an analysis allows one to determine the variation of Q_t as a function of $\hbar\omega$, as opposed to previous studies where Q_t was assumed to be constant for all states in the band. The measured Q_t-values are displayed as a function of $\hbar\omega$ in fig. 3b. As can be seen, the quadrupole moment Q_t, and hence the deformation, remain essentially constant ($Q_t \sim 20 \pm 2$ eb) over the entire frequency range. This result rules out centrifugal stretching as an explanation for the rise in $J^{(2)}$: this is illustrated by the dashed line in fig. 3b, where the values of Q_t have been derived assuming that the change in $J^{(2)}$ is entirely due to a gradual change in deformation.

It should be noted that the available mean-field calculations also find the SD minimum to be rather stiff, i.e. the Q_t values are calculated to be rather constant. This is illustrated by the solid line in fig. 3b as well as by the inset in this figure, where the calculated variations in the β_2 and β_4 deformation parameters as a function of frequency are shown to be very small. The calculations were performed within the framework of the cranked shell model, using the Woods-Saxon approach, and include effects due to static and dynamical pairing[29,31]. As discussed in ref. 29,32, these calculations also predict very small deformation changes in the SD bands of the A=150 region, in agreement with the measurements.

HIGH-N INTRUDER ORBITALS AND PAIRING

Differences in the variations of $J^{(2)}$ with $\hbar\omega$ from nucleus to nucleus have been attributed to changes in the occupation of specific high-N intruder orbitals (i.e. possibility (iii) above). These changes can have dramatic effects: in the case of the Dy isotopes for example, $J^{(2)}$ is seen to rise smoothly over the entire frequency range in ^{151}Dy (fig. 1), while a smooth decrease is seen in ^{152}Dy and an essentially constant value of $J^{(2)}$ is observed for the yrast SD band in ^{153}Dy (it is assumed that the most intensely populated band is the "ground" band in the second well). For these isotopes, the proton contribution has been assigned as $\pi 6^4$ (the nomenclature of ref. 30 is adopted and refers to $i_{13/2}$ intruder orbitals) and the observed differences have been attributed to changes in the occupation of the $j_{15/2}$ neutron orbitals ($\nu 7^1$, $\nu 7^2$ and $\nu 7^3$ for $^{151-153}$Dy, respectively), which result[30] in contributions of varying magnitude to $J^{(2)}$. While the calculations of ref. 30 were performed without pairing

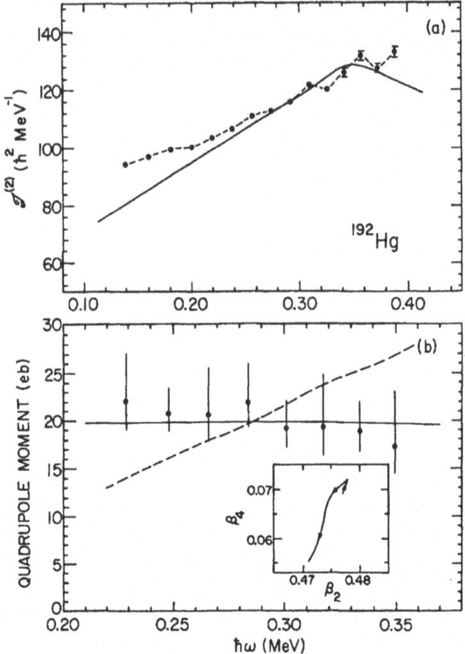

Fig. 3. Figure 3(a): Comparison between the measured and calculated (solid line) moments of inertia $J^{(2)}$ as a function of rotational frequency $\hbar\omega$ in the SD band of ^{192}Hg.
(b): Comparison between the measured and calculated transition quadrupole moments Q_t in the SD band of ^{192}Hg. The dashed line represents a calculation assuming that the rise in $J^{(2)}$ is due to a change in deformation only (see text); the solid line is the result of the cranked shell model calculation discussed in the text. The inset shows the calculated change in the deformation parameters β_2 and β_4 over the frequency range of interest. The arrows point towards increasing frequencies.

and at a fixed deformation, more recent Woods-Saxon cranked shell model calculations[29,32] have taken into account small variations of the shape with spin as well as changes in deformation from one nucleus to another. These calculations also treat pairing correlations self-consistently. The best indication for the importance of the inclusion of pairing correlations and shape effects in this mass region comes from the data on the Gd isotopes. In the calculations, the SD band of ^{150}Gd is assigned a $\pi 6^2 \nu 7^2$ configuration. This is in agreement with the assignment of ref. 30, but the alignment of a pair of $j_{15/2}$ neutrons at $\hbar\omega$ ~0.4 MeV (i.e. a band crossing) must be invoked to account for the sharp drop in $J^{(2)}$ at low $\hbar\omega$. The observation that the deexcitation out of the SD band in ^{150}Gd is extremely abrupt, with essentially all the intensity being lost over a single transition, has been interpreted as additional evidence for this band crossing, which in turn requires the presence of the pairing correlations[10]. Marked irregularities in the behavior of $J^{(2)}$ can also be seen in the yrast SD bands of ^{146,147}Gd as well as in an excited band in ^{148}Gd (fig. 1). At present, there is some argument regarding the exact orbitals involved in these crossings as well as concerning the role of pairing and/or octupole correlations[6,7,11,32]. The general conclusion, however, remains that in all yrast SD bands near A=150 the variations of $J^{(2)}$ with $\hbar\omega$ reflect the major role played by the few nucleons in high-N orbitals, and the adopted configurations are $\pi 6^{2,3,4}$ for Gd, Tb, and Dy respectively; $\nu 7^0$ or $\nu 7^1$ for N=82-85; $\nu 7^2$ for N=86 and $\nu 7^3$ for N=87. The number of occupied intruder orbitals varies in some of the excited bands.

Effects due to pairing are, perhaps, even more important in the A=190 region where mean-field calculations without pairing[33,34] give proton and neutron contributions to $J^{(2)}$ which remain essentially constant with $\hbar\omega$. The need to examine the effects of pairing (possibility (ii) in the discussion above) was originally pointed out by Ye et al[19]. This is done in the framework of cranked deformed shell model calculations already discussed above[22,29]. For ^{192}Hg, the relevant high-N intruder orbitals which are occupied in the SD configuration are four ($i_{13/2}$) protons and four ($j_{15/2}$) neutrons, i.e. the SD configuration in ^{192}Hg can be labelled ($\pi 6^4 \nu 7^4$). Figure 3a compares the calculated dynamic moment of inertia with the data. In the calculations, pairing correlations were treated self-consistently by means of the particle number projection procedure[29], but the neutron pairing interaction strength was reduced. The rise in the calculated $J^{(2)}$ can be ascribed to the combined gradual alignment of a pair of N=6 ($i_{13/2}$) protons and of a pair of N=7 ($j_{15/2}$) neutrons within the frequency range under consideration. The data are reproduced rather well. The calculations with pairing have also been very successful in reproducing the other SD bands of this region (see ref. 20-23, for example).

It is worth pointing out that the calculations are able not only to propose configurations for the yrast and excited SD bands, but also to account for details such as signature splitting between bands or relative changes in $J^{(2)}$: this is discussed in detail in refs. 5 and 35. As an illustration of the power of the calculations we discuss here a detailed comparison between the moments of inertia of ^{193}Tl and ^{191,192}Hg (see fig. 4). The occupation of the third $i_{13/2}$ orbital in ^{193}Tl ($\pi 6^3$) results in an increase in the value of $J^{(2)}$ with respect to ^{192}Hg at the lowest frequencies. In both ^{191}Hg and ^{193}Tl, $J^{(2)}$ is essentially constant at the lowest frequencies before exhibiting the characteristic rise described above. This feature is also present in the calculations and is proposed to be a signature for the occupation of these high-j intruder orbitals by a single nucleon[23]. In the ^{191}Hg band, the alignment of the $j_{15/2}$ neutron is blocked and the rise in $J^{(2)}$ is attributed to the alignment of the $i_{13/2}$ protons[20] while in ^{193}Tl the opposite situation occurs. The fact that the rise in $J^{(2)}$ with $\hbar\omega$ is very similar in both cases implies that neutron and proton alignments have contributions of comparable magnitude. Furthermore, the rise of $J^{(2)}$ in ^{193}Tl starts at lower frequency than that in ^{191}Hg. This suggests that the neutrons align at somewhat lower frequency than the protons, in agreement with the calculations.

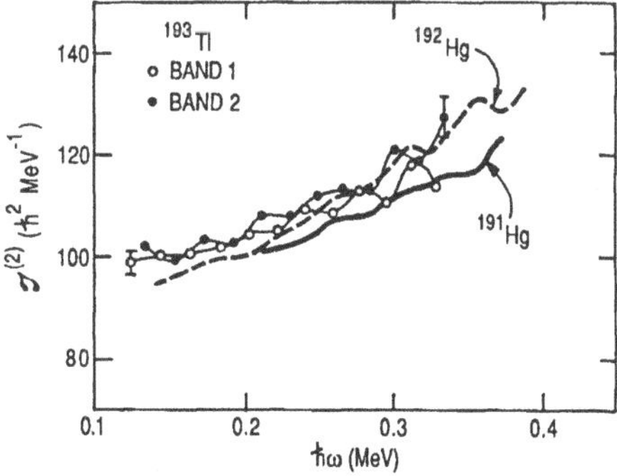

Fig. 4. Comparison between the dynamic moments of inertia $J^{(2)}$ measured in the SD of ^{193}Tl, ^{192}Hg and ^{191}Hg (yrast SD band).

From the discussion above, it can be concluded that a good description of the SD bands within the framework of cranked shell model calculations with pairing can be achieved. Three points need, however, to be emphasized. First, the transition energies in some of the SD bands in both the A=150 and A=190 regions are surprisingly close to those of bands in neighboring nuclei and several bands near A=190 can be related to ^{192}Hg[36]. This is discussed in section 4. Second, the inclusion of pairing is crucial for reproducing the data and, in particular, the smooth increase of $J^{(2)}$ with $\hbar\omega$ near A=190. In the proton system, pairing is reduced by the presence of the Z=80 shell closure. As was shown in refs. 22 and 31, the calculations require that the neutron pairing be reduced as well if one wants to reproduce the similarities in the behavior of $J^{(2)}$ with $\hbar\omega$ observed in all nuclei in this region. Reduced pairing is to be expected on the basis of general arguments. Pairing is sensitive to the overlap between orbitals of interest. At the very large deformations being considered here, states originating from different shells approach the Fermi level, and these states will only be very weakly coupled through the pairing interaction. Moreover, the coupling between the all important unique-parity levels (i.e. the various components of the high-N intruders) is also severely reduced because of their sizable energy splitting at large deformation. Third, it has been suggested that other effects may play an important role as well. Some pairs of orbitals such as the $i_{13/2}$-$f_{7/2}$ and the $j_{15/2}$-$g_{9/2}$ orbitals, which are responsible for strong octupole correlations in light actinide nuclei, also appear close to the Fermi level in SD configurations around ^{152}Dy and ^{192}Hg. Several recent calculations[21,37-39] have indicated that for many SD nuclei the minima in the total energy surfaces exhibit considerable octupole softness which is expected to persist even at the highest spins. Octupole correlations are also expected to reduce single-particle alignments, increase band interactions, and modify the deexcitation pattern of the SD states because of enhanced B(E1) rates. Most of the anticipated effects remain to be observed experimentally. However, first evidence for strong mixing of quasiparticle excitations with octupole vibrations may have been seen in ^{193}Hg[21] where two of the four SD bands are characterized by $J^{(2)}$ moments of inertia strikingly different from those of all other SD bands in this region (see fig. 2). Furthermore, there is indirect experimental evidence for enhanced E1 transitions linking one of the irregular bands with the yrast SD. These observations, together with the reduced alignments observed and the strong interaction between the crossing bands, have been interpreted as evidence for strong octupole correlations[21]. In the mass A=150 region, one of the proposed explanations for the behavior of one of the SD bands in ^{147}Gd invokes collective octupole excitations[7] as well, but other interpretations are also possible[11].

It has been suggested recently[40] that the residual n-p interaction may also play a role in SD nuclei. Such an interaction is expected to be strong in situations where protons and neutrons occupy rotationally aligned high-N intruder orbitals with large spatial overlap, such as the $\pi i_{13/2}$ and $\nu j_{15/2}$ orbitals involved in the SD bands of the A=150 and 190 regions. Such an interaction is expected[40] to lower the energy of SD bands based on these specific intruder configurations with respect to other excitations in the SD well. Firm experimental evidence for the importance of this interaction is currently lacking even though it has been invoked to explain the strong feeding of a SD band in ^{142}Eu[16] (interaction between N=6 odd proton and odd neutron) and the smooth rise of $J^{(2)}$ with $\hbar\omega$ in the lighter Hg isotopes[17] (interaction between 4 N=6 protons and 4 N=7 neutrons). The latter is illustrated in fig. 5 where the latest data on the ^{192}Hg SD band are presented. From an analysis of an enlarged set of coincidence data it has been possible to extend the SD band by 2, and possibly 3 transitions[41]. The $J^{(2)}$ values keep rising with $\hbar\omega$, in contrast with the expectations based on cranked shell model

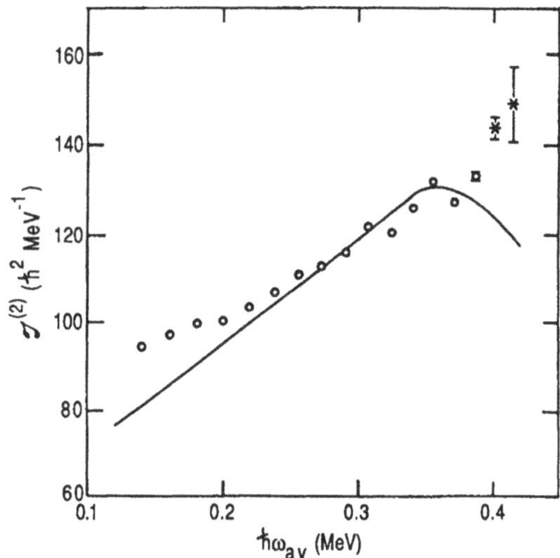

Fig. 5. Same as figure 3(a), but with the new data points at $\hbar\omega$ added (*). The figure illustrates that while the calculations predict that $J^{(2)}$ will decrease at the highest frequencies, the data keep rising.

calculations, and the agreement between experiment and theory is certainly not as satisfactory as one was led to believe from fig. 3a. The reason for the continued rise is not clear yet, but it is possible that some of the effects discussed above may account for this feature. Clearly, more work on the subject will be necessary.

"IDENTICAL" BANDS

The discovery of multiple SD bands within a single nucleus has made it possible to investigate the microscopic structure of both the ground and excited states in the second well. However, a greater impetus for studies of excited bands has been the discovery that several pairs of bands have almost identical transition energies. The first reported[42] cases consisted of the pairs ($^{151}Tb^*$, ^{152}Dy) and ($^{150}Gd^*$, ^{151}Tb) - the * denotes an excited SD band. Later, another similar pair ($^{149}Gd^*$, ^{150}Tb) was found[9] (the notation $^{149}Gd^{**}$ is used in ref. 9 as it is suggested that the second excited SD band in ^{149}Gd is involved). This is illustrated in fig. 6 where the difference between "identical" transitions energies ΔE_γ is plotted versus the transition energy. It can be seen that, on average, the deviation is less than 1 keV for the first pair and only slightly larger for the other two. This implies that transition energies are equal to better than 3 parts in 1000. This is a rather surprising equality: γ-ray energies should scale with the moment of inertia J, which is proportional to $A^{5/3}$ ($J \sim MR^2$), so that adjacent mass nuclei would have energies different by ~14 keV. Furthermore, the spins of corresponding transitions in each pair necessarily differ by $1/2\hbar$, leading to differences in E_γ of ~13 keV. For ($A-1^*$, A) pairs, with A even, these differences would reinforce each other. A word of caution is in order since it is assumed that each pair of transitions being compared has the appropriate spins (I+1/2,I), but the spins of the SD bands have not been measured. Since the spacing between consecutive transitions in each band is ~50 keV, the maximum difference in energy in a pair of bands is ~25 keV if no spin correlation is involved. However, with three cases and not just an isolated one, it is unlikely that these degeneracies are accidental. Furthermore, in all three cases the excited $(A-1)^*$ SD band is proposed to be characterized by a hole in the same specific orbital (see below).

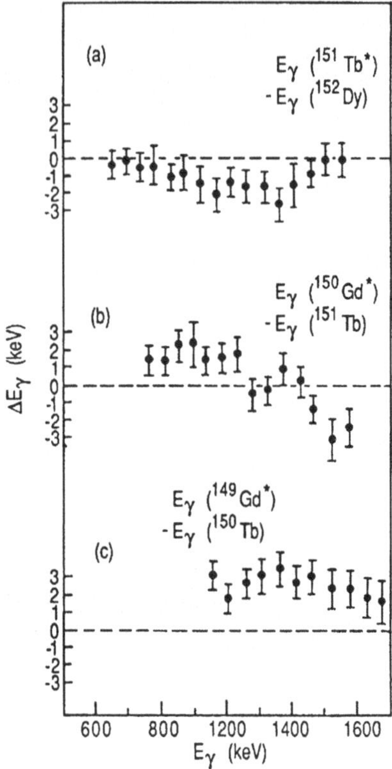

Fig. 6. Differences in the γ-ray energies ΔE_γ between the SD bands in (a) ($^{151}Tb^*$, ^{152}Dy), (b) ($^{150}Gd^*$, ^{151}Tb) and (c) ($^{149}Gd^*$, ^{150}Tb).

A related case of identical transition energies occurs in ^{153}Dy - the first case where excited SD bands were reported[14]. Here, two excited bands have been interpreted as signature partners and the averages of the transition energies in the partners reproduce the γ-ray energies in ^{152}Dy within 1-3 keV. Finally, the two SD bands of ^{147}Gd have been related to the yrast SD bands of ^{146}Gd and ^{148}Gd respectively, although the average ΔE_γ are somewhat larger in this case (~5 keV)[7]. In the case of ^{147}Gd, another relation applies as well: one of the bands has γ-ray energies following closely (1-4 keV) the average of two successive transition energies in ^{148}Gd, while the other band shows the same property when compared with the SD band of ^{146}Gd.

The current interpretation[43] of these observations is done within the framework of the strong coupling limit of the particle-rotor model in which one or more particles are coupled to a rotating deformed core and follow the rotation adiabatically. In the strong coupling limit, transition energies in an odd nucleus, relative to those in an even-even core obey simple relations (here the moments of inertia for all cases are assumed to be identical). When K≠1/2, the following relation applies:

$$1/2\ [E_\gamma\ (R+1/2) + E_\gamma\ (R-1/2)]_{odd} = E_\gamma\ (R)_{even} \quad (1)$$

This strong coupling relation provides a straightforward explanation for the ($^{153}Dy^*$, ^{152}Dy) pair. The two excited bands in ^{153}Dy have been interpreted[14] as a $^{152}Dy \otimes \nu[514]9/2$ structure with no signature splitting for which K≠1/2 and the relation above applies. For a K=1/2 band, the transition energies in the odd-nucleus are affected by the decoupling parameter, a, and obey the relation:

$$E_\gamma = \frac{\hbar^2}{2\mathcal{J}} [4I-2 + 2(-1)^{I+1/2}\, a\, \delta_{K,1/2}] \quad (2)$$

Transitions from both signatures form degenerate doublets, with the a=1 case giving energies identical to those of the core, while the a=-1 case has energies mid-way between those of adjacent transitions in the core. The three pairs ($A-1^*$,A) where identical energies have been observed can be interpreted as a case where a=1. In each case, the $(A-1)^*$ configuration is A$\otimes\pi([301]1/2)^{-1}$. If the decoupling parameter a is calculated from the Nilsson wave function of the [301]1/2 orbital, a value of 0.85 is obtained[43] very close to, but not exactly 1. On the other hand, for a $[Nn_3\Lambda]\Omega$ orbital, the decoupling parameter can be calculated from the asymptotic quantum numbers, by $a=(-)^N\delta_{\Lambda,0}$, which would result in a=0 for the [301]1/2 orbital. However, if one employed a different coupling scheme in terms of pseudo-spin, the appropriate quantum numbers are $[\widetilde{200}]1/2$, and a=1[43], hereby providing a natural explanation for the identical energies of the three pairs. In the pseudo-spin scheme[44], the natural parity orbitals form a pseudo-oscillator spectrum, labelled by the pseudo-asymptotic quantum numbers $\tilde{N}=N-1$, $\tilde{n}_z=n_z$. The pseudo-orbital angular momentum of the valence particles is strongly coupled to the core forming the total pseudo-orbital angular momentum, and the pseudo-spins are then added to form the total angular momentum $I=R+\tilde{s}$. The pseudo-Coriolis interaction is then expected to align the pseudo-spin with the total angular momentum[44]. In the pseudo-asymptotic limit, the decoupling parameter has the value $a=(-1)^{\tilde{N}}\delta_{\tilde{\Lambda},0}$, and a=1 follows for the orbital under discussion here.

It must be stressed again that the explanation of identical energies presented here follows only if the moments of inertia for odd and even nuclei are identical. It still is a puzzle why the moments of inertia $\mathcal{J}$ for adjacent nuclei appear to be constant to within ~0.2%. Since $\mathcal{J}$ depends on several factors (mass, deformation, polarization effects, alignment and pairing), the equality in $\mathcal{J}$ is very striking and leads one to wonder about the possibility of a fundamental explanation. The standard models are unable to reproduce transition energies with an accuracy of ~1 keV. We note that only a few identical bands are observed among the many SD bands in this region and that three of these pairs of bands involve a common orbital. This suggests that the phenomenon is associated with only a few specific orbitals. Ragnarsson[45] has used a simple harmonic oscillator model to show that for particles or holes in certain orbitals there can be cancellation among the different terms contributing to changes in $\mathcal{J}$. He has been able to reproduce the "identical" energies observed in (^{152}Dy, ^{153}Dy*), where the orbital occupied in ^{153}Dy* is either ν[402]5/2 or ν[514]9/2. These orbitals slope upward with increasing β_2 - so-called oblate orbitals - so that a particle here would tend to decrease the deformation of the (A+1) nucleus, compensating the increase in $\mathcal{J}$ due to the larger mass. Agreement for other cases calculated by Ragnarsson, including that involving the π[301]1/2 hole (responsible for the three pairs of bands with identical energies) is not as satisfactory, suggesting that other effects, such as changes in pairing, should be included.

There are even more examples of identical SD bands in the A=190 region[36]. However, two features distinguish the bands in this region from those near A=150: (a) many of the bands occur in pairs separated by two mass units, and (b) a large number of bands can be related to the SD band in ^{192}Hg, which appears to serve as a doubly-magic core. The proposed configurations for these SD bands are characterized by $K\neq1/2$[4,35]. Within the strong coupling limit, relation (1) should apply and can account for the observations[35]. However, there is again the requirement of equal moments of inertia in all nuclei. Another way of relating the energies of different bands to those of a reference, ^{192}Hg, has been proposed by Stephens[36], using the quantity $\Delta i=2\Delta E_\gamma/\Delta E_\gamma^{ref}$, where $\Delta E_\gamma = E_\gamma - E_\gamma^{ref}$ is obtained by subtracting the transition energy E_γ in a band of interest from the closest transition energy in ^{192}Hg (E_γ^{ref}), and ΔE_γ^{ref} is calculated as the energy difference between the two closest transitions in the reference. Δi has been called incremental alignment[36], but is not necessarily related to any physical alignment. In the strong coupling limit, $\Delta i = \pm 1/2$ for an odd nucleus referred to an even core. For even nuclei, bands with the same or different signatures have Δi = 0 or 1, respectively. Plots of Δi are shown in fig. 7, which illustrates that the SD bands of the A=190 region can be classified in two families. When additional particles (holes) with respect to ^{192}Hg occupy lower K members of the high-N intruder orbitals, values of Δi scatter significantly (fig. 7a). This is not surprising since particles in these intruder orbitals tend to both increase deformation and align with rotation, thus making the strong coupling scheme inappropriate. In contrast, when the orbitals involved do not show much variation with $\hbar\omega$, i.e. when there is little alignment, Δi values fall close to the limits of 0, $\pm$ 1/2, $\pm$1, which are expected in the strong coupling limit (fig. 7b).

Stephens et al.[36,46] suggested that many of the cases which exhibit Δi = 0, $\pm$1/2, or $\pm$1 also have integer alignment differences with respect to the core, which they refer to as quantized alignment. While the strong coupling limit corresponds to a zero alignment difference, they suggest alignments (with respect to ^{192}Hg) of 1$\hbar$ in the excited SD bands of ^{194}Hg, for example, which are then attributed to pseudo-spin alignment.

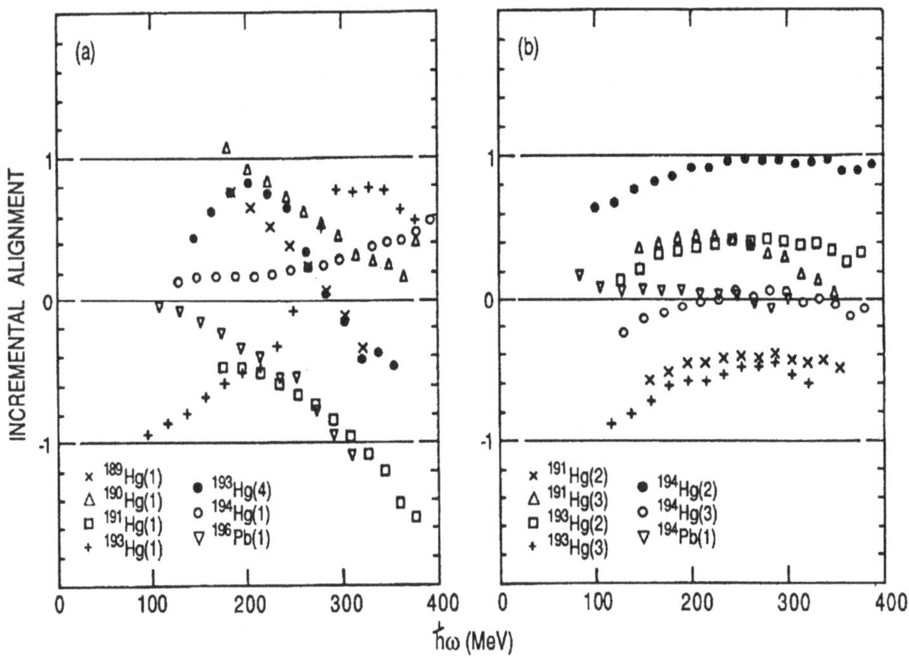

Fig. 7. Incremental alignment Δi as a function of rotational frequency for SD bands in $^{189-194}$Hg and ^{194,196}Pb with ^{192}Hg as a reference. As discussed in the text, the data are grouped [(a) and (b)] in order to illustrate the existence of two families; in (b) Δi clusters around integer and half-integer values. The data on SD bands in the Tl isotopes have been omitted for clarity; these are discussed in ref. 24.

An alignment of $1\hbar$ in ^{194}Hg is not easy to understand. However, a word of caution is in order. Whereas extraction of Δi does not require knowledge of spin, determination of alignment does. So far the spins of SD band members have not been assigned using the conventional, tested techniques of γ-ray spectroscopy. Methods to infer the spin have been suggested[47,48]. At present, it is not established whether these methods give the exact spin or have uncertainties of at least $1\hbar$[49]. Thus, until spins can be assigned firmly, one cannot be certain whether the alignments have the putative value of $1\hbar$[36]. It is quite possible that the strong coupling limit of the particle rotor model applies in the A=190 region as well as in the A=150 region. The caveat then remains that in this region the moments of inertia of neighboring nuclei have to be identical as well, a feature that present mean-field calculations cannot reproduce.

FEEDING AND DECAY OF SUPERDEFORMED STATES

Within the limited space available here, it would be impossible to review these subjects in their entirety. However, some very recent results should be discussed because they shed some new light on matters which have proven to be among the more difficult ones to study experimentally. Figure 8 presents very recent data[50] on the intensity with which the SD band in ^{152}Dy is populated in ^{32}S-, ^{48}Ca- and ^{82}Se-induced reactions. The SD intensity is given as a percentage of the total ^{152}Dy production and is plotted as a function of excitation energy in the compound nucleus. These data point towards the presence of a rather pronounced entrance-channel effect, i.e. the population of the SD band is enhanced by ~30% in the more symmetric fusion reaction. Similar results have also been obtained for the ^{149}Gd SD nucleus[50]. It was proposed several years ago[51,52] that entrance-channel effects might favor the SD shapes since projectiles and targets of essentially identical masses have a shape with an axis ratio close to 2:1 at the point of contact. This idea was originally proposed in attempts to account for the inhibition of neutron emission with respect to statistical model predictions observed in certain (H.I., xn) reactions. The result of fig. 8 may well represent one of the more intriguing examples of the interplay between nuclear structure and reaction dynamics.

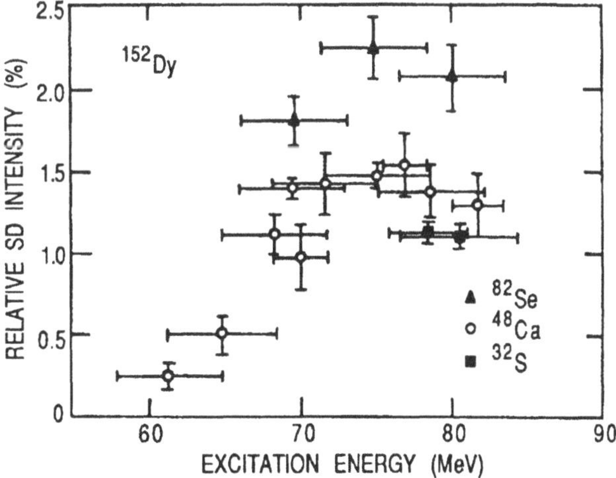

Fig. 8. Relative intensity for the feeding of the SD band in ^{152}Dy. The data are from the Liverpool-Strasbourg collaboration[50].

There is also new information regarding the feeding of the SD states. Important elements affecting the feeding are the level densities of both normal and SD states, the mixing between the two classes of states at moderate excitation energy, the electromagnetic decay rates in the normal and SD minima, as well as the barrier separating the two minima. Schiffer and Herskind[53] have performed an extensive series of calculations, which successfully described many of the observed feeding features in ^{152}Dy and which have provided insight into the population mechanism. In particular, it was found to be important that, as the excitation energy increases, the level density for the SD states increases less rapidly than for the normal states. This results in a limited region in excitation energy where the density of SD states is larger than that of normal states. This region corresponds to the highest partial waves and is located beyond the point where the SD band crosses the normal states and becomes yrast. Schiffer and Herskind suggest[53] that the entry states for the SD bands originate from this region and that this feature accounts for the "colder" feeding of these bands. From measurements of the fold and sum-energy distributions in coincidence with transitions in the SD band and in the yrast cascade of ^{192}Hg, there is now experimental evidence[54] for the decay picture outlined above. Figure 9 compares the entry distributions for the feeding of the SD and the yrast states in ^{192}Hg, and it is clear that the SD band is fed from the higher partial waves, in a rather limited region in excitation energy and spin. It should be pointed that it is important that the entire distribution has been measured rather than the average quantities which only give entry points.

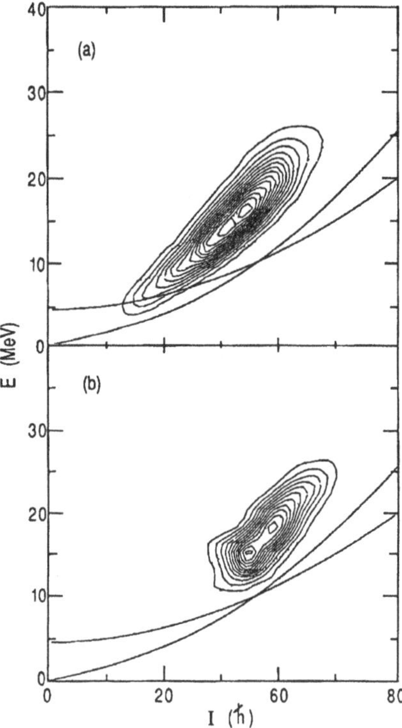

Fig. 9. Population of entry states in ^{192}Hg measured in coincidence with the yrast transitions (a) and with the SD transitions (b). The two lines in the figure represent the "normal" and the SD yrast lines. The excitation of the latter at spin 0 was assumed to be 4.8 MeV.

While most of the SD bands have been found to be in coincidence with transitions between low-spin yrast states in the same nucleus, none of the transitions linking the SD bands with the yrast and near-yrast states have been identified. It is likely that many decay paths share the intensity and that the decay-out process is statistical in nature. Transitions between coexisting low-spin states of moderate prolate and oblate deformations in A~190 Pt and Hg nuclei have been found[55] to exhibit unusually strong E0 admixtures. Since the known SD bands of Hg nuclei in this region extend to rather low spin ($I \sim 10\hbar$), it has been suggested recently[55] that E0 transitions could play an important role in the decay out of the SD states. If this were so, γ-ray techniques alone might be inadequate for studying the decay out of the SD bands, and electron measurements would also be needed. A limit on any E0 decay out of a SD band has been obtained recently[56] from a measurement of X-ray yields which relies on the fact that the presence of strong E0 decay is accompanied by intense K-shell ionization. Thus, by looking at the K X-ray yield in coincidence with SD transitions, one can probe for the presence of highly converted transitions. The K X-ray yields associated with the SD band in ^{192}Hg have been measured[56]. The measured value was compared with that expected from the K-conversion of known discrete lines in the band and of transitions following the deexcitation into the yrast states. The possibility of strong E0 decay out of the SD band can now be ruled out: an upper limit of 10% was found. The strength of the E0 decay between states of different shapes is governed both by the change in deformation and the mixing between initial and final states. Although there is a very large shape change in the present case, these results indicate that there is very little mixing between the SD and yrast states. Thus, the situation appears to be very similar to that found in the actinide nuclei where E0 decay between the 0^+ state in the second minimum and the 0^+ ground state was found to occur with low probability[57].

CONCLUSIONS

Here, I have attempted to review some of the many facets of SD nuclei. In the A=150 and 190 regions, it has been possible to perform spectroscopic studies in the SD secondary well of surprising detail and both "ground" and excited bands have been observed. The most unexpected discovery so far is that a number of the excited bands have energies identical to those of the lowest SD bands in adjacent nuclei. The degeneracies of better than one part in 500 require that the moments of inertia be identical and also that the strong coupling limit applies, both to a remarkable degree. It is not clear whether this is a result of an accident or is a consequence of a symmetry which has yet to be identified. Although there has been much progress in this area of research, many questions remain and, indeed, have actually been raised by the new discoveries. Certainly, a major challenge is to determine the excitation energies and spins of SD levels since there is not a single SD band in the A=150 and 190 regions for which these properties are known. The limits of the regions of superdeformation in the periodic table need to be established. The known excited SD bands are believed to correspond to particle or quasiparticle excitations, but the collective modes associated with states of large deformation, e.g. the beta, gamma or octupole vibrations have yet to be found. If the deformation is stiff with respect to quadrupole distortion, the beta vibrations may lie at high excitation energies, but soft octupole modes may exist at lower energies, possibly giving rise to exotic bending modes. Further studies are also necessary to shed more light on the cause of identical bands. At present experiments are limited by the detection capabilities of current γ-ray detectors. The next generation detector arrays currently under construction (EUROGAM in Europe and GAMMASPHERE in the USA) will improve the detection sensitivity by about two orders of magnitude and provide answers to the questions raised above. Many fascinating discoveries about superdeformation--and about nuclear structure in general--lie ahead.

ACKNOWLEDGEMENTS

The data and ideas reviewed above are the result of dedicated work by many colleagues and friends. It is impossible to name them all here. I would like, however, to acknowledge the contributions of my collaborators at Argonne, Notre Dame, Purdue, I.N.E.L., Stockholm and Warsaw. I also thank Mike Carpenter, Ian Bearden, and Teng Lek Khoo for numerous stimulating discussions and for carefully reading the manuscript. I am grateful to B. Haas and P. Fallon for communicating their latest results. This work was supported by the Department of Energy, Nuclear Physics Division, under contract no W-31-109-ENG-38.

REFERENCES

1. P. J. Twin et al., Phys. Rev. Lett. 57, (186) 811.
2. P. J. Nolan and P.J. Twin, Ann. Rev. Nucl. Part. Sci. 38, (1988) 533.
3. P. J. Twin, Nucl. Phys. A520, (1990) 17c.
4. R. V. F. Janssens et al., Nucl. Phys. A520, (1990) 75c.
5. R. V. F. Janssens and T. L. Khoo, Ann. Rev. Nucl. Part. Sci. to be published.
6. G. Hebbinghaus et al., Phys. Lett. 240B, (1990) 311.
7. K. Zuber et al., Nucl. Phys. A520, (1990) 195c, Phys. Lett. 254B, (1991) 308.
8. M. A. Deleplanque et al., Phys. Rev. Lett. 60, (1988) 1626.
9. B. Haas et al., Phys. Rev. Lett. 60, (1988) 503, Phys. Rev. C42, (1990) R1817.
10. P. Fallon et al., Phys. Lett. 218B, (1989) 137 and 257B, (1991) 269.
11. V. P. Janzen et al., Proc. Int. Conf. on High Spin Physics and Gamma-soft Nuclei, Pittsburgh (1990), World Scientific, p. 225.
12. M. A. Deleplanque et al., Phys. Rev. C39, (1989) 1651.
13. G.-E. Rathke et al., Phys. Lett. 209B, (1988) 177.
14. J. K. Johansson et al., Phys. Rev. Lett. 63, (1989) 2200.
15. R. M. Lieder et al., private communication.
16. S. M. Mullins et al., Phys. Rev. Lett. 66 (1991) 1677.
17. M. W. Drigert et al., Nucl. Phys. in press.
18. E. F. Moore et al., Phys. Rev. Lett. 63 (1989) 360.
19. D. Ye et al., Phys. Rev. C41 (1990) R13 and J. Becker et al., Phys. Rev. C41 (1990) R9.
20. M. P. Carpenter et al., Phys. Lett. 240B (1990) 44.
21. C. M. Cullen et al., Phys. Rev. Lett. 65 (1990) 361.
22. M. A. Riley et al., Nucl. Phys. A512 (1990) 178.
23. P. B. Fernandez et al., Nucl. Phys. A517 (1990) 386.
24. F. Azaiez et al., Phys. Rev. Lett. 66 (1991) 1030.
25. F. Azaiez et al., to be published.
26. J. Becker et al., private communication.
27. M. J. Brinkman et al., Z. Phys. A336 (1990) 115 and K. Theine et al., Z. Phys. A336 (1990) 113.
28. T. F. Wang et al., to be published.
29. W. Nazarewicz et al., Phys. Lett. 255B (1989) 208.
30. T. Bengtsson et al., Phys. Lett. 208B (1988) 39.
31. E. F. Moore et al., Phys. Rev. Lett. 64 (1990) 3127.
32. W. Nazarewicz et al., Nucl. Phys. A503 (1989) 285.
33. S. Aberg, Phys. Scr. 25 (1982) 23.
34. R. R. Chasman, Phys. Lett. 219B (1989) 227.

35. W. Satuła et al., Nucl. Phys. in press.
36. F. S. Stephens, Nucl. Phys. A520 (1990) 91c.
37. S. Aberg, Nucl. Phys. A520 (1990) 235c.
38. J. Dudek et al., Phys. Lett. 248B (1990) 235.
39. P. Bonche et al., Phys. Rev. Lett. 66 (1991) 876.
40. R. Wyss and A. Johnson, Proc. Int. Conf. on High Spin Physics and Gamma soft Nuclei, Pittsburgh (1990), World Scientific, p. 123.
41. R. V. F. Janssens et al., to be published.
42. T. Byrsky et al., Phys. Rev. Lett. 64 (1990) 1650.
43. W. Nazarewicz et al., Phys. Rev. Lett. 64 (1990) 1654.
44. A. Bohr et al., Phys. Scr. 26 (1982) 267.
45. I. Ragnarsson, Nucl. Phys. A520 (1990) 67c.
46. F. S. Stephens et al., Phys. Rev. Lett. 64 (1990) 2623 and 65 (1990) 301.
47. J. A. Becker et al., Nucl. Phys. A520 (1990) 188c.
48. J. E. Draper et al., Phys. Rev. C42 (1990) R1791.
49. C. Wu et al., Phys. Rev. Lett. 66 (1991) 1377.
50. P. Fallon and B. Haas, private communications.
51. W. Kuhn et al., Phys. Rev. Lett. 51 (1983) 1858.
52. R. V. F. Janssens et al., Phys. Lett. 181B (1986) 16.
53. B. Herskind and B. K. Schiffer, Phys. Rev. Lett. 59 (1987) 2416 and K. Schiffer et al., Z. Phys. A332 (1989) 17.
54. Ph. Benet et al., to be published.
55. E. Zganjar and J. Wood, Nucl. Phys. A520 (1990) 427c.
56. E. F. Moore et al., Phys. Lett. 258B (1991) 284.
57. J. Kantele et al., Phys. Rev. Lett. 51 (1983) 91 and J. Shirmer et al., Phys. Rev. Lett. 63 (1989) 219.

THALLIUM ISOTOPES: A SET OF "IDENTICAL" SUPERDEFORMED BANDS

F. Azaiez[+], C.W. Beausang[++], J. de Boer[**], M.A. Deleplanque, R.M. Diamond, J.E. Draper[#], W.H. Kelly[*], W. Korten, A. O. Macchiavelli, E. Rubel[#], M. Rohn[**] and F.S. Stephens

Lawrence Berkeley Laboratory, Berkeley, CA 94720 U.S.A.

J.A. Becker, M.J. Brinkman[§], J.A. Cizewski[§], E.A. Henry, A. Kuhnert, T.F. Wang, and S.W. Yates [≠]

Lawrence Livermore Laboratory, Livermore, CA 94550 U.S.A.

ABSTRACT

A total of eight superdeformed (SD) bands have been identified in two thallium isotopes (A = 194-195). Striking similarities were found between the transition energies of these bands and transition energies of the two SD bands in ^{193}Tl. This implies similarities between the dynamical moment of inertia of all the known SD bands in the thallium isotopes. The identical moment of inertia can in turn be interpreted as a constant alignment which was found, in all these cases, to be surprisingly close to 0h or 1h.

INTRODUCTION

The current interest in SD structures in the mass 190-region was inspired by the initial discovery [1)] of a SD band in ^{191}Hg. Since then more than two dozen discrete SD bands have been found in this region. The most startling feature of these rotational bands was revealed by F.S.Stephens [2)], when he pointed out the fact that a surprisingly large number of SD bands in the mercury isotopes are "identical" (they have identical transition energies within a few keV) or "equivalent" (they have within the same accuracy a trivial relationship between their transition energies). He also showed that the alignment of

[+]Institut de Physique Nucléaire - Orsay, F-91406 Orsay Cedex.
[++]Univ. of Liverpool, L69 3BX, UK
[*] Iowa State University, Ames, IA 50011.
[#] University of California, Davis, CA 95616.
[**] Universität München, Germany.
[§] Rutgers University, New Brunswick, NJ 08903.
[≠]University of Kentucky, Lexington, KY 40506.

Nuclear Shapes and Nuclear Structure at Low Excitation Energies
Edited by M. Vergnes *et al.*, Plenum Press, New York, 1992

these SD bands relative to the SD band in ^{192}Hg was quantized (with most often the value of 1h). Our contribution to this conference will be the systematics of SD bands in ^{193,194,195}Tl, isotopes[3,4,5] where a similar family of "identical" or "equivalent" bands was identified.

EXPERIMENTS AND RESULTS

The experiments were carried out at the Lawrence Berkeley Laboratory 88-Inch Cyclotron using the HERA (high-energy resolution array) facility. It consists of twenty Compton-suppressed Ge detectors and a 40-bismuth-germanate-element, 4π inner ball. Two different reactions were used to produce the thallium isotopes. First, the reaction ^{18}O + ^{181}Ta at beam energies of 95, 100 and 104 MeV was used to populate states in ^{195,194,193}Tl via the 4n, 5n and 6n evaporation channels. The second reaction, ^{15}N + ^{186}W at beam energies of 90 and 95 MeV, was used to populate ^{195}Tl and ^{196}Tl isotopes as the 6n and 5n channel, respectively. The targets of ^{181}Ta and ^{186}W consisted of three 0.5 mg/cm^2 self-supporting foils stacked together. The γ-rays from all threefold and higher coincidences between Ge detectors were recorded event by event on magnetic tape, together with the sum (H) and multiplicity (K) information from the inner ball. Twofold Ge coincidences were recorded only when they were in coincidence with at least six inner ball detectors. two hundred millions events per beam energy were recorded, with about 20% being threefold or higher coincidences.

Searches for SD bands were made, using gated projections from double-coincidence matrices with various multiplicity and total energy requirements.Analysis of the coincidence matrices from ^{18}O + ^{181}Ta reaction at 100 and 104 MeV revealed 6 SD bands. These are labeled 1a to 3a and 1b to 3b in figure 1. For all these bands, only very few lines were not contaminated by the intense normal γ-ray transitions. Therefore, extensive studies of the threefold and higher Ge coincidences were used to establish the mutual coincidences between all members of each band. The bands show properties characteristic of SD bands in the mass 190 region. The strongest band is 1a with an intensity of about 1.5% of the ^{194}Tl yield for $K > 10$ and $H > 5$ MeV. Under the same conditions bands 1b, 2a, 2b, 3a and 3b were found to have intensities of approximately 70%, 60%, 40%, 40% and 50% respectively of band 1a. The six SD bands were mainly observed in the data taken with ^{18}O at 100 and 104 MeV bombarding energies, where the dominant reaction products are ^{193}Tl and ^{194}Tl. We compared the yields of the normal and SD bands with a high multiplicity cut ($K \geq 16$) which did not affect the SD bands population but eliminated the lower spin population of normal bands. The excitation functions of the SD bands were found to match those of the normal ones in ^{194}Tl very well. We have therefore assigned the six bands to ^{194}Tl. The two SD bands in ^{193}Tl reported by the Argonne group[5] have been seen very weakly populated only in the 104 MeV data. Two other very weak SD bands have been identified in the ^{18}O data but only

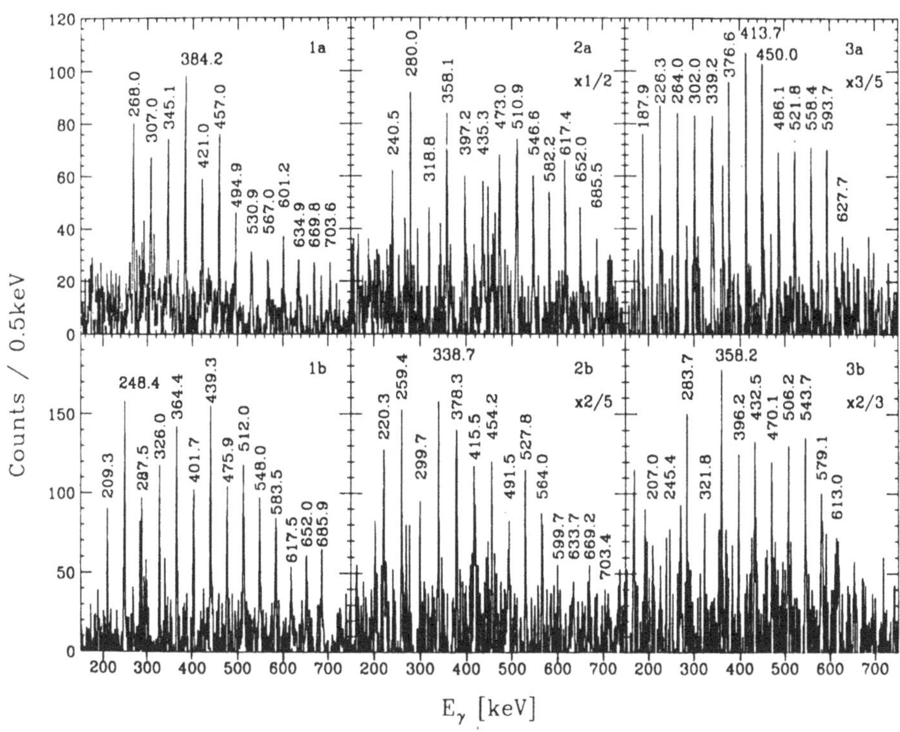

Fig.1. Triple coincidence spectra of the six SD bands in ^{194}Tl, obtained by adding spectra from different double-gate combinations in a given band. The uncertainties in the transition energies range from 0.3 to 1.0 keV. The ordinate scale should be multiplied by the factors 1/2, 2/5, 3/5 and 2/3 for spectra 2a, 2b, 3a and 3b, respectively.

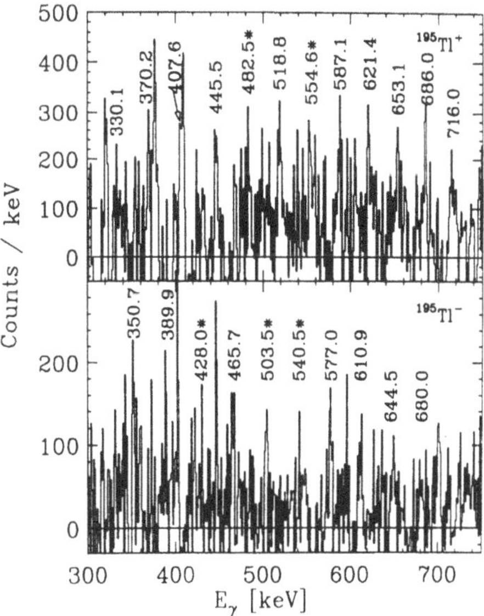

Fig.2. Twofold coincidence spectra of the two SD bands in ^{195}Tl obtained from ^{18}O data by gating on the starred transitions. All other lines than SD numbers are due to contaminations in the gates.

at the 95 MeV beam energy For each band a sum of single-gated spectra with a (K≥10) condition are displayed in figure 2. Under this condition, band 1 was found to represent about 0.5% of the ^{195}Tl intensity and to be about 2 times stronger than band 2. In the ^{15}N + ^{186}W reaction, the same two SD bands were observed only at the highest beam energy (95 MeV), and with approximately the same intensity as with the ^{18}O beam. Since the only common product seen in the two reactions is ^{195}Tl, we assigned the two bands to that nucleus.

Because of the weakness of the bands and the fact that very few clean gates were available , the angular correlations of only the strongest transitions in bands 1a and 1b in ^{194}Tl could be determined. The correlations indicated that the transitions involved have the same character, probably stretched quadrupole. The relative intensities of the transitions in all eight SD bands show the same behavior as those in the Hg nuclei, namely a slow increase of intensity with decreasing transition energy, followed by a region where the intensities stay roughly constant, and finally a rapid depopulation of the bands within one or two transitions at the low-energy end. Unlike the majority of the cases in the mass 190 region, the SD bands in ^{194,195}Tl are not seen unambiguously in coincidence with low-lying normal transitions, perhaps due to the large background subtractions involved and/or to the presence of isomers.

DISCUSSION

The dynamical moment of inertia, $J^{(2)} = dI/d\omega \simeq 4/\Delta E\gamma$, of all SD bands in the thallium isotopes are plotted in figure 3 as a function of $\omega = dE/dI = E\gamma/2$. They are remarkably similar. They all tend to start at ~ 100$\hbar^2$/MeV and increase smoothly by 20%, like almost all cases of SD bands in the mass 190-region. In order to study in more details the similarities or differences that these bands might have, we consider the notion of alignment relative to a reference band [1], which has been defined as the spin difference between a band and the reference band for at given frequency ω. We therefore need to

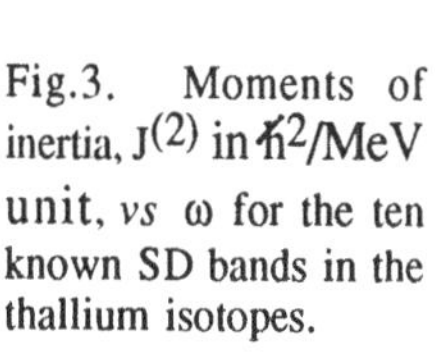
Fig.3. Moments of inertia, $J^{(2)}$ in $\hbar^2$/MeV unit, *vs* ω for the ten known SD bands in the thallium isotopes.

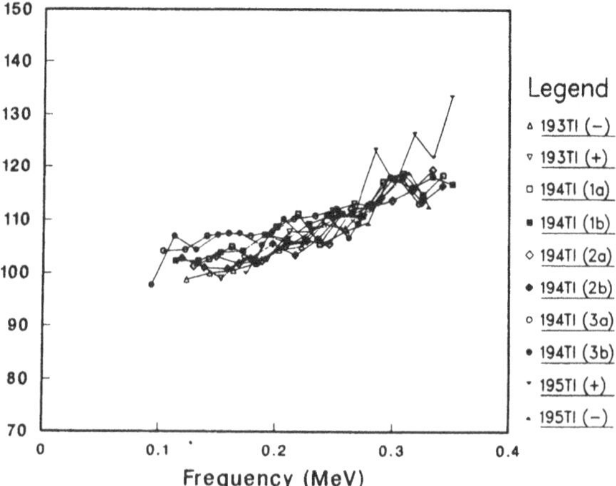

determine the spins of these SD bands. Because they are very good rotational cascades and go down to much lower transition energies than in the mass 150-region there is the possibility to determine their spins. This comes about because the transition energy of a rigid, axially symmetric rotor is proportional to its spin ($E\gamma = (4\langle I\rangle + 2)(\hbar^2/J)$, where $\langle I\rangle$ is the intermediate spin of a transition). Our collaboration has shown, in somewhat different developments, that one can express the dynamic moment of inertia, $J^{(2)}$, as a function of ω^2,

$$J^{(2)}/\hbar^2 = 2\alpha + 4\,\beta\omega^2 + \ldots \qquad (1)$$

and then evaluate the parameters α and β from the experimental values of $J^{(2)}$ at low spin. Integration of (1) gives the average or intermediate spin of the transition : $\langle I\rangle+1 \rightarrow \langle I\rangle-1$

$$\langle I\rangle + \frac{1}{2} = 2\,\alpha\,\omega + \frac{4}{3}\,\beta\,\omega^3 + \ldots \qquad (2)$$

as long as there is no discontinuous change in alignment in the extrapolated range from zero to the lowest spins observed such as a sharp backbend, and no frequency independante alignment. These seem not to be a problem, as the values determined so far are found very close to (half) integer for (odd) even mass nuclei. Also the values found are in general compatible with the average entrance spin into the ground bands wherever those have been observed in coincidence with the SD bands (as seen in some Hg or Pb cases).

Because the bands in the thallium isotopes are more weakly populated (less accurate transition energies) than the even-even cases in the Hg and Pb isotopes,the fitted values for the spins have larger uncertainties Nevertheless, spin assignments could be made in all the thallium nuclei, although for some cases their even (odd) mass nature had to be used to help assign the spin values to the closest integer (half-integer) values. In trying to understand superdeformed bands in the mass-190 region, plots of alignment relative to ^{192}Hg versus rotational frequency (half the γ-ray transition energy) have been discussed[1]. Figure 4 shows such a plot for the six bands in ^{194}Tl, the two in ^{193}Tl and the two in ^{195}Tl. It is apparent that the ^{194}Tl bands comprise three signature pairs and that those in ^{193}Tl and ^{195}Tl are signature partners that develop about the same signature splitting at frequencies above 0.2 MeV. This signature splitting is consistent with the

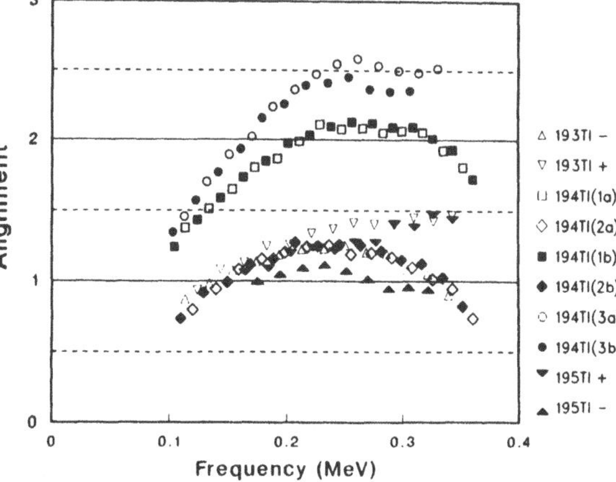

Fig.4. Alignment relative to the ^{192}Hg band, of the 2 SD bands in ^{193}Tl, the six SD bands in ^{194}Tl and the two in ^{195}Tl, the symbols for each band are indicated in the legend.

5/2 [642] configuration for the added proton when one goes from mercuries to thalliums. Above a frequency of 0.2 MeV only bands 3a and 3b in ^{194}Tl and ^{193}Tl(+) are rather constant and close to half-integer values. However, this kind of general behavior seen in the mercury isotopes is not really expected for this intruder proton orbital. A better classification is suggested by the similarity in shape of bands 1a, 1b, 2a, 2b in ^{194}Tl and ^{195}Tl(-) to ^{193}Tl(-) and bands 3a, 3b in ^{194}Tl and ^{195}Tl(+) to ^{193}Tl(+). The bands in ^{193}Tl are clearly better references for the bands in ^{194}Tl and ^{195}Tl than is ^{192}Hg. Thus in figure 5 we have used the ^{193}Tl bands suggested above as references for the bands in

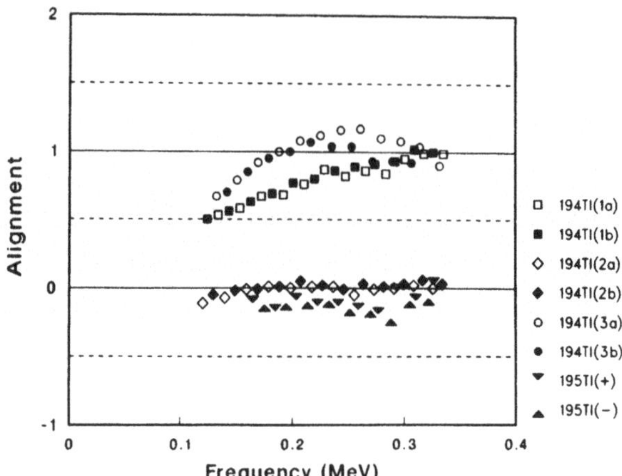

Fig.5. Alignment relative to the corresponding band in ^{193}Tl (see text) of all SD bands in ^{194}Tl and ^{195}Tl (^{193}Tl(+) is reference band for 1a, 1b, 2a, 2b and ^{195}Tl(+); and ^{193}Tl(-) is reference band for 3a, 3b and ^{195}Tl(-)).

^{194}Tl and ^{195}Tl. Now the bands in figure 5 show constant and near-integer relative alignments, indicating that the 113th and 114th neutrons are "better" spectators than the 81st proton. Each of two pairs in ^{194}Tl has alignment one against its respective ^{193}Tl reference band and one pair has alignment zero against the ^{193}Tl(-) band. This suggests that there might be still another pair of bands with alignment zero against the ^{193}Tl(+) band. These have been looked for, as well as crossing bands like those reported[7)] in ^{193}Hg, but none have been found yet.

ACKNOWLEDGEMENTS

One of us (F.A), for whom this conference coincided with the end of his visit to the Berkeley group, would like to thank them, and the Livermore group as well, for the pleasure he had working with them. This work was supported in part by the U.S. DOE under contract DE-AC03-76SF00098 (LBL) and contract W-7405-ENG-48 (LLNL).

REFERENCES

1 - E.F. Moore et al., Phys. Rev. Lett. 63 (1989) 360.
2 - F.S. Stephens et al., Phys. Rev. Lett., 64 (1990) 2623.
and Nucl. Phys. A 520 (1990) 091 c.
3 - F. Azaiez et al., Phys. Rev. Lett. 66, (1991) 1030.

4 - F. Azaiez et al., Z. Phys. A338 , (1991) 471.
5 - P.B. Fernandez et al., Nucl. Phys. A517, (1990) 386.
6 - J.A. Becker et al., Nucl. Phys., A520(1990) 187c and to be published
J.E. Draper et al., Phys. Rev. C42, R1791 (1990).
7 - D.M. Cullen et al., Phys. Rev. Lett., 65 1547 (1990).

MICROSCOPIC DESCRIPTION OF QUADRUPOLE AND OCTUPOLE MODES IN THE SUPERDEFORMATION REGION OF ^{192}Hg

J.P. Delaroche, J.F. Berger and M. Girod

Service PTN, CEA - Centre d'Etudes de Bruyères-le-Châtel
BP 12, 91680 Bruyères-le-Châtel, France

J. Libert and I. Deloncle

CSNSM, IN2P3-CNRS, Batiment 104-108
91405 Orsay Campus, France

L.M. Robledo

Universitad Autonoma de Madrid
E-28049 Madrid, Spain

INTRODUCTION

A large number of superdeformed (SD) bands has been observed in the mass 190-region, some of which showing nearly identical γ -ray transition energies. This has been illustrated for instance in ^{194}Hg by Cullen et al.[1] More recently, some measurements indicated that octupole correlations might exist in the superdeformed nucleus ^{193}Hg. Thus, both quadrupole and octupole collective modes may play a key role in explaining the observed SD band features in this mass region. These issues are discussed in the present work.

Constrained Hartree-Fock-Bogoliubov (HFB) calculations based on Gogny's force are performed to determine the potential energy surface (PES's) and tensors of inertia (TOI's) for ^{192}Hg. The range of quadrupole and octupole deformations under consideration is broad enough to cover the normal and superdeformation regions of this nucleus. The PES's and TOI's are used as inputs for Griffin- Hill- Wheeler equations which are solved in the Gaussian Overlap Approximation (GOA). Three π=+ superdeformed bands and the band head of a $K^\pi=0^-$ octupole band are predicted. Properties of these bands are discussed and compared with experimental information.

COLLECTIVE DYNAMICS

The SD levels are thought of as eigenstates of collective Hamiltonians. For the π=+ states, the Hamiltonian is defined in terms of the five quadrupole deformations: two shape coordinates (β ,γ) and three Euler angles ($\Omega = \theta_1, \theta_2, \theta_3$). On the other hand, a 2D Hamiltonian is solved to obtain the $K^\pi=0^+$ and $K^\pi=0^-$ excitations. In this case, the quadrupole (q_{20}) and octupole (q_{30}) moments are considered as collective variables. The difference in energies $[E^\star\ (0^-)\ -\ E^\star\ (\ 0^+)]$ then indicates the excitation energy of the $K^\pi=0^-$ band relative to that of the $K^\pi=0^+$ SD band taken as our reference energy in the following discussion.

Nuclear Shapes and Nuclear Structure at Low Excitation Energies
Edited by M. Vergnes *et al.*, Plenum Press, New York, 1992

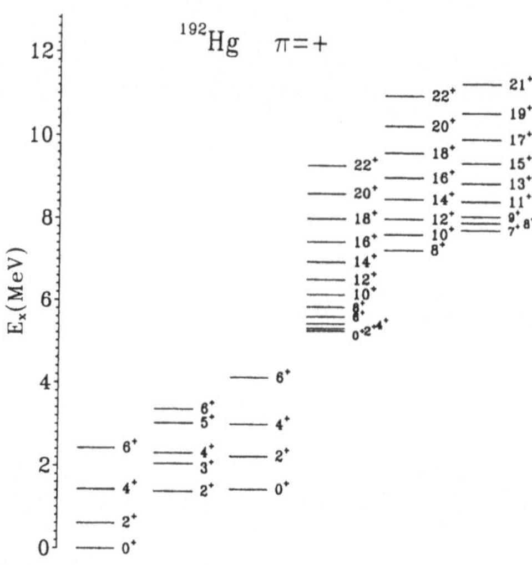

Figure 1. Predicted level scheme for ^{192}Hg. At normal deformation, only the first low energy states for ground, beta and gamma bands are shown. The three SD bands are displayed for spins up to 22.

The collective Hamiltonians are deduced from the Griffin- Hill- Wheeler equations treated in the Gaussian Overlap Approximation.[2] Inputs for the calculations are microscopic in nature. They include potential energy surfaces, zero point energy corrections, masses and moments of inertia that are obtained from constrained HFB calculations in which the external fields are: the moments $\hat{Q}_{20}, \hat{Q}_{22}, \hat{Q}_{30}$, and $\hat{Q}_{10}$, and proton and neutron number operators.[2,3] Gogny's force [2] is used throughout, and neither inert cores nor effective charges are considered.

RESULTS FOR $\pi = +$ LEVELS

The quadrupole Hamiltonian which describes quadrupole motion in ^{192}Hg has been solved by expanding the eigenstates on orthonormal basis vectors.[4] The calculations have been performed for excitation energies below 12MeV and spins up to I=22 to make possible a detailed analysis of the decay out of the SD region, and a comparison between our predictions and recent lifetime measurements.[5] In Fig.1 are shown the main results of our calculations: the ground state (gs), γ - and β -bands that take place at normal deformation ($\beta \sim 0.12$, $\gamma \sim 35^o$), and three SD bands ($\beta \sim 0.55$, $\gamma \sim 5^o$). The band head ($I_1^+ = 0^+$) of the lowest SD band has an excitation energy $E_X(1)$=5.3MeV. The next band head ($I_2^+ = 8^+$) is located at $E_X(2)$=7.3MeV, and the third one ($I_3^+ = 7^+$) at $E_X(3)$=7.8MeV. These SD bands labelled in the following as 1,2 and 3, respectively, may be interpreted as "gs", "β -" and "γ -" bands which develop in the isomeric potential minimum. In contrast to "normal" β - and γ -bands, the excited bands 2 and 3 do not extend down to very low spins. Bands 2 and 3 gain stability only with increasing rotational frequency.

When transformed into γ -ray spectra, the three predicted SD bands display remarkable features (see Fig.2). First, the energies $E_\gamma(I \to I-2)$ of the γ -ray transitions in band 1 are close to the energies $E_\gamma(I-1 \to I-3)$ in band 3. Second, the γ -ray energies $E_\gamma(I \to I-2)$ in band 2 fall mid-way between the energies $E_\gamma(I \to I-2)$ and $E_\gamma(I+2 \to I)$ in band 1. Theses features are similar to those observed [1] for ^{194}Hg, at least at low rotational frequency where the validity of our microscopic description is well established.[6]

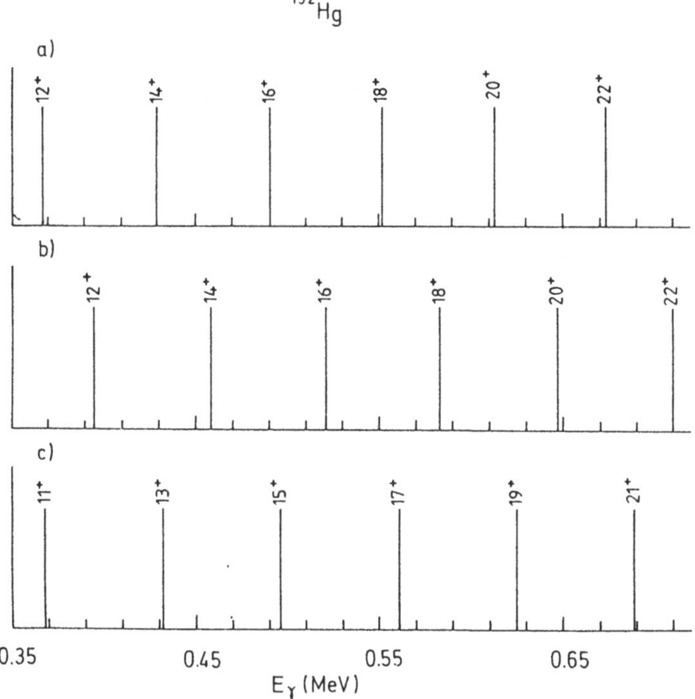

Figure 2. Predicted γ -ray spectra for a) band 1, b) band 2, and c) band 3.

Our predictions have been further tested in a comparison between calculated E2 reduced transition probabilities B(E2;$I_i \rightarrow I_f$) and B(E2)'s deduced for band 1 from lifetime measurements.[5] This comparison is shown in Fig.3. Good agreement between predictions and measurements is obtained for the $22^+ \rightarrow 20^+$ transition. The gradual decrease in magnitude that is seen in the calculated B(E2)'s as I gets lower than I=22 mainly stems from geometric factors entering the definition of B(E2)'s. Transition rates P_γ= B(E2;$I_i \rightarrow I_f$) E_γ^5 have also been calculated for electromagnetic transitions through which SD band 1 may decay. For this purpose, all the predicted $\pi = +$ levels (including those shown at normal deformation in Fig.1) have been considered. It was found that band 1 vanishes in four transitions. Since this band is observed [7] to disappear in one or two transitions, our results suggest that E2 transitions are not strong enough and/or that other decay modes should be considered.

RESULTS FOR THE $K^\pi = 0^-$ EXCITATION

The potential energy surface $V(q_{20}, q_{30})$, with $q_{20}=<\hat{Q}_{20}>$ and $q_{30}=<\hat{Q}_{30}>$, is shown in Fig.4 for q_{20} values in the vicinity of the isomeric potential minimum located at $q_{20} \sim 4500 fm^2$ (i.e. $\beta \sim 0.55$). This surface displays a minimum at $q_{30} = 0$, a feature similar to that deduced from cranking shell model calculations.[1] However, our predicted surface is stiffer in q_{30} than that reported by Cullen et al.[1] Solving the collective Hamiltonian in (q_{20}, q_{30}) coordinates shows that the excitation energy $E_X(4)$ of the K^π=0^-, band 4 head level, is 2.7MeV higher than the energy $E_X(1)$=5.3MeV found earlier for band 1 (i.e. $E_X(4)$=8.0MeV). Comparing the energies E_X found for the three $\pi = +$ and the

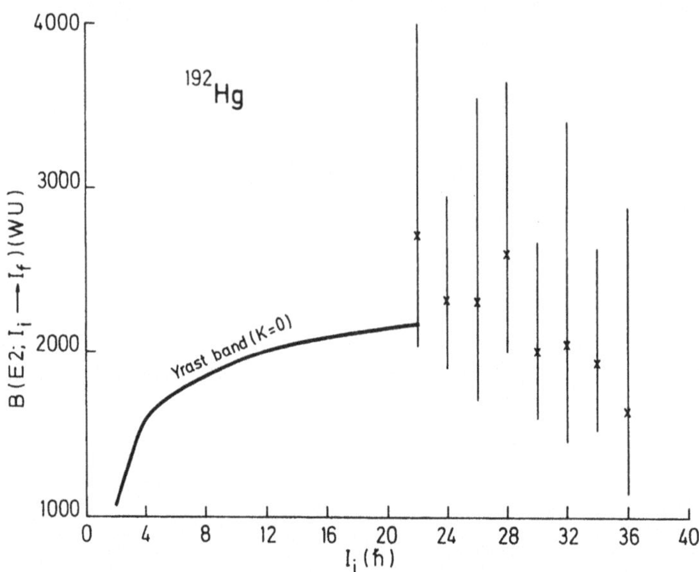

Figure 3. B(E2) values (in Weisskopf units) for the K=0 yrast SD band of ^{192}Hg. Comparison between present calculations and measurements.

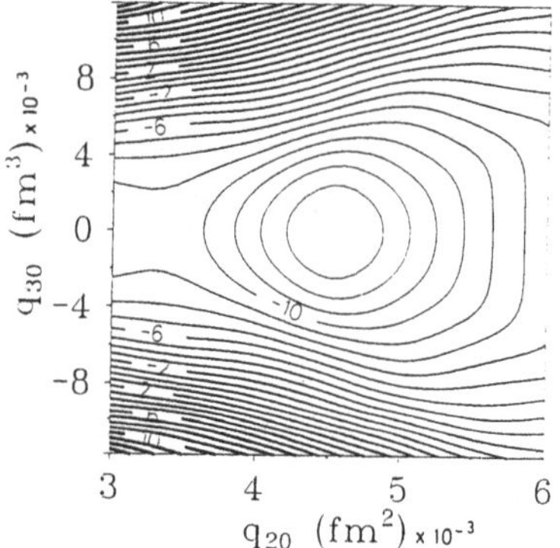

Figure 4. Potential energy surface $V(q_{20},q_{30})$ for ^{192}Hg. Zero point energy correction is included. Contours are separated by 1 MeV.

single $\pi = -$ band head levels indicates that band 4 is the highest. This result suggests that at low spins, the quadrupole mode associated with triaxiality is softer than the q_{30} octupole mode.

The calculated value for the E1 reduced transition probability is B(E1; $1^- \rightarrow 0^+$)= 0.024WU. This number is of the same order of magnitude as that inferred at higher spins (B(E1)$\sim$ 0.015WU) from measurements [1] in the SD bands of ^{193}Hg.

CONCLUDING REMARKS

Three positive parity SD bands and one $K^\pi = 0^-$ SD band head level have been predicted from solving microscopic collective Hamiltonians for ^{192}Hg. From the comparison between these predictions, it appears that the octupole mode q_{30} is stiffer than the quadrupole modes. Finally, the mode of triaxiality is soft enough so that it cannot be ignored in the description of excited bands in even–even SD nuclei of the A=190 mass region.

References

[1] D.M.Cullen, M.A.Riley, A.Alderson, I.Ali, T.Bengtsson, M.A.Bentley, A.M.Bruce, P.Fallon, P.D.Forsyth, F.Hanna, S.M.Mullins, W.Nazarewicz, R.Poynter, P.Regan, J.W.Roberts, W.Satula, J.F.Sharpey-Schafer, J.Simpson, G.Sletten, P.J.Twin, R. Wadsworth, and R.Wyss, Evidence for octupole softness of the superdeformed shape for band interactions in ^{193,4}Hg, Nucl.Phys. A520:105c(1990)

[2] M.Girod, J.P.Delaroche, D.Gogny, and J.F.Berger, Hartree–Fock–Bogoliubov predictions for shape isomerism in nonfissile even–even nuclei, Phys.Rev.Lett. 62:2452(1989)

[3] J.L.Egido and L.M.Robledo, A self-consistent approach to the ground state and lowest–lying negative–parity state in the barium isotopes, Nucl.Phys. A518:475(1990)

[4] J.P.Delaroche, M.Girod, J.Libert, and I.Deloncle, Collective structure of $I^\pi = 0^+$ shape isomers in the ^{190,192,194}Hg isotopes, Phys.Lett. B232:145(1989)

[5] E.F.Moore, R.V.F.Janssens, I.Ahmad, M.P.Carpenter, P.B.Fernandez, T.L.Khoo, S.L.Ridley, F.L.H.Wolfs, D.Ye, K.B.Beard, U.Garg, M.W.Drigert, Ph.Benet, P.J.Daly, R.Wyss, and W.Nazarewicz, Lifetime measurements in the superdeformed band of ^{192}Hg, Phys.Rev.Lett. 64:3127(1990)

[6] M.Girod, J.P.Delaroche, J.Libert, and I.Deloncle, Ab initio calculation of superdeformed bands in ^{192}Hg, submitted to Physical Review Letters (1991)

[7] J.A.Becker, N.Roy, E.A.Henry, M.A.Deleplanque, C.W.Beausang, R.M.Diamond, J.E.Draper, F.S.Stephens, J.A.Cizewski, and M.J.Brinkman, Observation of superdeformation in ^{192}Hg, Phys.Rev. C41:R9(1990). D.Ye, R.V.F.Janssens, M.P.Carpenter, E.F.Moore, R.R.Chasman, I.Ahmad, K.B.Beard, Ph.Benet, M.W.Drigert, P.B.Fernandez, U.Garg, T.L.Khoo, S.L.Ridley, and F.L.H.Wolfs, Superdeformed band in ^{192}Hg, Phys.Rev. C41:R13(1990)

VIII - EXOTIC NUCLEI

ISOSPIN FORBIDDEN β TRANSITIONS TO LOW-LYING 0^+ STATES AND THE EFFECT ON SUPERALLOWED DECAY RATES

E. Hagberg, V.T. Koslowsky, J.C. Hardy and I.S. Towner

AECL Research
Chalk River Laboratories
Chalk River, Ontario, Canada
KOJ 1J0

INTRODUCTION

Studies of superallowed $0^+ \rightarrow 0^+$ β decays furnish the necessary experimental data to test the Conserved Vector Current hypothesis and extract a precise value for the weak vector coupling constant, G_v. The V_{ud} quark-mixing element of the Kobayashi-Maskawa matrix is directly related to this coupling constant and the one extracted from muon decay. Consequently, its value is essential to establishing the unitarity of this matrix and the viability of the three-generation Standard Model as well as probing for the effects of right-handed currents and heavy neutrinos.

A major source of uncertainty in the extraction of the weak vector coupling constant from nuclear β-decay[1)] is the influence that small charge-dependent nuclear-structure effects have on the transition rate for superallowed $0^+ \rightarrow 0^+$ β decays. These effects alter the "structure-free" transition rates by about one percent. Unfortunately, they cannot be tested directly since there is no nuclear case that is free of charge-dependent effects. So far, only indirect tests of the calculated effects have been possible through systematic studies of $0^+ \rightarrow 0^+$ decays where the experimental variations in transition rates from nucleus to nucleus were compared with the calculated variations in charge dependence[1)].

One way of directly testing at least part of the calculations is to look for transitions to the non-analogue, excited 0^+ states. The transition operator, $\tau+$, for Fermi decays is the ladder operator of isospin algebra whose matrix element is non-zero only when the initial and final states are isospin analogues. Thus, Fermi decay to an excited non-analogue 0^+ state is possible only through an admixed component of the analogue state. Experiments designed to measure such transitions provide valuable tests for theories of charge-dependent mixing in nuclei.

The measurement of the transition strength to a non-analogue 0^+ state is very difficult since the branching ratios for even the strongest ones are expected to be in the 10^{-5} range. Only in one case relevant to the test of CVC, viz. ^{42}Sc, has such a transition been observed, and in that case the four reported measurements are not in very good agreement[1)]. In this paper, we report studies of ^{46}V, ^{50}Mn and ^{54}Co, where the aim was to observe weak, non-analogue transitions.

THEORY

We denote the Fermi matrix element for the ground state analogue transition as M_0 and for the non-analogue transition to the first excited 0^+ state as M_1. Then for states with J^π, T = 0^+, 1, we have

$$|M_0|^2 = 2(1-\delta_c^0)$$

$$|M_1|^2 = 2\delta_c^1$$

where δ_c^1 and δ_c^0 are small charge-dependent corrections that represent the difference in the amount of charge-dependent mixing that occurs in the parent and daughter nucleus[2]. Remembering that the square of the Fermi matrix element is inversely proportional to the ft-value, we express the branching ratio to the non-analogue excited state as

$$R \simeq \frac{t_0}{t_1} = \frac{f_1}{f_0} \cdot \frac{f_0 t_0}{f_1 t_1} = \frac{f_1}{f_0} \frac{2\delta_c^1}{2(1-\delta_c^0)} \simeq \frac{f_1}{f_0} \delta_c^1$$

where subscripts 0 and 1 again indicate the ground state and first excited 0^+ state, respectively, and where f is the statistical rate function and t the partial half life for the decays in question.

It remains to calculate δ_c^1. Towner and Hardy[3] first did this with the shell model by introducing charge-dependence into the effective interaction in the following way: (i) add two-body Coulomb terms in the proton-proton part of the Hamiltonian, (ii) increase the T=1 part of the proton-neutron Hamiltonian by 2% (justified by the charge dependence observed in nucleon-nucleon scattering data), and (iii) determine the one-body part of the Hamiltonian from the single-particle energies of the closed-shell-plus-proton and -neutron nuclei. More recently, following the lead of Ormand and Brown[4], Towner[5] refined the calculation still further. The strengths of the one-body and two-body parts of the Coulomb interaction and the charge-dependent nuclear interaction were further adjusted to reproduce the experimental b- and c-coefficients of the isobaric-mass-multiplet equation for the ground-state and excited-state multiplets, where they were known. The results for this calculation of δ_c^1 are given in the Table 1. The underlying shell-model interaction for the nuclei ^{46}V, ^{50}Mn and ^{54}Co was the renormalized interaction of Kuo and Brown[6] and the model space was truncated to configurations $(f_{7/2})^{n_1}(p_{3/2})^{n_2}$ with $n_2 \leq 2$. The results are quite sensitive to the choice of the underlying interaction[2] so an uncertainty of the order of a factor of 2 should be attributed to the calculated δ_c^1 values.

Table 1. The statistical rate functions for the ground-state and excited-state Fermi transitions, f_0 and f_1, the calculated charge-dependent mixing correction, δ_c^1, and the deduced excited-state branching ratio, R.

	f_0	f_1	f_1/f_0	δ_c^1(%)	R
^{46}V	7197.9	526.9	7.32×10^{-2}	0.033	2.4×10^{-5}
^{50}Mn	10727.0	249.1	2.32×10^{-2}	0.082	1.9×10^{-5}
^{54}Co	15743.0	2040.4	1.29×10^{-1}	0.017	2.2×10^{-5}

EXPERIMENT

The short-lived activities investigated in this work, ^{46}V, ^{50}Mn and ^{54}Co, were all produced at the TASCC facility in (p,n) reactions with enriched targets of ^{46}Ti, ^{50}Cr and ^{54}Fe, respectively. A row of 15 thin targets were positioned 5 mm apart in a helium-filled target chamber. The NaCl-loaded helium was continuously swept out of the chamber and transported in a 12.5 m long teflon capillary to a counting location in the next room. There the reaction products, clinging to the NaCl aerosol, were deposited onto the aluminized tape of a fast tape-transport system. The accumulated samples, barely visible to the naked eye as a small salt spot, were periodically moved through a vacuum interlock to a position in front of a shielded, 40% HPGe detector. The experimental arrangement is shown in Figure 1.

The strength of the samples must be kept quite high, with peak decay rates close to 10^6 s^{-1}, because the excited-state branches searched for are extremely weak. Consequently, the HPGe detector was shielded from direct exposure to the numerous, energetic positrons emerging from the sample by a 15 mm thick aluminum absorber. An additional layer of 15 mm thick lead shielding attenuated the low-energy portion of the bremsstrahlung produced when the positrons strike the aluminum absorber. However, the high-energy portion of the bremsstrahlung creates a continuous background in the spectra from the HPGe detector with an intensity that makes measurements at the 10^{-5} level all but impossible.

In our experiments a higher sensitivity was achieved by placing two thin plastic scintillators on either side of the sample on the tape. The scintillator on the same side of the sample as the HPGe detector is called the reject, or β_R, counter whereas the one on the opposite side is called the accept, or β_A, counter. The HPGe detector is operated in coincidence with signals from the β_A counter, thus guaranteeing that the detected positron emerged from the sample heading away from the HPGe detector. This coincidence selection removes most bremsstrahlung events since they originate from cases where the positron is emitted in the direction of the HPGe detector. However, some positrons strike the

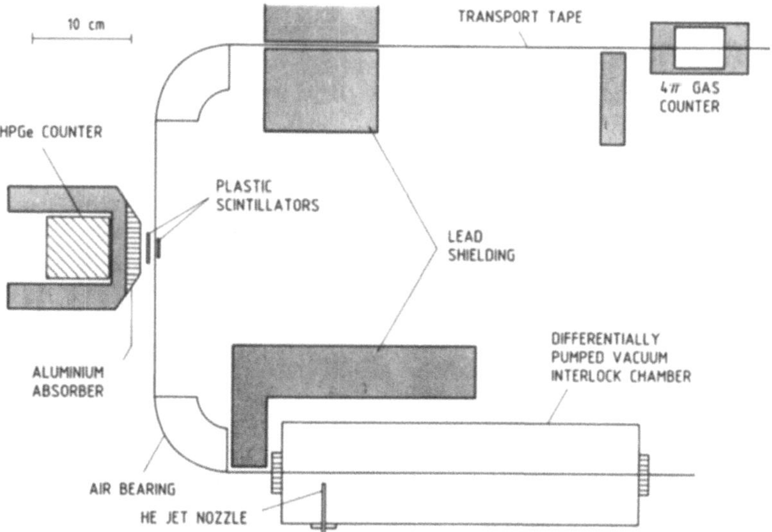

Fig. 1. Detector arrangement and tape transport system.

aluminium absorber, backscatter and traverse the β_A scintillator. Such events, resembling the ones in which the positron was initially heading away from the HPGe counter, are removed by operating the β_R scintillator in anticoincidence, since any backscattered event actually traverses this scintillator twice. The intensity of the continuous background seen in the γ-ray spectra with a $\gamma \cdot \beta_A \cdot \beta_R$ condition is about 400 times lower than that observed with only the passive shielding in place.

The excited state branch is observed with the HPGe - scintillator combination. The total number of decays is determined with a continuous-flow gas counter which has an efficiency of 92% for high energy positrons. Each collected sample is first assayed by the HPGe detector. After it has been observed for about 3 half-lives, the tape moves again and the same sample is brought into the centre of the gas counter. The events seen in this counter are multiscaled and the data is subjected to a decay-curve analysis. Two decay components were required to produce a good fit to the experimental data. In all cases the dominant component was found to exhibit the half-life of the activity under investigation. The second, small component was a constant background, or in the cases of ^{50}Mn and ^{54}Co, the long-lived isomeric state in these nuclides.

RESULTS

The gated γ-ray spectrum corresponding to 2.7×10^{10} decays of ^{46}V is shown in Figure 2. The first, excited 0^+ state in ^{46}Ti has an excitation energy of 2611 keV[7]. The known de-excitation mode of this state is a cascade of 1722 keV and 889 keV γ rays. Both are clearly seen in our spectrum and their intensities, when corrected for detector efficiency, are identical. Our deduced branching ratio for the non-analogue transition from ^{46}V to the 2611 keV, 0^+ state in ^{46}Ti is 5×10^{-5}. This transition has not been seen before.

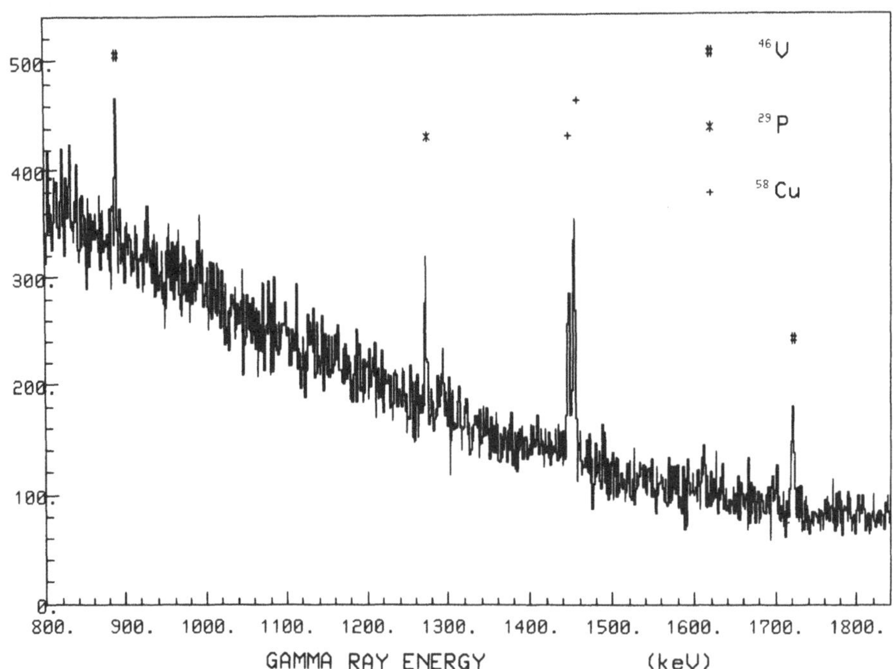

Fig. 2. Gated γ-ray spectrum observed in the ^{46}V experiment.

We have carefully searched for evidence of other known γ-ray transitions in ^{46}Ti, indicative of feedings to other excited states and found none. We do see γ rays from ^{29}P, ^{58}Cu and ^{90m}Zr with intensities comparable to those of the γ rays originating from the non-analogue transition of ^{46}V. These other activities originate from contaminants in the target at the 10^{-4} level of Si, Ni and Zr, respectively. The Si contamination most likely originates from the quartz tube of our NaCl aerosol generator which is kept at a temperature of 600°C. Zirconium is a likely contaminant in our titanium targets since both are group 4b elements.

For the cases of ^{50}Mn and ^{54}Co we have so far only spent a few hours on each nuclide in order to establish the optimum production conditions and the level of possible contaminants. However, some interesting results emerged from those short tests.

Gamma-ray spectra corresponding to 4×10^9 decays of ^{50}Mn are shown in Figure 3. These spectra are dominated by the numerous γ rays from the decay of ^{50m}Mn. The first excited 0^+ state in ^{50}Cr has not yet been located unambiguously[8] but the lowest tentative candidate is a level at 3694 keV. The most likely de-excitation mode of this state is a cascade of 2911 keV and 783 keV γ rays, the latter one also produced by the decay of ^{50m}Mn. There is no indication of a 2911 keV γ ray in our spectra and our present limit for the non-analogue branch from ^{50}Mn to a 3694 keV, 0^+ state in ^{50}Cr is 6×10^{-5}.

Two γ rays with energies of 2845 keV and 3628 keV are clearly seen in Figure 3. Their energies and relative intensities are in good agreement

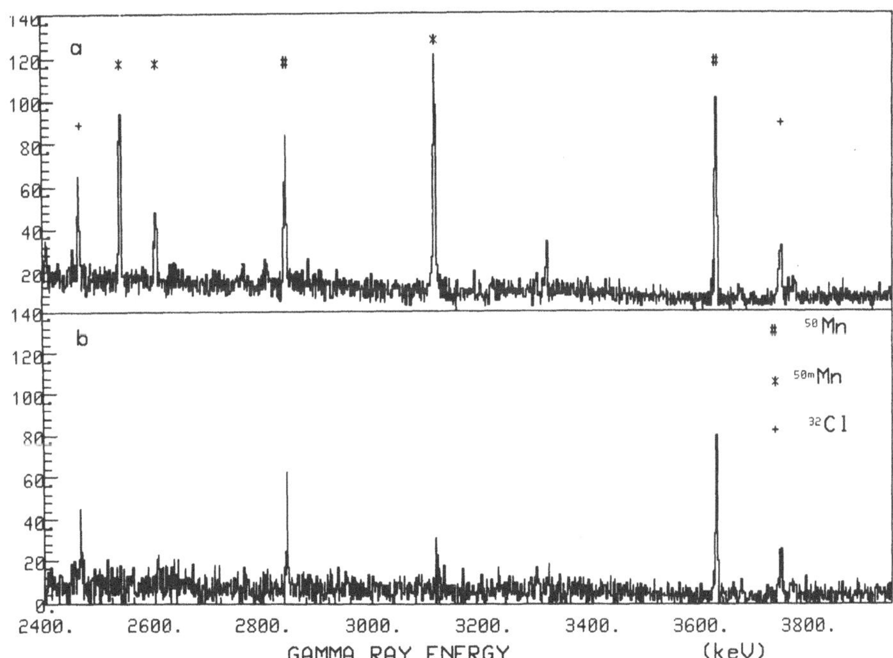

Fig. 3. a) Gated γ-ray spectrum observed in the ^{50}Mn experiment where each sample was counted for 1.2 s.
b) Spectrum where the long-lived activities have been removed. It was obtained by subtracting the spectrum measured during the last half of our counting period from that measured during the first half.

with those known to originate from the de-excitation of the lowest 1^+ state in ^{50}Cr[8]. The observation of these γ rays in Figure 3b confirms that they originate from a very short-lived activity. The allowed Gamow-Teller transition from ^{50}Mn to the 3628 keV excited state in ^{50}Cr has therefore been observed for the first time and our deduced value for this branch is about 7×10^{-4}. The existence of this branch will affect the value of the weak vector coupling constant deduced from the $0^+ \rightarrow 0^+$ ground-state transition of ^{50}Mn since the latter branch does not account for 100% of the decay intensity. Furthermore, no attempt has been made to account for the existence of beta branches to 1^+ states in the theoretical corrections to the $0^+ \rightarrow 0^+$ decay strengths because of the imprecise nature of the calculations for such branches. Only experiments can give us data on the intensity of these potentially vital, decay modes.

Our collected data on ^{54}Co correspond to 3×10^8 decays of this nucleus. No γ rays other than those of ^{54m}Co were observed. At this early stage, our deduced limit for the non-analogue transition from ^{54}Co to the 2561 keV 0^+ state in ^{54}Fe[9] is 1×10^{-4}.

CONCLUSION

An experimental arrangement has been developed that consists of a fast helium-jet transport system, a fast tape-transport system and a detector combination that is very sensitive to the observation of weak γ-ray transitions in a strong field of energetic positrons. It has made possible the observation, for the first time, of two beta-decay branches of interest to weak interaction studies.

Our deduced branching ratios for the non-analogue, $0^+ \rightarrow 0^+$ transitions of ^{46}V, ^{50}Mn and ^{54}Co are 5×10^{-5}, $<6 \times 10^{-5}$ and $<1 \times 10^{-4}$, respectively. The calculated values, shown in the table, for these nuclides are 2.4×10^{-5}, 1.9×10^{-5} and 2.2×10^{-5}, respectively, with uncertainties estimated to be a factor of 2. So far there is reasonable agreement between the theory and the data, but more extensive experiments on ^{50}Mn and ^{54}Co, which we plan for the near future, will provide more demanding tests. Note that in shell-model calculations: $\delta_c^0 = \Sigma_j \delta_c^j$, where j runs from 1 to n for the n excited 0^+ states in the model space. In most cases, the bulk of the depletion of strength in the ground-state analogue transition, δ_c^0, is found in the first excited non-analogue transition, i.e. $\delta_c^1 \simeq \delta_c^0$. This is true in the calculations described in section 2 for ^{50}Mn and ^{54}Co. The case of ^{46}V is somewhat unusual in that the calculations predict $\delta_c^1 \simeq 0.3\ \delta_c^0$.

It is worth noting that the only other basic theory of charge-dependent mixing effects in $0^+ \rightarrow 0^+$ beta decay[4] predicts a δ_c^0 value of only 0.01% for the ground-state transition in ^{46}V. Since $\delta_c^1 \leq \delta_c^0$, the predicted branching ratio to the excited non-analogue state must in this case be $R \lesssim 7 \times 10^{-6}$, a value that is not in agreement with our measurement. Additional data on non-analogue transitions should lead to a better understanding of theories of charge-dependent mixing in nuclei.

REFERENCES

1. J.C. Hardy, I.S. Towner, V.T. Koslowsky, E. Hagberg and H. Schmeing, Nucl. Phys. A509:429 (1990).
2. I.S. Towner, Phys. Lett. 43B:267 (1973).
3. I.S. Towner and J.C. Hardy, Nucl. Phys. A205:33 (1973).
4. W.E. Ormand and B.A. Brown, Phys. Rev. Lett. 62:866 (1989).

5. I.S. Towner, in Symmetry Violations in Subatomic Physics, Proc. of 6th Summer Institute in Theoretical Physics, ed. B. Castel and P.J. O'Donnell (World Scientific, Singapore, 1989) p. 211.
6. T.T.S. Kuo and G.E. Brown, Nucl. Phys. A114:241 (1966).
7. Nuclear Data Sheets 49:237 (1986).
8. Nuclear Data Sheets 61:1 (1990).
9. Nuclear Data Sheets 50:255 (1987).

NUCLEI FAR FROM STABILITY AND THE R-PROCESS WAITING-POINT CONCEPT

K.-L. Kratz[1], H. Gabelmann[1], P. Möller[1,2], B. Pfeiffer[1], H.L. Ravn[3], A. Wöhr[1] and F.-K. Thielemann[4]

[1]Institut für Kernchemie, Universität Mainz, Mainz, FRG
[2]Lund Institute of Technology, Lund, Sweden
[3]CERN, Geneva, Switzerland
[4]Harvard-Smithsonian Center for Astrophysics, Cambridge MA, USA

INTRODUCTION

The nucleosynthesis process by rapid neutron captures (the r-process) is responsible for the formation of about half of the nuclear species in nature beyond Fe. While the astrophysical site for the r-process is not yet unambiguously identified, its association with the cores of low-mass stars undergoing type II supernova (SN) events is strongly suggested (see, e.g., Refs.[1,2]).

R-process nucleosynthesis involves the progressive build-up of heavy isotopes via neutron captures on neutron-rich nuclei, interspersed by β-decays. Its abundance features clearly reflect nuclear properties - the maxima are related to the magic neutron numbers N = 50, 82 and 126 (see Fig. 1). This probes our knowledge of nuclear structure of isotopes very far from the valley of β-stability, even through the position of the neutron drip line. For the vast majority of nuclei close to or in the r-process path, no experimental information is available; hence, nu-

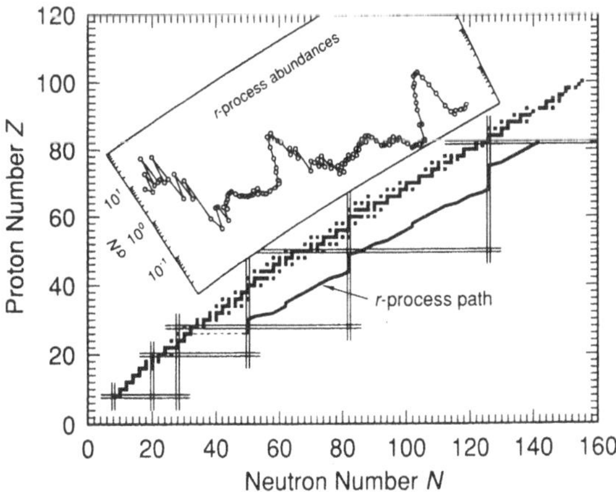

Fig. 1. Schematic illustration of r-process path and r-abundances. The abundance maxima are correlated with the positions of the N = 50, 82 and 126 closed shells.

Nuclear Shapes and Nuclear Structure at Low Excitation Energies
Edited by M. Vergnes *et al.*, Plenum Press, New York, 1992

clear properties have to be deduced from theoretical models. However, in the case of β-decay half-lives ($T_{1/2}$), even the most recent predictions[3,4] using the quasiparticle random-phase approximation (QRPA) still differ up to an order of magnitude far off stability. Therefore, it is of utmost importance to continue experimental efforts to measure the $T_{1/2}$ and β-delayed neutron (βdn) emission probabilities (P_n) of at least the key-isotopes in the r-process path. And in fact, only within the last few years has it been possible to determine these quantities for a few nuclei in the bends at N = 50 and 82 where the r-process path comes closest to β-stability.

BETA-DECAY PROPERTIES OF WAITING-POINT NUCLEI

From the β-decay properties of two of the most important 'classical' waiting-point nuclei $^{130}Cd_{82}$ (Ref.[5]) and $^{80}Zn_{50}$ (Refs.[6,7]) together with information on $^{131,133}In$ and $^{81,83}Ga$, respectively, where the r-process branches off from the neutron closed shells, strong evidence has been obtained that the solar-system r-abundances ($N_{r,\odot}$) of their stable isobars[8] result from a β-flow equilibrium (see, e.g., Ref.[9]). This so-called waiting-point concept implies approximate equality of progenitor abundance ($N_{r,prog}$) times β-decay rate for all isotopes in the r-process path. An additional test was performed for the - at that time unknown - N = 50 and 82 r-process isotopes ^{76}Fe to ^{79}Cu and ^{127}Rh to ^{129}Ag, respectively. From the known $N_{r,\odot}$ and predicted P_n-values, the $N_{r,prog}$ of the waiting- point nuclei were calculated; and with this also the $T_{1/2}$ for these isotopes were 'predicted' - provided that a β-flow equilibrium existed. Although these astrophysical requirements were found to agree closely with the nuclear-physics values (QRPA-predictions; see Tab. 1 in Ref.[9]), a final decision whether or not the steady-flow picture holds for the $A \simeq 80$ and $A \simeq 130$ $N_{r,\odot}$-peaks will only be possible from further experimental data on waiting-point nuclei. However, apart from the fact that these isotopes lie at the extreme borderlines of production and detection, isotopically clean beams are difficult to obtain at present separator facilities with 'chemical' ion sources.

Nevertheless, as a result of recent target/ion-source developments at the CERN-ISOLDE on-line mass separator, considerable progress could be achieved in the production of intense beams of neutron-rich isotopes of $_{29}Cu$ to $_{33}As$. In these experiments, a ^{238}UC-graphite target was irradiated with a beam of 600 MeV protons at a current of 2.2 μA from the Synchro-Cyclotron. The target was heated to about 2,400°C and connected to a plasma-discharge ion source via a temperature-controlled, heated transfer line. This line served as a kind of thermochromatography column allowing preferential extraction of volatile species, whereas non- or less-volatile elements were retained. Growth-and-decay curves, with collection and cycle times optimized for each isotope of interest, were measured by βdn-multiscaling. In this way, the β-decay half-life of the classical N = 50 waiting-point nucleus ^{79}Cu could be measured for the first time. Fig. 2 shows a typical multiscaling curve for A = 79, after subtraction of the 'longest-lived' component of ^{79}Ga. Its half-life was determined in a separate run as $T_{1/2} = (2,847 \pm 3)$ ms. Clearly, two 'short-lived' βdn-activities are seen, ^{79}Zn with a $T_{1/2} \simeq$ 1 s and the new waiting-point nucleus $^{79}Cu_{50}$ with a $T_{1/2}$ of (188 ± 25) ms. Furthermore, the P_n-values of the two N = 50 isotones ^{80}Zn (P_n= 1.0 ± 0.5) and ^{79}Cu (P_n= 55 ± 17) were determined in the same experiment. In addition, the $T_{1/2}$ of $^{74-78}Cu$ and $^{79-83}Ga$, and the $T_{1/2}$ and P_n-values of $^{84,85}Ge$ and ^{85}As were measured with high accuracy. Finally, two new isotopes lying 'beyond' the r-process path defined in Ref.[9], i.e. $^{81}Zn_{51}$ and $^{84}Ga_{54}$, were identified.[10]

It is interesting to note that the measured decay properties of ^{79}Cu are in good agreement with our earlier QRPA shell-model predictions of $T_{1/2} \simeq 193$ ms and $P_n \simeq 60$ %. As was discussed in detail in Ref.[11], correspondence of the nuclear-physics prediction of $T_{1/2}(^{79}Cu)$ with the astrophysical requirement could only be obtained with a large $P_n(^{79}Cu)$ - as, indeed, observed experimentally. With the $N_{r,\odot}(^{80}Se)$ from Ref.[8], the $T_{1/2}$ and P_n of ^{80}Zn and the experimental $P_n(^{79}Cu)$, one obtains a waiting-point request for $T_{1/2}(^{79}Cu)$ of (185 ± 29) ms. For details of the calculations, see Ref.[11]. We conclude from this excellent agreement for the ^{79}Cu waiting-point nucleus, that our hypothesis of a β-flow r-process[9] is further strengthened.

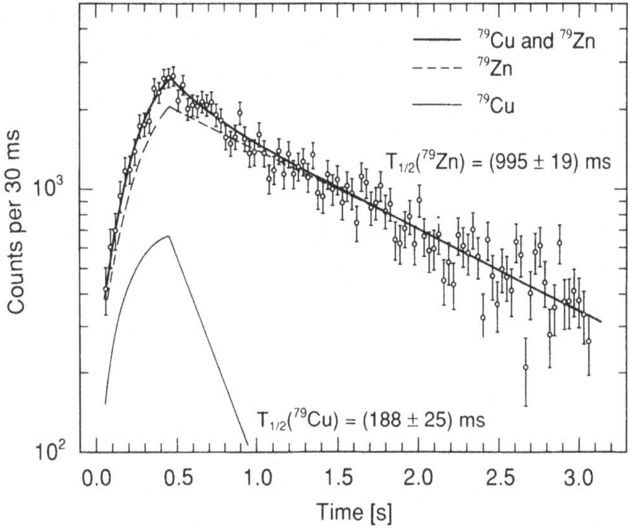

Fig. 2. Beta-delayed neutron growth-and-decay curve for A = 79, after subtraction of the independently determined 2.847 s component of ^{79}Ga. Besides ^{79}Zn with a known $T_{1/2} \simeq 1$ s, a short-lived component is observed which can be attributed to the new N = 50 waiting-point nucleus ^{79}Cu.

CONSTRAINTS ON R-PROCESS CONDITIONS

If we accept the implications of the above data, we can now ask whether the isotopic pattern found in the $N_{r,\odot}$ at A $\simeq$ 80 and A $\simeq$ 130 can give a clue to the stellar conditions under which the r-process has operated. In Ref.[9] the necessary time scale as well as stellar temperature (T) and neutron density (n_n) ranges have been determined which give the 'correct' steady-flow r-process path through the waiting-point nuclei ^{76}Fe to ^{80}Zn and ^{127}Rh to ^{130}Cd. With the additional information from the $N_{r,\odot}$ of ^{81}Br and ^{83}Kr which originate from ^{81,83}Ga, and the $N_{r,\odot}$ of ^{131,132}Xe which were formed by the progenitors ^{131,133}In, rather tight constraints for the T - n_n conditions could be given (see dashed area in Fig. 3).

Recently, Mathews and Cowan[1] have deduced a critical neutron density of $n_n \simeq 10^{20}$cm^{-3} from our measured $T_{1/2}$ and P_n-values and the above assumption that the r-process does indeed

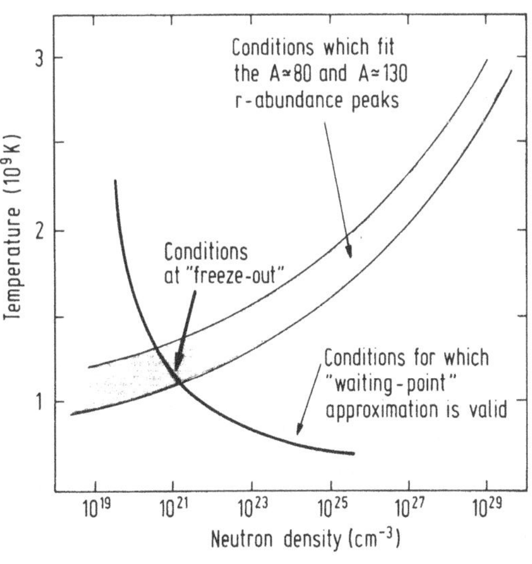

Fig. 3. Neutron densities (n_n) and temperatures (T) consistent with observed $N_{r,\odot}$-peaks and measured $T_{1/2}$ and P_n-values. Also shown are the conditions for which the 'classical' waiting-point approximation is valid.[12] The thick arrow indicates the freeze-out conditions for a β-flow r-process. (see text).

flow through the closed-shell waiting-point nuclei. For a classical r-process, this value corresponds to the minimum n_n required for the waiting-point approximation.[12] Combining our T - n_n conditions[9] with the above critical value for n_n, one can determine the freeze-out temperature for the r-process (see Fig. 3). One may argue that initially the r-process occurs at slightly higher temperature and significantly higher neutron density, involving isotopes further away from β-stability than the classical waiting-point nuclei. However, what one probably observes as final solar-system abundances seems to reflect the T - n_n conditions just at freeze-out. This then could also explain that the $N_{r,\odot}$-peaks are so sharp. Such narrow peaks may imply that most of the r-process material has been produced in a single well defined environment, rather than some combination of different environments.

PREDICTION OF R-ABUNDANCES IN THE $90 \leq A \leq 100$ REGION

Given the result that for the $N_{r,\odot}$-peaks there exists an $(n,\gamma)-(\gamma,n)$ and β-flow equilibrium, which implies that the ratio of progenitor abundance ($N_{r,prog}$) to progenitor half-life is constant, we then can use this requirement to also predict the $N_{r,prog}$ and - after correction for P_n-branching during β-decay back to stability - the $N_{r,\odot}$ of isotopes in between the magic neutron numbers, i.e. in between the r-abundance peaks.

As a first test, we have chosen the mass region $90 \leq A \leq 100$ for two reasons: (1) the stellar origin of ^{96}Zr is of special interest in connection with the recently observed correlated $^{50}Ti-^{96}Zr$ isotopic anomalies in several inclusions of the Allende meteorite (see, e.g., Refs.[13,14]), and (2) ^{100}Mo is an 'r-only' isotope which can - if necessary - be used for an absolute normalization of the calculated r-abundances to the observed $N_{r,\odot}$ (Ref.[8]).

For freeze-out conditions (see Fig. 3), we first have calculated relative r-abundances for each isotopic chain ($\sum N_{r,prog}(Z) = 1$). These relative $N_{r,prog}$ were then weighted according

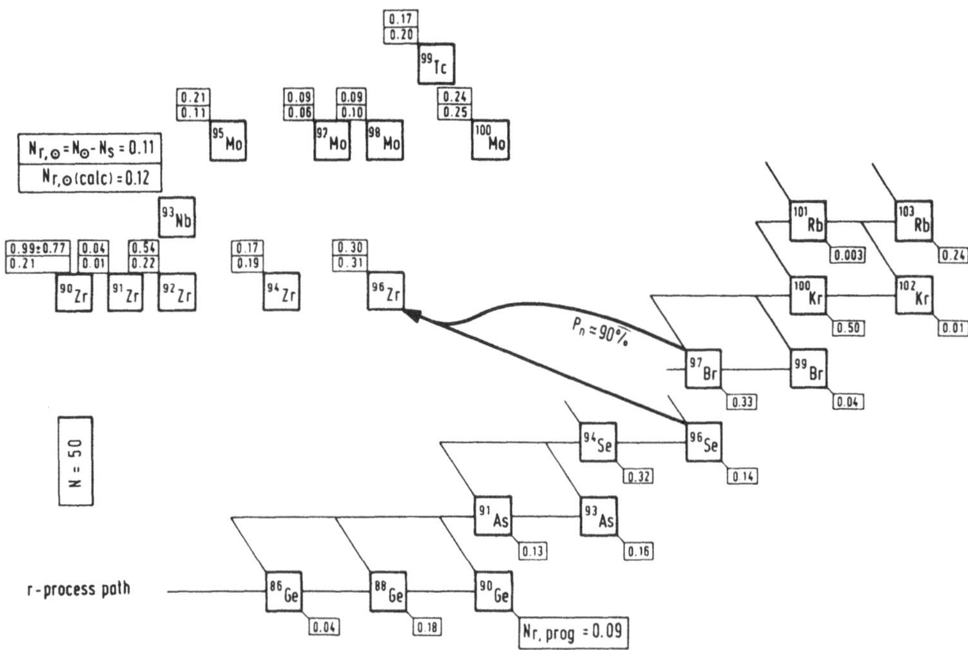

Fig. 4. Schematic view of the r-process path at freeze-out conditions and r-abundance features in the $90 \leq A \leq 100$ mass region. From the progenitor abundances ($N_{r,prog}$), after β-decay and P_n-branching, the final abundances ($N_{r,\odot}$(calc)) of the stable isotopes ^{90}Zr to ^{100}Mo were derived. They are in fair agreement with the observed r-process residuals from Käppeler et al.[8]. Under these conditions, ^{96}Zr would be an 'r-only' isotope, mainly originating from the precursor ^{97}Br. For further discussion, see text.

to their predicted $T_{1/2}$ (Ref.[4]) and normalized to the neutron closed shell isotopes $^{80}Zn_{50}$ and $^{130}Cd_{82}$. In both cases, P_n-branching is so small that one can assume approximate equality between initial $N_{r,prog}$ and final $N_{r,\odot}$. At this point it turned out that the values of $N_{r,\odot}(^{80}Se)/T_{1/2}(^{80}Zn) \simeq 42.5\ s^{-1}$ and $N_{r,\odot}(^{130}Te)/T_{1/2}(^{130}Cd) \simeq 11.5\ s^{-1}$ were not equal, as would be expected from a global waiting-point concept. This is a potentially important result which may indicate that - similar to what is established for the s-process[8] - also the r-process has different components (each with a β-flow equilibrium) to build up the $N_{r,\odot}$ from A $\simeq$ 70 to A $\simeq$ 240. This point will be taken up again in the next section.

A schematic view of the r-process path at freeze-out conditions and the r-abundance features in the $90 \leq A \leq 100$ region is shown in Fig. 4. The most interesting result is, that we obtain absolute r-abundances in fair agreement with the observed $N_{r,\odot}$ from Ref.[8] if we normalize to ^{130}Cd. This is most evident from the simultaneous reproduction of the abundance of the 'r-only' isotope ^{100}Mo and the $N_{r,\odot}$ of ^{99}Tc, which both originate from ^{100}Kr. Another important result is, that under the above $T - n_n$ conditions ^{96}Zr would also be an 'r-only' isotope, mainly produced from β-delayed neutron decay of the progenitor ^{97}Br. Although these results are rather sensitive to the neutron density assumed, even an order-of-magnitude agreement of our predictions with the (also rather uncertain) $N_{r,\odot}$-residuals[8] can put additional constraints on the r-process environment as well as on nuclear properties far off stability.

THE ASTROPHYSICAL SITE FOR THE R-PROCESS

The operation of an r-process is characterized by the fact that 10 to 100 neutrons per nucleus in the Fe-peak have to be available to form all heavier r-process nuclei by neutron capture. If only Fe-group elements and neutrons exist, this translates into a neutron excess of $\eta = (N - Z)Y = 0.4 - 0.7$, where Y represents isotopic abundances. Such a high neutron excess can only be obtained by converting protons into neutrons via capture of energetic electrons on protons or nuclei which have to overcome large negative Q-values. This can be achieved by degenerate electrons with large Fermi energies, which requires a compression to densities of $10^{11} - 10^{12}$g cm^{-3}, with an equilibrium between electron captures and β^--decays.[15] Such matter is located close to the neutrino sphere in type II SN; it can however also be convected out to lower densities in the so-called 'high entropy bubble' of delayed type II SN explosions.[16] Wilson and Mayle[17] report values of $\eta \simeq 0.6$ for these mass zones. This matter is initially composed of α-particles and free nucleons. In the expansion during the explosive ejection, a nuclear statistical equilibrium holds and leads to the dominance of Fe-group nuclei, with the highest binding energies per nucleon. At decreasing temperatures and after the freeze-out of charged particle reactions, rapid neutron captures will proceed to form heavy nuclei.

A different situation would emerge for maximum temperatures below freeze-out conditions for charged particle reactions with Fe-group nuclei. Then reactions among light nuclei which release neutrons, like (α, n)-reactions on ^{13}C and ^{22}Ne, can sustain a high neutron flux. The constraint of having 10 – 100 neutrons per heavy nucleus in order to attain r-process conditions, could then be met by small abundances of Fe-group nuclei. Such conditions were expected when a shock front passes the He-burning shell and enhances the (α,n)-reactions by orders of magnitude (see, e.g., Ref.[18]). This did, however, not materialize with realistic stellar models (see, e.g., Refs.[19,20]): even under explosive conditions, (α,n)-neutron sources are not strong enough to cause an r-process. Recent research, based on additional neutron release via inelastic neutrino scattering,[21] can also not produce neutron densities which are required for such a process to operate (see also Ref.[22]).

In the previous section, it was mentioned that the values of the $N_{r,\odot}/T_{1/2}$ = const were different for the $A \simeq 80$ and $A \simeq 130$ $N_{r,\odot}$-peaks, thus not supporting a global β-flow equilibrium for the whole mass region between $A \simeq 75$ and $A \simeq 135$. There exist two possible explanations for this discrepancy: (1) Within the nuclear statistical equilibrium scenario, the $N_{r,\odot}$ may represent a superposition of different zones with different neutron excess (in each case with a β-flow equilibrium). Then, for very high neutron excess the α-rich freeze-out could provide r-process seed abundances between A $\simeq$ 60 and 90 (Ref.[23]), and the abundance pattern beyond A $\simeq$ 90 would be decoupled from the A $\simeq$ 80 $N_{r,\odot}$-plateau. (2) Another option is the possible contribution of fission cycling from heavy nuclei (A $\simeq$ 240 – 280) down to A $\simeq$ 130 but not to smaller nuclei. This also would lead to a different constant of the β-flow coming from light nuclei and the one beyond A $\simeq$ 130.

CONCLUSION

While continued improvements have been evident in nuclear physics, the astrophysical site of the r-process is still not clearly established. However, from the new nuclear input data suggesting a β-flow r-process, together with recent astronomical observations (see, e.g., Refs.[1,2]), one can draw the following conclusions on the r-process environment: (1) Shock-induced explosive He or C burning in SN or He-core flashes in low-mass stars seem to be ruled out. All present calculations show that neither enough neutrons are produced by (α, n)-reactions nor is the available exposure time long enough to allow a β-flow process which can reproduce the observed $N_{r,\odot}$-pattern. (2) Only models assuming low mass type II SN sites seem to fit the astronomical data on the build-up of r-process matter during galactic evolution; and the evidence points to the core ejections of highly neutronized matter as the most likely site for the r-process.

ACKNOWLEDGEMENTS

This work was supported by grants from BMFT (06 MZ 106), DFG (Kr 806/1), and from NSF (AST-8913799).

REFERENCES

1. G.J. Mathews and J.J. Cowan, Nature 345:491 (1990)
2. J.J. Cowan and F.-K. Thielemann, Phys. Rep. (1991), in print.
3. A. Staudt, E. Bender, K. Muto and H.V. Klapdor-Kleingrothaus, At. Data Nucl. Data Tables 44:79 (1990).
4. P. Möller and K.-L. Kratz, "Table of QRPA-Predictions of $T_{1/2}$ and P_n-values" (1991), unpublished.
5. K.-L. Kratz, H. Gabelmann, W. Hillebrandt, B. Pfeiffer, K. Schlösser, F.-K. Thielemann and the ISOLDE Collaboration, Z. Physik A325:483 (1986).
6. B. Ekström B. Fogelberg, P. Hoff, E. Lund and A. Sangariyavanish, Phys. Scr. 34:614 (1986).
7. R.L. Gill, R.F. Casten, D.D. Warner, A. Piotrowski, H. Mach, J.C. Hill, F.K. Wohn, J.A. Winger and R. Moreh, Phys. Rev. Lett. 56:1874 (1986).
8. F. Käppeler, H. Beer and K. Wisshak, Rep. Prog. Phys. 52:945 (1989).
9. K.-L. Kratz, Revs. Mod. Astron. 1:184 (1988).
10. K.-L. Kratz, H. Gabelmann, P. Möller, B. Pfeiffer, A. Wöhr and the ISOLDE Collaboration, Verhandl. DPG (VI) 26:438 (1991); and subm. to Z. Physik A.
11. K.-L. Kratz, V. Harms, A. Wöhr and P. Möller, Phys . Rev. C38:278 (1988).
12. A.G.W. Cameron, J.J. Cowan and J.W. Truran, Ap. Space Sci. 91:235 (1983).
13. C.L. Harper, L.E. Nyquist, C.Y. Shih and W. Wiesmann, in: "Nuclei in the Cosmos", H. Oberhummer and W. Hillebrandt, eds., MPA/P4 138 (1990).
14. F.-K. Thielemann, K.-L. Kratz, B. Pfeiffer, P. Möller and W. Hillebrandt, ibid., 286 (1990).
15. A.G.W. Cameron, in: "Cosmic Abundances in Matter", C.J. Waddington, ed., AIP Conf. Proc. 183:349 (1989).
16. H.A. Bethe, Revs. Mod. Phys. 62:801 (1990).
17. J.R. Wilson and R.W. Mayle, Phys. Rep. 163:63 (1988).

18. F.K. Thielemann, M. Arnould and W.Hillebrandt, Astron. Astrophys. 74:175 (1981).
19. J.B. Blake, S.E. Woosley, T.A. Weaver and D.N. Schramm, Ap. J. 248:315 (1981).
20. J.J. Cowan, A.G.W. Cameron and J. W. Truran, Ap. J. 265:429 (1983).
21. R.I. Epstein, S.A. Colgate and W.C. Haxton, Phys. Rev. Lett. 61:2038 (1988).
22. S.E. Woosley, D.H. Hoffman and W.C. Haxton, Ap. J. 356:272 (1990).
23. S.E. Woosley, D.H. Hoffman and W.C. Haxton, submitted to Ap. J. (1991).

EMISSION OF NEUTRONS AND THE NEUTRON HALO OF ^{11}Li

M. Lewitowicz[a], R. Anne[a], S.E. Arnell[b], R. Bimbot[c], H. Emling[d], D. Guillemaud-Mueller[c], P.G. Hansen[e], L. Johannsen[e], B. Jonson[b], S. Mattsson[b], A.C. Mueller[c], R. Neugart[f], G. Nyman[b], F. Pougheon[c], A. Richter[g], K. Riisager[e], M.G. Saint-Laurent[a], O. Sorlin[c] and K. Wilhelmsen[b]

[a]GANIL, BP 5027, F-14021 Caen, France
[b]Fysiska Institutionen, Chalmers Tekniska Högskola, S412 96 Göteborg, Sweden
[c]Institut de Physique Nucléaire, BP 1, F-91406 Orsay, France
[d]GSI, D-6100 Darmstadt, FRG
[e]Det Fysiske Institut, Aarhus Universitet, DK-8000 Aarhus C, Denmark
[f]Institut für Physik, Universität Mainz, D-6500 Mainz, FRG
[g]Institut für Kernphysik, Technische Hochschule, D-6100 Darmstadt, FRG

INTRODUCTION

The matter distribution of the neutron drip-line nuclei has been a subject of intense interest since pioneering radioactive-beam experiments at Berkeley[1]. Unusually large interaction cross-sections measured at relativistic energies for ^{11}Li, ^{14}Be and ^{17}B nuclei were translated into effective matter radii. It was found that these were significantly larger than for other p-shell nuclei. In particular, the root-mean-square (rms) mass radius of ^{11}Li was found to be 3.2fm, which was considerably larger than the 2.4fm radius of ^{9}Li. At the same time, no signatures for significant deformation of ^{11}Li were found in measurements of electric quadrupole moment by means of laser spectroscopy[2].

The second qualitative feature of these extremely neutron-rich nuclei is that the binding energy of the last neutron(s) becomes very low. In the case of ^{11}Li the last pair of neutrons is bound by (250±80)keV[3] and ^{10}Li is unbound, thereby showing that a pairing is essential for the stability of ^{11}Li.

On the basis of these experimental results it was proposed[4] that the two last neutrons in ^{11}Li form a neutron halo or cloud around the ^{9}Li core, as is schematically presented in the inset of figure 1. One can estimate the main features of the neutron halo in a semi-quantitative way[4]. For a dineutron in an s-state the external wave function will be a Yukawa type:

Nuclear Shapes and Nuclear Structure at Low Excitation Energies
Edited by M. Vergnes *et al.*, Plenum Press, New York, 1992

$$\psi(r) \sim r^{-1}\exp(-r/\rho) \quad (1)$$

where the decay length ρ is related to the binding energy S_{2n} via the expression: $\rho=h/(2\mu S_{2n})^{1/2}$, with μ standing for the reduced mass. If we define x as the ratio R/ρ between the range R of the attractive nuclear potential and the halo range parameter ρ, the rms radius of the halo can be approximated by:

$$<r^2>^{1/2}= \rho((1+x)/2)^{1/2} \quad (2)$$

The total rms mass radius can be calculated from the expression:

$$<r_{tot}^2>=((M/(M+m))^{1/2}(<r_c^2>+m/(M+m)<r^2>)^{1/2} \quad (3)$$

here $<r_c^2>$ represents the mean square radius of the core of ^{9}Li with mass M and dineutron mass of m. If we adjust R so that for large binding energies $M<r^2>/(M+m)\cong<r_c^2>\cong2.5$fm , the value of the ρ parameter of ^{11}Li is 8.2fm which yields, from (2), $<r^2>^{1/2}$ equal to 6.4fm. The total rms radius calculated from equation (3) is 3.3fm for ^{11}Li and is in good agreement with the experimental value of Tanihata et al.[1)].

The broad spatial extension of the neutron halo via Heisenberg's uncertainty principle implies a narrow distribution in momentum. Kobayashi et al.[5)] have measured the transverse momentum distribution of ^{9}Li recoils from fragmentation of ^{11}Li on a carbon target at relativistic energies. The distribution obtained was interpreted as a superposition of a broad component coming from reaction with the core, and a narrow component associated with a breakup of the neutron halo. A similar effect should be observed in this reaction for the angular distribution of neutrons.

The dineutron model also predicts a large cross-section for the Coulomb dissociation of ^{11}Li in collisions with high-Z target nuclei. This effect was observed[6)] at 800MeV/u where, for a lead target, (15-30)% of the total cross-section is electrical. At 30MeV/u the cross-section is expected to be several times higher.

EXPERIMENT AND RESULTS

In a recent experiment performed at GANIL[7)] together with the total reaction cross section we also measured the angular distribution of neutrons emitted in reactions induced by 32.4MeV/u ^{11}Li, 31.6MeV/u ^{9}Li, 27.4MeV/u ^{8}He and 35.4 MeV/u ^{14}Be secondary beams on Be, Ni and Au targets. The secondary beams were produced in a reaction of a 55 MeV/u ^{18}O primary beam bombarding a 543mg/cm^2 Be production target. The outgoing fragments were separated by means of the LISE spectrometer[8)] and identified by their energy loss and time of flight. The reaction cross-sections were measured by a silicon telescope placed after a secondary target. Fast neutrons were detected in an array of 23 liquid scintillator counters.

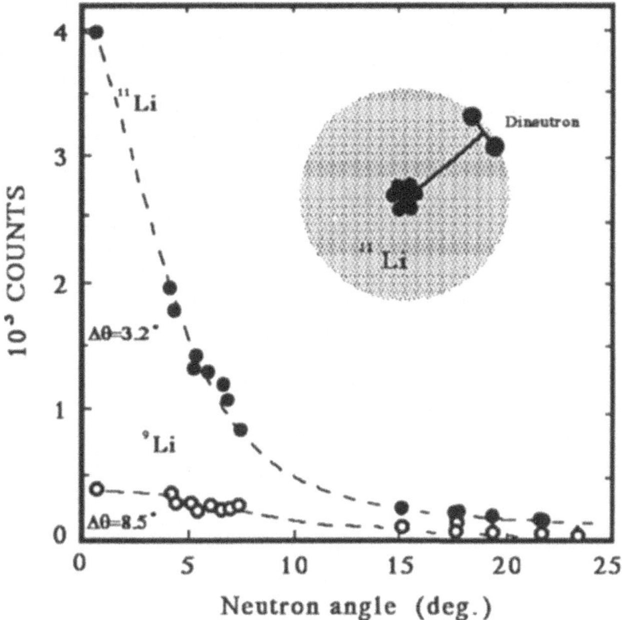

Fig. 1. The inclusive angular distribution of neutrons from secondary beams of ^{9}Li and ^{11}Li.

The raw data for the angular distribution of neutrons normalized to the same number of incident ions (for ^{11}Li and ^{9}Li) is shown in figure 1. Two features of these distributions are evident. The distribution for ^{11}Li is much narrower than that for ^{9}Li though at the same time the probability for emission of neutrons is much higher for the beam of ^{11}Li.

Another striking result comes from the angular distributions of neutrons measured for the reaction ^{11}Li→^{9}Li+2n on three different targets, presented in figure 2. All distributions are strongly forward-peaked (with a half angle of about 3°) and are very similar, suggesting that they reflect predominantly the distribution of the ground state of the halo. As can be seen from figure 2 the data are well described by a Lorenzian distribution which is obtained as the square of the Fourier transform of the wave function (1) transformed to the lab system. The assumed $\rho_{neutron}$ parameter of 12fm does not contradict the value of 8fm used for ρ since the former value includes a contribution from the size of the dineutron.

Table 1 summarizes measured reaction cross-sections and widths of the angular distributions of neutrons. The very large cross-sections obtained for ^{11}Li must arise essentially from peripheral reactions. The strong argument for this comes from the difference in the neutron-removal cross-sections presented in figure 1.

The cross sections measured for Ni and Au targets by the semiconductor telescope and by integration of the neutron distributions agree very well with each other if one assumes that for ^{11}Li+target→^{9}Li+xn+Y reaction channel, the mean multiplicity of neutrons is 2. This suggests the Coulomb dissociation as a dominant reaction mechanism. For the Be target it

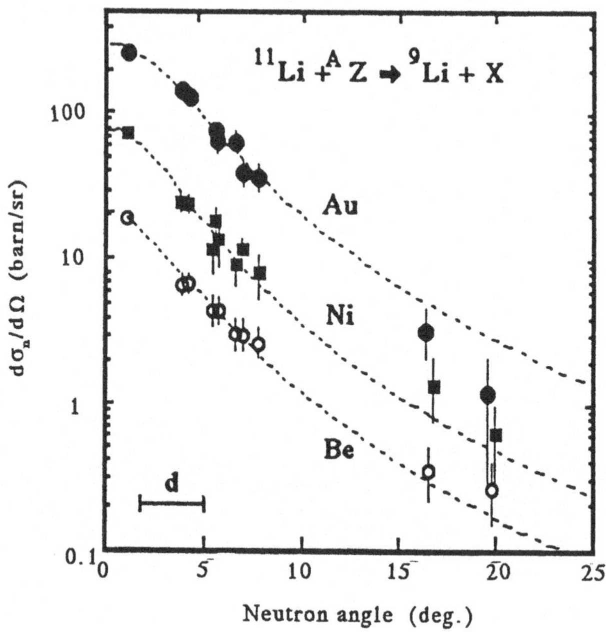

Fig 2. Differential cross-sections for observation of neutrons for a ^{11}Li beam on Be,Ni and Au targets.

Table 1. Cross-sections, widths of neutron distributions.

Beam	Observed fragment	Cross-section (barn) Be	Ni	Au	$\theta_{1/2}$ (°)[a] Be	Ni	Au
^{8}He	$^{6}He+n^{b}$	0.41±0.15	$1.5^{+1.0}_{-0.5}$	$2.9^{+5}_{-2.5}$	6.2±1.2	7.1±2.0	7±5
^{11}Li	$^{9}Li+n^{b,c}$	0.65±0.12	2.1±0.4	10.8±1.9	3.17±0.24	2.88±0.25	3.35±0.16
	$^{9}Li^{c}$	0.47±0.10	1.3±0.4	5.0±0.8			
	He^{d}	0.14±0.02	0.17±0.03	0.34±0.07			
	$^{4}He^{b}$				6.0±0.5		
^{14}Be	$^{12}Be+n^{b}$	0.37±0.08	$0.60^{+0.30}_{-0.21}$	$1.3^{+1.3}_{-0.6}$	4.8±0.5	6.4±1.8	6.6±2.3
	^{12}Be	0.44±0.10	1.0±0.3	2.1±0.6			
	Li^{d}	0.08±0.02	0.15±0.04	0.14±0.04			
	He^{d}	0.14±0.01	0.19±0.02	0.24±0.04			
^{9}Li	n^{e}					8.5±0.9[f]	

[a] Observed widths, see ref.9 for corrections.
[b] Calculated from the neutron distribution.
[c] From ref.7 .
[d] Sum over all isotopes.
[e] From the single neutron distributions.
[f] Average over all targets.

seems that absorption and diffraction dissociation play the most important role in yielding a neutron multiplicity of less than 2.

The data for ^{8}He and ^{14}Be (see also ref. 9) give some indications of the existence of the same reaction mechanisms as for reactions with ^{11}Li, but the effect is much less pronounced.

On the basis of the present results and many other measurements it seems that, at least in ^{11}Li, the existence of the neutron halo is well established. Of course, many other intriguing questions arise at the same time. What is the n-n momentum distribution in the dineutron ? Are there any bound excited states in ^{11}Li ? If yes, their large polarizability suggests that they will have fast E1 transitions of essentially single-particle strength. A measurement of an elastic scattering of loosely-bound nuclei on different targets should shed some light on the details of distribution of nuclear matter. As pointed out in ref.4, a precise determination of the mass of ^{11}Li is particulary important for setting a severe constraint on theoretical models. These and other theoretical and experimental problems should soon find clear answers.

REFERENCES

[1] I.Tanihata et al., Phys. Rev. Lett. 55, 2676 (1985)
[2] R.Neugart contribution to this conference
[3] C.Thibault et al., Phys. Rev. C12, 644 (1975)
J.M.Wouters et al., Z. Phys. A331, 229 (1988)
[4] P.G.Hansen and B.Jonson, Europhys. Lett. 4, 409 (1987)
[5] T.Kobayashi et al., Phys. Rev. Lett. 60, 2599 (1988)
[6] T.Kobayashi et al., Phys. Lett. B232, 51 (1989)
[7] R.Anne et al., Phys. Lett. B250, 19 (1990)
[8] R.Anne et al., Nucl. Instr. Meth. A257, 215 (1987)
[9] K.Riisager et al., Intern. Sympo. on Structure and Reactions of Unstable Nuclei, June17-19, 1991, Niigata, Japan, contributions

SEARCH FOR NEW RADIOACTIVITIES AT THE PROTON DRIP-LINE

F. Pougheon, V. Borrel, D. Guillemaud-Mueller, and A.C. Mueller

Institut de Physique Nucléaire, 91406 Orsay Cedex, France

R. Anne, C. Détraz, and M. Lewitowicz

GANIL, BP 5027, 14021 Caen Cedex France

D. Bazin, J.P. Dufour, A. Fleury, F. Hubert, and M.S. Pravikoff

CEN Bordeaux, Le Haut Vigneau, 33170 Gradignan, France

INTRODUCTION

A major achievement in the study of nuclei far from stability has been the observation of new radioactive modes [1]. Besides broadening the scope of reported nuclear processes, new radioactive modes provide specific structure information. Among them, direct emission of more than one nucleon carries unaltered informations on nuclear correlations. The simplest case is the long-predicted and not yet observed : direct two-proton radioactivity. At the proton drip-line, it might happen that a nucleus is bound by a one-proton emission but is able to decay by the direct emission of two protons. This process, discussed long time ago by Goldanski [2] and Jänecke [3] is governed by barrier penetration.

Favourable experimental conditions are available at GANIL. Some light nuclei which can be produced from GANIL projectiles are good candidates to exhibit new radioactivities. At the drip-line, isotopes like ^{22}Si, ^{31}Ar, ^{39}Ti, ^{42}Cr, ^{45}Fe and ^{48}Ni have already been observed as fragmentation products [4,5,6] or appear to be within reach of dedicated searches.

For a complete description and discussion of these experiments see ref. 4, 5, 6 and 7.

EXPERIMENTAL METHOD

The experimental method used at GANIL to produce, separate and identify nuclei far from stability is described in several papers (see e.g. 5 and 8). Basically, it uses the kinematic forward focusing of the products of fragmentation reaction to select those of interest through a magnetic separation. This is accomplished by the LISE spectrometer operated as an isotope separator [8]. Very recently, LISE

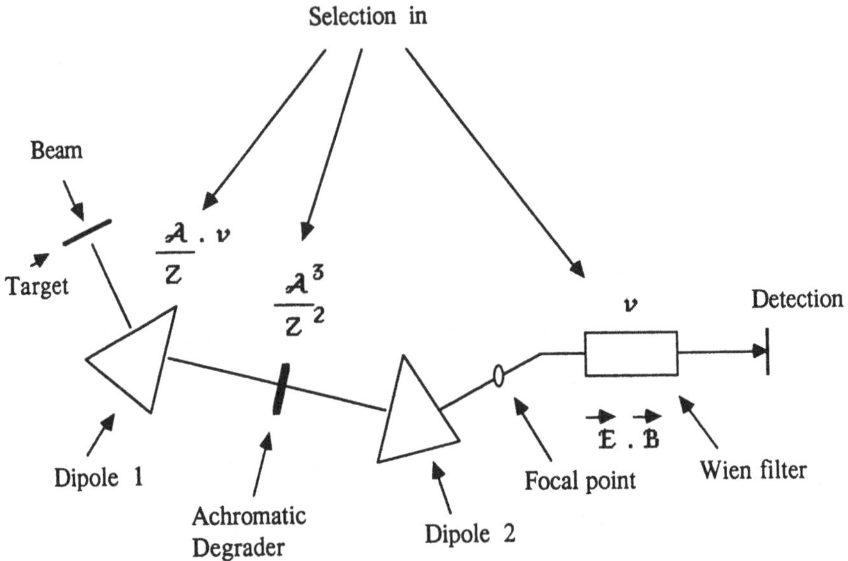

Fig. 1 Scheme of the LISE achromatic spectrometer

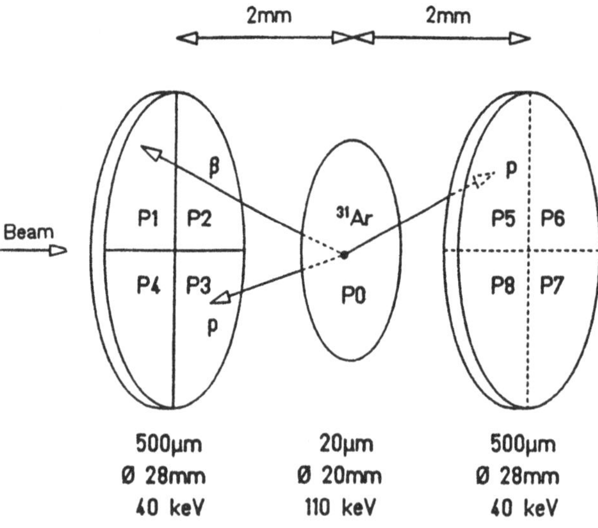

Fig. 2 Lay out of the silicon detector telescope. The numbers indicated below each detector are respectively their thickness, diameter and energy resolution

was upgraded, by adding a Wien filter, providing an additional selection [9] (fig.1).

After slowing down in an aluminium foil, the transmitted nuclei are identified by a ΔE versus time of flight (Tof) technique and stopped in a silicon detector telescope located at the final focal point of LISE. For a part of the ^{31}Ar decay study [7], a special telescope has been built (fig. 2). The ^{31}Ar nuclei were implanted into a thin central detector Po (20 μm). The two lateral thick detectors were segmented in four quarters. This special geometry provides at the same time good angular and particle-multiplicity sensitivities and is well adapted to the search for $\beta - 2p$ and $\beta - 3p$ radioactivity.

On line, two single-channel analysers examine the ΔE and Tof pulses. Each time that an isotope of interest is identified, the beam is interrupted for a time corresponding to four or five expected half-lives. During this time, the heavy-ion data-acquisition system is turned off, while a second data acquisition system records the charged particles, storing their time of arrival and the energy deposited in each of the detectors.

RESULTS

The decay of ^{31}Ar

Three different experiments were performed for the study of the ^{31}Ar decay. In the first one [5], several proton peaks were observed and attributed to the beta-delayed proton decay of this nucleus. The measured half-life of 15 $\pm$ 3 ms showed that ^{31}Ar does not exhibit an important branching ratio for direct two-proton decay.

The Q_β of ^{31}Ar amounts to almost 19 MeV and allows the feeding of the Isobaric Analog State (IAS) in the daughter nucleus ^{31}Cl with a relatively high probability. Then, the particle decay of this IAS is energetically possible through many ways (fig. 3) : $1p, 2p, 3p, \alpha$ and $p\alpha$ emission. Among them some of these processes are new and it is therefore interesting to know whether they occur or not.

Figure 4 displays the energy spectrum of the charged particles associated with the ^{31}Ar decay. The peak at 7580 $\pm$ 70 keV is attributed to the two-proton decay of the IAS in ^{31}Cl. Its two-proton nature can be established by :

- its width of 270 keV (FWHM) (for a beta-delayed two proton emission without angular correlation between two protons, a wide distribution for the ion recoil is obtained).

- The number of events which triggered two detectors at the same time.

Once the particle multiplicity of this peak clearly established, it is interesting to know whether the two protons are correlated or not. In the case of a $\beta -{}^2He$ process, the final interaction between the protons and the total energy of the pair, lead to a mean relative angle of about 30°. But the experimental points are fully compatible with an isotropic relative angle between the protons[7] ; thus it can be concluded to the non-observation of $\beta -{}^2He$ process.

The first hint of a $\beta - 3p$ decay mode arises from the observation of a peak at 4870 keV (fig. 4). This energy is fully compatible with the calculated one from the IAS decay. The deduced branching ratio for this decay is 2.5 $\pm$ 0.2 % compared to an estimation for this process of 2.1 $\pm$ 1 %. Furthermore, Monte-Carlo

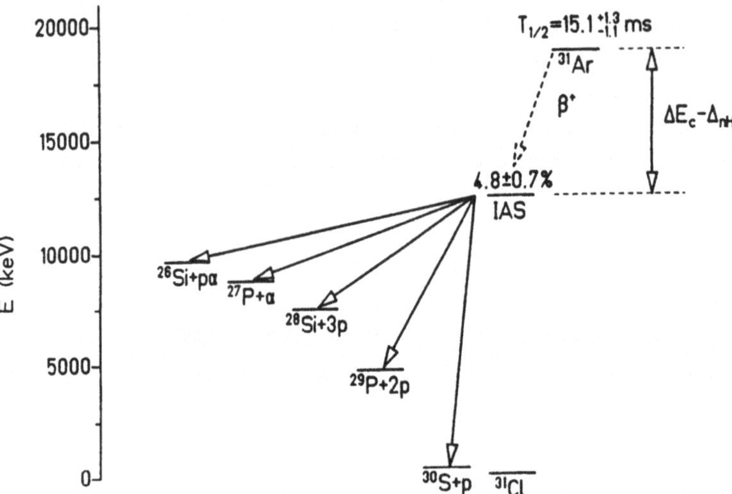

Fig. 3 Partial decay scheme of ^{31}Ar showing the different possible particle decay modes from the β feeding of the ^{31}Cl Isobaric Analog state

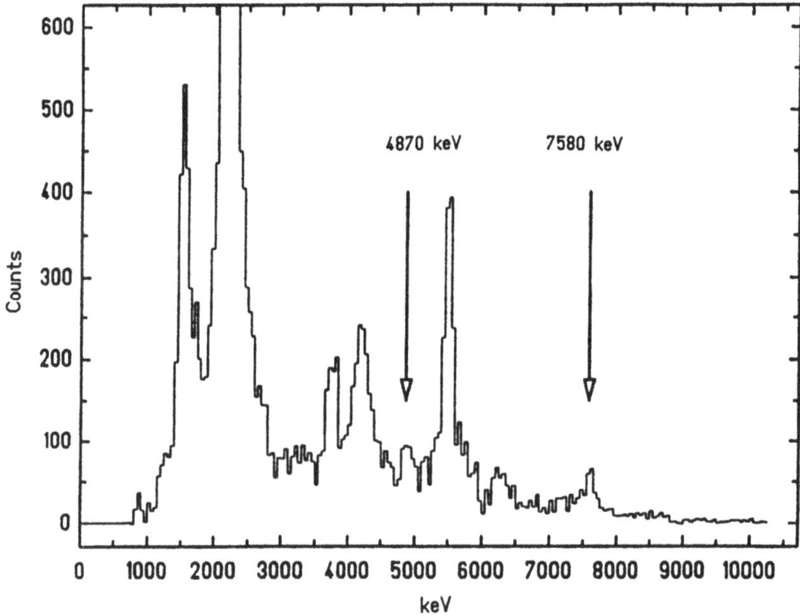

Fig. 4 Energy spectrum of the protons associated with the ^{31}Ar and with the multiplicity M=1. M=1 means Po and one of the quarter touched. Both observed two-proton and three-proton peaks are indicated by the arrows

simulations of the detector response are fully compatible with the experimental results [7]. Therefore there are strong evidences to assign this peak to the new radioactive mode : the β-delayed three protons emission.

The search of others possible decay modes $\beta - \alpha$ and $\beta - p\alpha$ led to conclude to the non observation of them.

The decay of ^{39}Ti

Due to its predicted S_{2p} of about $\sim$ -750 keV, the ^{39}Ti is a better candidate for the direct 2p emission than ^{31}Ar. About 75 counts, identified as ^{39}Ti nuclei, were collected during a beam time of 24 hours [6]. The mere observation of these counts shows that the half-life of this nucleus is at least of the order of the time of flight from the target of production to the detectors i.e. about 200 ns.

This a-priori leaves open the question whether it is a direct two-proton emitter. A half-life shorter than the expected β-decay half-life, a few 10^{-3} sec, would indicate a dominant hadronic 2p decay. But independant experimental proofs were obtained which rule out a significant 2p direct emission for ^{39}Ti :

- The time spectrum of charged particles correlated with a ^{39}Ti implantation leads to a half-life of 26 $\pm$ 7 ms close to the predicted value of 36 ms given by the β Gross Theory [10]. This provides an argument against a sizeable alternative decay channel.

- Another proof comes from the energy spectrum of the charged particles where only one peak at 3.60 $\pm$ 0.12 MeV is identified. Beta-delayed protons from ^{37}Ca produced as a daughter nucleus would also be observed in the case of direct two-proton emission and no counts are observed at the expected energy.

DISCUSSION

The negative experimental result obtained in the search for direct two-proton emission from ^{39}Ti was not expected since the predicted available energy for this process, $\sim$ 750 keV, appeared to correspond to a half-life short enough to favourably compete with β-decay and long enough to allow for the time of flight from target to detectors. An analysis of the parameters which govern direct 2p emission is presented in ref. 6 and will be just summarized here. A barrier penetrability calculation applied to an ^{2}He emitted particle has been used to obtain an order of magnitude of $T_{1/2}$. The main conclusion that can be drawn is the extreme sensitivity of the half-live value to the available two-proton energy. Therefore reliable prediction of the nuclear masses of interest is essential. A recent compilation of mass values at the proton drip-line [11] provides evidence for a deviation from predictions, when nuclei are one-proton or two-proton unbound. What happens at the proton drip-line, is that a single proton state becomes unbound which modifies its wave function from what it was when the same quantum state was bound. This effect, referred to as the Thomas-Ehrmann effect [12] leads to a stronger binding energy for that unbound individual particle state. The corrected S_{2p} value of ^{39}Ti is -570 keV, and is close enough to the threshold for experimental observation to provide a plausible explanation that ^{39}Ti preferably decays by β emission.

The next step would thus consist to look for $2p$ unbound nuclei with Q_{2p} values of 1 MeV or more in the same mass region which would simultaneously remain bound against one-proton decay. Two such cases have been identified : ^{45}Fe and ^{48}Ni.

In a very recent experiment, we have tried to produce ^{45}Fe ; only one count was observed during beam time of 48 hours. A very careful analysis of the data is necessary to be sure of this point.

REFERENCES

1. For a review see C. Détraz and D.J. Vieira, Ann. Rev. Nucl. Sc. 39 407 (1989) and references therein.
2. V.I. Goldanski, Nucl. Phys. 27 648 (1961).
   Sov. Phys. Usp. 8 770 (1966).
3. J. Jänecke, Nucl. Phys. 61 326 (1964).
4. M.G. Saint-Laurent et al, Phys. Rev. Lett. 59 33 (1987).
5. V. Borrel et al, Nucl. Phys. A473 331 (1987) and to be published in Nucl. Phys. 1991.
6. C. Détraz et al, Nucl. Phys. A519 529 (1990).
7. D. Bazin et al, submitted to Phys. Rev. C and Obernai Workshop edited by G. Klotz p. 21, nov. 1989.
8. R. Anne et al, Nucl. Inst. Meth. A257 215 (1987).
   J.P. Dufour et al, Nucl. Inst. Meth. A 248 267 (1986).
9. A.C. Mueller and R. Anne, Proceedings of the Int. Conference on the application of accelerators in research and industry Denton (1990), Nucl. Inst. Meth. (in press).
10. T. Tachibana et al, Proc. 5th Int. Conf. on Nuclei far from stability Rosseau Lake, Canada, sept. 87, p. 614.
11. E. Comay, I. Kelson and A. Zidon Phys. Lett. B 210 31 (1988).
12. R.G. Thomas, Phys. Rev. 80 136 (1950) ; 88 1109 (1952)
    J.B. Ehrmann, Phys. Rev. 81 412 (1951).

INTERNAL CONVERSION ANOMALY AROUND A = 187

B. Roussiere,[1] P. Kilcher,[1] F. Le Blanc,[1] J. Oms,[1] J. Sauvage,[1]
H. Dautet [2] and the ISOCELE Collaboration

[1]Institut de Physique Nucléaire
91406 Orsay Cedex, France
[2]Foster Radiation Laboratory
McGill University
Montréal, Quebec, Canada H3A 2B2

Abstract: γ-γ angular correlation measurements have been performed in ^{187}Pt in order to determine the spin value of the states connected by highly converted transitions independently of their multipolarity. For one of these transitions, we have determined that the spin values of the initial and final states are different, which implies abnormal M1 multipolarity for this highly converted transition.

INTRODUCTION

The region of Hg-Au-Pt nuclei is well known to exhibit shape coexistences [1]). Another important feature is the existence of highly converted transitions ($\alpha_K^{exp} > \alpha_K^{th}$(M1)) in odd nuclei of this region [2]). The large value of the conversion coefficient can be due to an anomaly in the M1 conversion or to an E0 component in the transition [3,4]). The nature (large E0 admixture or abnormal M1) of these transitions is a very debated question. For instance, we had pointed out that two of these highly converted transitions in ^{185}Au have abnormal M1 multipolarity [5]). Indeed, they deexcite the 330 keV level to the $5/2^-$ ground state and $9/2^-$ first excited state. Thus the 330 keV level must be assigned $7/2^-$, which rules out E0 contribution in these two transitions. This result was disputed by the UNISOR group [6]): they claimed that these two lines have multiplet structure, that the transitions involved in the 330 keV level decay have M1+E2 multipolarity and that the other transitions forming part of the multiplets are located elsewhere in the level scheme and have an E0 or E0+M1+E2 multipolarity.

To put an end to this contention, we have undertaken γ-γ angular correlation measurements with the aim of assigning spin value to the excited states irrespective of the multipolarity of the highly converted transitions. Thus if the spin values of states connected by a highly converted transition are found to be different, the multipolarity of this transition will not be allowed to be E0 since an E0 transition implies the same spin value for the initial and final states. We have performed these γ-γ angular correlation measurements in ^{187}Pt. Indeed ^{187}Pt is the nucleus of the Hg-Au-Pt region where the conversion anomaly is the strongest. Moreover, we quite agree with the UNISOR group on the location of the four highly converted transitions in the level scheme of ^{187}Pt [refs.[7,8])].

Nuclear Shapes and Nuclear Structure at Low Excitation Energies
Edited by M. Vergnes *et al.*, Plenum Press, New York, 1992

EXPERIMENTAL PROCEDURE

The ^{187}Pt excited levels were populated by the β^+/EC decay of ^{187}Au. Gold atoms were produced in the Pt(p(200 MeV),xn)Au reactions: a molten Pt-B alloy target, placed inside the ion source of the ISOCELE separator [9]), was irradiated by the I_{max}=2.5 μA proton beam delivered by the Orsay synchrocyclotron. The radioactive ions extracted from the ion source were then mass-separated and guided through a beam line to the experimental setup. Mass-separated radioactive sources were collected on a mylar/aluminum tape and moved to the counting station which consists of four hyperpure n-type Ge detectors (efficiency ~18%). The detectors were placed at 13 cm distance from the source, which makes the solid angle attenuation negligible. They were shielded by copper and lead in order to minimize scattering effects. The angles between the first detector and the three others were 50°, 173° and −100°. Thus, taking into account the symmetry properties of the correlation function, the following six angles are obtained for the different pairs of detectors: 7°, 30°, 50°, 57°, 80° and 87°. The data were accumulated for a week: for each angle, about 45 10^6 γ-γ coincidence events were obtained.

EXPERIMENTAL RESULTS

Off-line analysis of the data gives the number of counts $N_{ij}(E_1,E_2)$ for detector pair ij and energy gates around E_1 in detector i and E_2 in detector j. After correcting $N_{ij}(E_1,E_2)$ for the relative efficiency of detector i (and j) at energy E_1 (and E_2) and subtracting the accidental coincidences, we obtain the relative intensity $I(\theta)$ of the E_1-E_2 cascade at the angle θ. The intensities $I(\theta)$ were fitted by means of the least-squares method to the angular correlation function $W(\theta)=1+A_2P_2(\cos\theta)+A_4P_4(\cos\theta)$, $P_2(\cos\theta)$ and $P_4(\cos\theta)$ being the Legendre polynomials and A_2 and A_4 the γ-γ directional correlation coefficients. The theoretical A_2 and A_4 coefficients depend on the spin I_1, I_2 and I_3 of the states connected by the E_1-E_2 cascade and on the mixing ratios δ_1 and δ_2 of the two multipole components of the E_1 and E_2 transitions. In order to make spin assignments, the experimental A_2 and A_4 coefficients have been compared to the theoretical values calculated from the table of Hager and Seltzer [10]).

It is worth noting that the analysis of angular correlation functions in odd nuclei is very complex because, in these odd nuclei, few transitions have pure multipolarity. In many cases, δ_1 and δ_2 have to be taken into account in the A_2 and A_4 coefficient calculations. This increases the number of parameters and consequently the number of theoretical solutions presenting an overlap with the A_2 and A_4 experimental coefficients. However, in the case of ^{187}Pt, we could get round this difficulty because (i) the ^{187}Pt level scheme is well known [7,8,11]) and electron conversion measurements have already been performed in particular by means of a semi-circular magnetic spectrograph [11]), which provides us with the δ^2 value of the strong transitions, (ii) the 25.6 keV level is unambiguously $5/2^-$ [refs.[7,11]], (iii) the state located at 1340.9 keV, the most populated by the β^+/EC decay of ^{187}Au, is a positive parity state and decays by many E1 transitions to negative parity levels, which implies easier analysis (only one δ value) for the correlation functions involving this state. 35 correlation functions have been analysed: not only do the deduced spin values reproduce the experimental correlation functions, but they also agree with the whole decay scheme and in particular with the δ^2 values known from conversion electron measurements.

We present below only partial results: the spin assignment to the 1340.9 keV state and to the 507.8 and 9.3 keV levels which are connected by the 498.8 keV highly converted transition.

Spin value of the 1340.9 keV level

The spin of this state was attributed $3/2^+$ by Gnade et al. [8]). But, from its decay properties, the values $1/2^+$, $3/2^+$, $5/2^+$ and $7/2^+$ are allowed, although the $5/2^+$ and $7/2^+$ values are unlikely because of the 1340.9 keV log ft value. Fig.1 shows the comparison

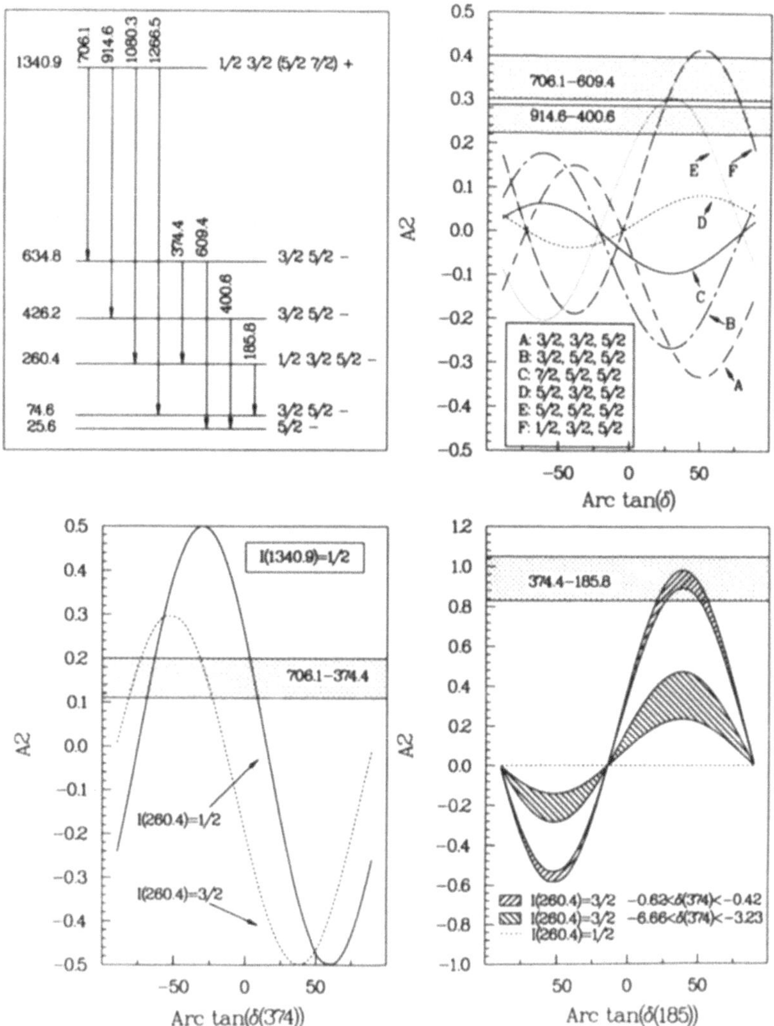

Figure 1. Comparison of the experimental and theoretical A_2 values for the 706.1-609.4, 914.6-400.6, 706.1-374.4 and 374.4-185.8 keV cascades.

between the experimental A_2 values obtained from the 706.1-609.4 keV and 914.6-400.6 keV cascades and the theoretical A_2 values calculated versus δ_2 (δ(609.4 keV) or δ(400.6 keV)) for all the possible spin sequences. The 3/2 and 7/2 spin value cannot be attributed to the 1340.9 keV level because there is no overlap between the experimental and theoretical values and the only possible spin sequences are $1/2 \rightarrow 3/2 \rightarrow 5/2$ and $5/2 \rightarrow 5/2 \rightarrow 5/2$. Let us assume now that the spin value of the 1340.9 keV level is 1/2: this implies that the 260.4 and 74.6 keV states cannot have 5/2 as spin value since they are linked by the 1080.3 and 1266.5 keV transitions to the 1340.9 keV level. Then from the overlap of the experimental and theoretical A_2 value for the 706.1-374.4 keV cascade (see fig.1), we can deduce the range of δ(374.4 keV) in the two cases I(260.4 keV)=1/2 or 3/2. Using these values of δ(374.4 keV), we can analyse the 374.4-185.8 keV correlation fonction and assign the 3/2 spin value to the 260.4 keV level, the [0.34,1.57] δ range to the 185.8 keV transition and the [−0.62,−0.42] δ range for the 374.4 keV transition. A similar analysis done by assigning the 5/2 spin value to the 1340.9 keV level fails to explain the high A_2 value of the 374.4-185.8 keV cascade. Thus the 1/2 spin value is firmly assigned to the 1340.9 keV level, which determines uniquely the spin value of all the levels displayed in fig.1.

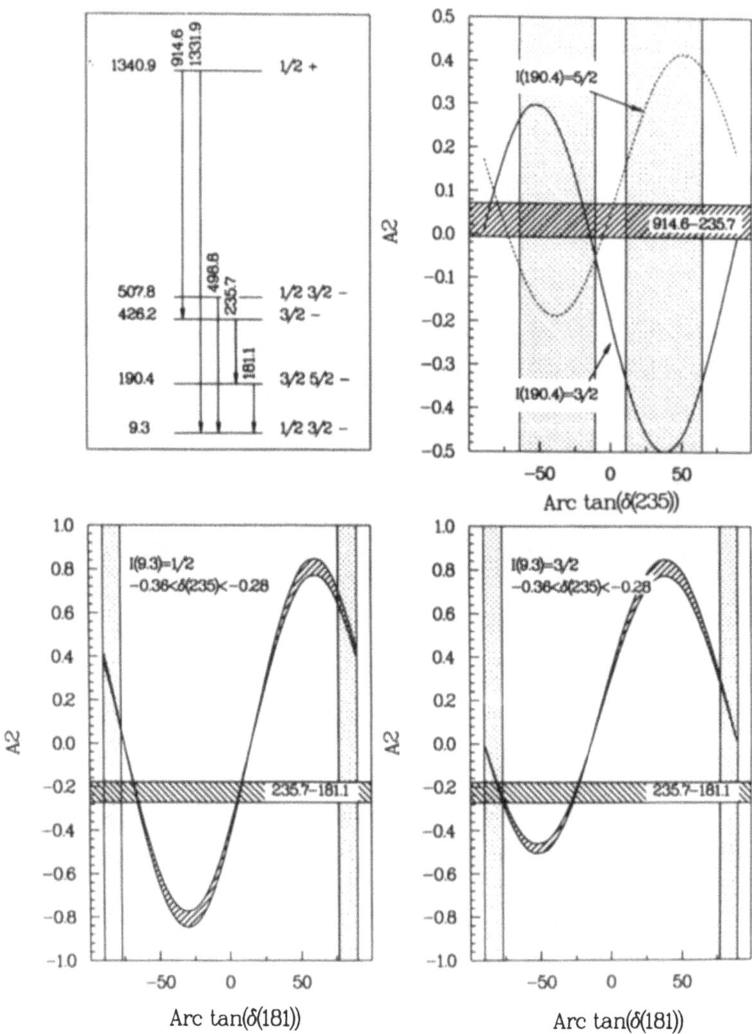

Figure 2. Comparison of the experimental and theoretical A_2 values for the 914.6-235.7 and 235.7-181.1 keV cascades.

Spin value of the 9.3 keV level

We have established above the spin of the 1340.9 and 426.2 keV levels, 1/2 and 3/2 respectively. If we consider the 914.6-235.7 keV cascade, we can see in fig.2 that the overlap between the experimental and theoretical A_2 values and the δ value of the 235.7 keV transition obtained from the conversion coefficient measurements implies that (i) the 190.4 keV level spin is 3/2, (ii) the δ value of the 235.7 keV transition lies in the $[-0.36,-0.28]$ range. Taking into account this δ range for the 235.7 keV transition, we can analyse the 235.7-181.1 keV correlation and deduce (since the δ^2 mixing ratio value of the 181.1 keV transition is well known) that the spin of the 9.3 keV level is 3/2 (see fig.2).

Spin value of the 507.8 keV level

The spin of the 507.8 keV level can be 1/2 or 3/2 since it is connected by the 833.0 keV E1 transition to the $1/2^+$ 1340.9 keV state. The simultaneous analysis of the 833.0-247.4 keV and 247.4-185.8 keV correlations allows us to rule out the 3/2 value for the 507.8 keV level. Indeed, if the spin of the 507.8 keV level were 3/2, the overlap between the theoretical and experimental A_2 values and the mixing ratio δ determined from the conversion

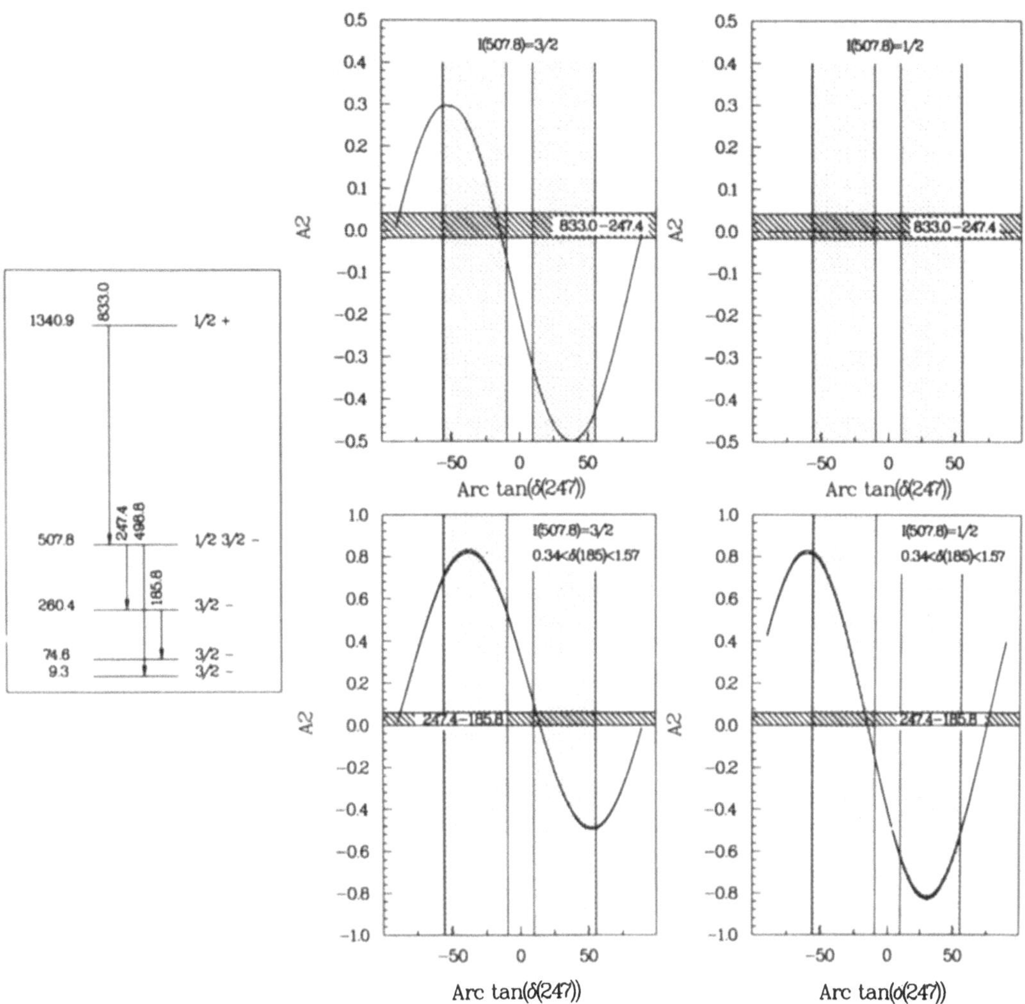

Figure 3. Comparison of the experimental and theoretical A_2 values for the 833.0-247.4 and 247.4-185.8 keV cascades.

coefficient of the 247.4 keV transition implies a negative value for δ(247.4 keV) from the 833.0-247.4 keV correlation and a positive one from the 247.4-185.8 keV correlation (see fig.3), which is impossible. On the contrary, assigning the 1/2 spin value to the 507.8 keV level, the 833.0-247.4 keV and 247.4-185.8 keV correlations are both consistent with the δ(247.4 keV) negative value (see fig.3).

Therefore the 498.8 keV highly converted transition, located between the $1/2^-$ 507.8 keV level and the $3/2^-$ 9.3 keV level is abnormal M1.

DISCUSSION

To summarize, the γ-γ angular correlation measurements performed in ^{187}Pt have allowed us to define the spin of most of the excited levels. In particular, the spin of the levels connected by three of the four highly converted transitions has been assigned uniquely: the 260.4 keV transition deexcites the $3/2^-$ 260.4 keV level to the $3/2^-$ ground state, the 262.6 keV transition connects the $5/2^-$ 288.2 keV level and the $5/2^-$ 25.6 keV state and the 498.8 keV transition lies between the $1/2^-$ 507.8 keV level and the $3/2^-$

9.3 keV state. Thus the 498.8 keV transition connecting states with different spin values cannot have an E0 component and consequently its nature has to be abnormal M1.

Conversion anomaly in odd nuclei seems to be a feature of the core. Indeed, the odd nuclei of the Hg-Au-Pt region in which highly converted transitions have been pointed out are generally next to even-even nuclei in which highly converted transitions are found with rather small energies [2]). The highly converted transitions in even-even nuclei have always been interpreted as transitions with E0 components. One can wonder whether M1 conversion anomaly can be ruled out in even-even nuclei as usually stated up to now, since we have pointed out clearly an abnormal M1 transition in ^{187}Pt. As far back as 1956, Church and Weneser have shown that M1 conversion anomaly could occur in the $2^+ \rightarrow 2^+$ transitions of heavy even-even nuclei [3]). Moreover, the results of e^--γ and γ-γ angular correlation measurements performed in ^{196}Pt indicate that an abnormal M1 component is possible in the $2^+ \rightarrow 2^+$ transition [12,13]). In order to point out abnormal M1 conversion in even-even nuclei next to ^{187}Pt, we have performed, for the 215.5 keV $2_2^+ \rightarrow 2_1^+$ transition in ^{186}Hg, precise measurements of conversion electrons in the K, L_I, L_{II} and L_{III} shells and γ-γ angular correlation measurements. The analysis is in progress.

REFERENCES

1 K. Heyde, P. Van Isacker, M. Waroquier, J.L. Wood and R.A. Meyer, Phys. Repts. 102 (1983) 291.
2 B. Roussière, C. Bourgeois, P. Kilcher, J. Sauvage, M.G. Porquet and A. Wojtasiewicz, Nucl. Phys. A485 (1988) 111 and refs. quoted therein.
3 E.L. Church and J. Weneser, Phys. Rev. 104 (1956) 1382.
4 R.J. Lombard, Phys. Lett. 9 (1964) 254.
5 C. Bourgeois, P. Kilcher, B. Roussière, J. Sauvage-Letessier and M.G. Porquet, Nucl. Phys. A386 (1982) 308.
6 C.D. Papanicolopulos, M.A. Grimm, J.L. Wood, E.F. Zganjar, M.O. Kortelahti, J.D. Cole and H.K. Carter, Z. Phys. A330 (1988) 371.
7 B. Roussière, thèse, Orsay (1986).
8 B.E. Gnade, R.W. Fink and J.L. Wood, Nucl. Phys. A406 (1983) 29.
9 P. Paris, C.F. Liang, D. Lecouturier, M. Arianer, J. Obert, A. Caruette, A. Ferro, J. Fournet, J.C. Putaux and J.L. Sarrouy, Nucl. Instr. Meth. 186 (1981) 91.
10 R.S Hager and E.C. Seltzer, Nucl. Data A4 (1968) 397.
11 A. Ben Braham, V. Berg, C. Bourgeois, P. Kilcher, J. Letessier, M.G. Desthuilliers-Porquet, C. Schuck, A. Huck, A Knipper, C. Richard-Serre and A. Hoglund, Nucl. Phys. A332 (1979) 397.
12 W.D. Hamilton, Nucl. Phys. A136 (1969) 257.
13 H.A. Doubt and W.D. Hamilton, Nucl. Phys. A177 (1971) 418.

EXPLORATION OF THE NEUTRON-RICH MASS SURFACE FROM ^{11}Li TO ^{66}Fe

D. J. Vieira[1], X. G. Zhou[2], X. L. Tu[2], J. M. Wouters[1], H. L. Seifert[1,3], K. E. G. Löbner[4], Z. Y. Zhou[1,5], V. G. Lind[2], and G. W. Butler[1]

[1]Isotope and Nuclear Chemistry Division, Los Alamos National Laboratory Los Alamos, NM 87545, USA
[2]Department of Physics, Utah State University, Logan, UT, 84322 USA
[3]II. Physikalisches Institut, Universität Giessen, D-6300 Giessen, FRG
[4]Sektion Physik, Universität München, D-8046 Garching, FRG
[5]Department of Physics, Nanjing University, Nanjing, PRC

INTRODUCTION

A small revolution in our ability to perform mass measurements of light-mass, neutron-rich nuclei has occurred during the last five years with the extension of the direct, total mass measurement method to fast recoiling nuclei. To date the work of two groups, one using the SPEG spectrometer at GANIL and the other using the TOFI spectrometer at LAMPF, has yielded over 85 mass measurements of which 62 had not been reported previously. Extending from the neutron-halo nucleus ^{11}Li up to ^{66}Fe, a region relevant to the astrophysical r-process, these measurements have provided a valuable first glimpse into the interesting nuclear structure present in many of these exotic nuclei.

Herein we highlight the nuclear structure insight which has been gained from these measurements, especially that learned from a comparison to recent shell model calculations. Attention is given to: (1) the binding of loosely-bound neutron halo nuclei; (2) the N=14-16 region in the neutron-rich isotopes of O, F, and Ne where the strong two-body interaction plays an important role; (3) the deformed intruder state region around ^{31}Na of long standing interest; (4) the neutron-rich isotopes of sulfur and chlorine; and (5) the question of the isospin dependence (or independence) of neutron and proton pairing energies in the fp shell. Only the briefest account of this work can be given here; emphasis is place on the most recent results.

DIRECT MASS MEASUREMENTS OF FAST RECOILS

As mentioned above two separate groups have pioneered slightly different approaches for determining the total mass of fast recoiling nuclei. In the mass measurements using the Energy-Loss Spectrometer SPEG[1-3], a combined two-parameter determination of velocity and magnetic rigidity is carried out, while at TOFI[4-7] a single-parameter measurement of the ion's time-of-flight through the Time-Of-Flight Isochronous spectrometer serves to determine the mass-to-charge ratio of the ion. Z and Q identifications were obtained in both approaches from additional measurements of stopping power, velocity, total kinetic energy, and/or range.

Nuclear Shapes and Nuclear Structure at Low Excitation Energies
Edited by M. Vergnes *et al.*, Plenum Press, New York, 1992

The SPEG experiments rely on the concentration of projectile fragmentation products at forward angles using 30–100 MeV/u heavy ion beams, while the TOFI measurements utilize fast (1-4 MeV/u) recoils produced in 800-MeV proton-induced target fragmentation and fission reactions. Given that a large variety of recoils are produced in these reactions and that both methods have reasonably large acceptances and are fast (with flight times typically of 1 μs or less), both groups have been able to make a wide range of systematic measurements which extend far from the valley of β-stability. Typical mass resolutions for both approaches are on the order of 3 x 10^{-4} with mass measurement accuracies ranging from 70 keV to 1.6 MeV depending on counting statistics. For the most part, good agreement between these two fast recoil methods has been found (see Fig. 1 caption for noteworthy exceptions).

A convenient way of displaying the nuclear mass surface without the complication of the odd-even staggering due to neutron pairing is to plot the masses in terms of two-neutron separation energies (S_{2n}) versus neutron number. Several interesting nuclear structure features can be gleamed from such a plot (see Fig. 1 and the discussion that follows).

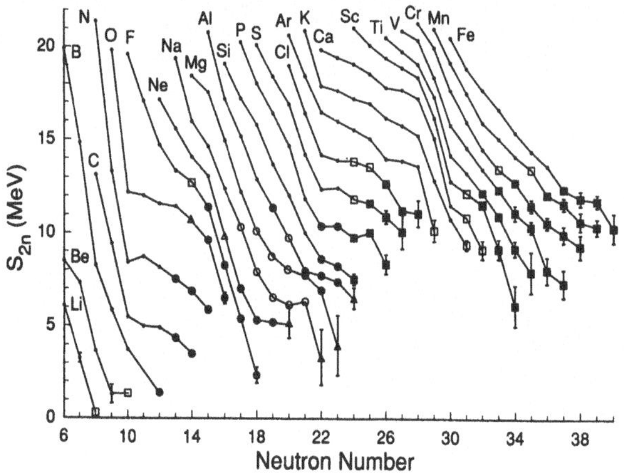

Fig. 1. Two-neutron separation energy versus neutron number for the neutron-rich isotopes of lithium to iron. Small data points (●) represent isotopes with well-known masses[8]. Open symbols indicate those isotopes for which masses have been remeasured and the solid symbols are given for those nuclei where the first mass determination was reported by either the SPEG or TOFI groups[1–7]. ● = SPEG + TOFI overlapping measurements; ▲ = SPEG measurements and; ■ = TOFI measurements. A weighted average of all reported masses is plotted with the exclusion of ^{27}Ne and ^{30}Na from Ref.[7], ^{38}P from Ref.[2], $^{31-34}$Na from Ref.[9,10], and ^{31}Mg from Ref.[11]. Error bars are given where they exceed the symbol size.

Z=3–15 NEUTRON-RICH NUCLEI

Of current excitement are those nuclei which are very loosely two-neutron (or single neutron) bound, such as ^{11}Li, (^{11}Be), ^{14}Be, ^{17}B, where enhanced interaction (both total and electromagnetic dissociation) cross sections have been observed. These results are being interpreted in terms of an increased neutron radius, i.e., "a neutron halo", the extension of which is directly related to how weakly bound the last neutrons are. In the case of ^{11}Li, for example, the rms radius of the last two neutrons is calculated[12,13] to be anywhere from 5 to 12 fm (i.e., roughly 2 to 5 times the size of the proton rms radius) depending on which model is used. To date three mass measurements[9,6,14] have been performed for ^{11}Li yielding S_{2n} values of 190 (110), 320 (120) and 340 (50) keV, respectively. The resulting weighted average is 315 (43) keV. Given that the size of the neutron halo is extremely sensitive to the S_{2n} value, additional higher precision measurements are needed to further constrain and test the validity of these models.

Other neutron–halo candidates worthy of attention are those of ^{19}B and ^{22}C which have been observed to be particle stable, but for which no mass determinations have been possible as yet due to their small production cross sections (roughly two orders of magnitude smaller than ^{17}B and ^{20}C). Using the latest SPEG and TOFI masses, the transverse Garvey–Kelson mass relationship[15] predicts S_{2n} values for ^{19}B and ^{22}C to be 360 and 110 keV, respectively. As such these nuclei should have neutron halos which are similar, if not larger, than that of ^{11}Li. Intensity and acceptance upgrades now being planned or built at several research facilities are likely to make the first measurement of such rare species possible.

Returning to Fig. 1, one prominent feature evident in this plot is the rapid S_{2n} falloff that occurs after N=8 which is then followed by a much slower decrease in S_{2n} values after N=10. This behavior is characteristic of a shell closure, in this case the completion of the p shell. Similar features, although less dramatic and not fully delineated here due to limitations in the vertical axis (i.e., several neutron–deficient isotopes have been excluded), are observed for the sd shell closure at N=20 and the completion of the $0f_{7/2}$ subshell at N=28. Analogous to the latter case, one might expect to see a sudden transition in the S_{2n} trend occurring at N=14 due to the completion of the $0d_{5/2}$ subshell. However, as is evident in the neutron–rich isotopes of O, F, and Ne, this transition occurs at N=15, not N=14. The explanation[6] of this effect came out of a detailed comparison to the shell model calculations of Wildenthal et al.[16] whose predictions, in contrast to most other mass models, were found to reproduce the observed S_{2n} trend in this region extremely well. In these calculations the single–particle energy spacing between the $0d_{5/2}$ and $1s_{1/2}$ levels is fairly small (~0.9 MeV), so the strong S_{2n} decrease observed in going from N=15 and N=16 results primarily from the interplay of two–body interactions, in this case the strongly attractive $[0d_{5/2}\text{–}0d_{5/2}]_{J=0}$ (~3 MeV) interaction is dominant. At N=14 and N=15 the contribution made by this $0d_{5/2}$–$0d_{5/2}$ interaction to the calculated S_{2n} value is sizable, while at N=16 its effect is minimal. In a strong sense this work has provided a challenging test of the shell model and verified the two–body interaction energies relevant to this region.

Moving up to the N=20 region one comes to the ^{31}Na region where enhanced binding energies relative to systematics and the predictions of most models has been noted for some time[9]. The SPEG and TOFI groups have now remeasured the masses of several Na isotopes and provided a number of new mass measurements for the adjoining Ne, Mg, and Al isotopes (see Fig. 1). The picture is now clear, given the good agreement between the SPEG and TOFI results, that the original[9] and subsequent[10] measurements of the Orsay group yielded masses which where consistently too bound with increasing deviations observed with increasing mass number, i.e., ^{31}Na (2.0 and 0.8 MeV, respectively), ^{32}Na (2.0 and 1.9 MeV), and ^{33}Na (– and 4.0 MeV). Although the reported Orsay error bars are large (0.6 to 1.1 MeV), most of these deviations fall between 2 and 3 standard deviations.

After discussions[17] with the Orsay group, it appears that there where many possible sources of error in this early work, such as interference due to hydrocarbons, a bad divider network, and sagging high voltage during the accelerator beam pulses. Determining the true source of these errors now appears difficult and for the most part academic. However, the suggestion[4] that these errors may have arisen from boot–strapping errors, analogous to those found[18] in the neutron–deficient Cs isotopes, is unfounded given that the different isotopic combinations used in the Na triplet (or double doublet) measurements never involved more than one unknown.

As can be seen in Fig. 1, an up-to-date view of the $^{31-32}$Na region no longer shows a dramatic S_{2n} upturn only a small increase, while the adjacent isotopes of Ne and Mg exhibit a slowly decreasing S_{2n} trend. However, evidence of enhanced binding in these nuclei is still indicated. These enhancements can be seen better in Fig. 2 where we contrast the experimental masses to recent shell model calculations. Concentrating on the solid line, i.e., the comparison to the normal $0\hbar\omega$ calculations of Wildenthal et al.[16] and Warburton et al.[19], a large deficiency in binding (2-3 MeV) is predicted in the N=20–21 isotones of Ne, Na, and Mg. By including two particle–two hole neutron excitations from the sd shell to the fp shell, so-called $2\hbar\omega$ excitations, Warburton et al.[19] has shown in the limit of weak coupling that increased binding in these nuclei occurs. In particular, a deformed intruder state which is dominated by neutron $(fp)^2$-$(sd)^{-2}$ configurations is found to be more bound (by 0.4 to 1.1 MeV) than the lowest $0\hbar\omega$ state in the localized Z=10–12, N=20–22 region. Moreover, Warburton et al. go on to show that the lowest $1\hbar\omega$ state in the odd neutron N=21,23 isotones competes with the lowest $2\hbar\omega$ state as being the most bound state.

Other shell model calculations have been performed by Poves and Retamosa[20] in the limit of $(0+2)\hbar\omega$ mixing. Even better agreement is noted in Fig. 2, however, there is considerable debate as to the validity of these calculations. For example, this calculation could suffer from the exclusion of $4\hbar\omega$ and higher order excitations which are known[21] to strongly mix with $0+2\hbar\omega$ configurations causing a binding energy shift in the ground state. None the less, both approaches result in the same basic finding – that a deformed intruder state is more bound than the lowest $0\hbar\omega$ state giving rise to enhanced binding of Z=10–12, N=20–22 nuclei. Given the very localized nature of this strongly prolate deformed region and the rich nuclear spectroscopy that these nuclei promise, additional investigations are clearly needed.

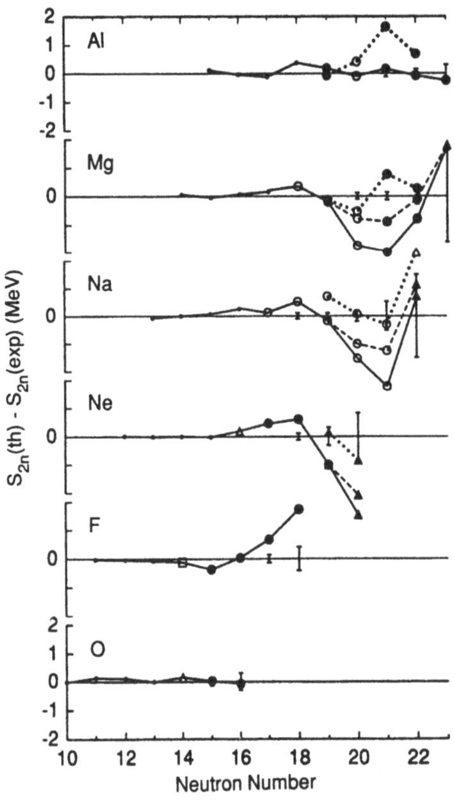

Fig. 2. A difference between the calculated two–neutron separation energies and the experimental S_{2n} values is plotted versus neutron number for the neutron–rich isotopes of Z=8–13. The symbols are as defined in Fig. 1. The solid line indicates the $0\hbar\omega$ shell model calculations of Wildenthal et al.[16] for N<18 or Warburton et al.[19] for N≥18. The dashed lines are for the $2\hbar\omega$ weak–coupling calculations of Warburton et al.[19] and the dotted lines indicate the mixed $(0+2)\hbar\omega$ shell model calculations of Poves and Retamosa[20].

As is noted in Fig. 1 several new mass measurements have recently been reported in the heavier nuclei by the TOFI group[4,5]. Two major features are worth mentioning here – one is a shell model comparison to several neutron-rich S and Cl isotopes and the second is the neutron and proton pairing energies for several Z=21–26 isotopes. Concerning the first topic, remarkably good agreement (rms deviation=280 keV) between the experimental mass of $^{40-43}$S and $^{41-45}$Cl and those predicted by the relatively restrictive $\pi 0d_{3/2}$-$\nu 0f_{7/2}$ shell model calculations of Hsieh et al.[21] was found. This agreement confirms the general observation that in the middle of both proton and neutron shells the shell model works quite well even within a limited basis space, while near shell closures problems often develop if the basis space is not large enough to include important cross-shell contributions.

Finally, we would like to turn you attention to pairing energies and the question of their isospin dependence. An evaluation[22,23] of pairing energy systematics has revealed an apparent neutron-excess ((N-Z)/A) dependence in both neutron (Δ_n) and proton (Δ_p) pairing energies. However, we suspect that the observed dependence could arise from the natural correlation of pairing energy with mass number rather than with neutron-excess since the masses used in the evaluation covered a large mass range and that their fits where dominated by those nuclei lying along the valley of β-stability (i.e., the average neutron-excess and mass number are strongly correlated along stability). Moreover, the recent pairing energy study by Möller and Nix[24] finds that no inherent neutron-excess dependence in the pairing interaction itself is required. To experimentally answers this question a localized test of pairing energies in a limited A region which covers a wide range of neutron-excess is needed. Given the strong increase in binding energy per nucleon and the individual character of light nuclei, the first region where such a test could be made is in the fp shell. Thus we have applied our latest mass measurements of neutron-rich $0f_{7/2}$ subshell nuclei[5] to examining the neutron-excess dependence of pairing energies (see Fig. 3).

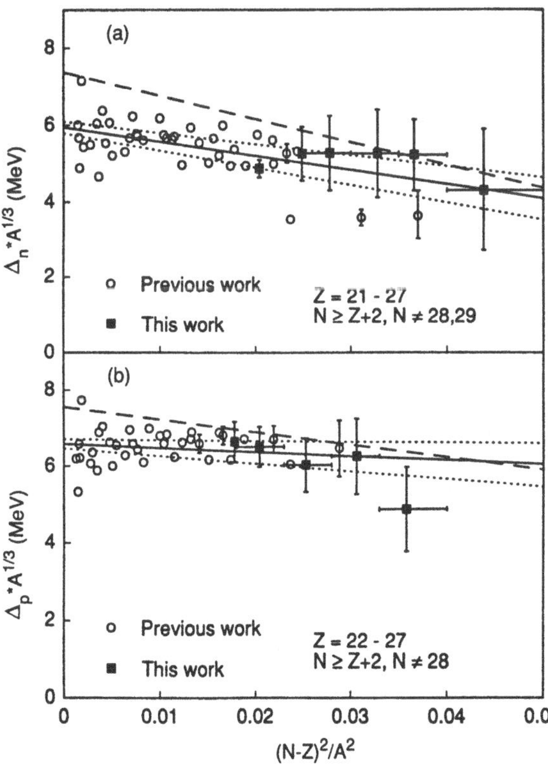

Fig. 3. A plot of (a) $\Delta_n A^{1/3}$ and (b) $\Delta_n A^{1/3}$ versus (N-Z)^2/A^2. The solid squares indicate the weighted average value of data points reported by Tu et al.[5] and the open circles indicate those data that were previously known[8]. The dashed lines are the global fits of Jensen et al.[23] while the solid lines are a fit to the data shown. The dotted lines indicate $\pm 1\sigma$ limits of the latter fit.

Although there is a considerable amount of scatter in the data and the latest measurements have larger error bars than are desired, the data show considerably less neutron–excess dependence than that obtained by Jensen et al.[23]. (Note that we are not so concerned with the offset difference, but rather the slopes of these fits.) In particular, for the neutron pairing energies the neutron–excess dependence is roughly half as large as that of Jensen et al. while the proton pairing energies are one quarter as large. More to the point, in the case of proton pairing energies no neutron–excess dependence at all is indicated at the 1σ level. Clearly additional measurements of this type and further theoretical efforts are needed to elicit the underlying nature of nuclear pairing interactions and to show to which degree they are isospin dependent.

ACKNOWLEDGEMENTS

One of us (DJV) wishes to acknowledge support from the Alexander von Humboldt Foundation and the warm hospitality of the Gesellschaft für Schwerionenforschung mbH, Darmstadt, while on leave from Los Alamos National Laboratory. This work was performed under the auspices of the U.S. Department of Energy.

REFERENCES

1. N. A. Orr et al., Phys. Lett. B 258:29 (1991).
2. A. Gillibert et al., Phys. Lett. B 192:39 (1987).
3. A. Gillibert et al., Phys. Lett. B 176:317 (1986).
4. X. G. Zhou et al., Phys. Lett. B 260:285 (1991).
5. X. L. Tu et al., Z. Phys. A 337:361 (1990).
6. J. M. Wouters et al., Z. Phys. A 331:229 (1988).
7. D. J. Vieira et al., Phys. Rev. Lett. 57:3253 (1986).
8. G. Audi and A. H. Wapstra, Interim Atomic Mass Adjustment (June 1989), private communication.
9. C. Thibault et al., Phys. Rev. C 12:644 (1975).
10. C. Thibault et al., in: "Atomic Masses and Fundamental Constants 6", J. R. Nolen and W. Benenson, eds., Plenum, New York (1980) p.291; A. H. Wapstra and G. Audi, Nucl. Phys. A432:1 (1985).
11. C. Dètraz et al., Nucl. Phys. A394:378 (1983).
12. W. R. Gibbs and A. C. Hayes, Phys. Rev. Lett. 67:1395 (1991), and references therein.
13. R. Anne et al., Phys. Lett. B 250:19 (1990).
14. T. Kobayashi et al., Spectroscopy of the Exotic Nucleus ^{11}Li via Pion–DCX Reaction ^{11}B(π^-,π^+)^{11}Li, KEK Preprint 91-22 (1991), submitted to Phys. Rev. Lett..
15. G. T. Garvey and I. Kelson, Phys. Rev. Lett. 16:197 (1966); G. T. Garvey, W. J. Gerace, R. L. Jaffe, I. Talmi, Rev. Mod. Phys. 41:S1 (1969).
16. B. H. Wildenthal, Prog. Part. Nucl. Phys. 11:5 (1984); B. H. Wildenthal, M. S. Curtin, B. A. Brown, Phys. Rev. C 28:1343 (1983); B. H. Wildenthal, private communication.
17. G. Audi, private communication.
18. H. Stolzenberg et al., Phys. Rev. Lett. 65:3104 (1990).
19. E. K. Warburton, J. A. Becker, and B. A. Brown, Phys. Rev. C 41:1147 (1990).
20. A. Poves and J. Retamosa, "A Theoretical Study of the Very Neutron Rich Nuclei Around N=20", submitted to Nucl. Phys.; A. Poves and J. Retamosa, private communication.
21. S. T. Hsieh, X. Ji, R. Mooy, and B. H. Wildenthal, AIP Conf. Proc. 142:357 (1986).
22. P. Vogel, B. Jonson, and P. G. Hansen, Phys. Lett. B 139:227 (1984).
23. A. S. Jensen, P. G. Hansen, B. Jonson, Nucl. Phys. A431:393 (1984).
24. P. Möller and J. R. Nix, Nuclear Pairing Models, Los Alamos National Report LA-UR-90-1629 (1990), Nucl. Phys. A (in press).

IX - CHAOS

STATISTICAL PROPERTIES OF COMPLETE LEVEL SCHEMES

T. von Egidy

Physik-Department
Technische Universität München
D-8046 Garching, Germany

INTRODUCTION

The atomic nucleus is a quantum system consisting of a limited number of protons and neutrons. Although the strong interaction between the nucleons is not yet precisely known, there exist many theoretical models which describe to a certain extent the structure of the nuclei at low excitation energies. However, already above 1 or 2 MeV there are experimentally and theoretically many open questions. Experimental investigations and theoretical models provide in addition to microscopic descriptions also statistical information. These statistical interpretations can reveal essential new features of the nuclei and of the various models. Such studies concern regular and chaotic dynamics which relate averages and fluctuations in nuclei[1], level densities, the K-quantum number, pairing and other problems. A basic prerequisite for these investigations are complete level schemes[2] including spin and parity information.

COMPLETE SPECTROSCOPY

The idea and necessity of completeness in nuclear level schemes has been discussed in several recent publications[2,3]. There are many modern and precise spectroscopic methods which provide very detailed information and which allow the construction of extensive and complete level schemes. Complete means that in a given spin window and in a given energy range all levels are known with spins and parities. The list of some most recent publications with extensive and complete level schemes in table 1 shows that (n,γ), (n,n'γ), (p,γ), (d,p), (d,pγ), (d,t), and (α,nγ) reactions are most useful in this context. Of course always the questions arises "How complete is complete and how to guarantee completeness?" To avoid this question, the problem of completeness is frequently not discussed in experimental publications. However, for comparison with theoretical models and for significant tests (IBM, shell model, VAMPIR model[21], level density, chaos[3]) the completeness information is essential and should be given with a confidence level (e.g. less than 5% missing levels) estimated by the careful experimentalist. As an example a fraction of the level scheme of ^{198}Au [19] is shown in fig. 1. In this case, as frequently, the theory (IBFFM) is tested by assuming a one-to-one correspondence of experimental and theoretical levels.

Table 1. Some most recent publications with extensive and complete level schemes

Nucleus	Number of levels	Experimental methods	Laboratories	Reference
^{25}Mg	83	(d,pγ),(α,nγ)	Freiburg	4
^{26}Al	156	(p,γ)	Utrecht	5
^{32}P	102	(n,γ),(d,p)	Göttingen,München,ILL	6,7
^{40}Ca	252	(p,γ) etc.	Utrecht	8,9
^{51,52}V	62,54	(n,γ)	Göttingen, ILL	10
^{87,88,89}Sr	140,108,68	(n,γ),(d,p)	Göttingen,München, ILL	11,12,13
^{130}I	46	(n,γ),(d,p)	Gatchina,ILL,München	14
^{143}Nd	87	(α,nγ)	Köln	15
^{154}Eu	103	(n,γ)	Riga,ILL,München	16
^{163}Dy	84	(n,γ),(n,n'γ), (d,p),(d,t)	München,Ill,Riga, Brookhaven	17
^{178}Hf	66	(n,γ)	Köln,ILL,Brookhaven	18
^{198}Au	140	(n,γ),(d,p)	München,ILL,Brookhaven	19
^{238}Np	82	(n,γ)	Livermore,Fribourg,ILL, Brookhaven	20

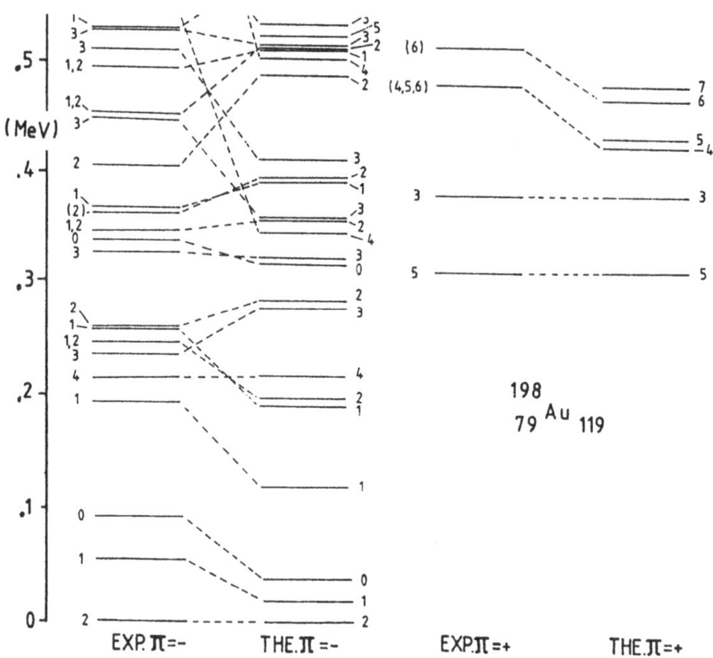

Fig. 1. The IBFFM energy spectrum of ^{198}Au compared with experimental levels (ref. 19).

LEVEL DENSITY

The level density and its energy, spin and parity dependence is a very fundamental property of nuclides and it is often necessary for the interpretation of experimental data. The level densities have been treated in many publications, but precise experimental values are available only for a fraction of the nuclides and the theoretical predictions are not yet sufficiently accurate. Therefore, new experimental and theoretical work is still necessary.

It has been shown that, in spite of the very different and complex modes of nuclear excitations, the exponential increase of level density with energy can be well reproduced with only two parameters[3,22]. The Bethe formula

$$\rho(E) = \exp(2\sqrt{a(E-E_1)})/(12\sqrt{2}\,\sigma\, a^{1/4}(E-E_1)^{5/4})$$

and the simple constant temperature formula

$$\rho(E) = \exp((E-E_0)/T)/T$$

reproduce experimental data equally well, if the level density parameters a, E_1 and T, E_0 are fitted to complete level schemes and neutron resonance densities. σ is the spin cut-off parameter.

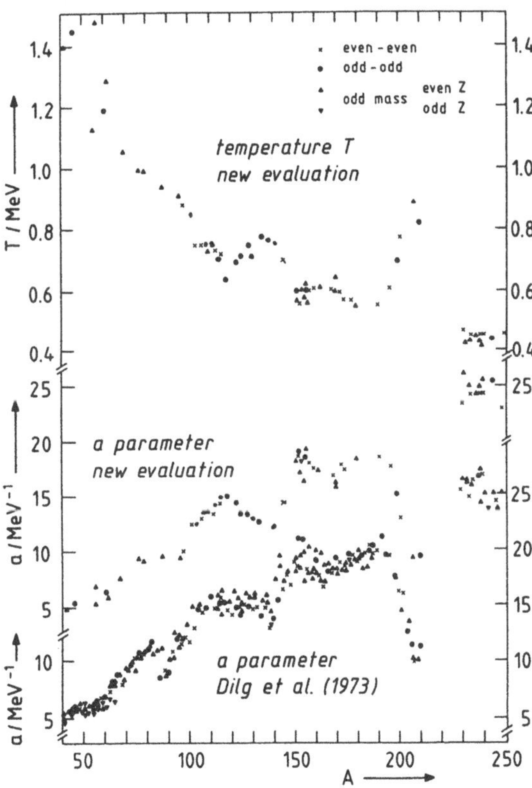

Fig. 2. The level density parameters "a" and T as a function of A.

While the previous evaluations showed large fluctuations of "a" as a function of the mass number A, new fits show smoother variations of a and T with A; the shell effects, especially near ^{208}Pb, are clearly visible (see fig. 2). It is a challenge for theoretical calculations to reproduce for instance the sequence of a or T from A = 100 to A = 140.

The spin distribution formula and the spin cut-off parameter σ

$$f(J) = \exp(-J^2/2\sigma^2)-\exp(-(J+1)^2/2\sigma^2)$$

were seldom tested. New experimental results[10] confirm the empirical formula[3] $\sigma = 0.98\ A^{0.29}$.

A new formula for the spin-dependent level density was suggested by Paar et al.[33] and should be carefully tested.

LEVEL SPACING DISTRIBUTION AND CHAOS

It has been suggested[1,24-26] that the spacing distribution of nuclear levels can be used as a measure for the chaoticity of the nuclear system, a Wigner distribution, $P(x) = (\pi/2)\cdot x\cdot \exp(-\pi\cdot x^2/4)$, corresponding to a chaotic system and a Poisson distribution, $P(x) = \exp(x)$, corresponding to a regular system ($x = S/D$, S = individual spacing, $D = 1/\rho$ = average spacing). The parameter ω in the interpolation formula of Brody[27], $P(x,\omega) = \alpha(\omega+1)\ x^{\omega} \exp(-\alpha\cdot x^{\omega+1})$ with $\alpha = \Gamma((\omega+2)/(\omega+1))^{\omega+1}$, is a convenient characterization for the chaoticity: ω = 1 for chaos and ω = 0 for regularity.

A recent analysis[28] of a total of 988 spacings from 60 different nuclides showed surprising results: ω decreases with increasing A (see fig. 3);

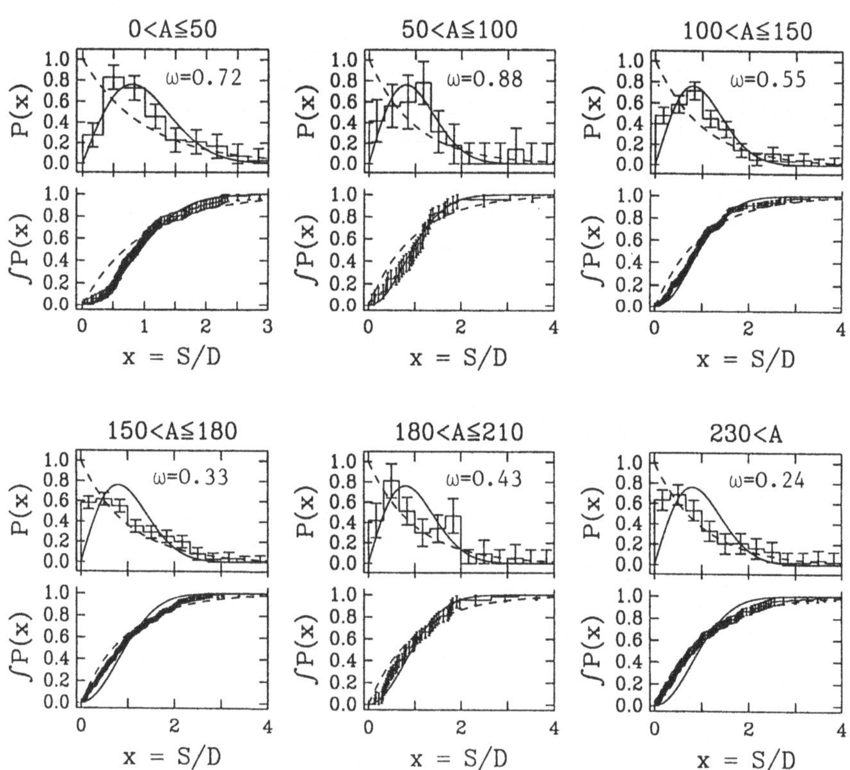

Fig. 3. Nearest neighbour spacing distributions P(x) (and cumulative distributions $\int p(x)dx$) with Poisson (dashed curve) and Wigner (solid curve) distributions (ref. 28).

ω = 0.60 for spherical and ω = 0.30 for deformed nuclei; and ω = 0.16 for 2^+ and 4^+ levels in even-even deformed nuclei and ω = 0.51 for levels other than 2^+,4^+ in these nuclei. $1/2^+$ levels at the neutron binding energy have a Wigner distribution[25,26]. Theoretical calculations with the IBM model[29,30] reproduced the experimental results: 2^+ and 4^+ levels in the SU(3) (deformed) and O(6) (γ-unstable) limit are regular, while 0^+ and 3^+ levels are chaotic. Fig. 4 of ref. 30 demonstrates that 8^+ levels at the SU(3) and O(6) limit are regular and that transitional 8^+ levels are chaotic. This observation shows that the phenomenon of chaos in nuclei has to be discussed in the context with the K quantum number.

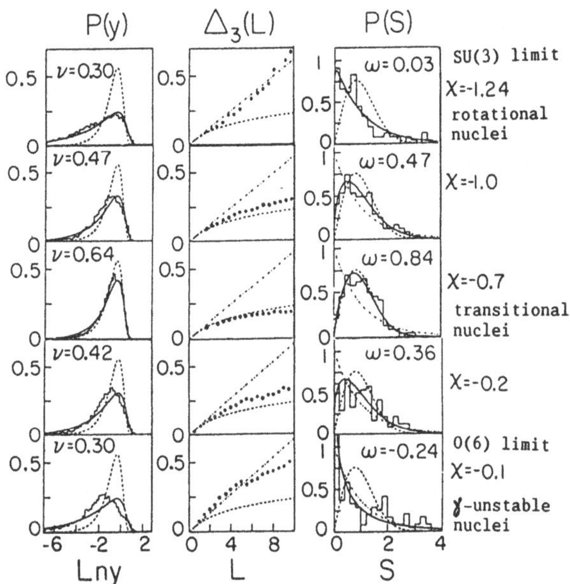

Fig. 4. The nearest-neighbor level-spacing distribution P(S) (histograms in the right-hand column), the Dyson-Metha Δ_3 statistics (dots in the middle column), and the B(E2) distribution P(γ) (histograms in the left-hand column) of the J = 8^+ states for several values of the parameter χ (ref. 30).

THE K QUANTUM NUMBER ABOVE 1 MeV

It is an open question up to what excitation energies the K quantum number is still a good quantum number for levels in deformed nuclei. It is generally assumed that Coriolis interaction mixes all K values if the level spacing is sufficiently small (few keV). A recent investigation[31] of J = 1 levels between 1 and 4 MeV in deformed even-even nuclei using nuclear resonance fluorescence[32] demonstrated that the γ-branching ratios R ($R=B(1\to 2_1^+)/B(1\to 0_1^+)$; R = 0.5 for K = 1 and R = 2.0 for K = 0 levels according to the Alaga rule) indicate that K seems to be in most cases a good quantum number (see fig. 5). The constant temperature level density formula with parameters of ref. 3 (A=168, T=0.609, E_0=-0.64, σ=4.33) give an average spacing of 1^+ or 1^- levels near 3 MeV of 41 keV.

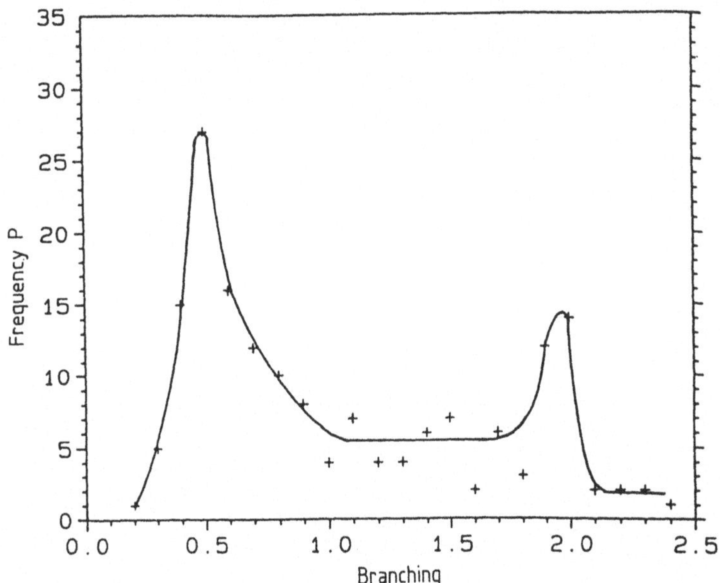

Fig. 5. Frequency distribution of experimental branching ratios $R_{exp} = B(1 \rightarrow 2_1^+)/B(1 \rightarrow 0_1^+)$. One notes two maxima, the first at $R_{exp} = 0.5$ for K = 1 and the second at $R_{exp} = 2.0$ for K = 0 levels (ref. 31).

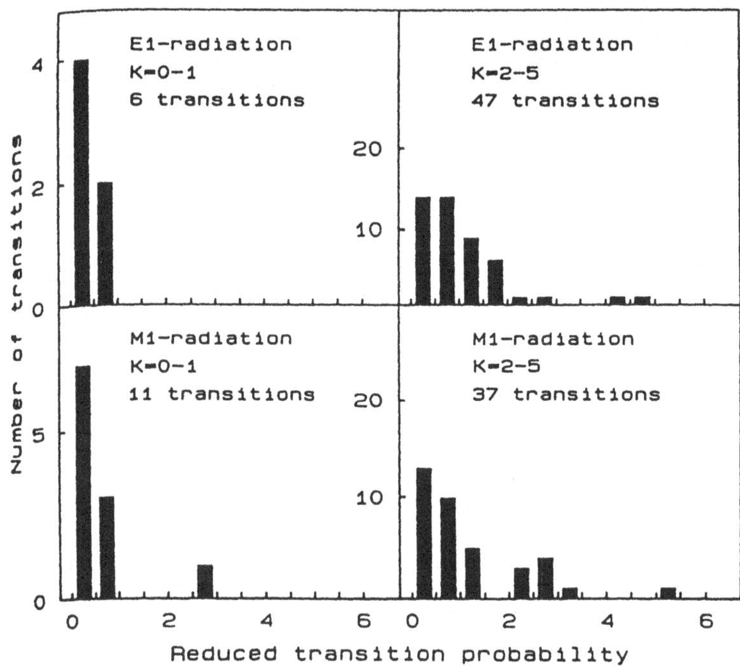

Fig. 6. Number of transitions vs reduced transition probability for E1 and M1 transitions to low-lying states with K = 0,1 and K = 2-5, respectively (ref. 33).

The partial conservation of the K quantum number is therefore not too surprising.

However another recent publication[33] compared the reduced γ-transition probabilities of the neutron capture state in ^{168}Er [34] and ^{178}Hf [18] to low lying levels with K = 0 or 1 and K = 2,3,4 or 5 and found that the transition probabilities to K = 2 to 5 levels are larger by a factor 3 (fig. 6). These authors conclude "The result contradicts the hypothesis that K is completely mixed in this energy region as expected for a chaotic structure". Since the average spacing of 4^+ levels in ^{168}Er at 7.77 MeV is 9 eV, conservation of K would be extremely surprising and why should the neutron capture state have K = 3 or 4. An explanation can be obtained if it is assumed that low K components are less frequent in the K mixture of the capture state. Table 2 gives estimated K compositions of levels in ^{168}Er at 7.77 MeV; the fraction with K = 0 or 1 is only about 1/3 for the 4^+ capture state.

Table 2. Number of levels per keV at 7.77 MeV in ^{168}Er calculated with level density and spin distribution formulae. The average K-components in the levels were estimated by assuming that all levels originate from the Coriolis mixing of rotational bands.

Spin	levels per keV one parity	Fraction of K quantum number in % K = 0	K = 1	K = 2	K = 3	K = 4
0	21	100				
1	61	34	66			
2	92	21	41	38		
3	109	15	30	29	26	
4	114	12	25	23	21	19

CONCLUSION

A wealth of very interesting nuclear properties and nuclear models can be investigated if complete level schemes are available. Therefore, the troublesome spectroscopic work of establishing these level schemes is worthy of appreciation. New experimental information on level spacing distributions, level densities, spin distributions and the K quantum number initiated and will continue to influence new theoretical considerations and models in order to promote our understanding of the nucleus.

REFERENCES

1. H.A. Weidenmüller, Comments Nucl.Part.Phys. 16 (1986) 199
O. Bohigas, H.A. Weidenmüller, Ann. Rev. Nucl. Part. Sci. 38 (1988) 421
2. P. von Brentano et al., Research Reports in Physics: Nuclear Structure of the Zirconium Region, eds. J. Ebert et al., Springer-Verlag, Berlin, (1988) p. 157
3. T. von Egidy et al., Nucl. Phys. A 481 (1988) 189
4. F. Heidinger et al., Z. Physik A 338 (1991) 23
5. P.M. Endt et al., Nucl. Phys. A 476 (1988) 333 and Nucl. Phys. A 487 (1988) 221
6. F.J. Eckle et al., Nucl. Phys. A 501 (1989) 413
7. S. Michaelsen et al., Nucl. Phys. A 501 (1989) 437
8. S. W. Kikstra et al., Nucl. Phys. A 512 (1990) 425
9. P.M. Endt et al., Nucl. Phys. A 521 (1990) 1
10. S. Michaelsen et al., Z. Physik A 338 (1991) 371
11. Ch. Winter et al., Nucl. Phys. A 460 (1986) 501
12. Ch. Winter et al., Nucl. Phys. A 473 (1987) 129
13. Ch. Winter et al., Nucl. Phys. A 491 (1989) 395

14. S.L. Sakharov et al., Nucl. Phys. A 494 (1989) 36
15. J. Wrzesinski et al., Nucl. Phys. A 515 (1990) 297
16. M. Balodis et al., Nucl. Phys. A 472 (1987) 445
17. H.H. Schmidt et al., Nucl. Phys. A 504 (1989) 1
18. A.M.I. Hague et al., Nucl. Phys. A 455 (1986) 231
19. U. Mayerhofer et al., Nucl. Phys. A 492 (1989) 1 and Thesis, TU München (1990)
20. R.W. Hoff et al., Phys. Rev. C 41 (1990) 489
21. K.W. Schmid et al., Nucl. Phys. A 452 (1986) 493
22. W. Dilg et al., Nucl. Phys. A 217 (1973) 269
23. V. Paar et al., Seventh International Symposium on Capture Gamma-Ray Spectroscopy and Related Topics, Asilomar 1990, to be published
24. T.A. Brody et al., Rev. Mod. Phys. 53 (1981) 385
25. R.U. Haq et al., Phys. Rev. Lett. 48 (1982) 1086
26. O. Bohigas et al., Lecture Notes in Physics 209 (1983) 35
27. T.A. Brody, Lett. Nuovo Cimento 7 (1973) 482
28. J.F. Shriner, Jr. et al., Z. Physik A 338 (1991) 309
29. V. Paar et al., Phys. Lett. B 205 (1988) 7
30. Y. Alhassid et al., Phys. Rev. Lett. 65 (1990) 2971
31. A. Zilges et al., Phys. Rev. C 42 (1990) 1945
32. D. Bohle et al., Phys. Lett. B 137 (1984) 27
33. J. Rekstad et al., Phys. Rev. Lett. 65 (1990) 2122
34. W.F. Davidson et al., J. Phys. G 7 (1981) 455

X - EXPERIMENTAL TECHNIQUES

NUCLEAR MOMENTS BY ORIENTATION METHODS

H. Postma

Department of Applied Physics
University of Technology
Lorentzweg 1
2628 CJ Delft
the Netherlands

INTRODUCTION

Static nuclear magnetic dipole and electric quadrupole moments provide valuable information about nuclear structure. In this paper the determination of such moments with the aid of low-temperature nuclear orientation (LTNO) techniques will be reviewed. With the development of these techniques the number of isotopes, which can be studied, has been increased very substantially; the shortest half lives used in LTNO are of the order of a few seconds.

HYPERFINE INTERACTIONS

The information about nuclear moments is contained in the hyperfine interactions of the nuclear magnetic dipole moment (μ) with a magnetic field ($\mathbf{B}$) and the electric quadrupole moment (Q) with electric field gradients ($\nabla^2 V$). The magnetic hyperfine interaction is presented by the hamiltonian:

$$H_{magn.} = -\mu.\mathbf{B} = -\mu_N g_N B I_z \ , \qquad (1)$$

with the z-axis chosen along $\mathbf{B}$; g_N is the nuclear gyromagnetic ratio. The magnetic field may be externally produced by a (superconducting) magnet, or it may be of internal origin, namely due to magnetic properties of transition elements (3d-elements, lanthanides, actinides). In addition nuclei of dia-magnetic atoms (ions) incorporated in ferromagnetic samples often experience large internal magnetic (hyperfine) fields $\mathbf{B}_{hf}$, which contain terms related to electron-orbitals and -spins and to the Fermi contact term. A large amount of data about magnetic hyperfine fields have been collected for iron, nickel, cobalt and gadolinium. An extensive survey is given by Rao[1] with recent additions from other authors (eg. ref.2). Notably the heavier elements may experience very strong magnetic hyperfine fields in the above-mentioned ferromagnetics. In first order the internal magnetic field is the same for different isotopes of the same element in a given host. However, the Fermi contact term of the hyperfine interaction

may be different for different isotopes since nuclear magnetic moments may not be distributed over the nuclear volume in an equal way. This leads to an anomaly factor $(1+{}^1\Delta^2)\mu_1/\mu_2$ in the hyperfine interaction. It is often negligible or small (order of 0.1 to 1%). However, there are some interesting exceptions in which it is of the order of 5 to 10 %; see ref.3.
In nuclear orientation experiments often an external magnetic field $\mathbf{B}_{app}$ (of order of ≈0.1 T) must be applied to magnetize a ferromagnetic sample. Therefore, the Lorentz field $(4\pi/3)\mathbf{M}$ and demagnetization field $-N\mathbf{M}$, where $\mathbf{M}$ is the magnetization of the ferromagnetic sample, must be added. These terms are usually very small compared to the hyperfine field and therefore often included in the other magnetic terms. The applied (external) field must be multiplied with 1+K, where K is the Knight shift related to magnetization of the conduction electrons, thus the final expression of the field at the nucleus can be written as:

$$\mathbf{B}_{tot} = \mathbf{B}_{hf}+(1+K)\mathbf{B}_{app}+(4\pi/3)\mathbf{M}-N\mathbf{M}. \qquad (2)$$

The Knight shift is often of the order of 0.1 to 1%; however, for some cases K is in the few per cent range[4].

The electric hyperfine interaction is due to strong internal electric field gradients in suitable (non-cubic) materials like the metals Zn, Cd, Hf, Re and Os, or in extreme "two-dimensional" materials like TaS_2 and graphite. It is not possible to create strong enough electric field gradients by laboratory means in order to achieve static nuclear orientation at suitable temperatures. Extensive compilations of electric field gradients have been made by Raghavan *et al.*[5], by Vianden[6] and more recently by Hagn[7]. To use electric hyperfine interactions for orienting nuclei, single crystals are necessary.
An electric field gradient has three components. The x,y,z-axes can be chosen such that $|\partial^2V/\partial z^2| \geq |\partial^2V/\partial y^2| \geq |\partial^2V/\partial x^2|$. These gradients are often denoted as V_{zz} etc. The electric hyperfine interaction hamiltonian can be written as follows:

$$H_{el.} = P\{I_z^2 - I(I+1)/3 + \eta(I_+^2+I_-^2)/6\}, \qquad (3)$$

with $P=3eQV_{zz}/4I(2I-1)$ and $\eta=\{V_{xx}-V_{yy}\}/V_{zz}$. The latter varies between 0 and 1. The field gradient V_{zz} is partly related to the lattice and partly due to the ion (atom):

$$V_{zz} = (1-\gamma_\infty)\{V_{zz}\}_{lattice} + (1-R)\{V_{zz}\}_{ion}, \qquad (4)$$

where γ_∞ is the Sternheimer factor[8,9], which can be very large (≈50-100) and negative (anti-shielding), and R is the shielding factor, which is a correction of the order of 10%.

Since hyperfine interactions depend on products of nuclear moments and electromagnetic fields, information on nuclear properties as well as on local properties in solid-state environments can be obtained. In this paper attention will be paid to the determinations of nuclear moments.
Both hyperfine interactions may occur simultaneously, but not necessarily with the same local z-axes. In such cases one of the two parts of the total hamiltonian $H = H_{magn.}+H_{el.}$ must be rotated; often one chooses to rotate the weaker electric part to the coordinate system of the magnetic hyperfine interaction.

Isotopes of nearly all elements have been oriented by low-temperature techniques using magnetic and/or electric hyperfine interactions.

In case of magnetic hyperfine interaction the sublevel energies are $E_m=-\mu_N g_N Bm$ in which m is the magnetic quantum number. Electric interaction removes the (2I+1)-degeneracy only partially ; with axial symmetry ($\eta=0$) there are doublets $\pm m$ plus one singlet m=0 if I is integer; further $E_m=P\{m^2-I(I+1)/3\}$. With $\eta\neq 0$ sublevels with $m'=m\pm 2$ are mixed and the expressions for sublevel energies become more complicated. In a sample with the nuclei in thermal equilibrium at the sample temperature T_L the sublevel populations p_m are determined by the Boltzmann distribution; that is:

$$p_m = \exp(-E_m/kT_L)/\Sigma_m \exp(-E_m/kT_L) \quad . \tag{5}$$

At low temperatures these populations differ distinctly; the nuclei become oriented with respect to the local z-axis. If local orientation directions are ordered (preferentially collinear), macroscopic orientation will occur. The p_m may be used to describe nuclear orientation. It is now common practice to use the following nuclear orientation parameters*:

$$B_k(\mu,I,T) = \{(2k+1)(2I+1)\}^{1/2}\Sigma_m (-1)^{I+m}\begin{pmatrix} I & I & k \\ -m & m & 0 \end{pmatrix} p_m \,, \tag{6}$$

where (:::) denotes 3j-symbols. The first order orientation parameter ,$B_1(I)$, concerns nuclear polarization; it depends on $\langle I_z\rangle/I=\Sigma_m mp_m/I$. The second order nuclear orientation parameter $B_2(I)$ describes alignment; it contains $\langle I_z{}^2\rangle/I^2=\Sigma_m m^2p_m/I^2$.
If we are dealing with oriented unstable nuclei, the emitted radiation will often show anisotropic directional distributions $W(\theta)$, which in general can be written as:

$$W(\theta) = 1 + f\Sigma_k B_k(I_o)U_k(I_oL_oI_i)A_k(I_iLI_f)Q_kP_k(\cos\theta), \tag{7}$$

where k is even for α- and γ-radiation, and odd/even for β-decay; k is limited to $\min(2I_o,2I_i,2L)$, where I_o is the spin of the oriented nuclei, I_i is the spin of the initial state of the radiation studied, L_o is the multipolarity of the intermediate transition ($I_o\rightarrow I_i$) and L is the multipolarity of the observed (studied) transition. The U_k-coefficients describe the change in orientation during preceding transitions, the A_k are related to the observed (studied) transition and $P_k(\cos\theta)$ are Legendre polynomials, θ being the angle of emission with respect to the orientation axis. The U_k and A_k can be calculated if the decay properties are known. Tables of these coefficients are given in the literature[10]. The Q_k are solid-angle corrections usually close to 1. The f-factor is introduced for cases where only a fraction f of the nuclei is oriented; thus fraction 1-f not. This is known as the two-component model, which works well in many cases. The unoriented fraction is due to atoms not properly located in the lattice. Whether atoms implant properly in a lattice or not depends on local solid state properties, size of the ions, solubility, concentration etc. Several detailed studies have been made; see e.g. ref. 11.

---

*note: In the more general case, in which m is not a good quantum number (no rotational symmetry),density matrix elements $\rho_{mm'}$ and statistical tensor elements $\rho_\lambda{}^q$ are used:

$$\rho_\lambda{}^q = (2\lambda+1)^{1/2}\Sigma_m(-1)^{I+m'}\begin{pmatrix} I & I & \lambda \\ -m' & m & q \end{pmatrix}\rho_{mm'} .$$

Thermal equilibrium exists due to spin-lattice interactions, which are responsible for equal temperatures of the spin-system and lattice. In a number of ways non-equilibrium conditions for the nuclear sublevel populations p_m can be introduced, namely:

i) by a sudden change of the lattice temperature, e.g. by pulsed heating,
ii) by applying an RF-field at a resonating frequency,
iii) by implantation of isotopes either continuously or in pulsed mode, and
iv) by reorientation during the life time of an intermediate state.

In order to study such situations it is necessary to consider the differential master equations for the level populations:

$$dp_m/dt = \sum_n (W_{nm}p_n - W_{mn}p_m), \tag{8}$$

where W_{nm} denotes the transition rates between sublevels n and m. The spin-lattice interaction, which is responsible for relaxation, is mainly between the nuclear spins and conduction electrons and is magnetic in nature. This interaction introduces transitions between adjacent sublevels m and m±1; that is:

$$W_{m,m+1} = (\Delta E/2kC_K)[I(I+1)-m(m+1)]/\{\exp(\Delta E/kT_L)-1\}, \text{ and}$$

$$W_{m+1,m} = (\Delta E/2kC_K)[I(I+1)-m(m+1)]/\{1-\exp(-\Delta E/kT_L)\}, \tag{8a}$$

where ΔE is the energy difference between the two sublevels and C_K is the Korringa[12] constant, given by:

$$C_K = (h/8\pi^2 k)(g_e/g_N K)^2 \approx 0.0827/(K\mu/I)^2, \tag{9}$$

where K is the earlier mentioned Knight shift parameter. In experiments which depend on the first order moment $\langle I_z\rangle/I$; that is on polarization, the relaxation time t_1 is given by*:

$$t_1 = C_K/T_L, \qquad \text{if } T_L \gg T_{int} = \Delta E/k \tag{10a}$$

$$\text{and } t_1 = C_K/IT_{int}, \qquad \text{if } T_L \ll T_{int}. \tag{10b}$$

In principle multi-exponential relaxation occurs in γ-ray and particle directional distributions[13]; that is, the following relaxation rates λ_n have to be taken into account:

$$\lambda_n = n(n+1)(\Delta E/2kC_K)\{1-\exp(-\Delta E/kT_L)\}, \text{ with } n=1,..k. \tag{11}$$

However, Templeton *et al.*[13] have shown that in good approximation an effective spin-lattice relaxation time t_{eff} can be used. Since the alignment parameter B_2 is the most significant orientation parameter in γ-ray anisotropies Klein[15] introduced the following definition for t_{eff}:

$$B_2(t_{eff}) - B_2(eq.) = \{B_2(t=0) - B_2(eq.)\}/e, \tag{12}$$

Shaw introduced the following empirical relations[16]:

---

*note: in order to avoid confusion the usual notation T_1 for the relaxation time is not followed in this paper, but instead t_1 is used.

$$t_{eff} = 4C_K/3T_L \qquad \text{if} \quad T_L>>T_{int} \qquad \text{or} \tag{13a}$$

$$t_{eff} = 3.3C_K/(I+1/2)T_{int} \qquad \text{if} \quad T_L<<T_{int}. \tag{13b}$$

For isotopes of the same element in the same host $(\mu/I)^2C_K$ is a constant.
For impurities in iron the following relation is approximately valid:

$$C_K\nu_o{}^2=6\times10^{16} \text{ K/s}, \tag{14}$$

where ν_o is the zero-field NMR frequency[15,17].

To resonate transitions between adjacent sublevels with m and m±1, the frequency of the RF-field should be $\nu_m=\Delta E/h =(\mu/I)\mu_N B/h$, where B is the total magnetic field experienced by the nuclei. In case of quadrupole interaction or a mixed magnetic/electric hyperfine interaction the level distances are not equidistant. For an $m\Leftrightarrow m+1$ transition the resonating frequency is:

$$\nu_{m,m+1} = \nu_M+3(m+1/2)\nu_Q/2I(2I-1), \quad \text{where } \nu_Q= eQV_{zz}/h. \tag{15}$$

Implanting a beam of unoriented nuclei the term $\Phi(t)/(2I+1)$ must be added to the master equation, where Φ is the flux of the implanted beam, which can be a function of time. Taking the mean nuclear life τ into account the master equations transform to:

$$dp_m/dt = \textstyle\sum_m(W_{nm}p_n-W_{mn}p_m) - p_m/\tau + \Phi(t)/(2I+1) \;. \tag{16}$$

On the basis of these equations Shaw and Stone[18] analysed the attenuation factors $\langle B_k\rangle/B_k(\text{eq.})$ as a function of T_{int}/T_L for a series of fixed values of the parameter $\tau T_{int}/C_K$.

BASIC METHODS TO DETERMINE NUCLEAR MOMENTS WITH LTNO

Through low-temperature nuclear orientation experiments strengths of hyperfine interactions and thus nuclear moments can be determined in the following ways:
i) temperature dependence of the anisotropy of emitted radiation of oriented nuclei. With decays sufficiently well known, the temperature dependence of the anisotropy $W(\theta)-1$ can be calculated as a function of $B(\mu/I)/kT$ or P/kT and used to determine the nuclear moments μ and Q from the measured temperature dependence if the magnetic fields or electric field gradients are known. The accuracy is of order of a few per cent.
ii) nuclear magnetic resonance (NMR) to determine the splittings $\Delta E=h\nu$ from the resonance frequencies. Anisotropies of γ-ray and particle emissions merely serve as indicators for resonances. Detailed knowledge of the directional distributions is not needed. Accuracy is of order of 0.1% for magnetic moments and about 1% to a few per cent for quadrupole moments depending on the knowledge of the hyperfine fields.
iii) Spin-lattice relaxation observed in time-resolved NO measurements may give the relaxation time t_1 and thus the magnetic moment using the quantity $C_K(\mu/I)^2$ presumably known from other isotopes. Accuracy is of order of 5 to 10%.
iv) Level mixing resonance at an intermediate state during the decay of oriented nuclei by a combination of magnetic and electric hyperfine interactions.

v) Energy reduction of the system with nuclear hyperfine splitting when cooling the sample; that is by increasing the population of the lowest nuclear sublevels.

In later sections mainly methods i) to iii) will be discussed. Method v), which has been used to determine magnetic moments of neutron resonances, is outside the scope of this paper and will not be treated.

SAMPLE PREPARATIONS

In early NO-experiments radioactive nuclei were grown into suitable single crystals with strong hyperfine interactions. These crystals were cooled into the mK region by adiabatic demagnetization. Hence, applications of nuclear orientation were initially very limited. Since the discovery of the induced hyperfine magnetic fields at impurity nuclei in iron and other ferromagnetic metals, samples for nuclear orientation can be made by melting and diffusion and also by implantation techniques using mass separators or recoil from heavy-ion nuclear reactions. Other developments, which have further enhanced applications of nuclear orientation, are the invention of the dilution refrigerator, superconducting magnets and improved γ- and particle detection. Particle detectors have been successfully used at ≈1K inside NO-equipment. Thermal nuclear orientation has been applied to isotopes of nearly all elements.

NUCLEAR ORIENTATION (NO) FACILITIES

The hardware for modern nuclear orientation experiments consists of a dilution refrigerator, a superconducting magnet (often a small one for magnetizing ferromagnetic foils in plane), and several γ-ray and particle detectors in fixed positions; usually at 0° and 90°, and opposite directions. The main recent expansion in the field of nuclear orientation has been achieved with semi and fully on-line systems allowing to implant short-living isotopes into suitable samples already cooled to temperatures well below 1 Kelvin. FOLBIS in Bonn is the pioneering NO-system on-line with a mass-separator, but not with an accelerator.

Nuclear orientation facilities on-line with mass-separators and coupled to isotope producing facilities are: i) KOOL from Leuven, ii) DOLIS-COLD in Daresbury, iii) NICOLE at CERN and iv) UNISOR/NOF in Oak Ridge and v) SPIN-2 in Dubna. Fig.1 gives an impression of the NICOLE cryogenic facility[19]. Its lowest base temperature is ≈6mK.

Special care is needed for the construction of the access tube in order to avoid a thermal radiation load into the sample region from warm surfaces. For this reason the NICOLE facility has a long side access tube with inner tubes connected to liquid nitrogen and liquid helium cooled shields.

With modern isotope production and source techniques large numbers of isotopes of many elements can be produced in considerable quantities[20]; up to 10^9 nuclei/sec in the beam. Fluxes of 10^4 are still useful for NO-experiments.

TEMPERATURE DEPENDENCE

The temperature dependence of NO-effects has often been used to derive nuclear magnetic dipole and electric quadrupole moments. The strength of the magnetic field or electric field gradient must be known. To circumvent the problem of good-site occup-

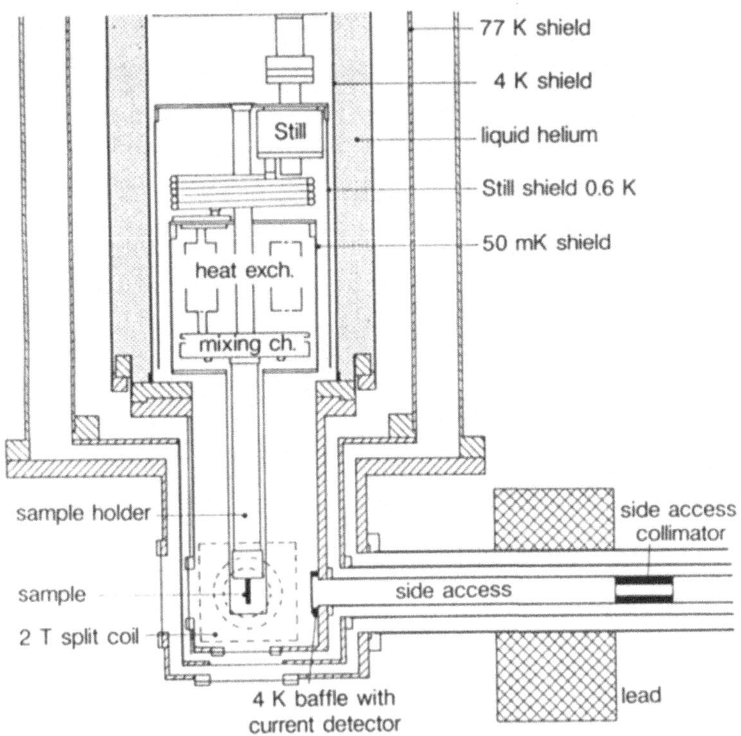

Fig.1.Cross section through the lower part of the NICOLE cryostat[19]

ation, it is necessary to obtain a temperature dependent curve in which saturation is at least approached. In that case the analysis of the temperature dependence provides both the strength of the interaction and the value of f, if the U_kA_k are known. If two or more isotopes are implanted in the same target under the same conditions (preferably simultaneously) ratios of nuclear moments can be obtained even if the temperature dependent curves are not well developed.

In principle the simplest way to orient nuclei is to use an external magnetic field acting directly on the nuclei. This "*brute-force*" type of nuclear orientation is not often applied to nuclear spectroscopy, the reason being that the combination of very low temperatures and strong external fields is still very difficult. An example of brute-force orientation is a γ-ray anisotropy experiment carried out with ^{178m2}Hf (31y) oriented in a magnetic field of 8.49 Tesla. This is a 16^+ isomer with a well-known decay scheme allowing to calculate the U_kA_k. A sample of Ta-powder with about 0.1 μCi of ^{178m2}Hf, solidly compressed with lead, has been cooled to about 10 mK. This sample has been obtained from old beam slits of the KVI-cyclotron. Fig.2 shows the temperature dependence of one of the γ-transitions[21]; the temperature has been determined with ^{57}Co<u>Fe</u> and ^{60}Co<u>Co</u> NO-thermometers. From the temperature dependence $\mu=7.26(\pm0.16+0.2/0)\mu_N$ has been determined. The systematic error is estimated on the basis of an estimated small temperature difference and a Knight shift of at most 2%. It is in good agreement with the value of $7.79(\pm0.02)\ \mu_N$ calculated by Ekström *et al.*[22] for the 4-quasi-particle configuration:

$$16^+\{p[514]9/2+p[404]7/2+n[514]7/2+n[624]9/2\}.$$

Using g_K values for the four quasiparticle states derived from neighbouring odd-A nuclei and g_R=0.3 the magnetic moment is estimated to be 8.6 μ_N. Only a small fraction of this moment is due to the particle spins. If, in the limit of large deformation the asymptotic Nilsson states are used, the magnetic moment is only due to the orbital parts of the uncoupled protons leading to a magnetic moment estimate of 7.8 μ_N.

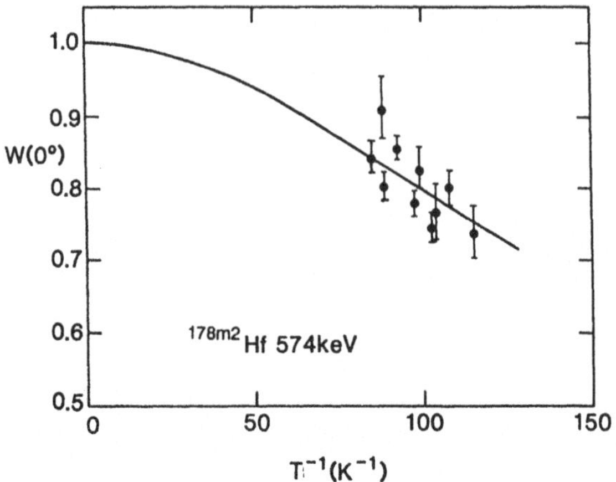

Fig.2.Anisotropy of the 574 keV γ-transition of $^{178m2}Hf$ as a function of inverse temperature[21]

Many other cases of temperature dependences of α, β, and γ emission from oriented nuclei and their use to get initial information about magnetic moments are available from the literature. Quite often such work is followed by NMR-ON experiments to get more precise magnetic moment values. As an example the temperature dependence of the 6059-keV α-emission of ^{199m}Po isotope, implanted in magnetized iron foils in the mK-region with the KOOL facility, is shown in fig.3. Fitting this curve with μB and fA_2 (A_4 is assumed to be very small) Wouters *et al.*[23] derived μB=235±7 μ_NTesla. With the hyperfine field of 238±16 Tesla, given by Herzog *et al.*[24], the magnetic moment $|\mu|$=0.99±0.07 μ_N is deduced.

Considerable effort has been devoted to orient nuclei with the aid of electric hyperfine interactions and to use this to

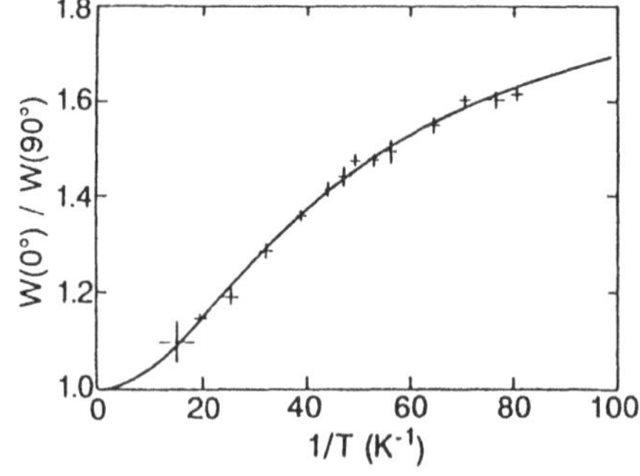

Fig.3.The anisotropy of the 6059-keV α-transition of ^{199}Po; μB and fA_2 were simultaneously fitted; A_4 is expected to be small[23]

extract nuclear quadrupole moments. After the first successful experiments by Kaindl *et al.*[25] a considerable number of nuclear quadrupole orientation experiments in non-magnetic non-cubic metals have been carried out. In this kind of work single crystals are needed. It is often difficult to derive precise nuclear quadrupole moments in this way since electric field gradients are not easily known. In several experiments ratios of nuclear quadrupole moments have been obtained. Calibration is possible by using quadrupole moments known from the electric hyperfine interaction of free atoms or muonic atoms. In these cases atomic wavefunctions are well-known, allowing to calculate interatomic electric field gradients precisely.

Examples for nuclear quadrupole orientation are ^{198}Au and ^{199}Au implanted at room temperature in Zn and Cd single crystals with the DOLIS facility[26] and ^{185}Pt and ^{189}Pt cold-implanted in Zn-single crystals using the NICOLE facility[27]. In both experiments ratio of quadrupole moments have been obtained.

Highly anisotropic layered materials (2H-TaS_2 and HOPG =highly oriented pyrolytic graphite) have been used to obtain nuclear quadrupole orientation of ^{182}Ta and some other isotopes[28,29]. From a recent experiment[30] concerning electric quadrupole orientation of ^{188}Ir and ^{189}Ir, implanted in HOPG foils, $V_{zz}=10.13(\pm0.38)\times10^{22}$ V/m^2, using the known quadrupole moment of ^{188}Ir, and $Q(^{189}\text{Ir})=0.80\pm0.10$ b are derived.

NUCLEAR MAGNETIC RESONANCE EXPERIMENTS

A very powerful method to study hyperfine interactions and to determine nuclear moments is the use of magnetic resonance techniques to change populations of nuclear sublevels and thus to alter the anisotropy of directional distributions of α-, β-, γ-rays or conversion electrons. Precise knowledge of decay properties is not necessary; NO-effects merily serve as indicators for observing resonances. To induce transitions between sublevels the resonance frequency must be equal to $\nu=\Delta E/h$, where ΔE is the distance between sublevels m and m±1. Frequencies for NMR-ON experiments are usually in the 100 to 1000 MHz range. RF-fields can penetrate only thin surface layers of the metallic samples; consequently the radioactive nuclei have to be in the outer surface layer, which is automatically the case with implanted sources. In case of pure

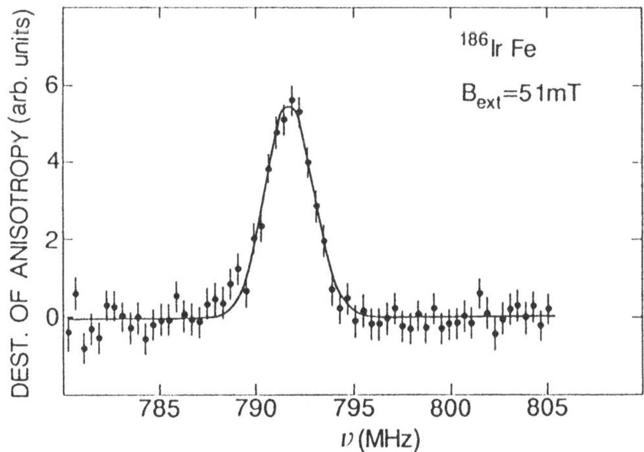

Fig.4.Nuclear magnetic resonance on oriented ^{186}Ir<u>Fe</u>; ref.31

magnetic hyperfine interaction a single sharp NMR-resonance line occurs with inhomogeneous broadening of the order of a few MHz. Fig.4 shows the NMR-ON curve observed with a ^{186}Ir<u>Fe</u> sample[31] by varying the frequency slowly. To destruct the orientation of a large number of nuclei FM-modulation is applied simultaneously. The RF-power also heats the sample. For this reason FM-modulated and unmodulated runs are subtracted to properly detect the NMR-ON effect. In addition increasing and decreasing frequencies are used. With cold implantations wider resonance curves are observed due to radiation damage frozen in at low temperature.

The NMR-ON resonance frequency $\nu_m = \mu_N(\mu/I)B/h$ makes a precise determination of $(\mu/I)B$ possible. To extract μ/I, the magnetic field must be well-known. This may not always be the case. Poor knowledge of the magnetic field value can be circumvented by varying the applied external magnetic field, which is used to magnetize the ferromagnetic sample; see fig.5 for an example[32]. The observed variation of the resonance frequency is given by:

$$\Delta\nu = (\mu/I)(1+K)\Delta B_{ext}/h. \qquad (17)$$

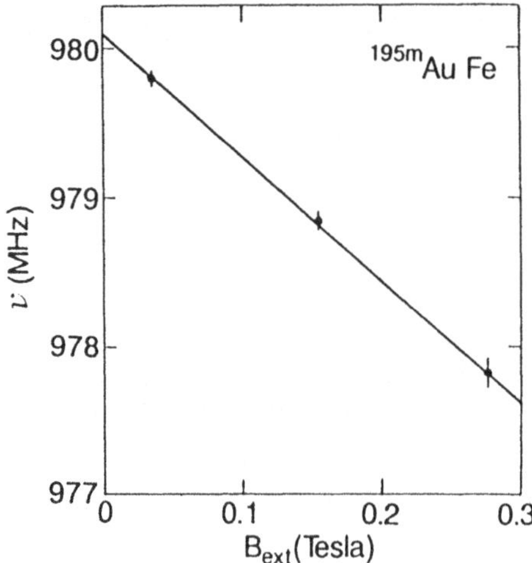

Fig.5.The field dependence of the NMR-ON frequency of ^{195m}Au<u>Fe</u>; ref.32

Often K is negligible small or known. In that case a reliable and precise value of μ/I can be obtained even if B_{hf} is not well known. By comparing $\Delta\nu$ and ν the strength and sign of the magnetic hyperfine field can be derived at the same time, which is very useful for other NO-experiments. Results of NMR-ON experiments have been collected by Herzog[33].

In the case of magnetic plus electric hyperfine interaction NMR-ON can be used to derive quadrupole moments from the structure of the NMR-resonance lines. Notably in hcp Co single crystals well separated quadrupole satellites occur as has been shown by Eder *et al.*[33] with ^{198}Au and ^{199}Au; see fig.6. Eq.15 can be used to derive ν_m and $\Delta\nu_Q$; the latter is the subresonance separation. Fig.6 clearly shows the resonances $\nu_M \pm \nu_Q/2$. These frequencies are related to the sublevel transitions m=±3/2 to ±1/2 for ^{199}Au(I=3/2). The experimental results are: for ^{198}Au ν_m=196.87±0.05 MHz, ν_Q=-14.92±0.2 MHz, and for ^{199}Au ν_m=126.19±0.07 MHz, ν_Q=-12.08±0.12 MHz. These results lead to Q(198)/Q(199)= +1.42±0.02. Comparing with quadrupole splittings of ^{197}Au and ^{199}Au in Fe Q(^{198}Au) =+0.62±0.2 b and V_{zz}=-0.90(±0.04)x10^{21} V/m^2 for gold isotopes in Co were derived by Eder *et al.*

So far continuous application of RF-power is implicitly assumed. Single-passage and pulsed NMR-ON experiments have been explored. With pulsed NMR sublevel populations are disturbed, whereafter they relax back to the Boltzmann distribution with an effective relaxation time t_{eff}; see eq.13.

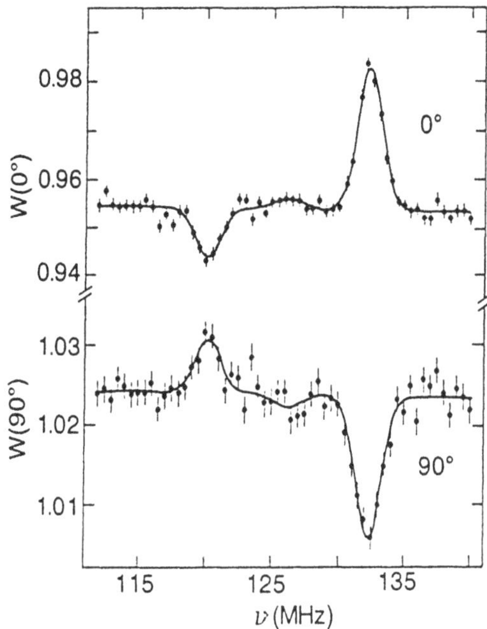

Fig.6.NMR-ON resonance curve of the 158-keV transition of ^{199}Au oriented in a single crystal of hcp cobalt[34]

Pulsed NMR-ON techniques have been significantly further developed by Wilson and coworkers in Canberra. Recent work of this group concerns Modulated Adiabatic Passage (NMR)-ON. This technique (MAPON) has revealed information about very weak electric quadrupole interactions and their distribution width in iron single crystals. From this work $Q(^{56}Co)/Q(^{58}Co)= +1.13\pm0.34$ was derived, leading to $Q(^{56}Co)=0.25\pm0.09$ b if the quadrupole moment of ^{58}Co is used[35]. In addition $\langle V_{zz}\rangle= 3.4(\pm0.5)\times10^{19}$ V/m^2 was deduced.

RELAXATION EFFECTS IN NUCLEAR ORIENTATION EXPERIMENTS

With shorter-living isotopes and lower temperatures relaxation times become of the same order of magnitude or larger than nuclear half lives. This has clearly consequences for NO-effects. If $t_{1/2}$ is of the order of the relaxation time only partial reorientation occurs. An example is the 8^- isomer (4.0 s) in the decay of $^{178m2}Hf$; ref.21. The observed anisotropies of transitions from this 8^- state are lower than expected on the basis of the observed orientation of the $^{178m2}Hf$ 16^+-isomer. Assuming a Knight shift of at most 2% the magnetic moment of the 8^--isomer is estimated to be between 1.4 and 5.3 μ_N. This shows that it is important to have accurate information about relaxation times for deriving nuclear magnetic moments on the basis of relaxation phenomena.

Another case of oriented nuclei decaying to a short-lived isomer is $^{89}Zr(78h)\rightarrow^{89m}Y(16.1s)$; ref.36. Both are $9/2^+$ nuclei of which the magnetic moments can be reliably estimated as 1.05 μ_N and 6.1 μ_N. The ^{89}Zr nuclei are produced with the Sr(α,xn) reaction and recoil-implanted into iron. The simple decay scheme and the known hyperfine field, determined by a DPAD measurement to be -27.4±0.4 Tesla, allows to calculate the NO-effect precisely. There are two extreme cases; a) complete relaxation, and b) no relaxation at all; see fig.7. The observed anisotropy of the 909-keV transition is in between the two calculated temperature dependent curves shown in this figure. An excellent fit could be obtained using f and C_K as free parameters and the relaxation calculations of Shaw and Stone[18]. The deduced values are f=0.76±0.03 and C_K=0.44±0.04 Ks.

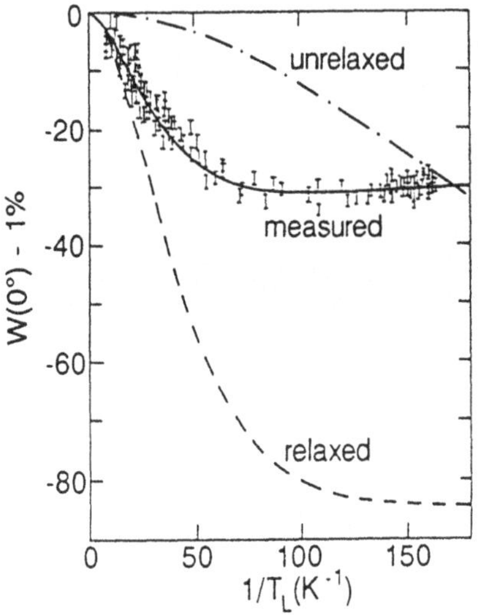

Fig.7.Anisotropy of 909-keV γ-transition of ^{89m}Y following the decay of ^{89}Zr showing a strong effect due to relaxation[36]

In implantation experiments chains of short-living isotopes may occur, which can lead to very complicated situations. An example are gold isotopes studied with the NICOLE facility using mass-separated Hg-isotopes as implants. Orientation can be inherited from preceding isotopes in competition with relaxation phenomena. If the nuclear mass number A is even the orientation is completely lost during passage of 0^+ ground-states of even-even nuclei. This makes the analysis simpler. In the case of odd-A nuclei this simplification does not occur. The situation becomes very complicated if also isomerism occurs; e.g. for ^{184}Au after implantation of ^{184}Hg; ref.27 and 37.

In time-integrated NO-experiments with short-living nuclei the influence of relaxation time is hidden in the observed effects making the analysis of the NO-data very difficult. It is therefore necessary to carry out time-resolved NO-experiments. Pulsed implantations followed by time-resolved anisotropy measurements reveal the effect of relaxation in competition with the half life. The implantation pulse is followed by several time steps during which the anisotropy is measured. After a few cycles a constant time-dependent curve of the anisotropy is expected allowing to sum the following cycles. This technique of pulsed implantation and time-resolved NO-measurements has been pioneered at the DORIS-COLD facility[38] and further developed in the NICOLE collaboration[26]. A disadvantage of this method is the loss of statistics due to the short implantation time compared to cycle time, but this is compensated by the fact, that the time dependence can be used to compare with computer simulations[37,39]. With the obtained relaxation effects time integral NO-measurements, which have better statistics, can be analysed properly.

Fig.8 shows the count rate of two γ-lines of ^{184}Au after 10 sec. pulsed implantations of ^{184}Hg. Note the shifted count rate of the 362-keV line with respect to the 161-keV transition[27]. This shift is related to an isomer of ^{184}Au, which has not been detected before. Electron spectroscopy has confirmed the ^{184}Au isomer. Fig.8 also shows the time-dependent anisotropy of the 272-keV line of ^{184}Au. Other time-resolved NO-experiments have been carried out by the NICOLE collaboration; results with computer simulations will soon become available.

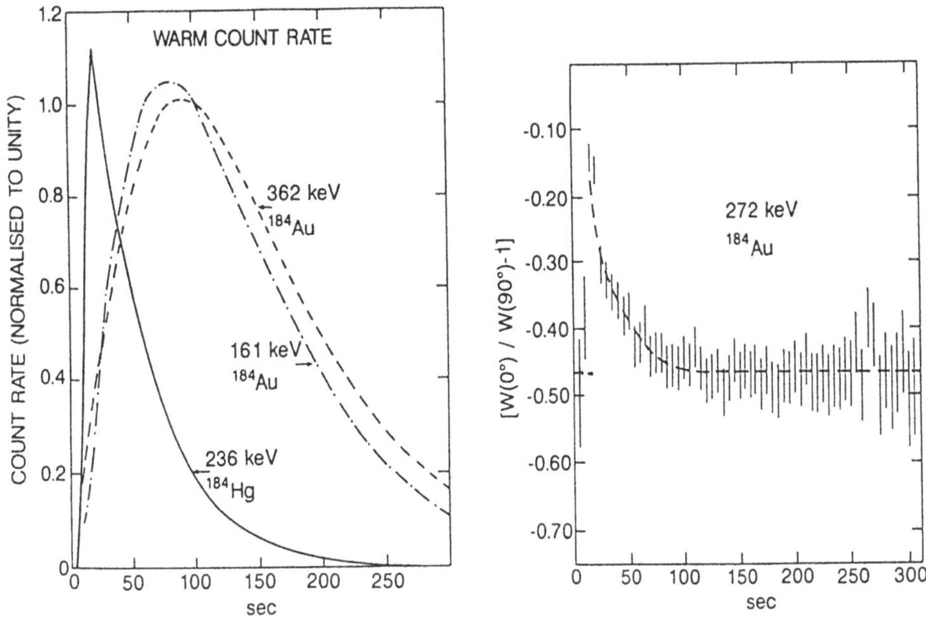

Fig.8. Warm count rates (left) and anisotropy (right) of ^{184}Au after ^{184}Hg implantation during 10 s and 300 s repetition periods[27]. For clarity no data points are shown in the figure on the left side.

LEVEL MIXING RESONANCE WITH ORIENTED NUCLEI

A novel technique to obtain information about hyperfine interactions has been invented by Coussement et al.[40] using mixing of intermediate state sublevels. Consider nuclear orientation by electric quadrupole interaction in a single crystal of which the orientation axis is making an angle $\beta \neq 0°$ with an external magnetic field. With an increasing magnetic field B or by varying β sublevels will cross at particular values of B and β and the sublevel populations mix. This level mixing is observed in the NO-anisotropy as a resonance behaviour. In this way the electric quadrupole interaction is compared with a magnetic interaction, and thus, this enables to determine the strength of the quadrupole interaction. Level Mixing Resonance on Oriented Nuclei (LMR-ON) has been used to determine quadrupole moments of $^{106, 107, 109}$Ag isomers.[41]

The combined determination of magnetic dipole and electric quadrupole moments can lead to interesting conclusions about nuclear structure, which cannot be reached by just one of them.

An interesting case concerns the asymptotic quantum number K of the 5^+ groundstate of ^{186}Ir. Its magnetic dipole moment is deduced to be $|\mu|=3.78\pm0.05$[31], 3.80(+0.12/-0.02)[42], and 3.88 ± 0.05[43] in excellent mutual agreement. However, this magnetic moment is much larger than expected by Ekström *et al.*[22], namely 2.0 μ_N. Emery *et al.*[44] and Hofstetter *et al.*[45] proposed that the groundstate of ^{186}Ir has a low value for K (0 or 1) essentially on the basis of the β-decay of ^{186}Ir. This nucleus is assumed to be the result of the coupling of the odd proton in a $1/2^-[541]$ Nilsson state and the odd neutron in a $1/2^-[510]$ state. Due to Coriolis effects the rotational band of this nucleus is thought be strongly disturbed with $I^\pi=5^+;K=0$ or 1 as lowest level (below the band head). A method to check this proposal is to measure the quadrupole moment, which depends strongly on K; that is:

$$Q=Q_o\{3K^2-I(I+1)\}/(I+1)(2I+3), \tag{18}$$

where Q_o is the intrinsic quadrupole moment.
The quadrupole moment of ^{186}Ir is derived to be -2.41 ± 0.20b[46] -2.54 ± 0.16b[47] and -3.00 ± 0.15b[42]. Taking the deformation into account for this region of nuclei Q_o is expected to be 6 to 7 barn. The experimental quadrupole moment of ^{186}Ir points to K=0 or 1 in accordance with the prediction in ref. 41 and 42. The magnetic moment of ^{186}Ir can be understood as a mixture of $|5^+0\rangle$ and $|5^+1\rangle$ giving a reasonable value of g_R; ref.42.

SIGN OF NUCLEAR DIPOLE MOMENTS

In the reported temperature-dependent NO-experiments, including NMR and relaxation measurements, it is not possible to derive the sign of nuclear dipole moments. In the case of quadrupole ratios the signs are incorporated, hence if the sign of one quadrupole moment is known, the other is known as well.

The sign of nuclear magnetic moments can be determined in NO-experiments if nuclear polarization aspects are included, that is:

i) β-asymmetry in case of allowed Gamow-Teller transitions,
ii) circular polarization of γ-radiation and X-rays,
iii) Mössbauer spectroscopy with polarized nuclei, and
iv) spin-neutrino correlation of polarized nuclei,
v) relaxation behaviour of the anisotropy as a function of time with a known relaxation time.

These experiments depend on the existence of the first order (polarization) nuclear orientation parameter $B_1(I)$. The observed asymmetries (i to iv) determine the direction of the polarization, and thus, the sign of μB. With the sign of B, the sign of μ is determined.

The determination of signs of magnetic moments with the aid of circular polarization of γ-rays has recently been explored by Scheidemann *et al.*[48] for the isotopes ^{193}Os(+), ^{191m}Ir(+), ^{192}Ir(+), ^{194}Ir(+) and ^{195m}Pt(-); the derived signs are indicated between brackets.
Berkes *et al.*[49] carried out a Mössbauer experiment with polarized ^{193}Os to determine the sign of the magnetic moment of this nucleus; a positive sign was obtained.

An experiment based on the nuclear-spin/neutrino correlation of the decay of polarized ^{56}Co has been carried out by Egorov *et al.*[50]; they determined a positive sign for the magnetic moment of ^{56}Co.

The last method v) is a new possibility depending on the relaxation process of a short-lived isotope after decay of oriented nuclei[36]. The relaxation curves are quite different for the two possible signs of a magnetic moment of the daughter nucleus. It takes much longer to reach equilibrium anisotropy if the sign of the magnetic moment of the daughter is opposite to the sign of the magnetic moment of the parent nucleus. The relaxation time of the short-lived isotope must be well known; e.g. from the systematics of the Korringa constant.

CONCLUDING REMARKS

Low-temperature nuclear orientation methods have been very successful in determining nuclear magnetic dipole and electric quadrupole moments of many nuclei. Several methods have been developed in the course of the years including NMR-ON techniques, relaxation phenomena, level mixing, implantation and recently time-resolved nuclear orientation. The shortest half lives, so far used in NO-experiments, are a few seconds. Especially implantation techniques into low-temperature targets have a great potential to explore in the near future many short-lived nuclei. Time-resolved NO-measurements will be very essential in this game. In-flight polarization techniques, e.g. the tilted-foil method, of short-lived isotopes before implantion in low-temperature targets are considered. The cooling of the target to moderately low temperatures is needed to have sufficiently long relaxation times.

REFERENCES

1. G.N.Rao, Hyp. Int. 24-26 (1985) 1119
2. N.Severijns *et al.*, Hyp. Int. 60 (1990) 889
3. E. Bodenstedt, Z.Phys. A322 (1985) 75 and references therein
4. L.E.Drain, Metallurgical Reviews 12 (1967) 195
   H.Ebert *et al.*, J.Phys.F: Metal Phys.16 (1986) 1133
5. R.S.Raghavan *et al.*, Phys. Rev. Lett.34 (1975) 1280, and R.S.Raghavan and P.Raghavan, Hyp. Int.9 (1981) 317,
6. R.Vianden, Hyp. Int. 15/16 (1983) 1081
7. E.Hagn, Chapt. 11 in *Low-Temperature Nuclear Orientation*, ed. N.J.Stone and H.Postma (1986) North-Holland Publ. Comp., Amsterdam; **"LTNO-book"**
8. R.M.Sternheimer, Phys. Rev. 80 (1950) 102
9. R.P.Gupta and S.K.Senn, Phys. Rev. A7 (1973) 850, and Phys. Rev. A8 (1973) 1169
10. K.S.Krane, Appendix 5 in LTNO-book; Report LA-4777 (1972), Ang. distribution of gamma-radiation from oriented nuclei
11. U.Dämmlich *et al.*, Hyp.Int. 43 (1988)169
12. J.Korringa, Physica 16 (1950) 601
13. F.Bacon *et al.*, Phys.Rev. B5 (1972) 2397
14. J.E.Templeton and D.A.Shirley, Phys.Rev.Lett. 18 (1967) 240
15. E.Klein,Chapter 12 in LTNO-book
16. T.L.Shaw, Ph.D.thesis, Oxford, 1987

17. N.J.Stone, in Hyperfine Interactions in Excited Nuclei, ed. G.Goldring and R.Kalish (1971) Gordon & Breach, New York.
18. T.L.Shaw and N.J.Stone, Atomic and Nuclear Data Tables 42 (1989) 339
19. K.Schlösser *et al.*, Hyp. Int. 43 (1988) 141
20. ISOLDE's users guide, CERN 86-05
21. H.Postma *et al.*, Hyp. Int. 52 (1989) 79
22. C.Ekström *et al.*, Phys.Scripta 14 (1976) 199
23. J.Wouters *et al.*, Phys. Rev. Lett. 56 (1986) 1901
24. P.Herzog *et al.*, Z.Physik A311 (1983) 351
25. G.Kaindl, F.Bacon and A.J.Soinski, Phys. Lett.46B (1973) 62
26. P.Herzog *et al.*, Z.Phys.A 314 (1983) 215
27. R.Eder *et al.*, Hyp. Int. 60 (1990) 83
28. S.Fabry *et al.*, Hyp. Int. 15/16 (1983) 863
29. B.Kastelein *et al.*, Hyp. Int. 35 (1987) 895; and Physica 165/166 (1990) 921
30. B.Kastelein, private communication
31. R.Spanhoff and H.Postma, Hyp. Int. 9 (1981) 99
32. H.J.Ligthart and H.Postma, Hyp. Int. 14 (1983) 125
33. P.Herzog, Appendix 3 in LTNO-book
34. R.Eder *et al.*, Phys. Rev. Lett. 61 (1988) 1301
35. P.J.Back *et al.*, Phys. Rev.B37 (1988) 4911
36. M.Fahad *et al.*, Hyp. Int. 43 (1988) 307
37. N.J.Stone, proc. *Exotic Nuclear Spectroscopy*, Amer. Chem. Soc. meeting, Miami, Sept. 1989
38. T.L.Shaw *et al.*, Phys. Rev. C36 (1987) 41
39. E.Zech, private communication
40. R.Coussement *et al.*, Phys. Lett. 97a (1983) 301
41. P.Put *et al.*, Phys. Lett. A103 (1984) 151; I.Berkes *et al.*, Phys. Rev., 30 (1984) 2026 and Phys. Rev. 33 (1986) 390
42. E.Hagn and E.Zech, Z.Phys.A 297 (1980) 329
43. A.L.Allsop *et al.*, J.Phys.G 8 (1982) 857
44. G.T.Emery *et al.*, Nucl. Phys. A211 (1973) 189
45. Hofstetter *et al.*, Phys. Rev.C8 (1973) 2442
46. H.Ernst *et al.*, Phys. Lett. 86B (1979) 154
47. D.W.Murray *et al.* Hyp. Int. 7 (1980) 481
48. E.Scheidemann *et al.*, Hyp. Int. 43 (1988) 493
49. I.Berkes *et al.*, J.Phys.G: Nucl. Phys. 11 (1985) 282
50. V.G.Egonov *et al.*, Nucl. Phys. A524 (1991) 425

HIGH-ACCURACY MASS DETERMINATION OF UNSTABLE Rb, Sr, Cs, Ba, Fr AND Ra ISOTOPES WITH A PENNING TRAP MASS SPECTROMETER

G. Bollen[1], H.-J. Kluge[1], H. Hartmann, T. Otto[1], G. Savard[1],
L. Schweikhard[1], H. Stolzenberg[1], G. Audi[2], R. B. Moore[3],
G. Rouleau[3], and the ISOLDE Collaboration[4]

[1]Inst. f. Physik, Universität Mainz, Fed. Rep. Germany
[2]Laboratoire René Bernas du CSNSM, Bât. 108, Orsay, France
[3]Foster Rad. Lab., McGill University, Montreal, Canada
[4]CERN, Geneva, Switzerland

INTRODUCTION

The majority of masses of radioactive isotopes has been measured by determination of Q-values in nuclear reactions or in nuclear decay. For a long time the use of direct mass determination has been limited to stable isotopes or isotopes close to stability. This changed in the 70's with magnetic spectrometers put on-line to isotope separators. The Orsay group (Audi et al., 1986) succeeded in measuring the masses in long isotope chains of alkali elements. They impressively demonstrated the possibilities embedded in direct mass determination of isotopes far from stability. The persisting demand for more precise masses of short-lived isotopes (or exotic particles) has prompted during recent years the development of new techniques. One powerful approach is the determination of the cyclotron frequency in a Penning trap, first demonstrated for light stable masses and now applied to unstable isotopes in the experiment reported here. We report on the technique and performance of a tandem Penning trap mass spectrometer at the on-line mass separator ISOLDE/CERN and summarize the very accurate results obtained on Rb, Sr, Cs, Ba, Fr and Ra isotopes.

THE PENNING TRAP MASS SPECTROMETER

Mass Determination via the Cyclotron Resonance

The principle used in the Penning trap mass spectrometer at ISOLDE/CERN is to determine the resonance of the cyclotron frequency

$$\omega_c = \frac{q}{m} \cdot B$$

of an ion with charge to mass ratio q/m in a homogeneous magnetic field B. The Penning trap allows the confinement of charged particles in a magnetic field for an extended period and the observation or excitation of their motions. The confinement is achieved by an electrode configuration that creates an axially symmetric electrostatic quadrupole field which prevents the ions from escaping along the magnetic field lines. In general this configuration consists of three electrodes; two endcaps and one ring electrode which are hyperboloids of revolution. Due to the additional electric force the ion motion is not a pure cyclotron motion with frequency

Nuclear Shapes and Nuclear Structure at Low Excitation Energies
Edited by M. Vergnes *et al.*, Plenum Press, New York, 1992

ω_c but a superposition of three harmonic eigenmotions; a slow drift around the trap axis called magnetron motion with frequency ω_-, a modified cyclotron motion with frequency ω_+ and an axial oscillation with frequency ω_z. An important relationship between the radial frequencies in a pure electric quadrupole field and a homogeneous magnetic field is

$$\omega_c = \omega_+ + \omega_-.$$

The ion motion in a Penning trap can be directly excited at this sum frequency by means of an azimuthal quadrupole field. This excitation couples magnetron and cyclotron motion so that a periodic conversion from one motion into the other can be achieved (Bollen et al., 1990). After an excitation time T_{RF} an initially pure magnetron motion is transformed into a pure cyclotron motion. In general this transformation leads to a large change in radial energy (since $\omega_+ >> \omega_-$) which is detected by a time-of-flight technique described below. The width of the cyclotron resonance curve $\Delta\nu(FWHM) \simeq 0.9/T_{RF}$ is determined by the Fourier-limit of the driving RF-field switched on for a time T_{RF}. For example, a singly charged ion with mass number $A = 100$ in a field of $B = 6\,T$ has $\nu_c \simeq 1\,MHz$. Exciting it for $900\,ms$ yields a line width of $1\,Hz$ corresponding to a resolving power of 10^6. Of course this theoretical limit can only be achieved if line broadening effects due to trap imperfections are much smaller.

Trap Imperfections

With regard to the high accuracy aimed for in this experiment it is important to find and understand sources of systematic errors which show up in a shift of the cyclotron resonance. Most of these frequency shifts arise from the trap itself and are due to deviations of the electric trapping field from a pure axial quadrupole, magnetic field inhomogeneites and misalignment of the trap relative to the magnetic field axis. (The effect of ion-ion interaction will be discussed later). In the case of the mass experiment a careful investigation allowed the construction of a Penning trap with very low magnetic and electric field imperfections (Becker et al., 1990).

Ion Capture and Cooling

In order to perform mass determinations on unstable isotopes delivered by an on-line mass separator a mechanism is needed to capture these ions in the trap. Two possibilities exist for capturing an ion from the outside world in a Penning trap. Either the trapping potential is just turned on when the ion has moved into the trap or the ion looses kinetic energy when passing through it. Both schemes are applied in the ISOLDE mass experiment. In the trap where the cyclotron resonance is determined, a short ion bunch is captured in flight by the first mechanism (Schnatz et al., 1986). This bunch is provided by another Penning trap which acts as an ion beam buncher and cooler for the continuous ISOLDE beam. In this trap, high temperature ions are stopped and cooled by buffer gas collisons. Buffer gas collisions in a Penning trap lead to the desirable decrease of the amplitudes of the axial and cyclotron motion but at the same time to an increase of the magnetron orbit due to the negative energy of this motion. In order to circumvent this process an additional oscillating quadrupole field at frequency ω_c is applied which re-centers the ions (Savard et al., 1991). This new cooling technique is mass selective since the cyclotron frequency is involved .

The Experimental Set Up and its Performance

Figure 1 shows the experimental set up at the on-line mass separator ISOLDE. It consists of two main parts, the lower Penning trap (trap ♯1) acting as ion beam buncher and cooler and the upper high precision trap (trap ♯2) in which the mass determination takes place. The first trap is placed in a vacuum chamber mounted in a $0.7\,T$ electromagnet. The continuous $60 keV$ ISOLDE beam is collected on a rhenium foil mounted in the lower endcap of this trap. The foil is rotated by 180° and current heated for $500\,ms$ so as to surface-ionize the collected material. By buffer gas collisions the ions are stopped and cooled by the mechanism explained above. Use of He gas at a pressure of $p = (10^{-4} - 10^{-3})\,mbar$ yields a capture

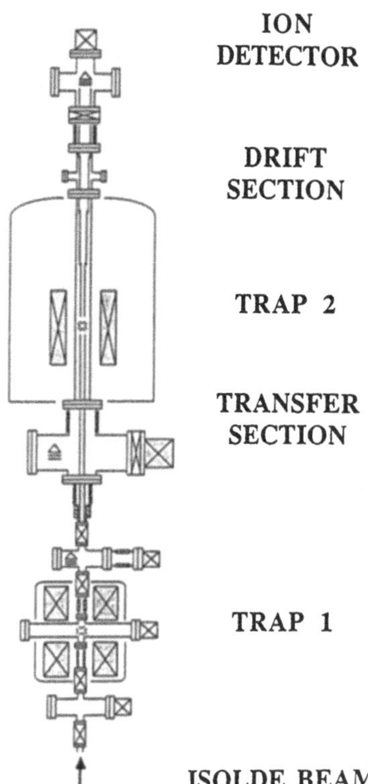

Figure 1. Experimental set up of the tandem Penning trap mass spectrometer at the ISOLDE on-line mass separator

efficiency of $\epsilon_{capture} > 10^{-4}$. A mass resolution of $m/\Delta m(FWHM) = 500$ is obtained in this trap, sufficient to remove isotopic contaminations present in the ISOLDE beam. After a cooling period of $200\,ms$ a voltage pulse is applied to the trap to eject the ions. The ion bunch is accelerated to 1 keV, transferred to the upper Penning trap with an electrostatic lens system and again retarded. The potential of the lower endcap is first switched to a low value allowing the ions to enter the trap and then raised again in order to capture them.

This high precision trap is constructed from oxygen-free copper and is gold plated. Glass ceramics are used for insulation. All parts are machined as thin as possible in order to minimize magnetic field inhomogeneities due to the susceptibility of the materials. Deviations from the ideal quadrupole field are compensated by correction electrodes installed between the ring electrode and the endcaps and at the holes required in the endcaps for injection and ejection of ions. The azimuthal quadrupole field for the excitation is created by splitting the ring electrode into four quadrants. The trap itself is placed off-center with respect to the axis of the magnetic field so as to to capture the ions at a magnetron orbit of $\rho \approx 1\,mm$ radius. After excitation of their motion by the RF-field the ions are gently kicked out of the trap and drift along the magnetic field lines to an ion detector placed in the weak fringe field of the magnet. The radial energy gained by the excitation is converted into axial energy in the inhomogeneous part of the magnetic field. Therefore the measurement of the drift time as a function of the applied radio-frequency yields a resonance curve with a minimum at $\omega_c = \omega_+ + \omega_-$, corresponding to a maximum of gained radial energy. As an example, figure 2 shows a resonance curve obtained for ^{120}Cs ($T_{1/2} = 64\,s$). As expected from the excitation time of $1.8\,s$ a Fourier-limited linewidth of $0.5\,Hz$ is achieved corresponding to a resolving

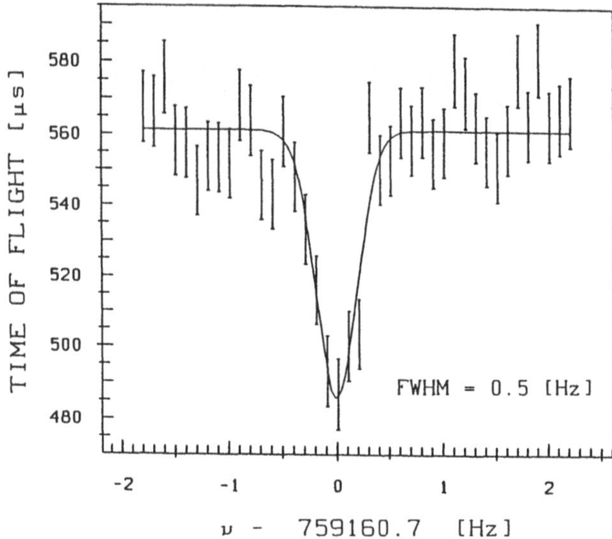

Figure 2. Cyclotron resonance of ^{120}Cs ($T_{1/2} = 64\,s$) with a resolving power of 1.5 million.

power of 1.5 million. This measurement took 7 minutes with 2000 ions detected. The overall efficiency (ratio of the number of detected ions to the the number of ions implanted in the foil) was determined to be $\epsilon_{total} = 5 \cdot 10^{-5}$. An accuracy of $\delta m/m = 2.1 \cdot 10^{-7}$ could be achieved for this isotope, corresponding to $\delta m = 24\,keV$.

MASS MEASUREMENTS

Several cyclotron resonances were observed for each isotope under investigation. The calibration of the magnetic field was performed via the cyclotron resonance of a stable reference isotope with well known mass. In order to achieve highest accuracy a careful search for systematic errors was performed before, during and after the runs. Within the quoted accuracy no systematic shifts due to field instabilities and imperfections were found. The only systematic shift of ν_c observed at a level of $\delta m/m \approx 10^{-7}$ is due to ion-ion interaction of simultaneously stored ions with different masses. Beside the isotopes in the ground state, isomers and isotopes of neighbouring elements are produced in the ISOLDE targets. Isobaric products are partly eliminated by the chemical selectivity of the ionization process either in the ISOLDE ion source or by the surface-ionizer in the first trap. In some cases with sufficient difference in half-lives between two isobars, the shorter lived isotope can be allowed to decay on the foil in trap ♯1 before starting the mass measurement. This method was also applied for isotopes with significantly different half-lives of ground and isomeric states. In cases of mass differences of $> 1\,MeV$, contaminant ions were removed from trap ♯2 by selectively exciting them to large cyclotron orbits by a dipole field at ω_+. This technique was applied to isobars. For isotopes with a half-life of the isomeric state comparable to that of the ground state none of the above techniques is applicable. There the only possibility is to work with a number of stored ions that is as small as possible in order to minimize the effect of the ion-ion interaction. This procedure and the high resolving power of the Penning trap mass spectrometer allowed for the first time in the history of mass spectrometry to resolve isomeric and ground state of a nucleus in the cases of ^{78}Rb and ^{84}Rb (fig. 3).

Table 1 summarizes the isotopes investigated so far with the Penning trap mass spectrometer. The stable isotopes ^{39}K,^{85,87}Rb,^{133}Cs listed in the table were used for calibrating the magnetic field. Masses could be determined for the first time in the cases of ^{78,79}Sr, ^{122m}Cs,

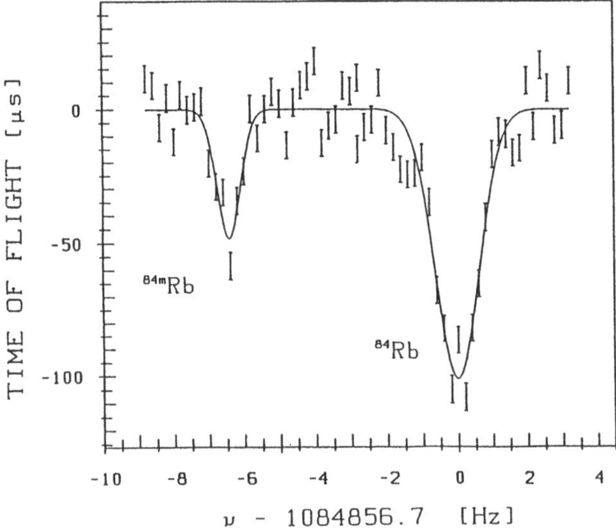

Figure 3. Cyclotron resonances for ^{84}Rb and ^{84m}Rb. The mass difference determined to be 463(10) keV is in full agreement with the known energy of the isomer of 463.62(10) keV.

^{124,126}Ba, and ^{230}Ra. The high performance of the Penning trap mass spectrometer allowed the determination of mass values with accuracies of $\delta m/m \approx 10^{-7}$ for all isotopes. The measurements performed on neutron-deficient Cs isotopes increased the accuracy of the masses far from stability by one order of magnitude (Stolzenberg et al., 1990). This has resolved the conflict between the mass values from Q_β-studies and values by a Mattauch-Herzog-spectrometer (Audi et al., 1986). In addition wrong mass values were found for ^{130}Cs and ^{135}Cs, both close to stability, due to an underestimation of the errors of Q_β-values. Presently, calculations with an extended macroscopic-microscopic model are performed for the Cs-region with the parameters adjusted to give the best fit to the accurate experimental masses. The model then allows the evaluation of spins, magnetic dipole and electric quadrupole moments which have to be compared with experimental data. The data evaluation for the other elements is nearly finished. In the case of the neutron-deficient Rb and Sr isotopes a comparison of theoretical mass values and experimental results reveals a reversed odd-even staggering in the experimental microscopic energy. We are presently studying this behaviour, taking into regard the large change of charge radii at (N=47) observed in isotope shift measurements by laser spectroscopy. The performed measurements not only improved the accuracy of the mass values of the investigated nuclei but also improved those of neighbouring isotopes of other elements due to the manyfold links between them.

OUTLOOK

The future aim of this experiment is to extend the mass measurements to elements which can not be surface-ionized. A first step forward has already been achieved with the development and successful test of a Paul trap beam collection system. This system can retard and collect the continuous ISOLDE beam. After cooling by buffer gas collisions the ions are extracted as an ion bunch. The buncher will allow us to investigate all isotopes of those elements which are not surface-ionizable and are produced relatively free of isobaric contamination. With the planned development of a highly efficient pulsed Laser-Induced-Surface-Ionization (LISI) source for positive and negative ions, the mass determination of isotopes of surface-ionizable elements can be extended to regions far away from stability. Fur-

Table 1. Isotopes for which cyclotron resonances have been determined with the Penning trap mass spectrometer. The stable isotopes ^{39}K, $^{85,87}Rb$, ^{133}Cs have been used for the calibration of the magnetic field.

Isotope	Mass Number
K	39
Rb	75, 76, 77, 78, 78m, 79, 80, 81, 82m, 83, 84, 84m, 85, 86, 87
Sr	[78] [79] 80, 81, 82, 83, 87
Cs	118, 119, 120, 121, 122+122m, [122m] 123, 124, 125, 126, 127 128, 129, 130, 131, 132, 133, 134, 135, 136, 137, 138, 139, 140
Ba	[124] [126] 138, 139, 140, 141, 142, 143, 144
Fr	209, 210, 211, 212, 221, 222
Ra	226, [230]

thermore the investigation of isotopes of halogenes will become possible. In addition, the LISI ion source would enable the creation of carbon clusters over a large mass scale. This clusters can be used for the calibration of the magnetic field. Hence not only direct but also absolute mass determinations will be performed. A future plan is to use laser-induced desorption and resonance ionization for the investigation of isotopes of refractory elements only available as daughter products. These techniques will extend the applicabilty of the Penning trap method to most of the isotopes produced at the new ISOLDE facility being constructed at CERN. They will allow a systematic study to very high accuracy of the nuclear binding energy as a function of a broad range of neutron and proton numbers.

ACKNOWLEDGEMENTS

This project is supported by the BMFT under contract number Hz-188-I

REFERENCES

Audi G., Coc A., Epherre-Rey-Campagnolle M., Le Scornet G., Thibault C., Touchard F., 1986, Mass-Spectrometic Measurements of Exotic Rb, Cs and Fr Isotopes, Nucl. Phys., A449:491

Becker St., Bollen G., Kern F., Kluge H.-J., Moore R. B., Savard G., Schweikhard L., Stolzenberg H., and the ISOLDE Collaboration, 1990, Mass Measurements of Very High Accuracy by TOF-ICR of Ions injected into a Penning Trap, Int. J. Mass. Spec. Ion Proc., 99:53

Bollen G., Moore R.B., Savard G., and Stolzenberg H., 1990, The Accuracy of Heavy Ion Mass Measurements, J. Appl. Phys, 68:4355

Savard G., Becker S., Bollen G., Kluge H.-J., Moore R.B., Schweikhard L., Stolzenberg H., and Wiess U., 1991, A New Cooling Mechanism for Heavy Ions in a Penning Trap, Phys. Lett., submitted

Schnatz H., Bollen G., Dabkiewicz P., Egelhof P., Kern F., Kalinowsky H., Schweikhard L. Stolzenberg H., and Kluge H.-J., 1986, In-Flight Capture of Ions into a Penning Trap Nucl. Instrum. Meth., A251:17

Stolzenberg H., Becker St., Bollen G., Kern F., Kluge H.-J., Otto Th., Savard G., Schweikhard L., Audi G., Moore R.B., 1990, Accurate Mass Determination of Short-Lived Isotopes by a Tandem Penning-Trap Mass Spectrometer, Phys. Rev. Lett, 65:3104

γ-RAY SPECTROSCOPY OF RELATIVISTIC FRAGMENTS: A NOVEL TECHNIQUE

R. Holzmann for the LAND Collaboration‡

Gesellschaft für Schwerionenforschung (GSI), Darmstadt
D-6100 Darmstadt, FRG

INTRODUCTION

During 1990 a new accelerator complex consisting of the heavy-ion synchrotron SIS and the experimental storage ring ESR came into operation at GSI. This facility provides beams of nearly relativistic velocities for ion masses up to uranium. After nearly one year of commissioning of the large-scale detector systems the physics programme has finally successfully taken off. This contribution addresses the relevance of near-relativistic heavy-ion beams to nuclear-structure studies and, in particular, on-line γ-ray spectroscopy. The production and storage of radioactive secondary beams with sufficient intensities at the SIS/ESR potentially gives access to a new field in the physics of nuclear structure and reaction dynamics. We will discuss the relevant physical and instrumental concepts, and finally present some preliminary results from our recent experimental work.

A NEW ROAD TO THE STUDY OF EXOTIC NUCLEI

Whereas the proton-rich side of the nuclear chart has been mapped in great detail over the last decades, the neutron-rich side is being explored only hesitatingly. Indeed for a long time, apart from neutron transfer, no experimental techniques have been available to investigate excited states in neutron-rich nuclei. Only quite recently have new methods been explored, based on the γ-ray spectroscopy of neutron-rich fragments from spontaneous [1] or in-beam [2] fission, and from deeply inelastic collisions [3]. Their applicability is largely determined by the A and Z distributions characteristic of these processes. Furthermore their selectivity critically relies on being able to discriminate unwanted nuclides, e.g. through γ-coincidence techniques.

We propose a novel experimental technique which can be applied throughout the chart of nuclides. It consists in producing excited exotic nuclei in a three-step procedure :

1. Relativistic projectiles from the SIS (with E/A = 100-1000 MeV/u) are fragmented on a production target yielding a distribution of mostly unstable nuclei.
2. Following a method pioneered at GANIL the fast-moving projectile fragments are mass- and charge-selected in a fragment recoil separator (FRS) [4], producing a mono-isotopic secondary beam.

3. Excited states in the separated fragments are populated through Coulomb and/or nuclear excitation in peripheral collisions on a second high-Z target. At the beam energies considered here the cross sections for these processes can be as high as a few barns [5]. Low-lying discrete states in the fragments, fed both by direct excitation and through statistical decay from high-lying collective modes (GDR, GQR), can be studied via their γ emission.

One of the bigger experimental devices installed at the SIS is the fragment recoil separator (FRS) built by a GSI-Bordeaux-Darmstadt-Gießen-Michigan collaboration [4]. The FRS allows to produce secondary beams by fragmentation of stable primary projectiles and in-flight isotopic separation. The fragments emerging from the production target are kinematically focused into a narrow cone ($\leq 1^o$) with a low momentum spread ($\Delta p/p \leq 1\%$). In the first section of the FRS nuclei are analyzed according to their magnetic rigidity, leading to a A/Z selection for fully stripped ions. In the second stage full A and Z separation is achieved exploiting the Z-dependent energy loss in a profiled degrader. With a primary beam intensity of $5 \cdot 10^{11}$ ions/s and a 1 g/cm^2 fragmentation target, one expects to realize production rates of exotic nuclei of up to 10^4-10^7/s [4].

One can now envision basically two different schemes for in-beam nuclear spectroscopic investigations :

- Isotopically separated radioactive beams are injected into the storage ring ESR, cooled down to reduce their emittance and improve storage times, to be finally decelerated to energies around the Coulomb barrier. Subsequently all of the standard spectroscopical techniques can be applied, either with internal targets, or, after re-extraction, on external targets. From a technical point of view this scheme is extremely demanding and will probably take a few more years to be fully implemented.

- In a more direct approach the separated, near-relativistic fragments are transported to a second target on which nuclear states are excited in peripheral collisions. These excitations can then be studied through a measurement of their particle and γ decay. The process of electromagnetic excitation in the 500-1000 MeV/u regime is expected to be well suited for this purpose [6, 7].

In a first step, before secondary beams of sufficient intensity become available, we propose to bypass the FRS and study directly the decay of the highly-excited nuclides produced in peripheral collisions with a stable beam. In the following sections some of the physical and experimental aspects of electromagnetic excitation with near-relativistic heavy ions will be discussed. It should be mentioned that the study of the γ decay of nuclei produced in the bombardment of various targets with relativistic heavy-ion beams has been used by Shibata et al. to measure fragmentation cross sections [8]. These authors have first demonstrated the feasibility of γ-ray spectroscopy in the presence of the very large background from atomic processes in the 100-1000 MeV regime [9].

ELECTROMAGNETIC AND NUCLEAR EXCITATION PROCESSES

Coulomb excitation at bombarding energies below and around the barrier is a powerful tool in nuclear spectroscopy. Because of its strength, in heavy-ion collisions multi-step excitations towards high-spin states occur in collective nuclei. Moreover, as the electromagnetic interaction is well understood, transition strengths can easily be extracted from measured cross sections. The same holds at relativistic bombarding energies, except that the adiabatic cut-off frequencies are much larger, allowing to excite much higher-lying states [6]. Thus the maximum excitation energy reachable in a single step at SIS beam energies amounts to about 30 MeV. The access to high excitation energies leads to very large cross sections for exciting giant resonances. Conversely, the short collision times imply that multi-step E2 excitations of high-spin states will be largely suppressed.

A semiclassical treatment (i.e. straight line approximation) of electromagnetic processes in heavy-ion collisions at relativistic velocities has first been proposed by Winther and Alder [6]; Bertulani and Baur have recently given a complete review [7]. The excitation of low-lying collective states and their coupling to high-lying resonances has also been treated by G. Baur [10]. Figure 1 illustrates the magnitude of excitation cross sections for both giant resonances and low-lying collective states, as calculated within the semiclassical approximation. Besides being excited directly, low-lying discrete levels are also populated in the decay of the giant resonances. Moreover, collective low-spin surface vibrations and rotations might get excited too. Such states can then be observed via their γ decay. Clearly this is of particular interest for exotic nuclei on the neutron-rich side, as already emphasized above. As a matter of fact Coulomb excitation of radioactive beams might be the only feasible way to access the giant resonance region in exotic nuclei and study the extreme isospin behaviour of these high-lying excitations.

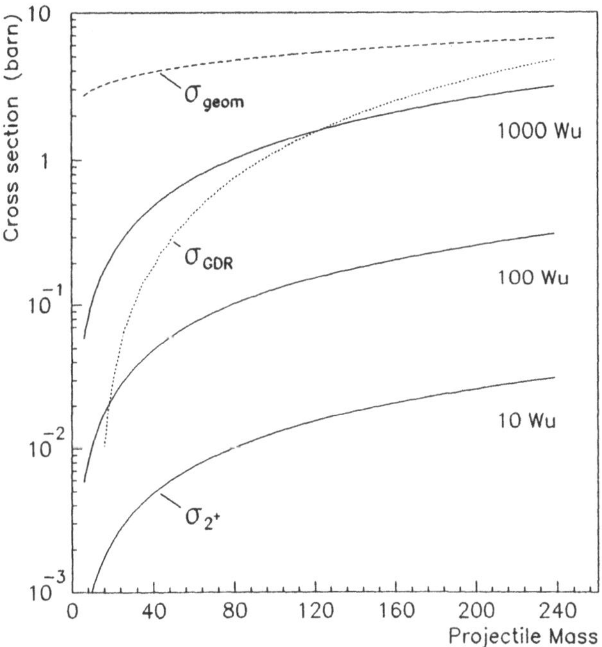

Fig. 1. Semiclassical calculation of the electromagnetic excitation of 700 MeV/u projectiles impinging on a ^{208}Pb target. Shown are the geometric cross section (dashes), giant dipole excitation (dots) and excitation of a 2^+ state (solid) with assumed B(E2) of 10, 100 or 1000 Weisskopf units.

Concurrently to the purely electromagnetic process, for close collisions which produce an overlap of the projectile and target, nuclear fragmentation leads to highly excited residual nuclei. The cross sections, as well as the excitation energies, linear momentum spreads and angular momenta of the fragments can be calculated in a two-step model, for example the abrasion-ablation model of Gaimard and Schmidt [11]. It describes the first step of the fragmentation process in peripheral collisions as a diabatic abrasion of the geometric overlap zone, followed in a second step by the evaporation of particles from the excited prefragments. This model predicts rather high angular momenta for the produced prefragments, e.g. for a prefragment ^{200}Au from fragmentation of ^{208}Pb one finds $\sqrt{< J^2 >} = 10\hbar$. This value is expected to be essentially preserved up to the end of the de-excitation process, because the centrifugal barrier favors the evaporation of particles with low angular momenta. Consequently the nuclear fragmentation process could be a promising mechanism for populating high-spin levels in exotic nuclei.

A POSITION-SENSITIVE GERMANIUM STACK DETECTOR

A major problem of the proposed experimental technique is the extremely high recoil velocity of the relativistic fragments, e.g. $v/c \geq 0.4$. Together with the large detector solid angles, required for very rare exotic nuclei, it produces an extreme Doppler broadening of the γ lines, destroying the intrinsically good energy resolution of germanium detectors. This problem can however be circumvented by measuring the angle of incidence of the photons and correcting the Doppler effects on an event-by-event basis. The necessary position sensitivity can be achieved by employing planar Ge crystals with a segmented contact. In this scheme each segment has to be read out separately, providing to some extent position information, while the energy information is obtained from the unsegmented opposite contact.

A prototype Ge stack detector consisting of 2 planar diodes (50x20 mm^2) has been designed and built [12]. The crystals are mounted back to back in a portable streamlined cryostat with a 65 mm diameter cap. They both have on one side a boron-implanted contact segmented into 8 parallel strips of equal area and on the other side a lithium-drifted unsegmented contact. The segmentation has been achieved by implanting through an appropriate ion mask. From the plain contacts energy and time signals are taken via two low-noise preamplifiers with a cooled FET, whereas the strips yield signals corresponding to the energies deposited in the individual segments of the crystals. These subenergies are fed to 16 additional hybrid-chip preamplifiers operating at room temperature.

The system has been tested with γ rays from various radioactive sources. The photopeak efficiency for the 1.33 MeV ^{60}Co line, totalled over both crystals, is 20 % (as compared to a 7.6x7.6 cm^2 NaI detector). The fwhm energy resolutions are 2.3 and 2.0 keV at 1.33 MeV on the front and rear diodes respectively, while the strip subenergies have resolutions of 2.5-3.0 keV. The intrinsic fwhm time resolution for photon energies above 500 keV is ≤ 3.5 ns for each crystal. A scan across the front face of the detector with a low energy source can be used to reveal the segmentation. Figure 2 shows such a scan in 1 mm steps for the 122 keV line of ^{57}Co.

From the strip signals one can compute the centroid of the energy deposited in each crystal. If for a given hit only neighboring segments are involved, this directly yields an estimate for the point of impact of the photon along the axis perpendicular to the strips. A more elaborate analysis of the subenergy pattern can be used to reconstruct full-energy events with Compton scattering between distant segments. The position resolution has been measured by irradiating the detector with sources collimated to 2x2 mm^2. For full-energy

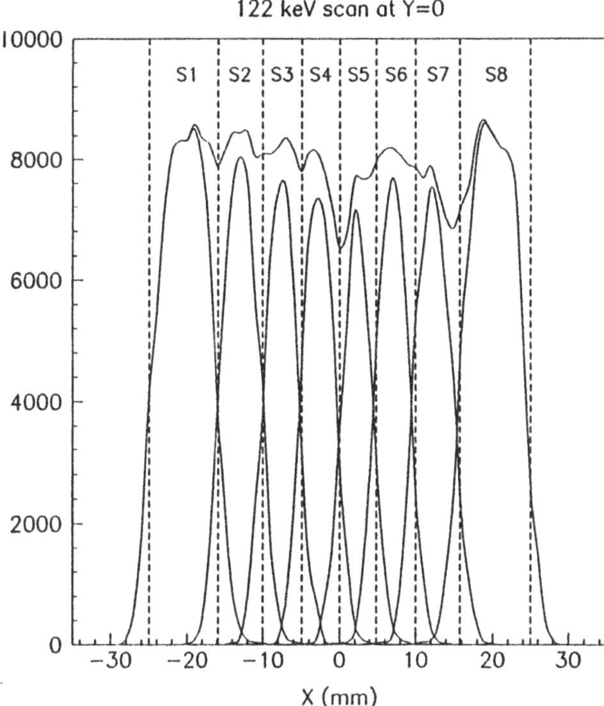

Fig. 2. Scan across the front face of the strip detector with a collimated ^{57}Co source. Plotted are the photopeak count rates on individual strips as a function of the source position. Also shown is the total, i.e. summed, rate. The two dips visible are due to absorption in the FET and HV decoupling capacity of the front diode. The dashed lines indicate the actual segmentation of the contacts.

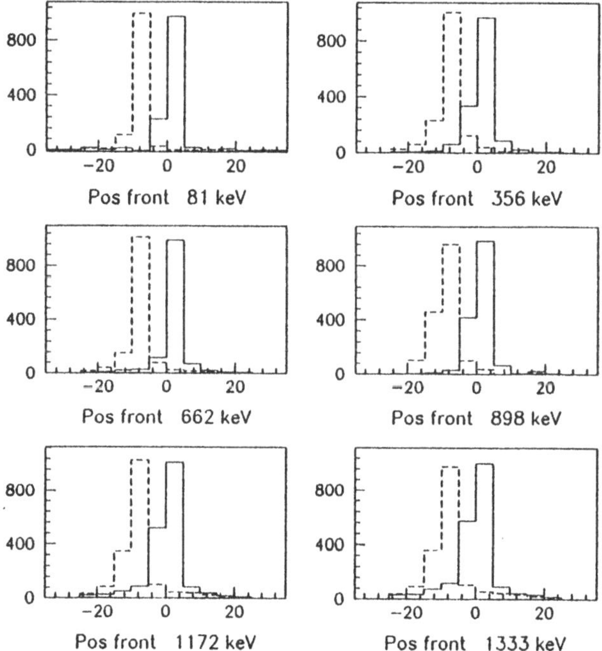

Fig. 3. Position spectra for γ energies of 356, 662 and 1333 keV measured with collimated sources irradiating the front face at the center (solid) and 10 mm off center (dashed). These spectra are obtained by gating on the full-energy peak.

events it is on the order of 5 mm (i.e. the strip width) at 356 keV, 7 mm at 898 keV and 10 mm at 1.836 MeV (see Fig. 3). These results agree well with Monte Carlo calculations done with the simulation code GEANT [13]. They show that the position resolution is good enough to allow for a 5-10 fold reduction of the Doppler broadening of γ lines emitted from a moving source.

The segmented Ge stack is a vital component of an experimental setup able to investigate the γ decay of fragments produced with very high recoil velocities. It appears that the necessary position sensitivity can be achieved over a range of γ energies large enough to make it a device of more general interest.

ELECTROMAGNETIC EXCITATION OF ^{136}Xe PROJECTILES

As mentioned above, the giant dipole resonance (GDR) is the most strongly excited mode in peripheral collisions with relativistic projectiles. A semiclassical estimate of the ^{136}Xe electromagnetic dissociation cross section gives $\simeq$2 barn on a lead target and a few tens of mbarn on a carbon target at a beam energy of 700 MeV/u. This has to be compared to nuclear fragmentation cross sections of $\simeq$ 500 and 250 mbarn, for Pb and C targets respectively. The cross section for direct excitation of the first excited 2^+ state in the closed-shell nucleus ^{136}Xe is only 16 mbarn. The prime population mechanisms for low-lying discrete levels are therefore the decay of the GDR, as well as fragmentation. We have undertaken an experimental programme with the aim to investigate projectile Coulomb dissociation and fragmentation processes in near-relativistic collisions [5].

In heavy nuclei the dominant decay mode of high-lying excitations is neutron evaporation and an efficient detector for high-energy neutrons is mandatory. Neutrons emitted from the projectile are kinematically focused into a narrow cone centered at 0°. They are detected in the highly segmented Large Area Neutron Detector (LAND) positioned 10-15 m downstream from the target. This position-sensitive detector covers 2x2 m^2 and allows to measure the time of flight and angle of incidence of the neutrons, yielding a momentum resolution of $\simeq$2 %. The beam is swept aside in a large dipole magnet (ALADIN). Incoming and outgoing projectiles are tracked with position-sensitive plastic scintillation counters with time and position resolutions of 50 ps and $\leq$1 mm respectively. The tracking detectors are complemented with a multiply sampling ionization chamber (MUSIC) and glass Cherenkov counters to guarantee complete A and Z identification ($\Delta A \leq$ 1 amu, $\Delta Z \simeq$ 0.3 e). A 36-element plastic scintillator box surrounding the target serves to veto violent central events. Finally a castle consisting of 48 hexagonal 5x20 cm^2 BaF_2 crystals, covering about 30 % of 4π, is used to measure the γ decay of the residual projectile fragments. The setup is completed with the segmented Ge detector described in the previous section.

First measurements have been done with Xe beams of 500, 700 and 800 MeV/u bombarding C, Cu, Sn and Pb targets. A preliminary analysis of the data reveals a Coulomb dissociation cross section on Pb of 2.1$\pm$0.2 barn at 700 MeV/u. From the energies released by neutron evaporation and γ decay, and taking into account the neutron binding energies, the projectile excitation spectrum can be reconstructed. We also observe γ rays from the decay of low-lying discrete levels populated in target fragments with rather high cross sections, namely a few hundred millibarn.

REFERENCES

[1] W.R. Phillips et al., Phys. Rev. Lett. **57** (1986) 3257.
[2] Y. Abdelrahman et al., Phys. Lett. **199B** (1987) 504 .

[3] J.X. Saladin in Proc. Conf. on High-Spin Nuclear Structure and Novel Nuclear Shapes, April 13-15 1988, Argonne.
[4] "The GSI Separator for Projectile Fragments", in GSI Nachrichten 11-87.
[5] H. Emling in Proc. of EPS-8 Meeting, September 2-7 1990, Amsterdam; and Nucl. Phys. **A520** (1990) 687c.
[6] A. Winther and K. Alder, Nucl. Phys. **A319** (1979) 518.
[7] C.A. Bertulani and G. Baur, Phys. Rep. **163** (1988) 301.
[8] T. Shibata et al., Nucl. Phys. **A308** (1979) 513.
[9] R. Anholt et al., Phys. Rev. **A39** (1984) 2234; Phys. Rev. **A33** (1986) 2270.
[10] G. Baur, Z. Phys. **A332** (1989) 203.
[11] J.-J. Gaimard and K.-H. Schmidt, submitted to Nucl. Phys. A.
[12] Intertechnique SA, France.
[13] R. Brun et al., GEANT3, CERN Report DD/EE/84-1 (1986).

‡ The Large Area Neutron Detector (LAND) Collaboration :
Th. Blaich[e)], Th.W. Elze[d)], H. Emling[c)], H. Freiesleben[a)], K. Grimm[d)], W. Henning[c)], R. Holzmann[c)], J. Keller[a)], H. Klingler[d)], W. Kneissl[a)], R. König[a)], J.V. Kratz[e)], R. Kulessa[b)], D. Lambrecht[e)], S. Lange[a)], Y. Leifels[a)], E. Lubkiewicz[b)], M. Proft[e)], W. Prokopowicz[b)], R. Schmidt[c)], C. Schütter[d)], H. Spies[d)], K. Stelzer[d)], J. Stroth[c)], E. Wajda[c)], W. Walus[b)], H.J. Wollersheim[c)], H. Xie[c)], M. Zinser[e)], E. Zude[e)]

[a] Univ. Bochum, [b] Univ. Cracow, [c] GSI, Darmstadt, [d] Univ. Frankfurt, [e] Univ. Mainz

LIFETIME MEASUREMENTS IN MEDIUM HEAVY NUCLEI USING THE GAMMA RAY INDUCED DOPPLER BROADENING (GRID) TECHNIQUE

J. Jolie and H.G. Börner

Institut Laue Langevin
156X, 38042 Grenoble Cedex
France

INTRODUCTION

Atomic nuclei are finite systems with, generally, too many interacting bodies to allow a few-body treatment, and too few bodies for a many (infinite) body treatment. Therefore, one finds a zoo of different ways to describe the wealth of phenomena showing up in the atomic nucleus. These can be classified in several ways. One very important classification relies on the total number of valence nucleons (particles or holes), present. Such a classification allows to describe the transition from the single-particle structure of nuclei near closed shells towards the very collective rotational structure of deformed nuclei. Here, the collective behavior builds up due to the increasing number of valence particles, while the other particles are considered to be inert at low energies.

Changes in structure can also be observed very clearly by increasing the excitation energy of the nucleus. At low energies the even-even nucleus has almost no excited states, due to the strong pairing force, and behaves very collectively with a dominant neutron-proton quadrupole-quadrupole force. Depending on the groundstate deformation the atomic nucleus can then be described as a 'semi-macroscopic' system, showing vibrations and rotations. With increasing excitation energy other energetically allowed possibilities show up and the paired off nucleons gradually can break up at energies higher than twice the pairing gap Δ ($\approx$0.8-1.0 MeV). Even when the pairing breaks down in the groundstate configurations it is still important for the excited states, the 2particle-2hole states, to also pair off. When such excitations happen, the consequence of this is that the number of interacting nucleons has increased. It is then not surprising that one finds still collective, but more deformed, states in the excited level scheme. These intruder states can then mix with the normal low-lying states.

Going further up in excitation energy the many quasi particle states, become much more numerous, and in general only the yrast band is made out of normal states up to the first backbending around 4 MeV in deformed nuclei. Before the yrare states exhibit the full complexity of single-particle and collective behavior and their mixing. In order to describe the yrare states above 3 MeV, both the collective and shell model approaches, are abandoned for statistical approaches, introducing the concept of temperature and/or chaos in order to describe the nucleus.

The study of the interplay between the single (quasi) particle and collective degrees of freedom in the nucleus is therefore ideally done at excitation energies around 2 MeV. It is there that one will be able to study in detail some of the assumptions used to describe collective behavior at lower energies. In particular we will focuss here on the validity of the concept of a phonon in the nucleus, by studying multiphonon states occuring around 2 MeV.

THE GRID TECHNIQUE

In order to study the nucleus at high excitation energies, one has to fulfil two conditions. The nucleus has to be brought into an excited state and one has to determine observables which characterize the structure of the state. At the Institut Laue Langevin, over the last years, a new experimental method has been developed which fulfills both conditions. This GRID technique allows to determine short lifetimes of excited states created after neutron capture. As lifetimes yield absolute transition rates and neutron capture is a non selective reaction, feeding the low spin yrare and yrast states, this technique is a very powerfull tool to study the twilight zone inbetween pure collective motion and the chaos of the many quasi particle excitations.

The essential ingredient of the GRID technique is the observation of small Doppler broadening of gamma rays (Börner et al., 1988): After neutron capture each emitted gamma-ray imparts a recoil to the nucleus and the energy of each subsequently emitted gamma ray is -in case the nucleus is still moving- Doppler shifted. As the recoil inducing gamma-rays are emitted isotropically this leads to a Doppler broadening ($\Delta E/E = 10^{-4}$-10^{-6}) of the gamma energy. This can be observed by ultra high resolving power gamma-ray spectroscopy as available with the spectrometer GAMS4 (NIST/ILL collaboration). The spectrometer uses subsequent gamma-ray diffraction on two flat crystals to achieve ppm resolution (Dewey et al., 1989).

Once the broadened lineprofile measured, the analysis of its shape allows to extract the lifetime, or limits thereon, of the nuclear level which is deexcited. The following simplified picture elucidates the mechanism. After neutron capture the nucleus is excited to a state at 8-10 MeV and will, generally, decay via a primary gamma ray. At the moment of the emission of this primary gamma ray the atomic nucleus also starts to recoil with a well defined velocity in the direction opposite to the one of the primary gamma ray. This defines a start time: at t=0 the recoil velocity is $v_R/c=E_{primary}/mc^2$, with mc^2 the rest mass of the atomic nucleus. The 'stop signal' is given by the moment of emission of a second gamma ray and is governed by τ, the lifetime of the intermediate excited state. The elapsed time t can be measured experimentally as, when the second gamma is emitted at time t the atomic nucleus will have slowed down in the target and will have a certain velocity $v(t) \leq v_R$ leading to a Doppler shift (and broadening for a set of secondary gammas) smaller than the one corresponding to t=0. Hence the form of the lineshape allows to extract the lifetime of the excited level in a direct way, provided the slowing down process is understood. Naturally, our simple example is not often realized in nature, as the levels under study are generally fed by many different gamma cascades.

ACCURACY OF THE LIFETIMES

The accuracy of the lifetimes determined with GRID depends on three factors: the counting statistics, the stopping theory and the knowledge concerning the feeding of the level.

Due to the fact that the Doppler broadened shapes rapidly change in the 3 to 100 fs region, the statistical error induces a very small error for short lifetimes (τ< 200 fs). For longer lifetimes statistics has a bigger influence on the error due to the decreasing recoil broadening. Moreover, this has to be distinguished from the broadening of the line due to thermal motion. Around 5ps the broadening is generally so small, that very good counting statistics is needed to deduce definite lifetimes.

The error induced by the stopping model is more difficult to evaluate. GRID accesses such small recoils (E_R< 20000/A eV) that no standard description of the slowing down is currently available. Up to recently the GRID data were analysed using a simple stopping model, which has the advantage of being completely parameter free and easily applicable to all kinds of targets (Jolie et al, 1989,1990). This model was extensively tested by comparing the results with known lifetimes and/or by checking its consistency when using different target compositions (Ulbig et al., 1991). Despite these successful tests its microscopic foundations are not obvious due to the different approximations which can cancel the effect of each other. Therefore, recently, two new approaches to describe the slowing down process were developed. The first relies on Molecular Dynamics simulations of the low energetic recoils (Kuronen, 1991), the other one on the Boltzmann equation (Janssen, 1991). First results of both approaches tend to find slightly smaller lifetimes (of the order of 10-20% for short lifetimes). These descriptions have just started to give results and are not yet fully tested in detail. In general, the error associated with the stopping description will be largest for short lifetimes and become relatively less important the longer the lifetime is. At the moment we can estimate roughly a 20% systematic error around 100fs down to 10% around 1ps.

The largest factor affecting the accuracy of the lifetimes in medium-heavy nuclei is the unknown feeding of the levels. While in light nuclei the feeding is often either direct or, at least quite well known, this is not the case for medium-heavy and heavy nuclei were the levels are fed via many unknown gamma cascades through levels situated between 2 and 8 MeV. Most of these levels are many quasi particle states and can only be described by statistical methods. In order to know exactly the initial feeding distribution one should not only know all transitions connecting a given level with the capture state but also the lifetimes of all intermediate levels which is clearly not the case. Therefore, several approaches are used to estimate the influence of the feeding on the lifetimes. Absolute limits can be obtained by taking extreme feeding assumptions: the upper limit is given by assuming 100% direct feeding from the capture state, the lower by considering all known feeding and adding the missing intensity to the largest known not primary feeding transition assuming all levels to be long lived. The lifetime windows obtained with these limits are, in general, reasonably small, allowing physical conclusions concerning the nature of the state to be drawn. In many cases additional experimental information helps to further reduce the width of the obtained lifetime window: For instance the relative intensity of the direct transition might be known, then one can get less extreme limits. A second method consists in trying to describe the feeding by simulation of the gamma cascades using statistical models. Finally a third, experimental approach, consists in trying to extract knowledge on the feeding from the line shapes directly. Recently a detailed set of simulations was started in order to study to which extent such an approach is feasible. The first results of those look very promising. When such an approach has proven to be reliable the present lifetime windows due to the unknown feeding might be reduced even further. At the same time this will allow to obtain direct experimental information on averaged feeding intensities which can be used to test models describing the nucleus above 3MeV.

GRID has helped to solve some long standing problems in nuclear structure -even in cases where only "lifetime windows" have been deduced. This because the (n,γ) reaction can reach low spin states at high energy where even an somewhat myope view reveals new information on nuclear structure.

RECENT RESULTS IN MEDIUM HEAVY AND HEAVY NUCLEI

The aim of first experiments in some heavy nuclei was to study the predictions of theoretical models at higher excitation energies. The nuclei looked at so far all represent characteristic examples of nuclear behavior. ^{196}Pt is the example of a nucleus described by the O(6) symmetry of the Interacting Boson Model; ^{168}Er is one of the best studied deformed nuclei; ^{114} Cd exhibits a rich play ground with it's vibrational structure and the well established intruder configurations around 1 MeV. Besides these studies, which will be discussed below, some GRID studies of negative parity states have been performed in recent experiments on ^{156}Gd and several Sm isotopes. These also led to large sets of lifetime values and will help to obtain a deeper theoretical understanding of the structure of negative parity states.

^{196}Pt

This γ-unstable nucleus is a very good example of the O(6) limit of the Interacting Boson Model. The low-lying excited levels and the E2 transitions between them follow nicely the O(6) rules up to rather high energy. Only the non-vanishing quadrupole moment of the first excited state is in contradiction with O(6) leading to some attempts to describe this nucleus as a strongly anharmonic vibrator. The low-lying states all belong to the maximal O(6) representation σ = N. Some years ago a 0^+ state at 1402 keV was identified to be a strong candidate for the lowest σ = N-2 state on the basis of its decay properties although absolute transition rates were unknown. In this case these rates are crucial for the identification of this state since the decay of σ = N -2 levels to σ = N levels should be strongly hindered as the E2 transition operator is a generator of O(6). With GRID a lower limit of $\tau > 1.86$ ps for the 0^+ state was measured (Börner et al., 1990). This corresponds to at least an order-of-magnitude hindrance for the E2 decay of the level and further confirms the O(6) character of ^{196}Pt.

^{168}Er

In ^{168}Er the aim of a recent experiment was to study whether or not two-phonon states in deformed nuclei exist. Some theoretical models do predict that the $K^\pi = 4^+$ double gamma vibration should be the lowest lying multiphonon state. For years a rotational band in ^{168}Er, with the J^π

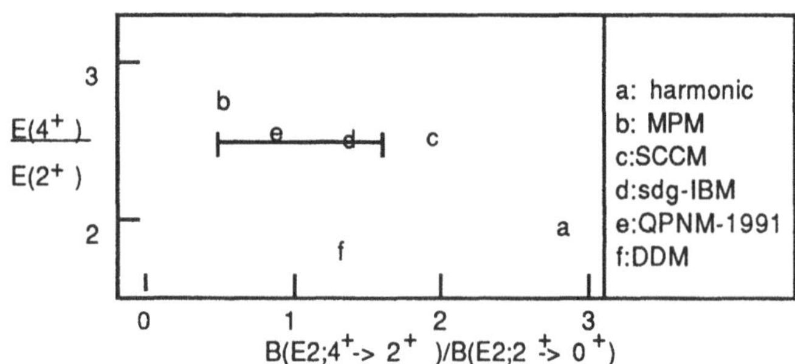

Figure 1. Comparison of the experimental ratio of energies and of B(E2) values connecting the K=4$^+$ and gamma bandhead and the groundstate in ^{168}Er with some theoretical values (all except e taken from Börner 1991).

= 4^+ bandhead at 2055 keV, was considered to be a candidate for such a two phonon state, but absolute transition rates, the most crucial test, were lacking. Using GRID the lifetime of this state was measured and $0.014e^2b^2 < B(E2;4^+ \to 2^+) < 0.041\ e^2b^2$ could be deduced for the decay to the gamma band (Börner et al.,1991). This collective B(E2) value nicely confirms the two-phonon character of the $K^\pi = 4^+$ band. It also solves the longstanding debate on whether the Pauli exclusion principle forbids the existence of multiphonon states in deformed nuclei, as was suggested by some theoretical predictions. It is only seldom that one is able to compare the experimental value to such an extensive set of theoretical predictions. Some are given in Figure 1 which gives the ratio of the deexciting B(E2) values versus the ratio of the energies of the $K^\pi= 4^+$ and $K^\pi= 2^+$ bands allowing a comparison with some of the theoretical values of which a to d are published before the experiment.

^{114}Cd

The experiments mentioned above focussed on very specific problems, connected to lifetimes of only a few states. GRID, however, can access many levels provided one has large neutron cross sections. This is the case for several medium-heavy nuclei. In a recent experiment on ^{114}Cd (ILL-BNL-Notre Dame-Clark collaboration) the deexcitation of all levels with spin lower than 4 up to 2.2 MeV in ^{114}Cd was investigated and definite lifetimes, as well as lower limits were obtained. This experiment shows the feasibility of very complete GRID-spectroscopy for low spin states, leading to quite complete sets of absolute transition rates.

The preliminary analysis of this study shows the existence of some very collective states around 2 MeV, including a candidate for a three phonon state in ^{114}Cd. The measured lifetimes of the states around this energy are listed in Table 1. They are obtained by taking the absolute upper limit for the lifetimes (only direct feeding) and for the lower limit by assuming the level to be only fed by a transition from an infinitely long lived state at 3.5 MeV (the level scheme as measured after neutron capture is complete up to 3.3 MeV (Mheemeed et al. 1984) and no relevant intensity from these levels feeds the ones under study). The measured lifetimes differs in two cases substantially from those obtained in a recent Coulomb excitation experiment (Fahlander et al, 1989). This might be related to large uncertainities in certain matrix elements needed to analyse the Coulex experiment as quoted in this paper.

Table 1. The lifetimes obtained with GRID after the $^{113}Cd(n,\gamma)$ reaction. Also given are the ones obtained by Fahlander et al. 1988.

J^+(E(keV)	2^+_4(1842)	0^+_4(1859)	3^+_1(1864)	2^+_5(2048)	2^+_6(2218)
τ(GRID)ps	>4.5	$0.42<\tau<1.1$	$0.96<\tau<2.58$	$0.42<\tau<1.71$	$0.073<\tau<0.263$
τ(Coulex)ps	0.91(17)	2.5(6)		0.55(16)	

From the lifetimes one deduces that both, the 3^+ level at 1864 keV and the 0^+ level at 1859.7 keV are very collective. The first one is connected with a large B(E2) value to the 2^+_2 state: $0.114 < B(E2;3^+_1 \to 2_2{}^+) < 0.307\ e^2b^2$ and extremely weakly to the groundstate: $0.00003 < B(E2;3^+_1 \to 0_1{}^+) < 0.00009\ e^2b^2$. Its E2 decay to the 4^+_1 state can not be determined due to the unknown mixing ratio, but we can obtain an upper limit of $B(E2;3^+_1 \to 4^+_1) < 0.066\ e^2b^2$ under the assumption of pure E2. If the 3^+_1, 2^+_2 and 4^+_1 states would belong to the three and two phonon vibrational excitations one would expect $B(E2;3^+_1 \to 2_2{}^+) = 0.218\ e^2b^2$ and $B(E2;3^+_1 \to 4_1{}^+) = 0.088\ e^2b^2$. One should, however keep in mind that the two phonon states are strongly mixed with intruding configurations (see for instance Jolie and Heyde, 1990). The decay of the other collective state, the 0^+_4 state at 1859 keV is different from the one of the 3^+_1 level. It decays mainly to the 2^+_3 level with $0.148 < B(E2;0^+_4\text{->}2_3^+) < 0.402\ e^2b^2$ and has a second branch to the 2^+_1 level with $0.0177 < B(E2;0^+_4\text{->}2_1^+) < 0.048\ e^2b^2$. The strong decay to the 2^+_3, which does not decay to the 2^+_1 state, suggests that it belongs as the 4^+_2 level to the intruding configuration. This is depicted schematically in Figure 2 where the B(E2) values as deduced from the lifetimes of the two higher lying 2^+ states are also included. In conclusion, GRID shows that in ^{114}Cd those states which are well separated from states with the same spin but have a different collective structure, do keep a rather pure character, e.g. the 3^+_1 state resembles a three phonon vibrational state and the 0^+_4 an intruder state.

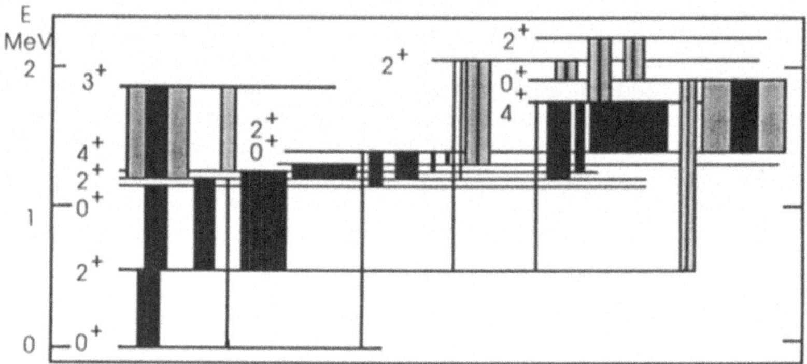

Figure 2. Part of the excited level scheme of ^{114}Cd showing the B(E2)↓ as obtained from the lifetimes given in Table 1. For the other states the B(E2)↓ obtained by Fahlander et al 1988 are also shown.

CONCLUSION

The GRID method provides a powerful new technique to study the structure of medium heavy and heavy atomic nuclei at relatively high energies. As show in the examples above, this region is appropriate to study the extent of collectivity and yields results which may help to elucidate the microscopic structure of collective motion.

ACKNOWLEDGEMENTS

We would like to thank S.J. Robinson, R.F. Casten, E.G. Kessler, M. S. Dewey, G. Greene, R.D. Deslattes, A. Aprahamian, D. Brenner, J. A. Cizewsky, B. Krusche, J. Klora, R. Piepenbrink, and W.T. Zhou for the many fruitfull discussions and numerous contributions to this work.

REFERENCES

Börner H. et al., Phys. Lett. **215B** (1988) 45.
Börner H. et al., Phys. Rev C. **42** (1990) R2271.
Börner H. et al., Phys. Rev. Lett. **66** (1991) 691 and erratum to be publ.
Dewey M.S.et al., Nucl. Instr. and Meth **A284** (1989) 151.
Fahlander A. et al. Nucl. Phys. **A485** (1988) 327.
Janssen D., private communication (1991).
Jolie J. et al, Europhys. Lett. **10** (1989) 231.
Jolie J. et al. Inst. Phys. Conf. Ser. No 105 (1990) 179.

Jolie J. and Heyde K., Phys. Rev. C **42** (1990) 2034.
Kuronen A., J. Phys.: Condensed Matter **3** (1991) 1363.
Mheemeed A. et al. Nucl. Phys **A412** (1984) 113.
Ulbig S.et al. Phys. Lett. **259B** (1991) 29

POST ACCELERATION OF RADIOACTIVE BEAMS: A NEW TOOL FOR NUCLEAR PHYSICS STUDIES

P. Decrock[2], Th. Delbar[1], P. Duhamel[3], W. Galster[1], M. Huyse[2], P. Leleux[1], I. Licot[1], E. Liénard[1], P. Lipnik[1], M. Loiselet[1], C. Michotte[1], G. Ryckewaert[1], P. Van Duppen[2], J. Vanhorenbeeck[3] and J. Vervier[1]

1) I.P.N. and C.R.C., U.C.L., B–1348 Louvain–la–Neuve, Belgium
2) I.K.S., K.U.Leuven, B–3001 Leuven, Belgium
3) I.A.A., U.L.B., B–1050 Bruxelles, Belgium

INTRODUCTION

The production of radioactive nuclear beams is a topic of great current interest. Possible applications for such beams can be found in nuclear astrophysics, solid state physics and nuclear structure physics[1]. Basically one has two schemes for the production of radioactive nuclear beams. A first scheme that is already used for a few years, bombards a target with a high–energy heavy–ion beam creating high–energy secondary (radioactive) beams. The kinematics of the reaction makes it possible to separate the recoil products from the primary beam[2]. In a second scheme the radioactive nuclear beams are produced by post–acceleration of ISOL (Isotope Separator On Line) beams. This scheme has been used, for the first time, at Louvain–la–Neuve (Belgium) by coupling two cyclotrons with an Electron Cyclotron Resonance (ECR) ion source.

In this contribution we will present details on the production of a medium–energy (~0.65 MeV/amu) ^{13}N(T1/2=9.96 m) beam, give the results of a cross–section measurement of the proton–capture reaction on ^{13}N and outline the future plans for the production and use of radioactive ion beams in Belgium.

PRODUCTION OF A RADIOACTIVE ION BEAM

The general layout of the production cycle is shown in Fig. 1. The first cyclotron, CYCLONE 30, is used to produce large amounts of radioactive atoms. The

Nuclear Shapes and Nuclear Structure at Low Excitation Energies
Edited by M. Vergnes *et al.*, Plenum Press, New York, 1992

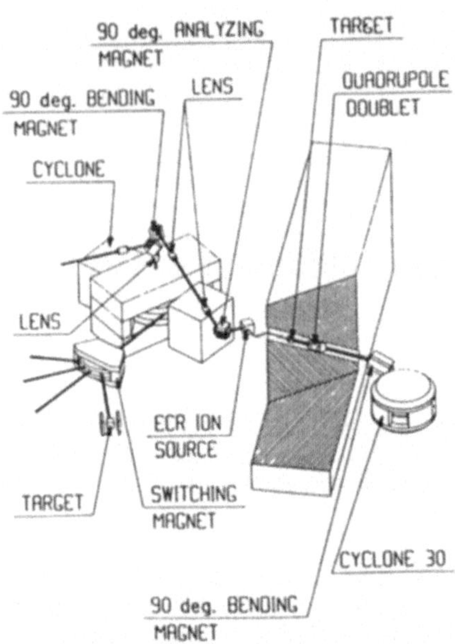

Fig. 1. *General layout of the radioactive ion beam facility.*

radioactive species diffuse out of the target and are ionized in the ECR ion source. The extracted low–energy ion beam from the ECR source, is injected into the main cyclotron, CYCLONE, and is accelerated to the desired energy. We will focus now on the relevant parts of this production cycle.

The CYCLONE 30 cyclotron

The CYCLONE 30 cyclotron can accelerate H^- ions up to 30 MeV and delivers beams with an intensity up to 500 μA. The proton beam is analyzed in a 90° bending magnet and at the same time tilted upward by 30°. Before the beam reaches the target – that is situated in the concrete wall separating the vaults of the cyclotrons – it can be steered by coils in order to get a uniform intensity over the target surface.

The ^{13}C target

The target consists out of a ^{13}C pellet, obtained by graphitizing 99% enriched ^{13}C powder under high temperature and pressure conditions. Large amounts of ^{13}N nuclei were produced by the $^{13}C(p,n)$ reaction ($2\ 10^{12}$ atoms per second) using a 200 μA proton beam on target. Due to the high temperature of the target the ^{13}N atoms diffuse out of the target and are transported to the ion source as $^{13}N^{14}N$ molecules. Therefore, a small nitrogen support gas flow is submitted to the target[3]. An overall release efficiency of 16% is obtained with a proton–beam intensity of 200 μA. Before reaching the ion source, the gas passes a liquid nitrogen trap in order to reduce the gas load for the source.

The ECR ion source

The nitrogen molecules are dissociated and ionized in a 1^+ charge state using a single–stage Electron Cyclotron Resonance ion source (ECR). This source is optimized with respect to the 1^+ ionization efficiency [4]. During off–line tests a maximum efficiency of 15% was obtained for ^{15}N atoms, however due to the relatively high gas load during on–line conditions, an efficiency of 5% was reached for ^{13}N. The ions are extracted at an acceleration energy of 8 keV and mass analysed in a 90° analyzing magnet. The low–energy mass 13 beam is then sent towards the CYCLONE cyclotron and axially injected. Apart from the ^{13}N atoms, the mass 13 beam contains large amounts of ^{13}C atoms (evaporation from the target) and ^{12}CH molecules: the $^{13}C/^{13}N$ ratio is over 1000.

The CYCLONE cyclotron

The desired energy for the first experiments of astrophysical interest is less than 1 MeV/amu. Therefore the central region of the cyclotron is modified to allow its use in the sixth harmonic mode. The overall acceleration efficiency is 4% and due to the excellent mass resolution of the cyclotron the $^{13}C/^{13}N$ ratio is reduced to a level lower than 1%. The 8.4 MeV mass 13 – charge state 1^+ – beam is extracted from the cyclotron and sent through a thin carbon stripper foil (25 $\mu g/cm^2$). There the ^{12}CH molecules are dissociated and are further separated by the switching magnet. After stripping the 5^+ charge state component is chosen by the switching magnet and an 8.2 MeV ^{13}N beam with an intensity of 50 ppA (3 10^8 particles per second) is sent towards the experimental set–up.

THE EXPERIMENT

The cross section of the proton–capture reaction on ^{13}N was measured: $^{13}N(p,\gamma)^{14}O$. This reaction is of crucial astrophysical interest, however it is beyond the scope of this paper to discuss the astrophysical implications. For further details the reader is referred to ref. 5. Here we will outline the experimental set–up used and the results of this cross–section measurement.

Experimental set–up

Fig. 2 shows the experimental set–up. The proton target was a 180 $\mu g/cm^2$ polyethylene $(CH_2)_n$ target that was mounted on a continuously turning ring. The thickness of the target was monitored with an alpha detector and an ^{241}Am source. A surface–barrier detector positioned at 17° with respect to the incident beam

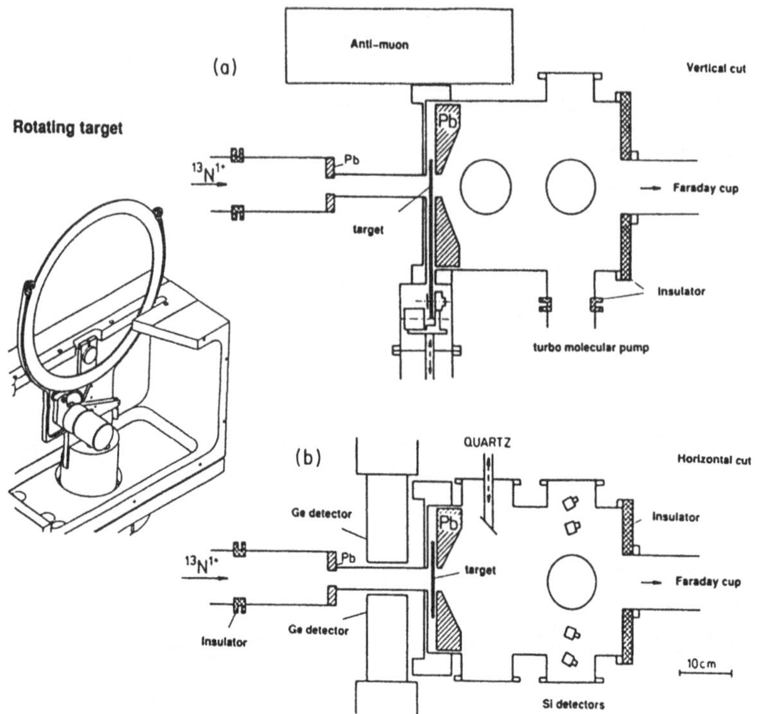

Fig. 2. *Schematic drawing of the reaction chamber – side (a) and top view (b) – the detector system and the target system (c)*

direction, detects the scattered ^{13}N projectiles and the ^{12}C and H recoils: the particle spectrum is shown in Fig. 3a. The ^{13}N peak yields the number of ^{13}N atoms incident on the target, assuming pure Rutherford scattering of the ^{13}N projectiles by the ^{12}C atoms. The proton–capture reaction on ^{13}N is studied through the 5.173 MeV capture gamma rays from the 1^- excited state of ^{14}O. These gamma rays are detected by two large volume Ge detectors (relative efficiency of 70 and 90%), positioned in the backward hemisphere. The time difference between the cyclotron beam pulses and the gamma rays was also recorded. The Ge detectors were vetoed by a large cosmic–ray umbrella, made out of plexiglas Cerenkov detectors.

Results

The prompt gamma–ray spectrum, after subtraction of the random events using time windows determined from fig. 3b, shows clearly the presence of the full energy, first and second escape peaks of the 5.173 MeV line (fig. 3c). A total of 85 full energy events were recorded. The total count rate in the Ge detectors was ~10 kHz and the correction for pile up and death time was 30%. From the ratio between the intensities of the full–energy gamma line and the scattered ^{13}N atoms, one can determine the

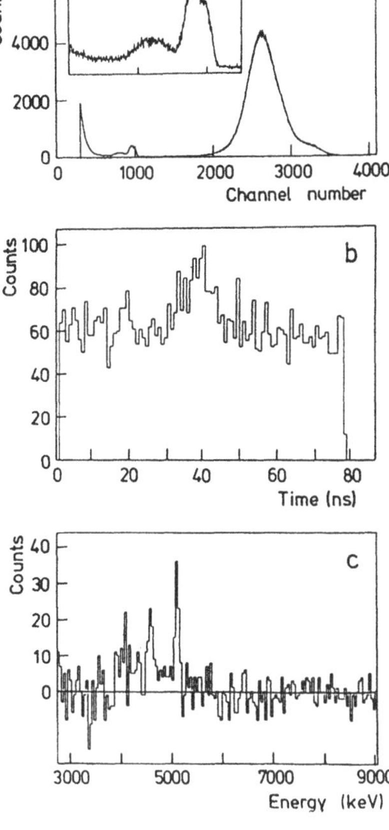

Fig. 3. *Charge particle spectrum (a). The peak to the right corresponds to the scattered ^{13}N projectiles and ^{12}C recoils (right shoulder), the peak to the left and in the insert, to the proton recoils. TAC spectrum (b) from the Ge detector and the cyclotron beam pulses. Prompt gamma–ray spectrum (c).*

cross section of the $^{13}N(p,\gamma)^{14}O$ reaction in an energy interval between 5.8 and 8.2 MeV: the result is 106(30) μbarn.

FUTURE DEVELOPMENTS OF RADIOACTIVE BEAMS

Future developments of the present radioactive nuclear beams project are proceeding in two direction: higher acceleration energies and new beams. Already beams of $^{13}N^{+2}$ at an energy of about 20 MeV are accelerated and intensities up to 20 ppA are obtained and higher energies up to 5 MeV/amu are foreseen. New beams that will become available in the near future are ^{19}Ne(T1/2=17 s), ^{15}O(T1/2=2 m) and ^{11}C(T1/2=20 m). These activities are already produced using LiF and B_2O_3 targets and extracted from the targets. The obtained intensities are comparable with the ^{13}N yields. The high energy beams, from a few MeV/amu on, will be used for nuclear–physics studies e.g. cross–section measurements with extreme neutron deficient beams like ^{14}O and ^{10}C, study of isospin mixing in the GDR,...

Unfortunately, the present set–up has its limitations. The CYCLONE 30 cyclotron can only deliver 30 MeV proton beams, this limits somewhat the different isotopes that can be produced. Furthermore the CYCLONE cyclotron has a low acceleration efficiency reducing the final beam intensity by at least one order of magnitude.

Based on the experience gained up to now, an alternative scheme for the production of radioactive isotopes and a new dedicated acceleration system for medium–energy radioactive beams is proposed. In this project either one of the two cyclotrons will be used for the production of radioactive elements – the CYCLONE cyclotron can deliver beams of protons (E=80 MeV), deuterons (50 MeV), ^{3}He (140 MeV) and alpha (100 MeV) with intensities up to 20 μA. The activity will be on–line ionized and mass analyzed and transferred to a new dedicated cyclotron whose acceleration energy ranges from 0.2–1.0 MeV/amu. These beams will then mostly be used for astrophysical experiments.

Meanwhile, an ECR ion source is installed at the LISOL facility [6]. High intensity, low–energy (typically 50 keV) beams will become available and will be used for solid–state studies.

These three radioactive ion beam facilities that are/will become available at Louvain–la–Neuve, will be converged in the **ARENAS^3** project: Acceleration of Radioactive Elements for Nuclear–, Astrophysical– and Solid State Studies.

ACKNOWLEDGEMENTS

This work has been made possible thanks to an Interuniversity Research Grant from the Belgian Government and thanks to the support from the National Fund for Scientific Research (Belgium).

REFERENCES

1. Proc. of the First Int. Conf. on Radioactive Nuclear Beams, U.S.A., ed. W.D. Meyers, J.M. Nitschke and E.B. Norman (1990)
2. R. Bimbot et al., Z. Phys. A322, 443 (1985)
   T. Tanihata et al., Phys. Lett. 160B, 380 (1985)
3. D. Darquennes et al., Nucl. Instr. and Meth. B47, 311 (1990)
4. P. Decrock et al., Nucl. Instr. and Meth. in press
5. C.E. Rolfs and W.S. Rodney, Cauldrons in the Cosmos (The University of Chicago Press, Chicago and London, 1988)
6. M. Huyse et al., Nucl. Instr. and Meth. B26, 105 (1987)

IN BEAM NUCLEAR POLARIZATION OF RADIOACTIVE ION BEAMS*

J. Wouters[a], W. Vanderpoorten[b], P. De Moor[b], P. Schuurmans[b], N. Severijns[a], R. Siebelink[b], J. Vanhaverbeke[b], L. Vanneste[b], and J. Vervier[a]

[a] Institut de Physique Nucléaire, U.C.L., Louvain-la-Neuve
[b] Instituut voor Kern- en Stralingsfysika, K.U.L., Leuven

INTRODUCTION

The polarization of radioactive nuclei has a lot of applications in the study of nuclear and solid state physics. The advent of intense radioactive ion beams with the recent astrophysical projects, opens new avenues for studies of other physics phenomena as well. A lot of unique and new experiments from applied physics to fundamental nuclear physics, can be carried out by polarizing these radioactive nuclei. Our first interests are : (i) condensed matter studies with nuclear magnetic resonance (NMR) on polarized ^{13}N, ^{15}O, ^{11}C, ^{19}Ne, implanted in different samples and which are available at the Cyclotron Research Center in Louvain-la-Neuve[1],(ii) weak interaction studies : search for the occurrence of right-handed currents in β-decay of polarized nuclei.
On line low temperature nuclear orientation is by now a well established technique to polarize shortlived radioactive species. The radioactive isotope beams from the Leuven Isotope Separator On Line (LISOL) are implanted in a magnetized ferromagnetic sample, which is attached to the cold finger of a ^3He-^4He dilution refrigerator with on-line base temperature in the order of 10 mK. The radioactive nuclei are polarized by the low temperatures together with the internal hyperfine fields in the ferromagnetic sample. The angular distributions of the nuclear radiations, α, β, γ, CE are observed, possibly in combination with NMR, and yield information about nuclear moments, nuclear decay, fundamental interactions and condensed matter properties[2]. This method is limited to isotope halflifes longer than the spin-lattice relaxation time T_1 which is a characteristic time for the hyperfine interaction of the implanted impurity in the host material.
A very important feature, however, is that the internal hyperfine fields giving rise to the nuclear orientation, are small for most Z < 25 elements, yielding small degrees of nuclear polarization for these elements. This prevents to obtain high polarization degrees for implanted impurities with $t_{1/2} < T_1$, which is particularly the case for light elements off the line of stability. These limitations can be circumvented by implanting at forehand polarized nuclei : in-beam nuclear polarization.

IN-BEAM POLARIZATION METHODS

In what follows we will mainly discuss polarization of low energy beams ; at isotope separator energies (about 50 keV) and slightly higher. In recent years some laser assisted techniques have been developed. At ISOLDE-CERN collinear fast-beam laser spectroscopy in combination with β-RADOP has been used to study in-beam nuclear polarizations of Li-

Nuclear Shapes and Nuclear Structure at Low Excitation Energies
Edited by M. Vergnes *et al.*, Plenum Press, New York, 1992

isotopes. Nuclear polarizations P_I of approximately $P_I \cong 7$ %, with $P_I = \langle I_z \rangle / I$, have been obtained for Li-isotopes[3]. Very recently, some new results with respect to the neutron halo of ^{11}Li have been obtained in this way[4]. Recently, at Argonne, a polarized $^{14}N^{4+}$ beam at 60 keV has been obtained by pick-up of polarized electrons from an optically pumped Na target. Nuclear polarizations of about $P_I \cong 4$ % were achieved[5].

In Louvain-la-Neuve, we recently developed in-beam polarization techniques based on beam-surface interactions, at the LISOL set-up. The radioactive ion beams are transmitted through tilted foils (TF) or reflected at grazing incidence from very flat polished crystals (IBSIGI : Ion Beam Surface Interaction at Grazing Incidence). After the beam-surface interaction, the in-beam polarized isotopes are implanted in a target material, which may be the hosts to be studied, and which can be mounted in the cryogenic apparatus, the ^3He-^4He dilution refrigerator. The low temperatures may be necessary to preserve the obtained polarization for a sufficiently long time. The TF and IBSIGI techniques have been chosen because we aim for an applicability as general as possible.

In both the tilted foil and scattering techniques, an asymmetric Coulomb interaction at the outgoing surface generates a large orbital angular momentum in a well defined direction (the **v** x **n** direction, where **v** is the velocity of the ions and **n** is the normal to the surface), which couples to the angular momenta of ground and excited atomic states. This atomic polarization, which can be measured by analyzing the angular distribution and the polarization of the emitted deexcitation light, is transferred to the nuclei by hyperfine interaction, thus creating nuclear polarization.

The tilted foil technique

Nuclear polarizations P_I up to several percents have already been obtained with a single tilted foil for beams with small nuclear spin I[6]. By using stacks of tilted foils, allowing for multiple interactions which are more efficient to build up polarization for high I, polarizations up to 20 % have been measured[7-10]. It has also been shown that in some cases P_I of the order of 1 % suffices to perform NMR experiments[9]. However, these experiments have all been carried out at energies well above our beam energy, i.e. from about 300 keV up to several tens of MeV. The radioactive ion beams from LISOL are accelerated to 50 keV. This puts severe constraints on the thickness of the foils to be used. Making these foils was the main difficulty concerning the application of the TF method. Tilted carbon foils have been used with thicknesses of 2-3 μg/cm2 (~80 A) and a surface area of the order of 0.2 cm^2.

The surface collisions at grazing incidence

By scattering ion beams from polished solid surfaces at grazing incidence angles (< 2°), nuclear polarizations up to 25 % have already been achieved by H. Andrä and H. Winter for stable isotopes, in particular for ^{14}N, but also ^7Li, ^{11}B, ^{13}C, ^{35}Cl [11-13]. Unlike the tilted foil method, up to now the scattering technique (IBSIGI) has not been applied to radioactive ion beams. Compared to TF this technique is more complicated, because the requirements for the quality of the scattering surface and the vacuum conditions are severe. In general, the polarizations obtained with IBSIGI are larger than for the TF method. We used Si(111) single crystals, with surface granulation better than 10 nm, as scattering surfaces in our experiments.

Energy dependence of the polarization

For both polarization techniques the 50 keV beam energy for singly charged ions at the separator is rather low. Beam energies above this value but below approximately 1 MeV may be ideal for low-Z elements. Not much is known about this question as yet because experimental data are scarce. However, the minimal beam energies that can be provided by existing accelerators are not well suited (for the CYCLONE cyclotron at Louvain-la-Neuve, 6 MeV is the minimal energy for an intense ^{13}N beam). Up to now, some data on the energy dependence of the nuclear polarization, obtained with the IBSIGI method, are available for only one isotope. For ^{14}N values of P_I = 23 % and 8.5 % have been recorded at 350 keV and 1 MeV respectively, decreasing further at higher energies[12-13]. Although the atomic polarization is probably saturated at energies above typically 4 keV per nucleon[13], the energy dependence of the nuclear polarization, which is experimentally unknown apart from these scarce data, reflects the variation of charge state distributions with energy after the beam-surface interaction.

Detection of polarization

We can measure both the atomic and the nuclear polarization in a direct way. The nuclear polarization is measured by detecting the anisotropically emitted nuclear radiation with semiconductor detectors installed inside the refrigerator. To measure the atomic orientation just behind the beam-surface interaction zone, an optical detection system has been built. The amount of circularly polarized light observed perpendicularly to the beam, is a direct measure for the atomic polarization, which is proportional to the nuclear polarization to be obtained. In this way, different surface conditions and interaction parameters, which play a role in the above described polarization techniques, can be studied and checked off-line with stable beams. The optical detector consists mainly of an interference filter, a rotatable quarter wave plate, a fixed linear polarizer and a cooled PM (Burle C31034). The intensity profile of the light as a function of the $\lambda/4$ plate setting is recorded and can reveal components with π periodicity refering to the amount of circular polarization of the emitted light. The circular polarization is defined through the S/I Stokes parameter. Neglecting very small alignment contributions, the S/I value of an optical transition is about equal to the atomic polarization ${}^{L}\rho$ of the originating (LS)J term. The nuclear polarization can be written $P_I = (I + 1/3I)^{1/2}\, {}^{I}\rho$ with ${}^{I}\rho = c\,{}^{L}\rho$. As defined in ref.13, the ρ are relative polarization parameters and the coefficient c depends on the angular momenta and spins in the coupling of nucleus with the most important atomic states. The values of the Stokes parameters can be determined from a Fourier transform of the intensity pattern as a function of the $\lambda/4$ plate setting[14]. The S/I value can be deduced straightforward from the recorded intensity as

$$S/I = \frac{I(\sigma^-) - I(\sigma^+)}{I(\sigma^-) + I(\sigma^+)}$$

where $I(\sigma^-)$ and $I(\sigma^+)$ are the intensities of the right-handed and left-handed circular polarized light, the handness being determined by the quarter wave plate setting.

Preservation of polarization

After the beam-surface interaction the polarized radioactive beam is implanted in a host material soldered onto the cold finger of the ${}^3He$-${}^4He$ refrigerator. Depending on the half-life of the polarized isotopes, the target will have to be cooled to low temperatures to preserve the obtained polarization for at least the halflife of the polarized isotopes. This is because the spin-lattice relaxation time T_1 for most materials is proportional to T^{-n} with $n > 1$. As an example, for 2sp impurities in Pt or Pd it has been measured that $T_1T \sim 300$ sK [15-16]. Thus for $T \sim 100$ mK one obtains $T_1 \sim 50$ min. Other materials may not have to be cooled down to such low temperatures.

Because the angular divergence of a 50 keV beam after passage through a foil is considerable and because of the non-ideal specular reflection (for IBSIGI) of the beam after interaction, the distance between beam-surface interaction and the target position (the implantation site) has to be minimal. As additionally the target material possibly has to be held at low temperatures, a very short cooled side access to the refrigerator has been constructed, which limits the heat irradiated onto the sample holder inside the refrigerator. With this mini side access, which has a length of only 92 mm, we still obtain a temperature of 15 mK in fully on-line mode.

More generally, the technical set-up and the experimental procedures are described in more detail in ref.17.

First test experiments have been carried out with singly charged ^{14}N ions accelerated to 50 keV, by the LISOL separator at Louvain-la-Neuve. Tests with TF as well as IBSIGI have been carried out. In this stage we are concentrating on the atomic polarization. The TF experiments were carried out with 2-3 $\mu g/cm^2$ C-foils at 60° and the IBSIGI measurements with a Si(111) single crystal and reflection at an angle of approximately 2.5°.

The ^{14}N ions are excited while passing through the foil and/or scattering at the single crystal. The atomic state populations of nitrogen after beam-foil interaction have been studied before, so that the most dominant optical lines are well known. The optical transitions NII $2s^22p3p\ ^2D - 2s^22p3d\ ^3F°$ at 500 nm and also NII $2s^22p3s\ ^3P - 2s^22p3p\ ^3D$ at 567 mm have been observed with a 500 nm, respectively 570 nm, interference filter (FWHM of 10 nm). The circular polarization Stokes parameter S/I has been deduced from the light intensity recorded as a function of the $\lambda/4$ plate setting. In both experiments the count rate was normalized to the transmitted beam current read from a faraday cup behind the interaction zone. For a single tilted foil we extract $|S/I|$ = 10.2(1.9) % from the 500 nm line. In figure 1 the light intensity of the 567 nm transition in NII, observed in the direction of specular reflection, is displayed as a function of the $\lambda/4$ plate angle for the surface scattering experiment. From these data we deduce $|S/I|$ =20.2(0.4) %. The fit shown in figure 1 is the result of the intensity pattern least square fitted to a Fourier series.

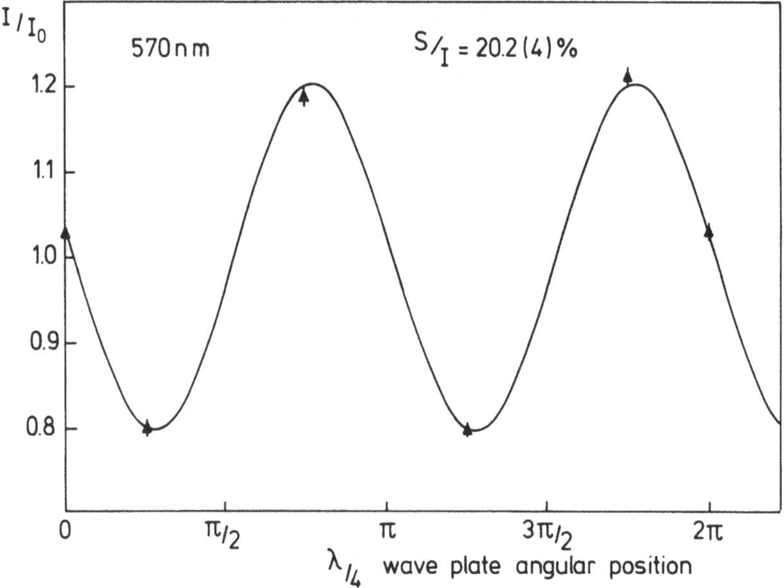

Fig.1. The intensity of the 567 nm transition in $^{14}N^+$ as a function of the $\lambda/4$ plate position after surface scattering at 50 keV at an angle of approximately 2.5° from a Si(111) crystal. A value $|S/I|$ = 20.2(0.4) % is deduced.

These first results are very encouraging. The TF polarization value agrees in magnitude with the polarization obtained for ^{13}N at higher energies around 300 keV with a single tilted foil[6]. Whereas these in-beam polarization methods are ready to be applied, the polarization values obtained with the scattering technique can still be enhanced. Recently, values of S/I = 60(3) % on 500 nm at 250 keV and S/I = 56(3) % on 567 nm at 150 keV have been obtained by H. Winter et al., with IBSIGI on a clean Fe(110) surface and scattered at angles inferior to 1° [18]. An S/I = 79.7(5) % on the 567 nm line at 300 keV has been reported

before[11]. We think that the difference is due to the non-ideal surface because up to now we did not apply any in-situ surface treatment. Tests will be done by cleaning the Si-surface with ion sputtering at low angles.
At 50 keV the atomic polarization should be saturated as described in ref.13. Coefficients for optimal polarization transfer to the nucleus have been calculated for ^{14}N to be of the order of 30 % [12]. This means that we are already able to generate nuclear polarizations of the order of more than 5 %.

ACKNOWLEDGEMENTS

We gratefully acknowledge stimulating discussions with H. Winter, R. Coussement, W. Galster. We also thank P. Decrock, P. Van Duppen and M. Huyse for operating the ECR-source at the separator. J.W. and N.S. are Senior Research Assistants and R.S. Research Assistant of the NFWO.

REFERENCES

* this work is part of the thesis of W. Vanderpoorten

1. P. Van Duppen et al., contribution to this conference
2. J. Wouters, N. Severijns, J. Vanhaverbeke, W. Vanderpoorten, L. Vanneste, Hyp. Int. 59 (1990) 59 and references therein
3. E. Arnold et al., Phys. Lett. B197 (1987) 311
4. R. Neugart, invited talk at this conference
5. C.J. Liu, N.B. Mansour, Y. Azuma, H.G. Berry, D.A. Church, R.W. Dunford, Phys. Rev. Lett. 64 (1990) 1354
6. B.I. Deutch, Hyp. Int. 24-26 (1985) 251
   Y. Nojiri and B.I. Deutch, Phys. Rev.Lett. 51 (1983) 180
7. Y. Niv, Hyp. Int. 24-26 (1985) 301
8. G. Goldring, Hyp. Int. 33 (1987) 19
9. W.F. Rogers, D.L. Clark, S.B. Dutta, A.G. Martin, Phys. Lett. B177 (1986) 293 and Nucl. Instr. Meth. Phys. Res. A253 (1987) 256
10. Y. Nojiri, Hyp. Int. 60 (1990) 599
11. H. Winter and R. Zimmy, Hyp. Int. 22 (1985) 237
12. H.J. Andrä and H. Winter, Hyp. Int. 5 (1978) 403
13. H. Winter, Habilitationnschrift (1987)
14. H.G. Berry, G. Gabrielse, A.E. Livingston, Applied Optics 16 (1977) 3200
15. R. Pirlot, M. Cyamukungu, L. Grenacs, J. Lehmann, K. Tompa, Phys. Stat. Sol. (b) 158 (1990) K197
16. K. Nakai, Hyp. Int. 21 (1985) 1
17. J. Wouters et al., to be published in Nucl. Instr. and Meth. in Phys. Res. B
18. H. Winter, H. Hagedorn, R. Zimny, H. Nienhaus, J. Kirschner, Phys. Rev. Lett. 62 (1989) 296

XI - CONCLUSION

CONCLUDING REMARKS

Philippe Quentin

CSNSM, CNRS/IN2P3

Bât. 104-108, 91405 Orsay-Campus, France

In the present remarks, I will collect my personal selection of what I liked to hear and see during this Conference. In view of the broad range of topics which have been covered, it is a rather impossible task to pack the various talks into an obvious succession of themes, I will therefore merely attempt to organize them in my own arbitrary way.

A very specific feature of this Nuclear Structure Conference has been the substantial role played by contributors coming from Atomic Physics and also to some extent from Condensed Matter Physics. To put it in the words of E. OTTEN, this does not mean that some of us "have attended the wrong Conference". Indeed this Conference has been specially designed for the purpose of allowing these disciplines to be fully participating. This is partly related to the fact that our meeting was meant to pay a tribute to our late colleague S. LIBERMANN whose enthusiasm and in-depth vision has been very much missed during this week. But, as E. OTTEN has stressed in his introductory remarks, atomic and nuclear spectroscopy are indeed, unseparable domains in many respects. This is not a general and old statement. Indeed as we have seen, it is still true in many aspects of current research. Out of the talk of E. OTTEN, I will only mention the speculation on an E1 polarization in parity doublets of octupolly deformed nuclei for instance.

One unexpected consequence of such wide spread origins of the attending physicists, has been the unusual number of rather pedagogical presentations of up to date results, taking stock of the fact that a substantial part of the audience in any given talk, was not supposed to be very familiar with most of what was discussed. Altogether this feature has been rather positive, even though it has often resulted in a two regime type of invited talks where after a quiet presentation of the problems at stake, the most original part was indeed present but discussed at quite another pace. However, I think that this was an unavoidable price to pay for the above recalled benefits of such a conference program.

Nuclear Shapes and Nuclear Structure at Low Excitation Energies
Edited by M. Vergnes *et al.*, Plenum Press, New York, 1992

J. BAUCHE has set for us the stage, by clarifying the theoretical basis for the use of hyperfine structure and isotopic shift studies in nuclear spectroscopy, providing us with some useful guidelines on what is to be expected and what is unlikely to provide accurate quantitative informations.

One striking aspect of the Atomic Physics works which we have heard of during these days, is their impressive systematic character. G.D. ALKHAZOV and collaborators after their pioneering work in resonance ionization spectroscopy have obtained and presented here results for more than 100 isotopes in the Xe, Cs, Ba and Nd, Sm, Eu regions. G.D. ALKHAZOV has highlighted the rather high Z-dependence of the isotopic shift behaviour below the N=82 neutron magic number.

The comparison of ^{40}Ca and ^{48}Ca radii has always constituted a crucial test for theoreticians of various obediences. From R. NEUGART's talk we have learned that they will have to do as well with the radius of ^{50}Ca...

For decades, the region of transitional nuclei below the doubly magic ^{208}Pb nucleus, has attracted a lot of attention. Let me mention at this time the seminal experimental work of the late R. FOUCHER which has been rather inspirating for many of us. From the isotopic shift measurements of the Mainz group and much later of the Orsay group , has emerged more and more clearly the picture of a shape transition for some ground state isotopic series and more generally of a well established shape coexistence. As we have learned from E. OTTEN's talk, the body of data is now very impressively complete.

In R. NEUGART's talk, we have had also a report on some spin and quadrupole moment measurement in ^{11}Li, ruling out a static deformation. M. LEVITOWICZ reviewed for us also, the current explanation of reaction cross section data and of his own neutron emission data for this nucleus in terms of a "neutron halo". Obviously more is still to be known there in particular on the possible correlation between the two last neutrons giving rise or not to a dineutron stabilized in the vicinity of the remaining nucleons.

Coming back now, to isotopic shift data, both G.D. ALKHAZOV and G. MEISEL have insisted on the fact that most of the odd-even staggering which is observed is still to be explained. Another standing problem is the behaviour near magic numbers of the isotopic shift. One may mention as an example NEUGART's Strontium data, but this is quite a general fact indeed. Mocking up the vibrational states by a statically deformed density might appear as a rather short sighted ersatz. On the other hand, implementing various correlations one by one so to speak seems unsufficient as well, as it has been alluded to by P.H. HEENEN in a discussion.

A.M. MARTENSSON-PENDRILL has depicted for us the very intricate situation into which one is caught when dealing with parity non-conservation in atomic physics. One has indeed to struggle there at the same time with particle, nuclear and atomic aspects of that problem.

The anomalously converted M1 transitions constitute a rather controversial subject. The UNISOR group has ruled out their occurence in ^{185}Au on the sole basis of the existence of an E0 transition corresponding to the same transition energy. Now in ^{187}Pt, if the spin assignments presented by B. ROUSSIERE are to be accepted, one is facing again with a rather challenging puzzle.

H. POSTMA has made a very comprehensive survey of the study of nuclear moments by orientation methods. I might have been particularly biased there but I have been very much interested by the data on the $^{178}Hf^{m}{}_{2}$ magnetic moment. I would certainly consider that making such studies for as many isomers as possible in this region, would be a very useful task in view of the rather simple single particle character of such states allowing a direct comparison of theoretical estimates with experimental data.

+-+-+-+

Let us turn now to what is generally considered as nuclear spectroscopy in its own right. To put some kind of order in my personal records, I will make use of a picture proposed by K. HEYDE at the end of his introductory talk.

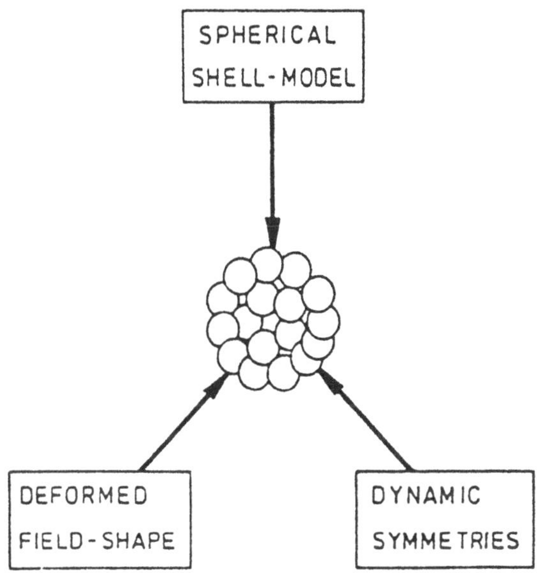

Fig. 1 K. HEYDE's schematic presentation of three classes of approximation schemes to Nuclear Structure properties.

As for Algebraïc methods we have had rather short but illuminating surveys by K. HEYDE and R. CASTEN. From the latter's talk, I will single out the in-depth discussion of the difference between γ-soft (à la JEAN-WILETS or O(6) IBA limit) and γ=30° rigid potential landscapes à la DAVIDOFF.

Let us turn now to contributions related to the Spherical Shell Model Approach.

S. GALES in his talk, has compared two competing ways of

experimentally assessing the single particle response function :

- the electronic probe, as in (e-e'p) reactions
- the hadronic probe, as in transfer reactions.

On the particular example of ^{207}Tl, he has highlighted a difference by a factor of two between the two classes of data. This type of results, and a similar quenching for (e,e') longitudinal sum rule might indicate (from my probably too naïve point of view) that the interpretation of (e,e'p) data might gain a lot by a refinement of the currently practiced CDWIA treatment.

C. PAPANICOLAS has recalled to our attention the Saclay results on the proton 3s1/2 density in the lead region. As it has emerged from the discussion, it is very dangerous and may be not very useful to be quantitative in the assessment of the validity of nuclear mean fields from such results. Indeed, from the rather scarce evidence provided by a number of reasonably good effective forces, it is hard to a priori rule out a force (yet to come) wihch could, may be, do even better than what is presently computed. Moreover, this validity strongly depends, of course, on what one is looking at. As pointed out by C. PAPANICOLAS, high-momentum transfer data are most likely to be badly reproduced by mean field calculations.

D. ROWE has exposed in a very clear fashion, a point of view which seems -at least to me- rather unorthodox, on the proper representation of quadrupole-quadrupole correlations which would need quite a large number of spherical shells. Part of the answer to the necessary reconciliation of such a statement with the current practice of shell model calculations of deformed nuclei (e.g. in s-d shell nuclei) might be hidden in the somewhat mysterious "effectiveness" of the interactions in use.

K. HEYDE has shown a very suggestive result for ^{116}Sn, where he has made restricted shell model calculations, yet using the same extended Skyrme force for generating the single particle properties and for the residual interaction with a rather good qualitative agreement.

Many experimental and theoretical contributions could be considered (more or less arbitrarily) as belonging to approaches where the concept of a deformed mean field plays an important role. This is summarized in the Table, where the relevant talks are assigned to various Nuclear Structure phenomena.

The crucial importance of **intruder states** in the occurence of nuclear deformation has been recalled by R. BENGTSSON. This is especially true, in the lead region. N. TAJIMA has presented the results of a self-consistent calculation where the quasi-particle vacuum and two 2-qp states (involving $\{1h9/2, K = 9/2\}^2$ and $\{1h9/2, K = 1/2\}^2$ configurations) have been properly mixed in a GCM approach, in the same region .

M. HUYSE has reinterpreted the results obtained at LISOL for α-hindrance factors in the Pb region in terms of the presence or the absence of intruder states in the relevant nuclear states.

Nuclear **shape coexistence** is a very old topic whose origin may be

Table 1

Topics	Speakers
Intruder states	R. Bengtsson, M. Huyse, N. Tajima
Shape coexistence	J. Hamilton, A. Korichi
Superdeformation	F. Azaïez, P. Bonche, J.P. Delaroche, R. Janssens, J. Scharpey-Schaeffer
Parity-breaking modes	P. Butler, P. Kleinheinz, I. Mikhailov, W. Nazarewicz, R. Piepenbring
Multiphonon states	J. Jolie, P. Kleinheinz, R. Piepenbring, V. Soloviev
Odd-odd systems	A. Kreiner
"Scissor" mode	D. Zawischa

traced back to some remarks formulated years ago by G. BROWN, concerning the 0^+ level at about 6 MeV in ^{16}O. As already mentioned, a genuine playground for shape coexistence is to be found in the Os-Pt-Hg region. J. HAMILTON in his very rich introductory talk has presented some new results on ^{186}Hg from which he has proposed the first evidence of a γ-band built on the coexisting 0^+ state. Of particular interest also, is the 8^- isomeric state present in ^{178}Hf, for instance, which is built on the prolate equilibrium solution for lighter nuclei and does therefore correspond to an excited prolate band in ^{186}Hg.

A. KORICHI has revisited the well-known case of coexistence in ^{187}Au and showed that it could be pushed up to 57/2 $\hbar$ where it still exists.

Ideally, in this Conference, we were only supposed to deal with **superdeformation (SD) at low spin**. However, such rules are made to be violated when this is done à propos. That is why R. JANSSENS has made a clear presentation of the entire SD set of data, highlighting the similarities and the differences (as in the $\mathfrak{J}^{(2)}$ moment of inertia data for instance) between the A~150 and A~190 regions. He presented also some very interesting new results. Two further transitions in ^{192}Hg (contradicting a theoretical prediction for a decrease of $\mathfrak{J}^{(2)}$ supposedly due to a pairing variation) are another illustration of the very true sentence of an unknown (at least to me) nuclear chemist :"the terrible tragedy of science is the murdering of beautiful theories by ugly facts". He has also discussed the recent claim that the yet unobserved decay of SD states to "normal" states could as well proceed through an E0 transition. Then, the electrons so produced should induce some extra X-ray emissions which seems not to have been found in the particular case of ^{192}Hg.

J. SHARPEY-SCHAEFFER has shown some data, hinting at the existence of alignment effects also for SD states in ^{197}Pb. F. AZAIEZ has displayed quite a generous crop of SD bands in Thallium isotopes (10 in total, of which 6 in the odd-odd ^{194}Tl isotope). All speakers of course have insisted on the startling feature of identical transition energies for which theoretical explanations that are fully convincing, are still to be produced. The strong coupling arguments are very interesting. Yet one has to cope with the fact that for $|K| = 1/2$ bands one has not $|a| = 1$ exactly and, even if so, the constancy of $\mathfrak{I}$ is at any rate, quite mysterious. The variation $\delta\mathfrak{I}$ of the moment of inertia $\mathfrak{I}$ as :

$$\delta\mathfrak{I} = \delta\mathfrak{I}_A + \delta\mathfrak{I}_\beta + \delta\mathfrak{I}_\Delta + \delta\mathfrak{I}_\omega + \ldots$$

(where A, β, Δ, ω, ... stand for the variation due to the change in nucleon number, deformation, pairing strength, angular velocity,...) is currently performed. Such a linearization is rather ambiguous when adding contributions that are not linearly independent. After all there are only two such independent variables : the chemical potential λ and the angular velocity ω, or equivalently the particle number A and the total spin I (neglecting here for the sake of simplicity the difference in charge states).

Concerning the superdeformation phenomenon we have had two theoretical talks presenting two complementary aspects of the current research :
- the GCM by P. BONCHE
- a microscopic BOHR Hamiltonian approach by J.P. DELAROCHE.

From an a priori quantum mechanical point of view, the GCM seems better. Yet, the bottom line is (despite of the incredible achievement of 2 dimensional GCM calculations) that one approach can only deal with 2 (may be 3) collective variables so far, whereas the other can cope with 5.

The **parity-breaking modes** have been widely discussed here. W. NAZAREWICZ made a very comprehensive survey of the data and their interpretation with suggestive connexions with Molecular Physics. I found very exciting his comments on rational harmonic oscillator magic numbers. Whether such consideration are to survive in actual SD realistic mean fields is of course a question. However, it surely provides some very valuable guidelines.

I. MIKHAILOV presented some theoretical evidence for the presence at low energy of isoscalar dipole strength in his FERMI liquid dynamical approach. This was substantiated by the results of a 3-parameter fit of 53 experimental levels, where, in some limit of an IBM approach with s, p, d, f bosons, the role played by the p-bosons has been found to be essential.

As N. NAZAREWICZ has noted, parity-breaking modes in nuclei could ideally be found in the two limiting "ideal" cases, namely the rotational and the vibrational limits. However, he insisted, they are more likely to belong in real cases to an intermediate situation which is much richer, which means harder to describe theoretically (see figure).

In such "mixed case" situations, R. PIEPENBRING has demonstrated that either of the two limiting cases were indeed equivalently appropriate for a discussion of the prominent spectroscopic properties. In the same context, P. BUTLER has convincingly demonstrated from his Coulomb excitation data in ^{148}Nd, that one should not follow too hastily, some simple-minded theoretical reasonings. For instance, going

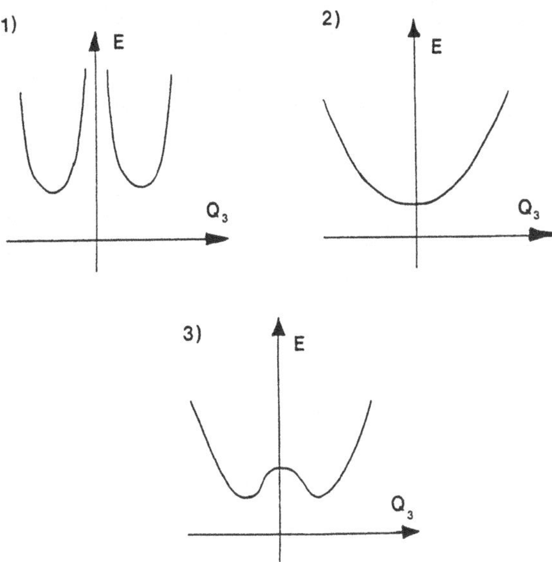

Fig. 2 Potential energy curves for parity-breaking modes in the rotational (1) and vibrational (2) limits and in the actual "mixed" situation (3).

up in the spectrum, the observed level sequence looks much closer to a typical rotational pattern. One could infer that upon increasing the spin, for instance, the symmetrical ($Q_3=0$) barrier is raised. However this should change for instance the intrinsic $\langle Q_3 \rangle$ value which is in contradiction with what is found experimentally.

Let us turn now, towards a heavily disputed subject : the evidence of **multiphonon states**, particularly in the well-documented case of the 4^+ state located at about 2 MeV in ^{168}Er. V. SOLOVIEV summarized the experimental and theoretical situation for this nucleus. Its personal conclusion out of his own calculations is that there is no evidence for a 2-phonon state there. However, H. BORNER, J. JOLIE and collaborators have measured a strong E2-deexcitation of this state on the 2^+ γ-vibrational band head, concluding thus to a 2 γ-phonon character of this 4^+ state. This is fully in accordance with the results obtained by R. PIEPENBRING in his multiphonon model approach.

If I dare to involve myself in such a risky situation, I have to give now a personal point of view. I would, first, stress that we are dealing here to a large extent with a semantic problem. Collectivity, in my opinion, should refer to a given transition (and thus not to a wawefunction) in terms only of a suitably chosen sum rule. Then, as a second statement, I would suspect that much of that problem depends on the particular interaction in use. Let us make an assumption and in doing so I cautiously underline that I do not assert that this has necessarily been done. Let us assume thus an interaction which has been fitted in its hexadecapole-hexadecapole part such as to reproduce exactly the energy of the relevant 4^+ state as a one-phonon state of hexadecapole character. It is likely that one will not find it as a two-phonon state of the quadrupole type ! Beyond this trivial statement, we can say more. J.P. BLAIZOT and D. GOGNY, some years ago, found in their self-consistent RPA calculations that the collectivity of any given state depends dramatically on the consistency between the single-particle wawefunctions and the residual interaction in use. Lacking such a self-consistency, one uses the relevant multipole-multipole force to build up the correctly deformed mean field, mocking up thus as a correlation effect what is indeed purely a mean field effect. In both V. SOLOVIEV's or R. PIEPENBRING's calculations, it is not very easy to assess the fulfilment of this selfconsistency requirement between the Nilsson (or Woods-Saxon, or the like), wawefunctions and the multipole-multipole force which has been considered.

Less controversial, may be, but nevertheless very interesting, has been the report of the longstanding quest by P. KLEINHEINZ on one-phonon, two-phonon states and one-phonon states coupled to one or two particles around ^{146}Gd. He has indeed identified two two-octupole-phonon states deexcited through a stretched E3-E3 cascade. As in ^{168}Er these states substantially deviate from the harmonic limit.

A. KREINER has demonstrated, for us, various geometrical coupling situations playing with various configurations for both proton and neutron valence states in some **odd-odd nuclei**. For high-j orbitals, depending on the K-value, one may find oneself in a strong coupling or in a decoupled situation. In other cases as in heavy Thallium isotopes (A~198) one finds a semi-decoupled structure where one nucleon is coupled while the other is not.

Turning now to what is currently dubbed as **"scissor" modes**, D. ZAWISCHA has proposed an alternate description for these low energy magnetic dipole excitations. Quantum Mechanics is, as we all know, full of surprises : we have seen in his talk a nice current flow chart of "scissor" that do not open...

+-+-+-+-+

I will say now some words about some contributions which, despite some decent efforts on my side, I found definitely too difficult to squeeze into the previously discussed theoretical scheme.

We have had two nice talks on mass measurements. D. VIEIRA reported on the results obtained at TOFI (Los Alamos) on neutron rich isotopes up to ^{66}Fe. Such results together with those obtained at GANIL have considerably increased our knowledge concerning the outskirts of this part of the nucleidic chart. G. BOLLEN has summarized the current endeavours and the impressive achievements already attained in the Penning trap measurements of nuclear masses.

From neutron-rich nuclides to astrophysical r-or s-process, the transition is quite natural. K. KRATZ has presented the astrophysical consequences (for the r-process here) of his studies of "waiting point" isotopes like ^{80}Zn, ^{130}Cd and ^{195}Tm. E. HAGBERG reported on an attempt to constrain the value of the V_{ud} KOBAYASHI-MASKAWA matrix elements from super-allowed $0^+ \to 0^+$ beta decay. The search for new radio-activities (of the 2p, 3p type) near the proton drip line has been pursued during the last years at GANIL. F. POUGHEON has shown its significant progress, particularly about the β-3p radioactivity found in ^{31}Ar.

Even though, I should probably be among the less competent persons in this audience to do so, I will briefly mention the outcome of the nice technical session which we have had here. J. SHARPEY-SCHAEFFER has made a lively presentation of the γ-multidetector arrays presently under construction : EUROGAM, GAMMASPERE to which one must add the current actvity of the italian and nordic groups for instance. From this talk, a clear picture of a very active field emerged undoubtedly, whose physical impact has been nicely stated. In particular J. SHARPEY-SCHAEFFER made the important statement that even though superdeformation studies constitute an obvious and fascinating domain of study for these arrays, there are many other topics which can valuably take full advantage of these instrumental developments. I am thinking in terms of a full exploration of the (ω,λ,T) space - T being the temperature, while ω and λ have been already defined. In particular within the (ω,λ) plane a full study of the diabolic pair transfer would be very nice. It is also clear that these arrays will permit to get spectroscopic informations as complete as possible. Such a completeness, as always, could provoke our reflection and even keep some surprises in store. One often have a blasé attitude towards such systematical approaches. However, one must remember, as somebody noted, that to have a DARWIN, on needs to have had some LINNE.

C. PAPANICOLAS has presented the prerequisite of a γ-multidetector device needed for experiments of the $(e,e'\gamma)$, $(e,e'X\gamma)$ or $(e,e'\gamma_1\gamma_2\gamma_3)$, etc... type. R. HOLZMANN has summarized a currently developped technique of γ-ray spectroscopy for neutron-rich isotope studies with relativistic ion beams. Very impressive results obtained by the so-called GRID method have been presented by J. JOLIE. This method appear to be particularly well-suited for the measurement of lifetimes ranging from the femtosecond to the picosecond, which therefore makes it a very useful tool for many nuclear deexcitations.

P. VAN DUPPEN and M. LEVITOWICZ have presented results and projects in the blooming field of relatively high intensity secondary beam. The potentiality of these projected factories (Louvain and GANIL) of unstable isotopes is indeed very high, entailing valuable consequences for Astrophysical as well as current Nuclear Structure studies, and possibly also for the Materials Physics community. More or less in connection with these efforts, J. WOUTERS has presented the first and interesting results of in-beam polarization of light radioactive elements.

The pioneering role of Nuclear Physics concepts, Nuclear Physics data and nuclear physicists as well in the understanding of what could be the fingerprint of classically chaotic systems in their quantal analogs, has been very well presented by M.J. GIANNONI. This is, as she said, one issue where the "trade balance" of Nuclear Physics versus the rest of Physics, is nicely located in the positive region. The conjecture connecting GOE level spacing distributions with systems where classical analogs are S-systems, proved to be highly important in many areas of current Theoretical and Experimental Physics. T. VON EGIDY illustrated these concepts in the region of transitional nuclei and at an energy range (up to some MeV in excitation energies) where one may expect to decipher a coexistence or a transition between regular and chaotic dynamical regimes.

Let us conclude now. Restricting the Conference themes to low-excitation, low-spin Nuclear Structure studies, the Organizing Committee could be thought a priori as having made a rather risqué choice. The perspective could have come out as being too much limited and the topical domain unduly narrowed. From what we have heard and seen here, I should conclude that this has not been so. The field defined within the Conference boundaries remains indeed very active, lively and full of projects (let us mention for instance γ-multidetector arrays, Penning traps, exotic beams and targets). It is also a domain which is widely and deeply connected with many other disciplines, with very fruitful links with Theoretical Physics (as examplified by its contribution to Quantum Chaos).

I feel consequently obliged (and in that, I certainly reflect the desire of many of us) in proposing to renew this stimulating experience. I think we should consider another such Conference in about three years. I have personally no doubts that we should have, then, a Conference as profitable and lively as the present one, in view of all the valuable experimental and theoretical works currently undertaken which we have heard of during this week.

PARTICIPANTS

Participants of the Advanced Research Workshop on Nuclear Shapes and Nuclear Structure at Low Excitation Energies
June 3-7, 1991
Cargèse, France

PARTICIPANTS

Dr. G.D. ALKHAZOV
Nucl. Physics Leningrad
Department of Physics
Gatchina
188350 LENINGRAD
U.S.S.R.

Dr. Faical AZAIEZ
Div. Recherche Expérimentale
Institut de Physique Nucléaire
F-91406 ORSAY Cedex
FRANCE

Dr. Vittorio BARCI
Laboratoire de Radiochimie
Université de Nice
28, avenue Valrose
F-06034 NICE Cedex
FRANCE

Dr. Jacques BAUCHE
Laboratoire Aimé Cotton
CNRS - Campus Orsay
Bât. 505
F-91405 ORSAY Cedex
FRANCE

Dr. Ragnar BENGTSSON
Oak Ridge National Laboratory
Bldg. 6000-MS6374
P.O. Box 2008
TN 37831 OAK RIDGE
U.S.A.

Dr. Istvan BERKES
Institut de Physique Nucléaire
43, bd. du 11 Novembre 1918
F-69622 VILLEURBANNE Cedex
FRANCE

Dr. Georg BOLLEN
Institüt für Physik
Universität Mainz
Postfach 3980
D-6500 MAINZ
F.R.G.

Dr. Paul BONCHE
CEN Saclay SPTh
ORME DES MERISIERS
F-91191 GIF/YVETTE Cedex
FRANCE

Dr. Hans BORNER
Institut Laue Langevin
BP 156 X
F-38042 GRENOBLE
FRANCE

Dr. Chantal BRIANCON MIKHAILOVA
CSNSM Orsay
Bât. 104-108
F-91405 ORSAY Cedex
FRANCE

Dr. Peter BUTLER
Oliver Lodge Laboratory
University of Liverpool
PO. Box 147
L69 3BX LIVERPOOL
UNITED KINGDOM

Dr. R.F. CASTEN
Brookhaven National Laboratory
Department of Physics
Building 510
NY 11973 UPTON
U.S.A.

Dr. John E. CRAWFORD
Mc Gill University
Foster Radiation Laboratory
3610 University Street
H3A 2B2 MONTREAL
CANADA

Dr. Joao DA PROVIDENCIA
Universita de Coimbra
Departamento De Fisica
P-3000 COIMBRA
PORTUGAL

Dr. Jean-Paul DELAROCHE
C.E.N. Bruyeres-le-Châtel
Physique théorique
BP 12
F-91680 BRUYERES-LE-CHATEL
FRANCE

Dr. Philippe DESSAGNE
Centre de Recherche Nucléaire
BP 20 CRO
F-67037 STRASBOURG
FRANCE

Dr. Hong Tuan DUONG
Laboratoire Aimé Cotton
CNRS - Campus Orsay
Bât. 505
F-91405 ORSAY Cedex
FRANCE

Dr. Sydney GALES
Div. Recherche Expérimentale
Institut de Physique Nucléaire
F-91406 ORSAY Cedex
FRANCE

Dr. Maria José GARCIA BORGE
Insto Rocasolano
C.S.I.C.
Serrano 119
E-28006 MADRID
SPAIN

Dr. Marie-Joya GIANNONI
Division de Physique Théorique
Institut de Physique Nucléaire
F-91406 ORSAY Cedex
FRANCE

Dr. Michel GIROD
C.E.N. Bruyeres-le-Châtel
Physique théorique
BP 12
F-91680 BRUYERES-LE-CHATEL
FRANCE

Dr. Andrée GIZON
Institut Sciences Nucléaires
53 av. des Martyrs
F-38026 GRENOBLE Cedex
FRANCE

Dr. Dominique GOUTTE
DPhN-SEPN
CEN Saclay
Orme des Merisiers
F-91191 GIF/YVETTE Cedex
FRANCE

Dr. Erik HAGBERG
Atomic Energy of Canada Lted
Nuclear Laboratories
CHALK RIVER Ontario
K0J 1J0 CHALK RIVER
CANADA

Dr. Dennis HAMILTON
University of Sussex
Physics Division
BN1 9QH BRIGHTON
UNITED KINGDOM

Dr. Joe HAMILTON
Vanderbilt University
Dept of Physics & Astronomy
P.O. Box 1807-B
TN 37235 NASHVILLE
U.S.A.

Dr. Fazia HANNACHI
CSNSM Orsay
Bât. 104-108
F-91405 ORSAY Cedex
FRANCE

Dr. Paul-Henri HEENEN
Université Libre Bruxelles
Physique Théorique
Campus de la Plaine-CP 2050
B-1050 BRUXELLES
BELGIUM

Dr. Kris HEYDE
Nuclear Physics Laboratory
University of Gent
Proeftuinstraat 86
B-9000 GENT
BELGIUM

Dr. Romain HOLZMANN
G.S.I. Darmstadt
Postfach 110552
D-6100 DARMSTADT
F.R.G.

Dr. Friedrich HOYLER
Physikalisches Institut
Universität Tübingen
Auf der Morgenstelle 14
D-7400 TUBINGEN
F.R.G.

Dr. Max G. HUBER
Institüt für Physik
Universität Mainz
Postfach 3980
D-6500 MAINZ
F.R.G.

Dr. Mark HUYSE
Katholieke Universiteit Leuven
Inst. Kern-en Stralingsfysika
Celestijnenlaan 200D
B-3030 LEUVEN (Heverlee)
BELGIUM

Dr. R.V.F. JANSSENS
Physics Division
Argonne National Laboratory
IL 60439 ARGONNE
U.S.A.

Dr. Jan JOLIE
Institut Laue Langevin
BP 156 X
F-38042 GRENOBLE
FRANCE

Dr. Jean KERN
Université de Fribourg Suisse
Institut de Physique
Pérolles
CH-1700 FRIBOURG
SWITZERLAND

Dr. Peter KLEINHEINZ
Institüt für Kernphysik
Kernforschungsanlage Jülich
Postfach 1913
D-517O JULICH 1
F.R.G.

Dr. Amel KORICHI
Div. Recherche Expérimentale
Institut de Physique Nucléaire
F-91406 ORSAY Cedex
FRANCE

Dr. Karl Ludwig KRATZ
Institut für Kernchemie
Johannes Gutenberg Univ.
Postfach 3080
D-6500 MAINZ
F.R.G.

Dr. Andres J. KREINER
Dept. de Fisica
TANDAR CNEA
Av. del Libertador 8250
1429 BUENOS AIRES
ARGENTINA

Dr. Thomas KUHL
G.S.I. Darmstadt
Postfach 110552
D-6100 DARMSTADT
F.R.G.

Dr. Reinhard KULESSA
Jagellonian Univ.
Institut of Physics
Ul. Reymonta 4
PL-30059 KRAKOW 12
POLAND

Dr. François LE BLANC
Div. Recherche Expérimentale
Institut de Physique Nucléaire
F-91406 ORSAY Cedex
FRANCE

Dr. Jonathan LEE
Mc Gill University
Foster Radiation Laboratory
3610 University Street
H3A 2B2 MONTREAL
CANADA

Dr. Marek LEWITOWICZ
GANIL
BP 5027
F-14021 CAEN Cedex
FRANCE

Dr. Jean LIBERT
CSNSM Orsay
Bât. 104-108
F-91405 ORSAY Cedex
FRANCE

Dr. Peter LIEVENS
CERN/Isolde
CH-1211 GENEVE 23
SWITZERLAND

Dr. Pertti LIPAS
University of Jyvaskyla
Department of Physics
P.O. Box 35
SF-40351 JYVASKYLA
FINLAND

Dr. Ann-Marie MARTENSSON-PENDRILL
Chalmers Tekniska Högskola
Fysika Institutionen
Dept. of Physics
s-41296 GOETEBORG
SWEDEN

Dr. Gerd MEISEL
Karlsruhe GmbH
IK III
Postfach 3640
D-7500 KARLSRUHE
R.F.A.

Dr. Vincent MEOT
DPhN-SEPN
CEN Saclay
Orme des Merisiers
F-91191 GIF/YVETTE Cedex
FRANCE

Dr. Igor MIKHAILOV
J.I.N.R. Dubna
Laboratory Theoretical Physics
P.O. Box 79
101000 MOSCOW
U.S.S.R.

Dr. Vladimir MOROZOV
J.I.N.R. Dubna
Laboratory Theoretical Physics
P.O. Box 79
101000 MOSCOW
U.S.S.R.

Dr. Witold NAZAREWICZ
Oak Ridge National Laboratory
Bldg. 6000-MS6374
P.O. Box 2008
TN 37831 OAK RIDGE
U.S.A.

Dr. Valentin NESTERENKO
J.I.N.R. Dubna
Laboratory Theoretical Physics
P.O. Box 79
101000 MOSCOW
U.S.S.R.

Dr. R. NEUGART
Institüt für Physik
Universität Mainz
Postfach 3980
D-6500 MAINZ
F.R.G.

Dr. Gerda NEYENS
Katholieke Universiteit Leuven
Inst. Kern-en Stralingsfysika
Celestijnenlaan 200D
B-3030 LEUVEN (Heverlee)
BELGIUM

Dr. Ernst W. OTTEN
Institüt für Physik
Universität Mainz
Postfach 3980
D-6500 MAINZ
F.R.G.

Dr. Costas PAPANICOLAS
University of Athens
Nuclear&Particle Physics Section
Panepistimioupoli-Kouponia
GR-15771 ATHENS
GREECE

Dr. Robert PIEPENBRING
Institut Sciences Nucléaires
53 av. des Martyrs
F-38026 GRENOBLE Cedex
FRANCE

Dr. Jacques PINARD
Laboratoire Aimé Cotton
CNRS - Campus Orsay
Bât. 505
F-91405 ORSAY Cedex
FRANCE

Dr. Hans POSTMA
TU Delft
Department of Applied Physics
Lorentweg 1
2628 CJ DELFT
NETHERLANDS

Dr. Françoise POUGHEON
Div. Recherche Expérimentale
Institut de Physique Nucléaire
F-91406 ORSAY Cedex
FRANCE

Dr. Philippe QUENTIN
CSNSM Orsay
Bât. 104-108
F-91405 ORSAY Cedex
FRANCE

Dr. Nadine REDON
Institut de Physique Nucléaire
43, bd. du 11 Novembre 1918
F-69622 VILLEURBANNE Cedex
FRANCE

Dr. Mario RENTERIA
Div. Recherche Expérimentale
Institut de Physique Nucléaire
F-91406 ORSAY Cedex
FRANCE

Dr. Brigitte ROUSSIERE
Div. Recherche Expérimentale
Institut de Physique Nucléaire
F-91406 ORSAY Cedex
FRANCE

Dr. David J. ROWE
Department of Physics
University of Toronto
M5S 1A7 TORONTO
CANADA

Dr. Krzysztof RYKACZEWSKI
G.S.I. Darmstadt
Postfach 110552
D-6100 DARMSTADT
F.R.G.

Dr. Jocelyne SAUVAGE
Div. Recherche Expérimentale
Institut de Physique Nucléaire
F-91406 ORSAY Cedex
FRANCE

Dr. Henri SERGOLLE
Institut de Physique Nucléaire
F-91406 ORSAY
FRANCE

Dr. Claude SERRE
CERN/Isolde
CH-1211 GENEVE 23
SWITZERLAND

Dr. John F. SHARPEY-SCHAFER
Oliver Lodge Laboratory
University of Liverpool
PO. Box 147
L69 3BX LIVERPOOL
UNITED KINGDOM

Dr. V.G. SOLOVIEV
J.I.N.R. Dubna
Laboratory Theoretical Physics
P.O. Box 79
101000 MOSCOW
U.S.S.R.

Dr. Henry STROKE
New York University
Department of Physics
4 Washington Place
NY 10003 NEW YORK
U.S.A.

Dr. Naoki TAJIMA
Division de Physique Théorique
Institut de Physique Nucléaire
F-91406 ORSAY Cedex
FRANCE

Dr. Olof TENGBLAD
CERN
Divison PPE
CH-1211 GENEVE 23
SWITZERLAND

Dr. Andrzej WOJTASIEWICZ
University of Warsaw
Institute for Nuclear Studies
Ul. Hoza 69
PL-00 681 WARSAW
POLAND

Dr. Piet VAN DUPPEN
Katholieke Universiteit Leuven
Inst. Kern-en Stralingsfysika
Celestijnenlaan 200D
B-3030 LEUVEN (Heverlee)
BELGIUM

Dr. Jan WOUTERS
Institut de Physique Nucléaire
Chemin du Cyclotron 2
B-1348 LOUVAIN-LA-NEUVE
BELGIUM

Dr. Michel VERGNES
Div. Recherche Expérimentale
Institut de Physique Nucléaire
F-91406 ORSAY Cedex
FRANCE

Dr. Dietrich ZAWISCHA
Theoretische Physik Inst.
Universität Hannover
Appelstrass 2
D-3000 HANNOVER
F.R.G.

Dr. David J. VIEIRA
Los Alamos Nat. Lab.
Isotope and Nuclear Chemistry Div.
P.O. Box 1663
NM 87545 LOS ALAMOS
U.S.A.

Dr. Nissan ZELDES
The Racah Institute of Physics
The Hebrew University of Jerusalem
Givat Ram Campus
91904 JERUSALEM
ISRAEL

Dr. T. VON EGIDY
Technische Universität München
Fakultät für Physik
D-8046 GARCHING München
F.R.G.

INDEX

Abnormal M1, 359, 438
Algebraic models, 203, 439
Alignment, 123, 275, 302, 315, 441
Alpha decay, 220, 440
Anapole moment, 88
Angular correlations, 359
Angular distributions, 349, 429
Anharmonicity, 179
Astrophysics, 425, 429, 444
Atomic beams, 70, 93
Atomic physics, 55, 69
Atomic spectroscopy, 3, 87, 437
Axial asymmetry, 210, 247, 442

B.C.S calculations, 154, 182
Beta-delayed 3 proton decay, 357, 445
Bethe-Goldstone equation, 58
Bogolyubov-Valatin transformation, 180

Chaos, 373, 416, 445
Charge radii, 5, 26, 56, 72, 81, 87, 93
Complete level schemes, 373
Coriolis interaction, 144, 307, 377, 396
Coulomb dissociation, 77, 348
Coulomb excitation, 163, 267, 408
    419, 442
Cross sections, 111, 347, 408

Damping width, 102
Deep hole states, 112
Deformation of the Fermi surface, 272
Deformations, 17, 42, 233
Deformed nuclei, 3, 143, 164, 179, 239
    415
Deformed shell gaps, 23
Diabatic effects, 169
Diabolic pair transfer, 445
Doppler broadening, 70, 410, 415, 445
Droplets of Fermi liquid, 273, 442
Dye lasers, 82, 93

Dynamical groups and algebras, 233
Dynamical symmetries, 47

E.C.R ion source, 423
Electric dipole moment, 87
Electric quadrupole moment, 18, 58, 269
Electromagnetic moments, 81, 93, 269

Fast neutrons, 412
Fermi surface, 107, 127, 144, 242, 247, 301
Final state interaction, 106
Fine structure, 3, 219
Fission, 230, 407
Fluorescence spectroscopy, 72
Four Quasi-particle state, 163
Fragmentation, 104, 353, 365, 408
Fragment separator, 12, 408
Fusion - evaporation, 228

Gamma ray spectroscopy, 227, 268, 299
    407, 416
Gaussian overlap approximation, 323
Generator coordinate method, 45, 170, 286
    440
Giant resonances, 6, 216, 233, 409
Gold isotopes, 132, 227, 359
Grazing surface scattering, 430

H.F.B calculations, 47, 323
Hartree-Fock calculations, 39, 58, 79, 84
    101, 248, 281
Heavy ions, 267, 355, 365, 408, 423, 429
Heavy nuclei, 6, 102, 185, 247, 271
HF+BCS calculations, 45, 164, 169, 249
    281
Hidden symmetries, 248
High resolution, 69, 107, 402
High resolution (e, e' p), 101, 438

High spin states, 47, 72, 143, 214, 227
247, 274, 299, 324, 408
Hill-Wheeler equation, 107, 286, 323
Hyperfine structure, 3, 55, 69, 81, 87, 93
383, 438
Hyperpure Ge detectors, 333, 360

"Identical" bands, 143, 216, 305, 315, 442
In-beam nuclear polarization, 429, 445
Interacting boson model, 40, 166, 185, 203
275, 377, 418, 442
Internal conversion, 359
Intruder orbitals, 41, 123, 219, 259, 301
320, 365, 418, 440
Iridium isotopes, 158
Isomers, 72, 163, 169, 402
Isotope series, 14, 63, 78
Isotopic shift, 5, 55, 69, 81, 88, 438

King plot, 63
Knight shift, 384
Kobayashi-Maskawa model, 89, 331, 444
K quantum number, 31, 127, 146, 183, 206
249, 306, 323, 377, 396, 418, 444

Laser spectroscopy, 5, 69, 81, 93, 163, 403
429
Lead region, 129, 169, 219, 440
Lie algebra, 233
Lifetimes, 268, 299, 355, 416
Lithium isotopes, 72, 347
Low-lying J = O levels, 28, 41, 131, 169
224, 331
Low spin states, 124, 143, 281, 331
Low temperature, 383, 429

M1 sum rule, 51
M2, M3, strength distribution, 198
Magic numbers, 17, 41, 75, 101, 339
Magnetic analysis, 107, 353, 365, 399
Magnetic dipole, 18, 58, 239
Mass determination, 22, 69, 365, 399, 444
Mass spectrometer, 399
Mean field, 76, 101, 169, 248, 267, 282
440
Medium/heavy nuclei, 101, 143, 239, 267
299, 340, 399, 417
Mercury isotopes, 6, 125, 281, 299, 323
359, 441
Microscopic calculations, 40, 79, 116, 164
215, 240, 247, 281, 323
Mixed symmetry states, 32, 48
Multicluster model, 253
Multi detector, 227, 268, 274, 316, 349
412, 445
Multiphonon, 179, 203, 416, 443

Neutron drip line, 76, 339, 347
Neutron halo, 11, 18, 76, 347, 365, 430
438
Neutron-rich nuclei, 76, 347, 365, 407, 444
New radioactivities, 353
Nilsson model, 22, 74, 125, 146, 172, 183
215, 237, 242, 248, 307, 396, 444
Noble gases, 74
Nuclear charge-dependent effects, 331
Nuclear collective motion, 179, 299
Nuclear dipole modes, 17, 272, 444
Nuclear magnetic resonance, 14, 59, 76
391, 429
Nuclear masses, 10, 69, 402
Nuclear models, 93, 101, 373
Nuclear moments, 72, 93, 383, 429
Nuclear orientation, 33, 74, 163, 383, 429
Nuclear polarization, 5, 429
Nuclear shapes, 17, 39, 233, 267
Nuclear spectroscopy, 123, 383, 408, 437
Nuclear structure, 3, 39, 69, 93, 101, 195
365
Nuclei far from stability, 69, 81, 339, 353
365, 403-407

Occupation numbers, 102, 147
Octupole deformations, 6, 34, 78, 83, 143
188, 247, 267, 293, 323, 437
Octupole vibrations, 187, 203, 271, 304
Odd-even staggering, 76, 83, 93, 143, 366
438
Odd-odd nuclei, 143, 444
On-line isotope separation, 69, 219, 340
353, 360, 399, 423, 429
Orbital M1 excitations, 239
O (6) symmetry, 203, 377, 418, 438

Pairing, 43, 129, 163, 170, 196, 237, 243
248, 289, 299, 347, 366, 415, 441
Parity non-conservation, 87
Pauli principle, 180, 196, 254, 419
Penning trap, 69, 399, 444
Phase inversion, 143
Platinum isotopes, 131, 281, 299, 359, 441
Polarized beams, 72, 113, 429
Polonium isotopes, 93
Population of oblate and prolate systems
227
Position sensitive Ge detectors, 410
Proton drip-line, 86, 353, 444

Proton-neutron residual interactions, 40
143, 304
Proton rich nuclei, 353, 407
Pseudo-spin scheme, 307
Pseudo SU (3), 203, 238, 249

Quadrupole deformations, 72, 83, 281, 323
Quadrupole vibrations, 196
Quenching, 108, 440

Radioactive beams, 11, 45, 69, 340, 347
407, 423, 429, 445
Radioactive nuclei, 69, 81, 360, 399
Radium isotopes, 14, 78, 248, 271
Rare earth isotopes, 51, 72, 81, 132, 163
179, 196, 236, 239, 248, 267, 299
Rational harmonic oscillator, 249, 442
Reaction models, 105
Relativistic fragments, 4, 347, 407
Resonance ionization spectroscopy, 14, 72
81, 438
Rotational structures, 127, 144
R-Process nucleosynthesis, 339, 365, 444
R.P.A calculations, 196, 239, 255, 340
444

Scissor modes, 32, 49, 239, 444
Selenium region, 139
Shape coexistence, 6, 27, 47, 123, 169
219, 227, 359, 438
Shape isomer, 169, 287
Shape transition, 6, 84, 132, 438
Shell corrections, 247
Shell model, 9, 40, 75, 166, 219, 233, 332
365, 438
Shell structure, 101, 299
Shielding, 58
Signature splitting, 135, 144, 203, 303, 319
Single particle response function, 101
Solid state telescope, 349, 355
Spectroscopic factors, 101
Spin-flip excitations, 243
Spins and parities, 17, 114, 359, 373
Standard Model, 10, 88, 331
Statistical properties, 373, 415
Strength distributions, 104
SU (3) symmetry, 47, 234, 377
Superallowed beta decay, 331
Superdeformation, 23, 51, 233, 247, 281
299, 315, 323, 441
Supernovae, 339
Supersymmetry scheme, 215
Symplectic models, 203, 234

Tamm-Dancoff approximation, 181, 240
Thallium isotopes, 315
Time of flight technique, 353, 365
Tin region, 134
Tilted foil, 430
Toroidal moments, 273
Transfer reactions, 111, 163, 440
Triaxiality, 149, 166, 326
T violation, 87
Two Phonon states, 179, 195, 418
Two Quasi-particule - plus - rotor model, 144

U (5) symmetry, 212

Waiting-point concept, 339, 444
Weak interaction, 87, 336, 429
Weak vector coupling constant, 331
Wilets-Jean model, 438
Woods-Saxon potential, 107, 126, 183, 248
270, 301, 444

The manufacturer's authorised representative in the EU is Springer
Nature Customer Service Centre GmbH, Europaplatz 3, 69115 Heidelberg,
Germany. If you have any concerns regarding our products, please
contact ProductSafety@springernature.com

Printed and bound by CPI Group (UK) Ltd, Croydon, CR0 4YY
15/07/2026
02167631-0009